Cheng Zhi Huang, Jian Ling, Jian Wang
Elastic Light Scattering Spectrometry

Also of Interest

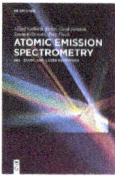

Atomic Emission Spectrometry.
AES – Spark, Arc, Laser Excitation
Golloch, Joosten, Killewald, Flock, 2019
ISBN 978-3-11-052768-1, e-ISBN 978-3-11-052969-2

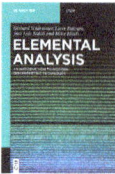

Elemental Analysis.
An Introduction to Modern Spectrometric Techniques
Schlemmer, Balcaen, Todolí, Hinds, 2018
ISBN 978-3-11-050107-0, e-ISBN 978-3-11-050108-7

Organic Trace Analysis.
Niessner, Schäffer, 2017
ISBN 978-3-11-044114-7, e-ISBN 978-3-11-044115-4

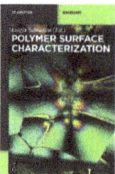

Polymer Surface Characterization.
Sabbatini (Ed.), 2014
ISBN 978-3-11-027508-7, e-ISBN 978-3-11-028811-7

Cheng Zhi Huang, Jian Ling, Jian Wang

Elastic Light Scattering Spectrometry

—

DE GRUYTER 科学出版社

Authors
Prof. Dr. Cheng Zhi Huang
Southwest University
College of Pharmaceutical Sciences
400715 Chongqing
PR China
chengzhi@swu.edu.cn

Dr. Jian Ling
Yunnan University
School of Chemical Science and Technology
Kunming
PR China

Prof. Dr. Jian Wang
Southwest University
College of Pharmaceutical Sciences
400715 Chongqing
PR China
wj123456@swu.edu.cn

ISBN 978-3-11-057310-7
e-ISBN (PDF) 978-3-11-057313-8
e-ISBN (EPUB) 978-3-11-057324-4

Library of Congress Control Number: 2018954461

Bibliographic information published by the Deutsche Nationalbibliothek
The Deutsche Nationalbibliothek lists this publication in the Deutsche Nationalbibliografie; detailed bibliographic data are available on the Internet at http://dnb.dnb.de.

© 2019 Science Press Ltd. and Walter de Gruyter GmbH, Beijing/Berlin/Boston
Cover image: luckyvector/iStock/Getty Images Plus
Typesetting: Integra Software Services Pvt. Ltd.
Printing and binding: CPI books GmbH, Leck

www.degruyter.com

Contents

List of Contributing Authors

Prof. Dr. Cheng Zhi Huang
Southwest University
College of Pharmaceutical Sciences
400715 Chongqing
PR China
chengzhi@swu.edu.cn

Dr. Jian Ling
Yunnan University
School of Chemical Science and Technology
650091 Kunming
PR China

Prof. Yuan Fang Li
Southwest University
College of Chemistry and Chemical
Engineering
400715 Chongqing
PR China
liyf@swu.edu.cn

Prof. Dr. Yun Fei Long
Hunan University of Science
and Technology
School of Chemistry and Chemical
Engineering
411201 Xiangtan
PR China
l_yunfei927@163.com

Qie Gen Liao
Jiangxi Academy of Agricultural Sciences
Agricultural Product Quality Safety and
Standards Institute
330200 Nanchang
PR China

Prof. Dr. Jian Wang
Southwest University
College of Pharmaceutical Sciences
400715 Chongqing
PR China
wj123456@swu.edu.cn

Dr. Yue Liu
Army Medical University
College of Basic Medical Sciences
Department of Chemistry
400038 Chongqing
PR China

Dr. Peng Fei Gao
Southwest University
College of Pharmaceutical Sciences
400715 Chongqing
PR China

Dr. Li Qiang Chen
Yunnan University
College of Chemical Science and Technology
650091 Kunming
PR China

Dr. Zhong De Liu
Southwest University
College of Pharmaceutical Sciences
400715 Chongqing
PR China

Dr. Li Zhang
Nanchang University
Department of Chemistry
330031 Nanchang
PR China

Dr. Sai Jin Xiao
East China Institute of Technology
Jiangxi Key Laboratory of
Mass Spectrometry
and Instrumentation
330013 Nanchang
PR China

Prof. Dr. Ke Jun Tan
Southwest University
College of Chemistry and Chemical
Engineering
400715 Chongqing
PR China
tankj@swu.edu.cn

https://doi.org/10.1515/9783110573138-201

Prof. Dr. Xiao Bing Pang
Nanjing Uinversity of Information Science &
Technology
School of Atmosperic Physics
210044 Nanjing
PR China
pangxbyuanj@163.com

Dr. Yan Yin
Nanjing Uinversity of Information Science &
Technology
School of Atmosperic Physics
210044 Nanjing
PR China

Cheng Zhi Huang, Jian Ling, Yuan Fang Li

1 Introduction to light scattering

1.1 Synopsis of light scattering

1.1.1 Light scattering phenomenon

Why is the sky blue? Why is the deep sea mazarine? And why is the lake green? All these questions are concerned with light scattering and its related optical phenomena.

Light scattering is an optical natural phenomenon that exists extensively. Literally, light scattering refers to a phenomenon that scatters light in all directions. When the light passes through an inhomogeneous medium, photons of different wavelengths interact and collide with the inhomogeneous area of the medium, scattering the light in different directions. Therefore, the light can be observed in other directions besides the direction of the incident light. For example, in a dark night, when you turn on the electric torch, you can observe the light cross, and this is because the electric torch light is scattered by the suspended particles in the air; hence, the light is visible to our eyes. The larger the particles, the clearer would be seen for these floating particles. It is because of the existence of these particles resulted in scattering of the electric torch light that the flashlight could not spread to infinity.

1.1.1.1 Light scattering acting as an important form of light decay

The same as the light absorption, light scattering also can make the light passing through a medium decay. For example, when a white light passes through water-diluted milk, in the direction where the light advances, the milk appears pink, while from the side and above, it appears light blue. This is because the white light collides with small particles in milk emulsion, causing variation of different degrees for the light of different wavelengths from the original incidence direction of the white light. A short-wavelength light has more variation, while a long-wavelength light has less variation.

In optical propagation, the flow of photons interacts with the inhomogeneous area of a medium, such as an atom, a molecule, molecular aggregates, or particles of different sizes, and so light intensity changes in terms of spatial distribution, polarization state, or frequency. In other words, when the light propagates through a medium, if there is any inhomogeneity, it can lead to light scattering. Thus, light scattering is a form of interaction between the flow of photons and the inhomogeneous area, which widely exist in nature and our daily life.

https://doi.org/10.1515/9783110573138-001

1.1.1.2 Propagation of sunlight

Propagation of light can be seen in atmosphere, such as blue sky, white cloud, rosy cloud, and fog. Either blue sky, beautiful morning light, pretty sunset glow, or a rainbow after the rain, all these phenomena are the combined result of light scattering, light reflex, and light refraction, produced by the interactions between sunlight and atmospheric suspended solids, and gas molecules. Because of this, on cloudy rainy days or in case of fierce dust storm, sunlight hardly reaches the ground. In addition, because of this, even superstrong laser beam cannot propagate to infinity.

When sunlight passes through the atmosphere, rays of different wavelengths of the sunlight react with the atmosphere, that is, with floating particles and gas molecules in it. For the fact that the light at different wavelengths has different degree of scattering, the shorter the wavelength, the larger the degree of scattering is. A short-wavelength light has a larger degree of scattering; it scatters all over the sky, resulting in pretty blue color of the sky. As a long-wavelength light has a smaller degree of scattering, it directly propagates to the ground, and the color of the light is croci or orange red. For this reason, on a sunny noon, the sun shines directly at the ground with strong short-wavelength light scattering, and we can see the blue sky and nearly white sun. The sun appears white because the sunlight is too strong, so that the scattered intensity distributed all over the wave band (white light) is not enough to change the color of the sun. While in both the morning and at night, sunlight reaches the earth through atmosphere with a large dip angle, there are much more interactions between the sunlight and floating particles and air molecules in the atmosphere, and so the shorter-wavelength including blue light and yellow light are side scattered, and only the longer-wavelength including red light reaches the earth, making the sky blue and the sun yellow to red.

When compared with the wavelength of a visible light, dust, droplet, and other suspended particles have larger sizes. For example, $PM_{2.5}$ refers to a particle with a size smaller or equal to $2.5\,\mu m$; PM_{10} refers to a particle with a size smaller or equal to $10\,\mu m$. Diameters of $PM_{2.5}$ and PM_{10} are far larger than the wavelength of the sunlight of visible region (360–760 nm). So, when the sunlight passes through the atmosphere and collides with these larger particles, the sunlight is reflected in different directions. Since this kind of reflected light has no selectivity for light of different wavelength, so these particles reflect white light. On foggy and cloudy days, as there are lots of mist particles, the light of different wavelengths is reflected by these particles, and so the light that reaches the earth appears white, and the light intensity is rather low.

In fact, an analogical observation can also be seen when the white light passes through milk. It has known that milk appears white because it contains a number of 100 nm casein and fat globule particles of large diameters. These particles can equally scatter and reflect the visible light of all wavelengths. If the milk is well-diluted with water, however, large particles are less and so the reflected light is weaker; in such a case, we can see scattering of different colors in different directions of the container.

Consider another example, the motorcycle exhaust is generally light blue in color, especially for a motorcycle with a two-stroke engine. The reason is that gasoline combustion produces some particles. In fact, these particles are not light blue, and the color we see is a result of the scattering of white light.

1.1.1.3 Cognitive process of light scattering

In Chinese history, there are many literary quotations on light scattering. *Two Kids' Debate* recorded in the book of Tang questions in Liezi Series is an example. It is a story about Confucius who encountered a debate of two kids during a study tour with his students. The debate of the kids focused on: (1) when the sun rises, it appears closer to us, while it gets farther away by the middle of the day. This is because the sun appears as a hood when it first rises; however, it appears as large as a plate by the middle of the day. (2) The sun is far when it first rises and gets closer by the noon. This is because it is cold in the morning and gets hot by the noon. Confucius could not settle the debate because of his lack of knowledge about light scattering and thus, he was mocked by the two kids: "Who says that you know about everything?"

People's knowledge about light scattering comes from the cognition of life such as dust, smoke, fog, droplet, crystal, suspension liquid, emulsion, colloid, and other common particulate matters in life. In the nineteenth century, people's cognition of light scattering started from "blue sky phenomenon" or blue sky physics, and finally found that the essence of light scattering is the interaction of light wave electromagnetic field and refractive index inhomogeneous area, led by medium particles, molecules, or density fluctuation. During this period, people mainly focused on scattering by small particles and molecules as well as on scattering intensity variation, but not so much on energy change due to scattering. The scattering involved only included Rayleigh and Mie scattering.

After the twentieth century, people started to focus on the light scattering of particles smaller than the molecules, such as chemical bonds, quasiparticles, atoms, and free electrons, including Compton scattering, Thomson scattering, and so on. People started to pay attention to its energy change. Scientists were especially attracted to different scattering mechanisms involving wavelength and energy changes. They classified the scattering and applied them to molecular structural analysis. These researches mainly include Raman scattering, Brillouin scattering, and so on.

1.1.2 Classification of light scattering

Light scattering includes three participants: photons, homogeneous medium, and particles. A photon has energy, it is a kind of particle that flows with momentum, and it collides with suspended particles during transmission in a homogeneous medium and results in light scattering. A medium is a carrier of particles and also collides with particles (except for vacuum), while particles are normally object of the study wherein. The photon serves as a probe. On the basis of the phenomenon of their interactions and their relations between photons and particles, scattering can be divided into different types.

1.1.2.1 Classification based on scattering particle size

Light scattering is closely related to the size of the particles and can be classified based on that. Table 1.1 lists the general relationship between the scattered light and incident light, their energy changes, and refractive index.

Table 1.1: Types of light scattering.

Types of elastic scattering	Scatter particles size (d)	Relative refractive index (ρ)
Rayleigh, Debye	$\leq 0.05\lambda$	<0.3
Rayleigh–Gans–Debye	$>0.1\lambda$	<<0.3
Mie	$\approx\lambda$	≥ 0.3

Notes: d is the size of the scattering particle and λ is the wavelength of the incident light. In equation $\rho = 2\pi d(\eta\text{-}1)/\lambda$, η is the ratio of the refractive index of the scattering phase and medium, and it can be expressed as $\eta = \eta_1/\eta_0$.

1.1.2.2 Classification based on the energy changes in scattering process

After the photons interact with the medium and floating particles within, based on the basis of change in photon energy (frequency or wavelength), light scattering can be divided into elastic scattering and inelastic scattering. A photon with no energy loss or change in frequency or wavelength during collision mainly includes Tyndall, Rayleigh, and Mie scattering. In inelastic scattering, a photon has slight change in energy or displacement of frequency or wavelength; this mainly includes Raman and Brillouin scattering[*]. Related scattering energy changes and intensity changes are shown in Figure 1.1.

---

[*] For discussion convenience, we separate medium with floating particles in this book. However in fact, on the basis of the density fluctuation theory, scattering also exists in a homogeneous medium.

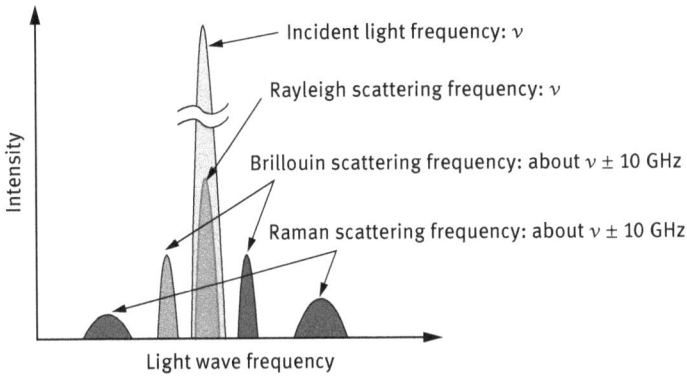

Figure 1.1: A distribution scheme of light scattering frequency and intensity.
Notes: On the x-coordinate are the frequency values, rather than the wavelength.

1.1.3 Reasons for ubiquitous light scattering

1.1.3.1 Optical inhomogeneity

Optical inhomogeneity of a substance is because there are other substances with different refractive index in the homogeneous substances, inhomogeneity of the substance itself, or time-related density fluctuation of the homogeneous substances.

Theoretically, except for vacuum, all the media have inhomogeneity to some degree, which in turn results in light scattering. Because when a light beam, with a certain intensity and wavelength, passes through a medium, forced vibration of the electrons of medium molecules or particles occurs under the influence of photon, forming dipole of certain vibrational frequency. The two ends of a dipole have slight positive and negative charges, and thus the dipole makes different dielectric constants, causing a different refractive index of a medium. As the refractive index is different from that of its surrounding medium, optical inhomogeneity occurs. At the same time, photons interact with the inhomogeneous area, radiating electromagnetic wave, and producing the secondary radiation, which is actually the scattered light. Different from dipole polarization of small particles, forced vibration of large particles, activated by the incident electromagnetic wave, is multipolar, causing complex scattering light waves. In such a case, the existence of particles changes the direction of propagation of the electromagnetic wave with no energy loss. If an electromagnetic wave is propagated inside the substance, the molecules vibrate and part of the electromagnetic wave energy transfers into heat or has molecular transition, causing light absorption and energy loss. Thus, the wavelength of the scattered light changes.

1.1.3.2 Elementary excitation and light scattering

Elementary excitation is an important concept in the theory of solid-state physics. A crystal in the state of ideal ground state is perfect with all lattice atoms in balance. However, it is impossible that a crystal in real practice could be perfect. There are always some atoms in the excited state – most of them are just close to the ground state. Thus, according to solid-state physics, the low excited state, close to the ground state, consists of some independent basic excitation units, which are just elementary excitation. It is sometimes called quasi-particles, which include phonon, exciton, magneton, plasmon, and so on. These elementary excitations under photoelectric field produce vibrations, resulting in disturbing the lattice from its balanced state and also producing inhomogeneity, and sometimes light scattering.

Thus, we can say that not only common particles can produce scattered light, but also atoms, molecules, crystal and condensed state, plasma, even electrons, chemical bonds, and total or partial state of these substances can produce scattered light. For this reason, light scattering is an important pathway in the study of molecules (solid) variation or rotary property, solid elementary excitation (phonons, excitons, and magnetons), and their interactions.

1.1.3.3 Particle flow and particle scattering

Except for the scattering of a visible light, there are scatterings of X-ray and other rays on the basis of the energy of photons. These scatterings are because of the interactions of the flows of photons and an inhomogeneous medium. Since there are different types of particle flows and variety of inhomogeneous mediums, there are different kinds of particle scattering.

Besides the light scattering owing to the interaction of photon flows with inhomogeneous media, there are also many other particle scatterings, such as neutron scattering, electron scattering, and so on. It should be noted that particle scattering of different kinds should be applied to different objects that should be investigated. For example, a visible light scattering is mainly applied to the study of molecule vibration and elementary excitation in a condensed state, while X-ray is applied to substance's microstructure and spatial symmetry.

By studying light scattering spectrum in terms of frequency shift, intensity, and linear and polarization state, people can obtain the structural and functional information of different scattering areas (such as elementary excitation), and it can also be combined with absorption and fluorescence spectrometry to characterize molecular interacting system.

1.1.3.4 Light scatterometry and light scattering spectrometry

In general, all light scattering concepts mentioned above have been applied in the analytical chemistry to some extent but their applications depend on the demands of the subjects and research fields. This book discerns related light scattering concepts to help us understand these phenomena and their applications in terms of analytical chemistry.

Light scatterometry has been widely used in atmospheric remote sensing and air quality monitoring, which is comparatively narrow. So, in this book, we discuss the light scattering in terms of spectrometry, belonging to the field of light scattering spectrometry, so as to express its application prospects in quantitative analytical chemistry. For this, in the following, we will introduce a few vital light scattering phenomena and their developing history, briefly discuss their arguments.

1.2 Tyndall scattering

1.2.1 Discovery of Tyndall scattering

In 1869, a British physicist, John Tyndall (1820–1893), found that dispersoid particle can scatter visible light during his study of colloids. Later, the phenomenon was named after him as Tyndall scattering. Tyndall's early study (1850–1856) was bene-fited from the famous R. W. Robert Wilhelm Bunsen (1811–1899), the founder of spectrometry. Later, he mainly focused on fields involving magnetism and diamagnetic polarity, as well as the function of radiation and air.

Tyndall scattering was discovered during his study of air thermal radiation, where in floating dust and other particles must be removed. In a sunny afternoon of nineteenth century, Tyndall accidentally found that when the sunlight passes through curtain gaps and falls in a dark room, the floating particles in the light crosses can be seen with human's naked eyes, and most visible light is from the larger particles floating in the air. The light of these large particles is quite different from that of small particles.

Tyndall further found that if the dark background is lighted with a specific light beam, even without a microscope, low-concentration particles that are hardly visible under normal conditions can be observed and counted owing to the scattering light of these particles. On the basis of this, Tyndall used the electric-powered lights as the source of light, equipped with light concentrators, and developed nephelometer, turbidimeter, and other similar facilities, which are able to exhibit solution's aerosol and colloidal properties through the spotlight beam. Combined a microscope, he even if successfully made an ultramicroscope.

Tyndall is the founder of light scattering equipments, and later generations made a lot of improvement. By far, people have developed ultramicroscope and turbidimeter and other commercial equipment with many modern functions and high sensitivity to determine the particle size and density in aerosol or other colloidal substances. These light scattering based equipments have been widely used in the production and research of polymer science, drug development, water, and atmospheric environment.

Tyndall scattering is the earliest recorded discovery and research on light scattering, revealing scattering of light in homogeneous suspension particles or colloidal ones, and reflecting the optical properties of light scattering. Tyndall scattering is the early foundation of the blue sky physics, and it is believed that the blue color of atmosphere is from all-direction scattering of the blue light, which are blocked by small particles suspended in the air. However, today, Tyndall scattering is usually defined as the light passes or crosses in the direction vertical to the direction of the incident light when a light passes through the colloidal sol.

1.2.2 Benefits and drawbacks of Tyndall scattering

The intensity of Tyndall scattering light depends on biquadrate of the frequency of the incident light, and so the scattering intensity of blue light is stronger than that of the red light. With the increase of the radius of a particle, an increase in light scattering is availabe. When white light falls on suspension, blue light scattering could be observable, together with partial polarization. Because of this, when the light beam passes through a suspension, as the particle size in a disperse system is rather wide, and the sizes of most particles are far larger than the wavelength of the visible light, light refraction, reflection, and scattering occur simultanouly, making the system appear turbid (nearly white).

If the floating particles are in the range of 40–900 nm, as its distribution strides range from visible light to near-infrared wavelength (360–760 nm), Tyndall scattering can be seen easily. In other words, when the light beam passes through a colloidal solution with the particle size in the system from dozen to hundred nanometers, Tyndall scattering occurs with strong scattering light. For example, in Figure 1.2, a red beam formed when the red light passes through 30–50 nm gold colloid, which is

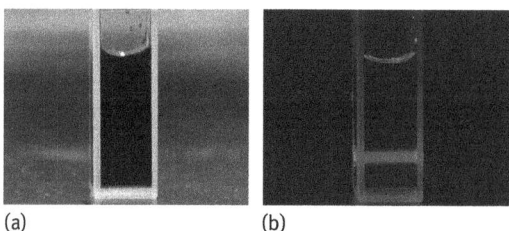

(a)   (b)

Figure 1.2: The Tyndall effect of gold colloid (refer to colored postscript picture): (a) gold colloid in white light; and (b) gold colloid with a red laser pen in a dark room (photo by Mingxuan Gao).

the classic Tyndall scattering. Thus, the Tyndall effect is especially suitable for the identification of a colloidal mixture.

The Tyndall effect only established the relationship between the particle size and light scattering intensity and can be used for distinguishing from suspension, colloid, and solution in a simple way. However, in terms of the concept of modern physics and chemistry, Tyndall scattering has drawbacks since it is only the discovery of light scattering and its simple applications. No further research and discussion on the basis of optics and electromagnetic theory, and the theory of light scattering is deficient. For example, molecules or ions in a clear solution scatter the incident light very weakly, and the scattered light can hardly be observed. In such case, the Tyndall effect does not exist, which does not mean that the scattered light is not present.

It should be noted that the explanation about the blue sky phenomenon is not perfect with the Tyndall effect. It is widely believed before that the blue sky is just scattering of floating particles in the air without considering scattering of a large amount of smaller particles, such as gas molecules. In fact, gas molecules, including oxygen and nitrogen with diameter about 0.3 nm, can also scatter the light at different frequencies, where in the light at higher frequency can be more easily scattered. This phenomenon was discovered by a famous English physicist, Rayleigh, and was later called Rayleigh scattering.

1.3 Rayleigh scattering

1.3.1 Discovery of Rayleigh scattering

A famous English physicist, Lord Rayleigh the Third (1842–1919), made a great contribution and promotion to the theory of light scattering. Rayleigh, first known as John William Strutt, was interested in physics from his very young age. His contribution is considerable in the field of acoustics, optics, electromagnetics, hydraulics, hydrodynamics, and other physics and related fields. He put forward the law of Rayleigh scattering, and predicted the existence of a surface wave [1]. By careful measurement of gas density and analyzing data difference, Rayleigh found a noble gas element Argon, owner of the "third decimal success". He won the Nobel Prize in physics in 1904, with English chemist, William Ramsay (1852–1916), who independently found Argon element.

As described earlier, Tyndall scattering seems perfectly to explain the blue color in the sky in terms of the scattering of sunlight by tiny dust, droplets, crystals, and other particles in the air, producing a shorter wavelength scattered light including blue, purple, indigo, and so on. As the blue light is most easily scattered, the sky appears blue. However, later it was found that Tyndall scattering does not give a complete explanation, that is to say, the concept of blue sky physics in early stage had drawbacks. According to Tyndall scattering, if the blue color is owing to the scattering of large particles the color and shade should change with the air components and humidity, as the temperature and humidity change with the number and size of water drops and crystals floating in the air. But, in reality, in areas with considerable climatic changes, such as humid areas and desert, the blue sky can be seen everywhere. Tyndall scattering could not explain the same color of blue sky in different meteorological condition. So by the end of the nineteenth century, people started to question this explanation since they were not sure if aerosol particulate matter are the true reason for the blue sky.

In 1871, 29-year-old Rayleigh published a paper, *On the light from the sky, its polarization and colour*, to address this problem [2] and reported a series of funny phenomena: (1) When the light falls on an inhomogeneous medium such as emulsion or colloidal solution, scattering exists. The primary cause for its existence is the refractive index of the inhomogeneous medium; (2) even for a homogeneous medium, molecular particles have continuous thermal motion, and disturb the fixed relationship in an intermolecular position with the occurrence of scattering. In 1880, Rayleigh further found that inorganic molecules in air such as oxygen and nitrogen, could produce scattering, and blue light is more easily scattered. Thus, he believed that the blue sky was a result scattering of sunlight by gas molecular particles, smaller than the wavelength of the light, and there is no change in the wavelength when the incident light scatters particles with a smaller size of irradiation wavelength.

To solve the observed problems, Rayleigh used a dimensional method to couple with light scattering with factors such as wavelength of the sunlight, molecular property, scattered light angles measurement, and so on. Through repeated experiments and calculations, Rayleigh believed the following:

(1) Scattering of particles smaller than the wavelength is the same as Tyndall scattering, namely, inverse biquadrate relationship followed between scattering light intensity of incident light wavelength in detection direction.

(2) Short-wavelength light is more easily scattered as compared to, long-wavelength light. The shorter the wavelength, the stronger is the scattering intensity.

(3) The intensity of scattered light of a certain wavelength is in direct proportion with the scattering angle (θ, included both angle of scattered light and incident light) related $(1+\cos\theta)$.

All scattering phenomena following the aforementioned rules are called Rayleigh scattering.

In 1899, Rayleigh concluded all his discoveries and wrote a groundbreaking paper, "Origin of the blue sky"[3]. The paper infers that "We will still have beautiful blue sky even without foreign particles". The "foreign particles" here actually refer to the solid or liquid aerosol particles of 0.001–100 μm in Tyndall scattering, such as dust, smoke, droplet, crystal, suspension, emulsion, and so on.

1.3.2 Rayleigh scattering law

1.3.2.1 Rayleigh scattering equation

In 1899, Rayleigh put forward the Rayleigh scattering equation for providing an explanation for light scattering under ideal conditions. The intensity of light scattering of a spherical molecule or particle depends on the particle size (a), incident wavelength (λ_0), particle relative refractive index (m), environment refractive index (n_{med}), test angle (θ), and other parameters:

$$I = \frac{8\pi^4 a^6 n_{med}^4 I_0}{d^2 \lambda_0^4} \cdot \left|\frac{m^2-1}{m^2+1}\right|^2 \cdot (1+\cos^2\theta) \tag{1.1}$$

where I is the intensity of light scattering, a is the particle size, n_{med} is the refractive index of the medium, I_0 is the intensity of the incident light, d is the distance from the scattering particle to the observer (detector), λ_0 is the wavelength of laser light in vaccum, and m is the relative refractive index of particles at the excitation wavelength, with the following equation:

$$m = \frac{n}{n_{med}} = \frac{n_{rel} + i n_{im}}{n_{med}} \tag{1.1a}$$

where n is the refractive index of particles; i is the complex number, composed of two components of the refractive index, where n_{rel} is related to nonphotoabsorption behavior of particles, called the real part of the refractive index, and n_{im} is related to photoabsorption behavior of particles, named the imaginary part of the refractive index.

From equation (1.1), it can be seen that the main factors to determine the intensity of single-particle scattered light are as follows: (1) sixth power of particle size (a); (2) biquadratic of the incident wavelength (λ_0); (3) distance (d) and angle of the observer (θ); and (4) refractive index. In further study, it was found that the refractive index was rather vital and complex, which will be discussed in later chapters. With the Rayleigh scattering equation, we can calculate that a relationship between the particle size (a), incident wavelength (λ_0), and angle of observer (θ) when a visible light interacts with particles far less than the incident wavelength, which can be given as shown in Figures 1.3–1.5.

On the basis of equation (1.1), dawn, sunset, rainbow, or blue sky depend on two factors, the first is the scattering degree of atmosphere and the floating particles for the light of different wavelength in sunshine, and the other is the observer's angle.

From equation (1.1) we can know that the intensity of scattered light is related to the distance and angle of observers. In other words, as shown in Figure 1.5, the intensity of Rayleigh scattering and angle dependence relationship demonstrates

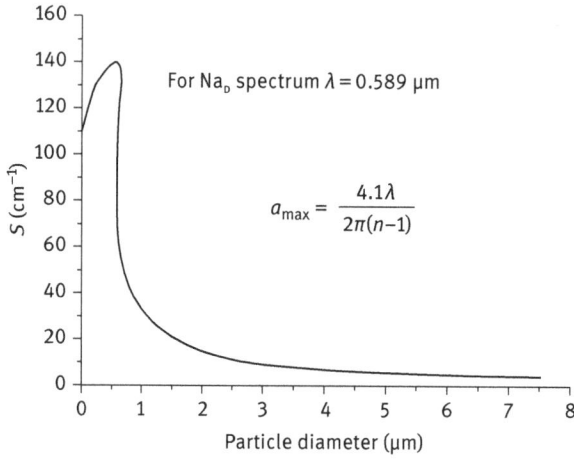

Figure 1.3: Influence of the particle size (a) on the scattering coefficient.

For Na$_D$ spectrum $\lambda = 0.589\ \mu m$

$$a_{max} = \frac{4.1\lambda}{2\pi(n-1)}$$

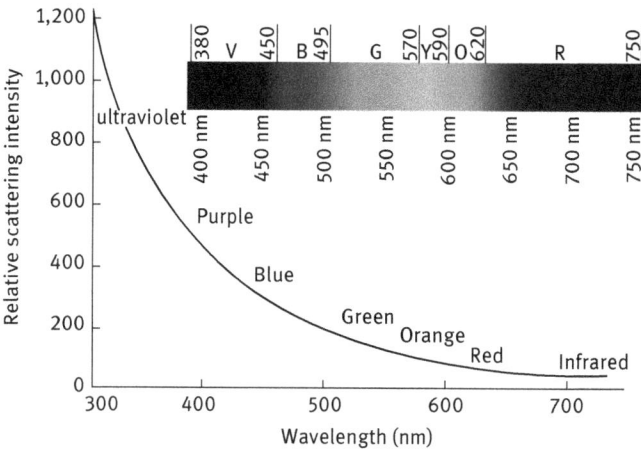

Figure 1.4: A frequency diagram of Rayleigh scattering.

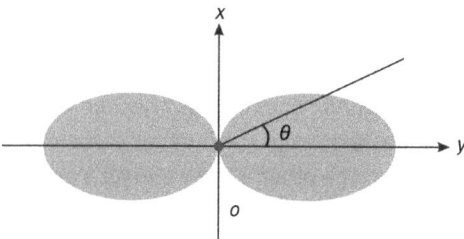

Figure 1.5: Angle distribution of the light intensity of Rayleigh scattering.

that Rayleigh scattered light has directivity, and both the right forward and backward scattering are the strongest and almost identical, while that vertical to the incidence direction is the minimum.

1.3.2.2 Light scattering examples

As sunlight has a continuous spectrum that is almost consistent with the radiation of absolute blackbody. The range from near ultraviolet to intermediate infrared has the most concentrated and rather stable, while that in X-ray, far ultraviolet and microwave band, energy is low but has great variation (Figure 1.6). Thus, the discussion about blue sky can be done by taking into consideration of following two aspects: (1) There are much less large particles in the air, making the ultramarine light scattering with short wavelength more outstanding. (2) Most electromagnetic radiation of atmosphere is nontransparent except for the visible light and a few other radio communication bands, that is to say, atmosphere would absorb electromagnetic of these wave bands.

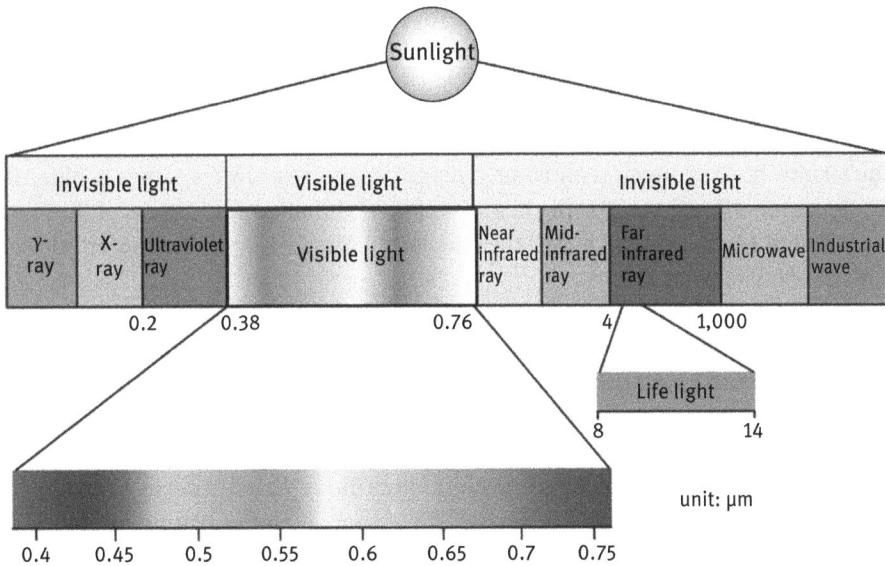

Figure 1.6: Sun radiation spectrum.

So, at noon, the sun is direct above us, and the sunlight passes through atmosphere and interacts with air molecules, resulting in Rayleigh scattering. Short-wavelength scattering is all over the sky with very few part of long-wavelength light scattering, making the sky blue. However, the sun itself and its surrounding area are white or yellow, and for this reason we mostly see the direct light rather than scattered light.

In such case, the light of long wavelength, red and yellow light, and a small amount of scattered light of short wavelength, blue and green light, get mixed, and the color basically would not change, and remain white.

When the sun sets or rises, the surrounding area of the sun looks red because the sunlight passes through a rather long way in atmosphere, during which most blue lights are scattered, and only that lights remain are red and orange light. At this time, the cloud looks red with the reflection of sunlight, while the sky still appears blue. If we observe from the moon, around which no atmosphere, the sky appears black even if during the day.

For the same reason, the energy of the sunlight reached to plateau region is considerably decreased as compared to that of highland. People living there would normally have "red plateau" on their faces for long-time ultraviolet irradiation. Caution lights in daily life such as for traffic are usually red in daily life because the long-wavelength of red light is not easily scattered by floating particles in the air so that it has strong penetrating power as compared with blue and green light. The reason is that people are more sensitive to red light. Therefore, the red light as the traffic light is to prevent road accidents.

We could also explain the blue sea with the same reason. Water molecules and small particles floating in the sea that are much smaller than the visible light wavelength have large degree of Rayleigh scattering for sunlight, and 450–550 nm light has good water penetrability with the maximum penetrating wavelength about 480 nm. This explains why the deeper water body or clearer water body has much more blue or blue-green color. Therefore, the sea always appears blue. It should be mentioned that for the blue sea, Rayleigh mistakenly assumed that the sea reflected blue sky, which mistake was later corrected by an Indian physicist, Raman (see Section 1.6.1).

It is noticeable that it is hard to observe the Rayleigh scattering when a light transmits through a nearly pure and homogeneous substance (such as a low-temperature pure crystal), molecules or atom particles produce scattered light coherent with the incident light, destructive in almost all directions except for the transmission direction of the incident light. To observe it, an important pre-condition would be an inhomogeneous refractive index distribution in a substance or a medium. If there is impurity (as small suspended matter or tiny bubbles), a slight inhomogeneous area becomes the scattering center with incoherent scattered light, and so the intensity of each beam must be directly added to observe scattering of light.

The Rayleigh theory assumes that all electrons and incident lights have the same phase and frequency oscillations , and the collective oscillation would surely lead to a large oscillating electron dipole and scattering of light. A precondition for the Rayleigh assumption is that the size of spheroidal particle (d) is far smaller than the wavelength of the incident light (λ). So, we can see that the Rayleigh scattering is actually mainly about molecule scattering.

The Rayleigh scattering equation can describe the phenomenon of molecules in a solution and molecular aggregates. However, from the point of view of modern physics, the Rayleigh scattering equation is too simple with boundedness. It mainly refers to scattering of mass point, far smaller than the wavelength of the incident light, especially for molecule scattering. When the size of the particles gets enlarged as comparable to the wavelength, especially when particle size is near or even larger than wavelength, different electrons in particles could oscillate at different phases, causing interfering scattering light of electrons to different degree. In such case, the intensity and angle distribution of scattered light deviate oscillating electron dipole, the Rayleigh scattering equation is no right again.

1.4 Mie scattering

1.4.1 Discovery of Mie scattering

For particles larger than the wavelength, there is a complex relationship between the intensity of scattered light and particles size and shape, then the Rayleigh scattering law is no longer followed (Table 1.1). The electromagnetic wave scattered by uniform, isotropic particles is called Mie scattering.

In 1908, Mie scattering was put forward by a German physicist, Gustav Adolf Feodor Wilhelm Ludwig Mie (1868–1957). Its concept is based on Mie's complete solving theory using the Maxwell equation for scattering of arbitrary magnitudes homogeneous spherical particles [4]. Mie first learned mathematics and physics, but he was also interested in chemistry, zoology, geology, mineralogy, astronomy, logics, and philosophy. In 1908, he used the classic Maxwell equation to study gold particles related color effect, and published the famous paper *Contributions to the optics of turbid media, particularly of colloidal metal solutions* in Germany *Annalen der Physik*. This paper gave him an idea as how to calculate a small spherical particle scattering, which was one classic paper in the field. He considered scattering particles as conductive balls. By calculating the polarization of the particles in light electric field which radiate electromagnetic wave, he put forward the scattering theory of floating particles with the size close to the wavelength of the incident light. He also creatively applied the Maxwell electromagnetic wave equation into plasma resonance for theory processing.

Mie's calculation results show that the following:

(1) Large particle scattering is different from Rayleigh scattering. Only when the radius and wavelength (λ) of a spheroidal particle satisfy the equation $r < 0.3\lambda/2\pi$, the electromagnetic scattering of spheroidal particles conforms to the Rayleigh law.

(2) If r is large, the intensity has no inverse biquadrate relation with the wavelength (as shown in Figure 1.7).

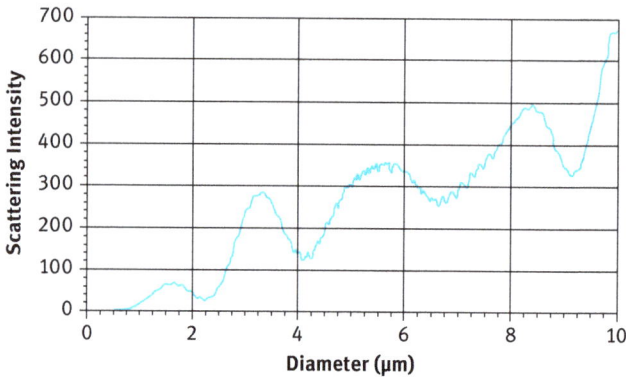

Figure 1.7: The intensity of scattered light of different droplets. The incident wavelength was 633 nm, and scattering angle (θ) 4°–12°. Data source: http://www.app17.com/tech/infodetail/157035.html.

Because of this, when the white light falls on large particles, as cloud in the sky, the scattered light is still white. Although Mie theory could only solve scattering of spherical particles, it was a good start for particle scattering research in modern physical theory, making a way for systematic research in the field of light scattering.

Compared with Rayleigh scattering, Mie scattering theory assumes that scattering of light of large particles is overlaid by electron dipole, quadrupole, octopole, and higher electron polypole oscillator, formulation of random size of spherical particles by the Maxwell equation. When the particle size is far smaller than the wavelength of the incident light, the electron dipole in the Mie equation plays a leading role and the equation can be simplified to the Rayleigh equation. Although the Mie theory could solve light scattering of any size, in light concept description, people still called scattering of molecules or particles (a few nanometers), far smaller than the wavelength of scattered light, as Rayleigh scattering, while large particle scattering that was unsuitable for the Rayleigh scattering theory is called Mie scattering. That is, to say, Mie scattering is suitable for particles whose size are of the same order or larger than light wave, and so Mie scattering is also called particle scattering.

Different from scattering intensity distribution of Rayleigh scattering, Mie scattering has asymmetric scattering light intensity, where the forward light scattering is larger than the backward light scattering. With the increase of the particle size, the ratio of the forward light scattering intensity and backward light scattering intensity gets increased, and the lobe of forward light scattering increases (Figure 1.8). When a particle is larger than the wavelength, the scattering process has no obvious reliable relationship on the wavelength.

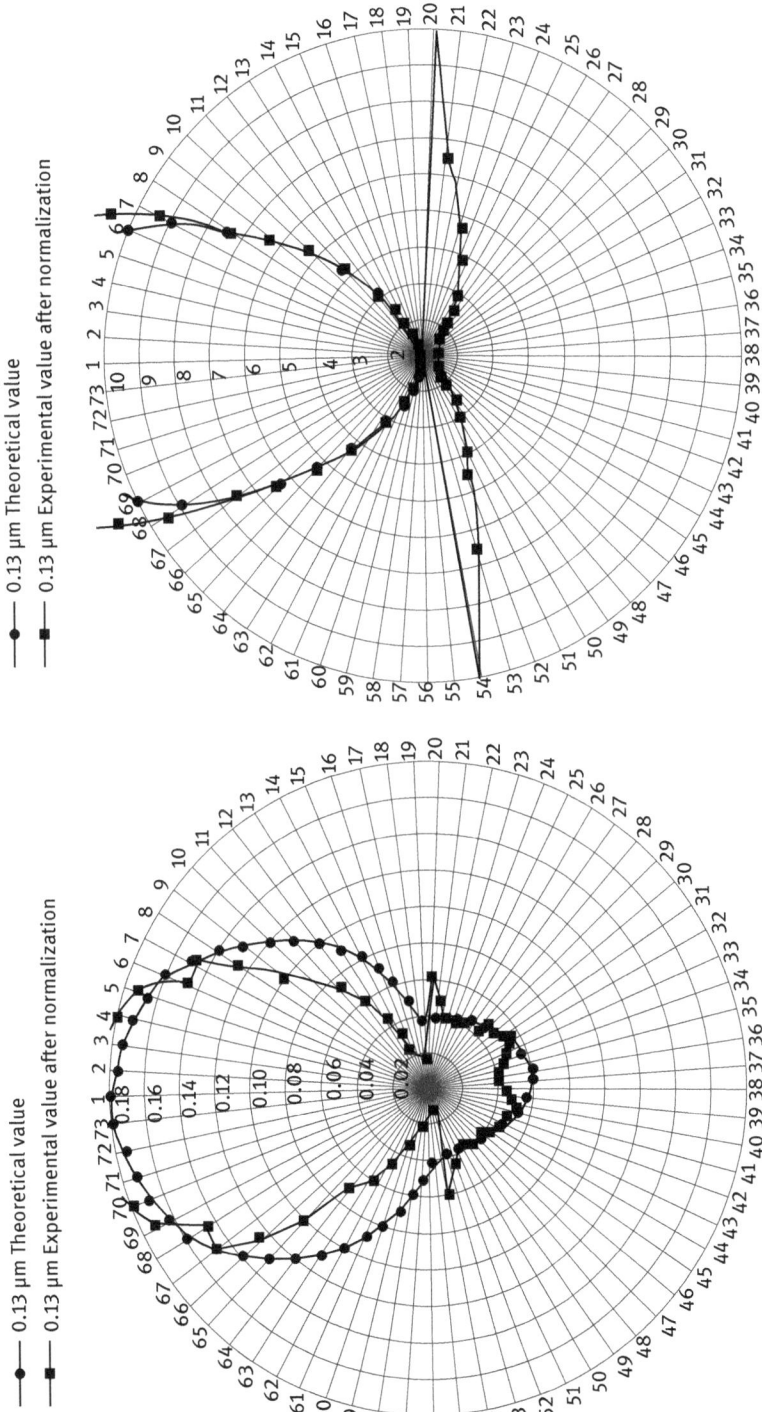

Figure 1.8: The light angle distribution of Mie scattering. Data source: http://www.app17.com/tech/infodetail/157035.html

1.4.2 Features of Mie scattering

Compared with Rayleigh scattering, Mie scattering has the following features:
(1) The light intensity of Mie scattering is much larger than that of Rayleigh scattering, while its variation with the incident wavelength is not as obvious as that of Rayleigh scattering.
(2) With the change of scattering angle, the light intensity of Mie scattering has many maximum and minimum values. When size parameter increases, the number of extremum value also increases.
(3) With the increase of scattering particle size, the total energy of Mie scattering increases rapidly, and then it comes to a constant value in vibrating form.
(4) When scattering particle size is rather small, Mie scattering can be simplified to Rayleigh scattering. With an increase of the particle parameter, the ratio of forward light scattering and backward light scattering increases, and forward light scattering increases. When the particle has a large size, the Mie scattering result is same as that by geometrical optics, namely there is light reflection or diffraction. In the moderate range of a size parameter, only Mie scattering could obtain the correct result. That is, Mie scattering calculation mode could widely describe scattering features of homogeneous spherical particles of any size.

By using the Mie scattering theory, we could well understand why cloud at noon appears white or grey. Scattering particles in the cloud such as water drops or crystals are large enough to the wavelength of sun lights, and so Mie scattering occurs with no obvious relation between the intensity and wavelength of the incident light. The scattered light has the color of sunshine, and so cloud in the sky appears white. If the cloud is too thick, the scattered light is unable to pass through the cloud, and the cloud appears grey or even black to our eyes.

It should be noted that Mie scattering in essence is not an independent theory, but solutions of Maxwell equations for a spherical medium. However, as the solution to Maxwell's equation is rather complex, the first perfect solution provided by Mie became classic, so this solution algorithm was called the Mie theory. Through Mie theory, people obtained many regular things, for example, rules of scattering related anisotropy coefficient changing with the relative diameter of medium balls, scattering of complex particles and particle groups.

1.4.3 Lorenz–Mie–Debye scattering

In the history of light scattering study, Mie was not the first scientist to formularize electromagnetic scattering. Before him, a German mathematician Rudolf Friedrich Alfred Clebsch (1833–1872) used the potential energy function to solve the scattering issues of perfectly rigid sphere as a elastic point source, while a Danish mathematician and physicist Ludvig Valentin Lorenz (1829–1891) independently put forward a similar theory in 1890, almost 20 years before the Mie theory. Although Lorenz's study was earlier than Mie's, his paper was published in Denmark and was less concerned. Comparatively, the Mie theory was drawn wide attention and was further developed, and so the phenomenon is called the Mie scattering theory.

In 1908, when Mie put forward a scattering solution to a homogeneous spherical particle of any size with the Maxwell equation, a Dutch-American physicist and chemist Peter Joseph William Debye (1884–1966) also studied the light scattering influence of surface reflection, penetrating particles, or inside refraction, and he also introduced the key Debye series to the Mie theory.

In 1909, Debye used two scalar potential functions to discuss radiation pressure of a spherical particle. Because of this, the Mie theory, discussing the related homogeneous isotropic sphere plane wave scattering, became Lorenz–Mie theory or Lorenz–Mie–Debye theory.

1.5 Essence of light scattering

1.5.1 Smoluchowski and density fluctuation theory

As described earlier, to observe the scattering signal, the main precondition is the existence of an inhomogeneous distribution of refractive index. A pure substance (or a medium) has density fluctuation everywhere, which constitutes the inhomogeneity of the refractive index distribution, and so it can be concluded that a normal, pure, and transparent material could also have Rayleigh scattering. As density fluctuation exists even in a pure substance, Rayleigh scattering is as also called as molecule scattering. As for the scattering signal produced by density fluctuation, a Poland statistical physicist Marian Smoluchowski (1872–1917) and a Jewish-American scientist Albert Einstein (1879–1955) reported their own research result in 1908 and 1910, respectively.

An Austrian physicist Ludwig Eduard Boltzmann (1844–1906) developed statistical mechanics to explain and predict physical properties, such as how mass, charges, and structures determine the viscosity, thermal conductivity, and diffusion of a substance. As deeply influenced by the study of Boltzmann, Smoluchowski started a fundamental research about the theory of object dynamics. He put forward the existence of density fluctuation in the gas phase. In 1908[*], he first concluded the critical

---

[*] When the substance is at a critical point, there is large density fluctuation, and its microcell is larger than the wavelength and stays for a long time. The liquid density microcell close to the critical density is quite like large particles. If we cast light on it, there is a strong molecule scattering in the substance, and it is called the critical opalescence. In fact, the fluid is transparent at higher temperatures, but if it gets muddy suddenly, it could scatter most or all of the incident light.

opalescence* to density fluctuation. He believed that gas is in a critical state in liquidation, making a small area much larger than the wavelength of light, similar to large particles, while strong light scattering of large particles changes the originally transparent substance into a muddy state, producing critical opalescence. Smoluchowski also focused on blue sky. He believed that the light dispersion is a result of photon interacting with the atmospheric waves. He also explained the Brownian motion of particles. In 1906, he described the Brownian motion and put forward stochastic processes theoretical equation.

1.5.2 Einstein and density fluctuation theory

As mentioned earlier, Rayleigh scattering could successfully explain the blue sky, but the precondition of the theory is that "air is an ideal gas." However, air is not an ideal gas, so Rayleigh scattering has shortcomings. Thus, in 1910, Einstein used just-developed entropy statistical thermodynamics theory to prove that "even the most pure air has fluctuation." Einstein proved that density fluctuation of air itself could scatter, producing the blue sky. By far, Einstein has completed the *Blue Sky Physics*, showing that there is an ineradicable self-impurity because of fluctuation, resulting in density fluctuation and scattering of sunlight that form the ineradicable blue sky. Only after this finding, *the debate about the blue sky came to an end*.

The relative studies of both Smoluchowski and Einstein not only perfectly explained the reason for the blue sky, but was also quite important for later studies. For example, a vital application of *Blue Sky Physics* is the optical fiber communication. A study of Nobel Prize winner in physics, a Chinese physicist Charles Kuen Kao (1933.11-2018.9), was made based on the Smoluchowski–Einstein equation [5] to prove that the most ideal glass, without a bubble or flaw, also has ineradicable self-impurity, which is the existence of ineradicable fluctuation in the glass itself. This kind of fluctuation is from the light wave signal in optical fiber transmission, a main physical factor in its signal loss. Thus, in optical fiber communication, only long-wavelength light can be chosen to transmit signal.

1.6 Raman scattering

1.6.1 Discovery of Raman scattering

In 1928, an Indian physicist Sir Chandrasekhara Venkata Raman (1888–1970) found that the scattering light frequency changes [6] in gas and liquid, which was later known as the Raman scattering or Raman effect. Raman scattering refers to the inelastic scattering produced by molecular vibrations when a certain frequency incident light interacts with molecules, which has different scattering frequency from the incident light. It should be noted that, before Raman scattering, investigations only include problems of light intensity and direction changes, not involving the changes of light energy (frequency or wavelength).

In 1921, 33-year-old Raman represented Calcutta University and went to Oxford for Commonwealth Universities meeting, British Royal Society, and some scholar tours. When the passenger ship of S. S. Narkunda that he went back to India through Mediterranean, the blue sea shining on the summer day greatly attracted him, and made him doubt Rayleigh's idea that "the blue sea is not the color of water, but the reflection of the blue sky."

To find the source of the color of the sea, Raman used a simple optical instrument to observe the sea level on the deck. By using the Nicol prism to diminish the blue light from the sky, he found that the color of the sea itself is even deeper than that of the sky. Then, he further used optical grating to observe for the color of the sea, finding that the maximum absorption of the seawater is bluer than that of the sky. In such case, he believed that the blue color of seawater was from the light scattering of the seawater, rather than the reflection of the sky. In such case, he wrote two papers on the ship and sent to England when S. S. Narkunda stopped at wharf, which were later published in two magazines of London. Once he went back to India, he immediately started a series experimental and theoretical researches to find out light scattering rules in each transparent medium.

With his students together, Raman found that if sunlight was used as the source of light to set pure water or alcohol flask, there is a quite weak green component of light observed from side with a purple filter (Figure 1.9). As the component has polarization, they ruled out the possibility of fluorescence.

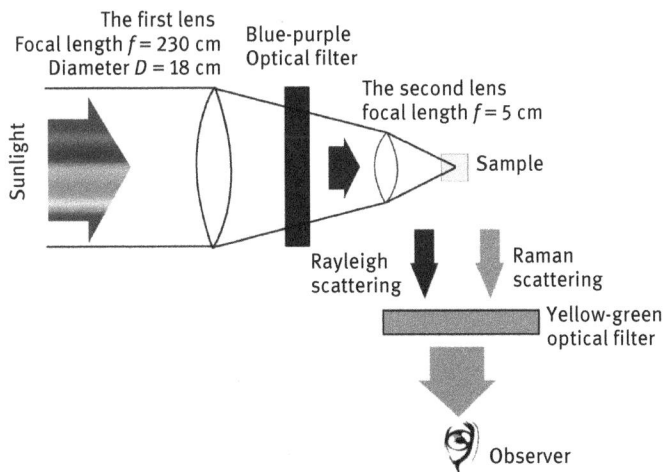

Figure 1.9: Experiment setup designed for Raman scattering observation by Sir. Raman.

Inspired by Compton effect[*], Raman named it as modified scattering, and wrote a short assay titled "New-type second-level radiation" on 16 February 1928 and published in *Nature* on 31 March 1928. The signal was later called Raman scattering. In 1930, an American spectroscopist Robert Williams Wood (1868–1955) named the modified scattering with lower frequency as the Stokes line and modified scattering with higher frequency as the anti-Stokes line.

1.6.2 Raman scattering spectrum

Raman's discovery proved that, except for some light components of the original incident light (Rayleigh scattering) there are some weak light scattering signals with different wavelength from that of the incident light (Raman scattering). Since the scattering of light is related to the intensity of the incident light, these weak light scattering signals can only be observed with a strong incident light. That is, developing strong light source is an effective way to study the weak signal and weak interaction.

Raman's discovery proved the prediction of the Austrian theoretical physicist Adolf Gustav Stephan Smekal (1895–1959) in 1923 [7]. Later, the former Soviet Union physicist Grigory Samuilovich Landsberg (1890–1957) and Leonid Isaakovich Mandelstam (1879–1944) also observed the scattering in a quartz crystal [8]. Raman's discovery provided new evidence for the light quantum theory and quickly won the public recognition, and thus Raman won the Nobel Prize in physics in 1930, only two years after his short note in Nature published. After X-ray Compton effect was found in 1920, Heisenberg predicted a similar effect in visible light in 1925. Thus, the British Royal Society named that Raman Scattering is "one of the most excellent discoveries of experimental physics in the twenty century." Raman's discovery greatly promoted the field of spectroscopy and also reflected deep influence to scientific history of India and even Asia.

In 1930–1934, a Czechoslovakian physicist George Placzek (1905–1955) researched the first Raman spectrograph, greatly promoting Raman Effect research. The spectrograph used the mercury arc as the source of light and a camera for analysis and test. However, the Raman spectrum, owing to weak signal and easily interfered by the strong Rayleigh scattering was limited to the study of chemical molecule vibrational spectra. With the appearance of intense light source laser until 20st century 60th year, Raman spectroscopy started its splendid history and now becomes a newly-developed study field [9–10]. As different chemical structures and different physical states, different molecules have their specific Raman spectra. This

---

[*] The Compton effect was discovered by the American physicist Arthur Holly Compton (1892–1962) in 1920. He found that there is longer wavelength besides the scattering light identical to the original wavelength when X-ray is scattered by a crystal.

information plays an important role in the study of molecular structure, similar to the fingerprint region of an infrared spectroscopy, which then is named as the Raman fingerprint.

The Raman spectrum is produced by the collision of photons and the outer electrons of molecules. During the collision, overlapping of molecular vibration or rotation energy and photon energy occurs, and the electron rises to an imaginary energy level and goes back to a higher or lower energy level, making the frequency of the scattered light changed, and the photon either obtains energy from molecules or passes energy to the molecules. Since the changes of the frequency of scattered light are closely related to the molecular or atomic microstructure, so the Raman spectrum is a useful tool for the investigation of molecular structure. People could understand molecular or atomic structural characteristics through the scattering spectrum.

In the Raman scattering spectrometry, on the both sides of the incident light with a frequency v_0, there are many Raman scattering spectral lines with a frequency of $v_0 \pm v_i$ (i=1, 2, 3, ...), where $v_i = \Delta E_i/h$. Lines on the side of the long wavelength is called as Strokes line, which is ascribed to the decrease of the scattering light energy owing to the energy transfer from the photon to a molecule. Anti-Strokes line, on the other hand, is located at the short-wavelength side with an increase in the energy of scattered light owing to the energy acquisition from molecules. Thus, the frequency difference (v_i) of the Raman scattering spectrum is irrelevant to the frequency of the incident light (v_0), but is decided by the structure of the substance.

The Raman scattering peak has certain correspondence with its infrared spectrum peak and can be used for qualitative and quantitative analysis of a substance. In the fluorescence spectrum scanning with a fluorospectrophot-ometer, the solvent Raman scattering peak could be detected. By changing the excitation wavelength we can observe the changes of the location of emission peak, which helps to judge whether the peak is Raman scattering peak of the test solvent or not.

1.7 Brillouin scattering

1.7.1 Brillouin scattering discovery

Brillouin scattering is one of scattering phenomena, which is caused by the sound speed propagation pressure fluctuation in the substance, and was first put forward by a French physicist Léon Nicolas Brillouin (1889–1969) in 1922. Brillouin studied physics in Paris in his early years and took part in lattice X-ray diffraction study. In 1914, he studied scattering light frequency spectrum. Although he joined the army for the First World War in 1914–1919, he went back to Paris University afterward and continued his study in quantum theory of solids, and became the doctor of science. In his PhD paper, he mentioned solid-state equation based on atomic vibration (phonon). In 1922, Brillouin calculated scattering light frequency distribution of density fluctuation of a sound wave in a scatterer, and predicted that there should be symmetrically distributed lines with different wavelengths around the incident light frequency.

Brillouin also made great contributions to quantum mechanics, air radio wave propagation, solid-state physics, and information theory. He studied monochromatic light wave transmission and wave band function, and found that scattering light has frequency shifts much smaller than the Raman scattering frequency, which are related to the frequency shift and scattering angle and can be used for acoustic vibration researches in gas, liquid, and solid.

1.7.2 Essence of Brillouin scattering

When the light interacts with the areas in a medium (such as air, water, or crystal) with the time-dependent light density to produce the changes of light energy (frequency) and propagation path, Brillouin scattering occurs. The density changes may be caused by the acoustic mode such as phonons, or magnetic modes such as magnetic oscillators, or temperature gradient. As described in classic physics, when a medium is compressed, the refractive index changes, conforming to Doppler frequency shift (refer to Section 1.8.1).

Brillouin scattering is an phenomenon that discloses the interaction among light wave, sound wave, magnetic wave, or any other vibration waves to slightly change the photon propagation direction and vibrational frequency. Any change in photon vibrational frequency is comparable to the frequency of the interacting sound wave or magnetic wave, and so the Brillouin scattering spectrum could be applied to analyze and determine the speed of the sound, temperature, and other parameters in the medium.

Similar to Raman scattering, Brillouin scattering is an inelastic scattering of elementary excitation in a medium, and the frequency changes show the energy of elementary excitation. However, Brillouin scattering is different from Raman scattering, which only involves in the small-energy elementary excitation, such as phonons and magnetic oscillators. As a practical study method, although Evgenii Fedorovich Gross (1897–1972), the former Soviet physicist, claimed to observe the Brillouin–Mandelstam light scattering during their observation of the fine structures of Rayleigh scattering in a condensed matter in 1930, the Brillouin scattering signal is weak, and it did not develop quickly until the appearance of intense light source laser.

To some degree, Brillouin scattering is in the field of Raman scattering, namely, the inelastic scattering of elementary excitation in a medium. Even though, Brillouin scattering is because of the vibration of sound, magnetic, or other vibration waves, while Raman scattering is because of molecule vibration with different sound waves and molecule vibration frequencies. The vibration frequency of Brillouin scattering is below 500 GHz, while Raman scattering could reach THz range. Besides, Brillouin scattering and Raman scattering have obvious differences in terms of equipment structure and applications.

1.8 Dynamic light scattering and static light scattering

1.8.1 Dynamic light scattering

In terms of time concept, the detection mode of the light scattering signals, there are dynamic light scattering and static light scattering. The concepts are originated from the ignorance of the movements and time-dependance of the scattering particles, namely, the dynamics. In fact, particles in the solution have Brownian movement, which produces the Doppler shift.

In 1842, Christian Johann Doppler (1803–1853), an Austrian physicist and mathematician, proposed the concept of Doppler shift. He believed that the radiation wavelength of a substance changes with the radiation source and observer's relative movements. That is, the wave in front of a moving radiation source is compressed (blue shift), while later the wave is stretched (red shift). The faster the speed of the radiation source, the larger is the shift effect. Thus, when the light falls on the moving particles, the irregular Brownian movement makes the particles sometime move away from the radiation source of the incident light and sometime close to it. So, fluctuation of the scattering light wavelength (λ) and excitation wavelength (λ_0) exists, which can be given as

$$\lambda = \lambda_0 \left(1 + \frac{v}{c}\right) \tag{1.2}$$

where v is the instant speed of the particle in a solution and c is the speed of light. The Brownian motion causes the Doppler shift. Scattering light intensity of particle move fluctuates with time, so it is named as dynamic light scattering. Generalized, dynamic light scattering refers to the signal that vibrates with time. Narrowly, dynamic light scattering refers to dynamic light scattering technology, which refers to those to analyze the particle informations including the size and state by measuring the light scattering signals of sample varying with time.

The fluctuations of dynamic light scattering signal are originated from the Brownian motion of molecules or particles in the medium. The Brownian motion is related to particle size, medium viscosity, temperature, and so on. The smaller the particle, the faster the Brownian motion is, and then the faster the scattering light intensity fluctuation is. So, by using the dynamic changes of the scattering light intensity with the aids of laser to excite the solution and other spectral technologies, people could obtain the informations of particle or molecule size, molecular weight, dispersibility, uniformity etc, and these light scattering technologies have been widely used in polymer chemistry, drug development for molecular weight measurement and particle size distribution study.

1.8.2 Static light scattering

Static light scattering is a generalized concept as compared with the dynamic light scattering, which refers to the light scattering without considering the movements of particles with time going on. The previously mentioned Rayleigh, Mie, and Brillouin scatterings could all be considered as static light scattering. Therefore, static light scattering can be developed for physiochemical techniques, which make use of the scattering intensity in any direction or angle to study and analyze the scattering particles. The particle size, mass, component, and other parameters can be measured based on the relationship of intensity and angle according to the Mie theory.

Static and dynamic light scattering are concepts of scattering signals and their changes with time without referring to the essence of the light scattering. For example, most static and dynamic light scattering facilities coupled with near-infrared laser as the source of light, trying to avoid fluorescence or other scattered. They seldom coupled with splitting system, directly with photomultiplier as signals detector. So, the detected light scattering signal could not be simply classified as Rayleigh or Mie scattering, elastic scattering, or inelastic scattering.

References

[1]   Rayleigh L. On waves propagated along the plane surface of an elastic soil. Proc. Lon. Math. Soc., 1885, 17(3): 4–11.
[2]   Rayleigh L. On the light from the sky, its polarization and colour. Phil. Mag., 1871, 41(271): 107–120.
[3]   Rayleigh L. On the transmission of light through an atmosphere containing small particles in suspension, and on the origin of the blue of the sky. Phil. Mag., 1899, XLV(II): 375.
[4]   Mie G. Beiträge zur Optik trüber Medien, speziell kolloidaler Metallösungen. Ann. Phys., 1908, 25(3): 377–445.
[5]   Kao C. Proc. IEE, 1966, 113(7): 2010.
[6]   Raman C V, Krishman K S. A new type of secondary radiation. Nature, 1928, 121(3048): 501–502.
[7]   Smekal A G. The quantum theory of dispersion. Nature, 1924, 114(2861): 310–311.
[8]   Landsberg G, Mandelstam L. A novel effect of light scattering in crystals. Naturwiss, 1928, 16: 557–772.

[9]   Li J F, Huang Y F, Ding Y, et al. Shell-isolated nanoparticle-enhanced Raman spectroscopy. Nature, 2010, 464(7287): 392–395.
[10] Zhang R, Zhang Y, Dong Z C, et al. Chemical mapping of a single molecule by plasmon enhanced Raman scattering. Nature, 2013, 498(7452): 82–86.

Yuan Fang Li, Jian Ling

2 Electromagnetic wave and light scattering theory

2.1 Overview

According to the electromagnetic theory, light makes the molecules vibrate and forms dipole oscillator that creates a secondary wave coherent with the incident light. So, in a homogeneous pure gas or liquid system, the total intensity of scattering light is zero, because for any differential element of a homogeneous medium, people can always find one correspondent differential element to offset at observing point owing to the interference of scattering electromagnetic field.

In order to make that happen, it should assume that scattering system is static in an ideal state. As molecules are in a state of constant random thermal motion, it is difficult to find a completely fixed or regular unit, so scattering, even if becomes very weak or even negligible, but exists objectively. Even in a homogeneous medium, molecules are in constant random thermal motion, which induce the fluctuations of partial density or concentration and so as to make instant changes of partial refractive index or dielectric constant, producing a weak inductive dipole moment that is different in various thermal motion areas, and thus leading to a certain light scattering. Because of this, we can still observe light scattering of gas or liquid of homogeneous pure substance. Obviously, if a medium, such as sol, is light inhomogeneous, an there are suspended particles in the system and there are great differences of the refractive index between the dispersion phase and the medium, producing a large induced dipole moment, in which secondary waves cannot be diminished, causing a significant light scattering.

Light scatterometric methods discussed in this book are mainly quantitative analysis based on elastic scattering without considering photon energy changes before and after scattering, and only considering scattering light intensity and direction. Even so, related theories are still needed to discuss the properties of a scatterer (including intensity, spectrum range, and influencing factors of the environment, such as dielectric constant and enviromental temperature). It needs related scattering theory for instruction and reckoning. In this book, the basics of scattering theory involves in Rayleigh scattering theory, Mie scattering theory, discrete dipole approximation algorithm, and so on. This chapter will simply introduce these theories and calculation methods.

As described in Chapter 1, both Rayleigh scattering theory and Mie scattering theory are the most basic. Rayleigh was the first to put forward scattering theory equations, which is mainly used to solve light scattering properties of an ideal spherical scatterer of small particles, whereas Mie theory describes ideal spherical scatterers of any size. Because Mie theory has no limit on the particle size, it includes

https://doi.org/10.1515/9783110573138-002

Rayleigh scattering theory, whereas Rayleigh scattering theory can be seen as a simplified equation of Mie theory when particles are small [1].

We can use a parameter, α, which can express the size of scattering particle, to judge which theory should be used for explanation and calculation the light scattering properties, and the parameter is as follows:

$$\alpha = \frac{\pi r n_m}{\lambda_0} \tag{2.1}$$

where r is the scattering particle radius, λ_0 is the environment wavelength, and n_m is the environmental refractive index. When $\alpha \ll 1$, equation (2.1) can be simplified into Rayleigh scattering equation; when $\alpha \gg 1$, it can be explained by macrogeometric reflection and refraction.

For easily understanding scattering theory, we will explain the related theories of electromagnetic waves in the following sections. Meanwhile, we will also discuss the related problems of scattering light as a secondary radiation.

2.2 Maxwell equations and electromagnetic wave

2.2.1 Maxwell equations

As described earlier, Mie theory discusses ideal spherical scatterers of any size, theoretical solutions of light scattering based on Maxwell equation.

James Clerk Maxwell (1831–1879), the English physicist and mathematician believed that, on the basis of three experimental laws of electromagnetism (Faraday's law of electromagnetic induction, Coulomb's law, and Ampere's law), changing the electric field can induce the changes of magnetic field surrounding it, whereas the changes of magnetic field induces the changes of the distant electric field. Electric field and magnetic field induce each other, appear alternately, and are mutually perpendicular, transmitting from the near to the distant in space with limited speed (Figure 2.1), and forming electromagnetic waves. Properties of electromagnetic waves, such as electric field, magnetic field, charge density, and current–density relationship can be described with Maxwell's equations[*].

---

[*] In the history of scientific development, Maxwell was a great figure to unite electricity and light, after Newton's union of air and ground moving rules. Maxwell put forward 20 equations and 20 variations in 1865, and in 1873, changing the equation into a quaternion expression. In 1884, Oliver Heaviside (1850–1925), a British physicist, and Josiah Gibbs (1839–1903), a American chemical physicist, put forward the calculus and vector analysis, changing the differential equation into an ordinary algebraic equation, and realizing the re-expression of Maxwell's equations' vector form, which is still commonly accepted even if up to now.

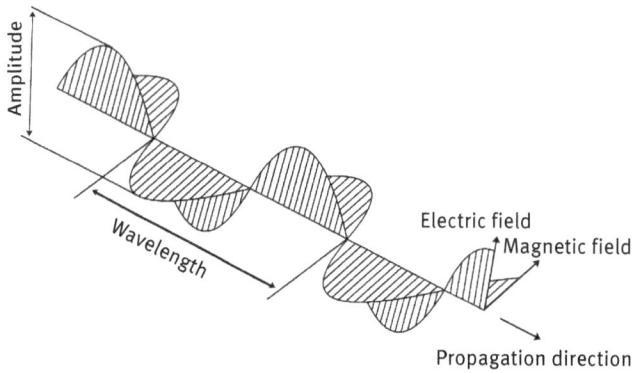

Figure 2.1: Electromagnetic wave formation and propagation.

Maxwell's equations have differential form, integral form, plural form, and so on. We will discuss an in-depth relationship between the electric field and the magnetic field in differential form and integral form in the following sections.

2.2.1.1 Differential form

(1) Defines what is a electric field:

$$\nabla D = \rho \tag{2.2}$$

It demonstrates how charge generates an electric field and describes its properties, where D is the electric displacement, ∇ is the divergence operator, and ρ is the charge density.

(2) Defines what is a magnetic field

$$\nabla B = 0 \tag{2.3}$$

That is, a magnetic monopole has no Gauss's law for magnetism. In the equation, B is the magnetic induction intensity.

(3) Explains how the change of the magnetic field (vortex magnetic field) induces electric field:

$$\nabla \times E = -\frac{\partial B}{\partial t} \tag{2.4}$$

Its essence is the Maxwell–Ampere law, where $\times$ is the rotor operator.

(4) Explains how the changes of the electric field (displacement current) induces magnetic field:

$$\nabla \times H = j + \frac{\partial D}{\partial t} \qquad (2.5)$$

The essence is the Faraday's law of induction, where j is the total current density.

2.2.1.2 Integral form

Above partial differential equations usually involve a closed system, so that can be expressed in an integral form to reflect the relation of area quantity of electromagnetic field (**D, E, B, H**) and field source (charge q, current I). Thus, these integral equations can be expressed as the following:
(1) Electric field properties correspond to the Gauss theorem:

$$\oiint_s D dS = \sum q_i \qquad (2.2')$$

Generally, the electric field is related to the Coulombian field and induced electric field owing to the changes of magnetic field. The former one is active field, and the latter is eddy field. As the electric displacement line of eddy field is closed, it has no contribution to a curve flux of closed surface.
(2) Magnetic field properties obey the Gauss theorem:

$$\oiint_s B dS = 0 \qquad (2.3')$$

Namely, the magnetic field can be excited by transmit current and also can be induced by the displacement current of the changes of electric field. It is noticeable that all of these magnetic fields are eddy fields with closed magnetic induction line, which has no contribution to the closed curve flux of a close surface. In addition, electric fields, either are induced by a transmit current or a displacement current, are all passive fields.
(3) Circuital theorem of electric field

$$\oint_L E dl = - \iint_s \frac{\partial B}{\partial t} dS \qquad (2.4')$$

All electrostatic fields are conservative fields, changing only the magnetic field can induce electric field and produce vortex electric field.
(4) Ampere circuital theorem

$$\oint_L H dl = - \iint_s \frac{\partial D}{\partial t} dS \qquad (2.5')$$

The change of magnetic field can induce an electric field, demonstrating that the transmit current and the changes of electric field can induce vortex electric fields.

Therefore, magnetic field and electric field are interdependent. That is, the changes of a magnetic field can induce a vortex electric field and the changes of a electric field can induce a vortex magnetic field. So, both the electric field and the magnetic field are closely related, inducing each other and forming a uniform electromagnetic field. That is the core of the Maxwell electromagnetic field theory.

2.2.1.3 Solution of Maxwell's equations

As mentioned earlier (Section 1.4.1), Mie scattering theory is strict mathematical solutions to the Maxwell equations under the boundary conditions of a homogeneous single scattering particle in a medium exposed to parallel homochromatic waves, which are the most common and basic arithmetic. As there is no other theory capable of exceeding the precision of Mie scattering theory when treating the scattering of particle at the size of wavelength scale there are lots of software available up to now for Mie scattering treatment in order to solve the solutions conveniently.

Mie scattering theory is mainly used for light scattering of particles within the range of 100 nm to 1 μm. For particles below 100 nm, scattering can be approximated into a simple Rayleigh scattering law; large particles over 1 μm can be utilized in a Fraunhofer diffraction pattern.

2.2.2 Generation and propagation of electromagnetic waves

A wave source is required for the generation of an electromagnetic waves. Candle, light, chemiluminescence of chemical reaction, and triboluminescence of mechanical friction are common wave sources*. Different wave sources produce different electromagnetic waves. As there is no objective with temperatures equal to or lower than absolute zero, any object above absolute zero can emit electromagnetic waves. That is, any object in the world is a wave source of electromagnetic waves, so electromagnetic waves widely exist.

Heinrich Rudolf Hertz (1857–1894), the German physicist, experimentally confirmed that electromagnetic wave is a transverse wave in 1886–1888 (Figure 2.1) and published his experimental results in 1888. He also confirmed Maxwell electromagnetic

---

* Kilohertz, kHz, 10^3 Hz; megahertz, MHz, 10^6 Hz; gigahertz, GHz, 10^9 Hz; terahertz, THz, 10^{12} Hz; petahertz, PHz, 10^{15} Hz; exahertz, EHz, 10^{18} Hz. visible light: 380–790 THz; frequency telemetering: 200 GHz; millimeter wave radar frequency: 94 GHz, common 300 MHz to 40 GHz; Wi-Fi frequency: 2.4 GHz; Wide band Code Division Multiple Access (WCDMA): 1,920–2,100 MHz; ultrasonic: above 20 MHz; sound wave: 20 Hz to 20 MHz; infrasonic wave: below 20 Hz.

wave theory and properties through an experiment. He believed that electromagnetic field can be expressed in partial differential equations (2.6) and (2.7):

$$\frac{\mu}{c}\frac{\partial H}{\partial t} = -\frac{\partial E}{\partial x} \tag{2.6}$$

$$\frac{\varepsilon}{c}\frac{\partial E}{\partial t} = -\frac{\partial H}{\partial x} \tag{2.7}$$

where ε is the medium relative dielectric constant, μ is the relative permeability, and c is the speed of light.

Afterward, Guglielmo Marchese Marconi (1874–1937), an Italian electrical engineer and inventor, performed a series of researches to prove that light is not just one electromagnetic wave, but has many forms, such as X-ray and γ-ray. Even though, all these forms have same essence and the only difference is in their wavelength or frequency. Based on this, Marconi invented radiotechnics, and he became the inventor of the first useful radiotelegraph system in the world.

The two partial differential equations (2.6) and (2.7) show fluctuation in the electric field and the magnetic field, as well as their close relationship demonstrating that electromagnetic wave transmits in the form of electromagnetic field. Magnetic field, electric field of electromagnetic wave and their propagation directions are mutually perpendicular. Namely, electromagnetic waves propagates along a straight line at the speed of light (c) in vacuum. In its propagation process, amplitude has periodic alternate changes in a vertical direction with the intensity inversely proportional to square of the distance. It should be noted that the propagation of electromagnetic wave, when its frequency is low, has to rely on the physical conductivity of the body, and can be carried out in free space or in the physically conductive body when its frequency is rather high. Electromagnetic wave has energy during its propagation, and the power of any position is in direct proportion to the square of the amplitude.

Electromagnetic wave energy is decided by the Poynting vector:

$$S = E \times H$$

where S is the Poynting vector, which is an energy flow density vector in the electromagnetic field; E is electric field intensity; and H is the magnetic field intensity. Either in static field or constant field, energy transfer depends on Poynting vector properties.

2.2.3 Electromagnetic wave properties

According to the electromagnetic wave equations, it can be predicted that electromagnetic wave speed is the same as light speed, so light wave is actually an electromagnetic wave, which is the basis of Maxwell's electromagnetic wave theory. As electromagnetic field has energy, wave propagation process is just the propagation process of electromagnetic energy. That is, electromagnetic wave, namely, the light, has energy. Max Karl Ernst Ludwig Planck (1858–1947), a famous German physicist and quantum mechanics founder, found in his study of object's thermal radiation rules in 1900 that if emission and absorption of electromagnetic wave is inconsistent, but performs one by one, the calculation results are identical with the experimental ones. This "one by one" energy unit is named as energy quantum or light quantum, where the energy of one photon is as follows:

$$E = h\nu = \frac{hc}{\lambda} \tag{2.8}$$

where h is the Planck's constant, 6.63×10^{-34} J·s, assuming emission and absorption of electromagnetic wave is inconsistent.

2.2.3.1 Wave–particle duality

Wave–particle duality means that light has both undulatory and corpuscular properties. Wave–particle duality is one of the important concepts in quantum mechanics, which is a leap in our understanding of the universe, laying foundation to the development of wave mechanics. In classical mechanics, subjects are divided into wave and particle and describe the macro-mass point state by using mass point position and momentum (or speed), highlighting the corpuscular property of mass point. Equation (2.8) perfectly unifies the wave–particle duality of light, wherein $h\nu$ highlights its corpuscular property and hc/λ highlights its undulatory property.

When an electromagnetic wave passes through a medium, regardless of the medium component and homogeneity, there is always refraction, reflection, scattering, absorption, and so on. Electromagnetic wave of long wavelength has less reduction and it is easier to continue its propagation by rounding the obstacles. As all types of waves have wave–particle duality, mechanic wave, the same as electromagnetic wave, also has refraction, reflection, diffraction, and interference, where refraction and reflection belong to the corpuscular properties, while diffraction and interference belong to undulatory properties.

Although equation (2.8) appears simple, its birth in fact has experienced a complex process of two extremes to a combination. Francesco Maria Grimaldi (1618–1663), the Mathematics professor of University of Bologna, Italy, found in his observation in 1655 that bright and dark stripes resembling a water wave could be found when a light beam falls on the screen through two holes in a dark room, so he concluded that (1) light is a fluid with undulatory movements and (2) lights with different colours come from different wave frequencies. In 1663, British chemist Robert Boyle (1627–1691) also proposed that color of the object is due to light irradiation, rather than the nature of the body itself. The finding of both Grimaldi and Boyle were in dominant position at that time, and the both science giants were staunch supporters and defenders of light undulatory property.

As compared, Sir Isaac Newton (1643–1727), the famous apple-revealed British physicist, found that a colorful spectrum can be obtained when sunlight passes through a prism in a dark room through a hole, then he believed that light composition and decomposition is similar to a mixture of different colour particles and they can be separated. He published a paper titled *New theory about light and colors* in 1672 to describe his experimental results and put forward a light particle theory. Newton's theory was contradictory to the popular undulatory theory of the day, causing a debate more than one century, which was not solved until the establishment of wave–particle duality of quantum mechanics in the beginning of the twentieth century. Quantum mechanics believes that all particles in nature, either photon, electron, or atom can all be described in one complex differential equation, and the solutions to the differential equation are wave functions.

Einstein introduced photon concept on the basis of Planck's concept of energy quantum in 1905. He believed that photon is some particle without mass in static state and has the energy of $E = h\nu$ and the momentum of $P = h/\lambda$ or $h\nu/c$. Photon propagation in space is inconsistent and in the form of unit by unit. Based on this, Einstein further put forward the photoelectric effect theory, believing that if one photon has high enough of energy, it gets off from a metal surface to produce light current. His photoelectric effect theory theoretically solved an unexplainable experimental phenomenon of wave theory. So, Einstein was the first scientist to settle the debate about the undulatory and corpuscular properties of light, defining the wave–particle duality of light. At the same time, British mathematician and physicist William Henry Bragg (1862–1942) and American physicist Arthur Holly Compton (1892–1962) also put forward similar views. For example, Compton proved that there is energy and momentum exchange in the interaction of electrons and photons. In 1923, French theoretical physicist Louis Victor de Broglie (1892–1987) promoted Einstein's wave–particle dualism to microscopic particles and put forward the material wave hypothesis, demonstrating the microscopic particle's undulatory properties, which was later identified by an electron diffraction experiment.

2.2.3.2 Reflection and refraction of light

When light falls on the surface of two mediums, both reflection and refraction occur, namely, one incident light beam is divided into reflection and refraction light beams. The relationship between the reflection and refraction beams is determined by the reflection and refraction law. On the basis of the transverse wave of light, and by dividing the incident light beam into linearly polarized lights, whose vibration planes are parallel and vertical to the incident plane, respectively, Augustin-Jean Fresnel (1788–1827), the physical optics founder and French civil engineer and physicist, derived Fresnel formulae, which can show refraction–reflection ratio relationship. Fresnel's formula has two groups, four equations (2.9)–(2.12), and the discussion on electric vector and incident electric vector of refraction and reflection instance is given as follows.

Suppose electromagnetic vibration vector of incident light, reflection light, and refraction light is divided into two components, one parallel to and the other vertical to an incident plane, and expressed by p and s. A_1, A_1', and A_2 are also adopted, respectively, to express the instant electric vectors of incident, reflected, and refractory light beams, then their amplitude components on p and s directions are A_{p1}, A_{p1}', A_{p2} and A_{s1}, A_{s1}', A_{s2}. Light energy can be written as follows:

$$\frac{E'_{p1}}{E_{p1}} = \frac{n_2 \cos i_1 - n_1 \cos i_2}{n_2 \cos i_1 + n_1 \cos i_2} = \frac{tg(i_1 - i_2)}{tg(i_1 + i_2)} \tag{2.9}$$

$$\frac{E'_{s1}}{E_{s1}} = \frac{n_1 \cos i_1 - n_2 \cos i_2}{n_1 \cos i_1 + n_2 \cos i_2} = \frac{\sin(i_1 - i_2)}{\sin(i_1 + i_2)} \tag{2.10}$$

where E is energy of the light, E'_{p1} and E'_{s1} are the partial energy of the reflected light in p and s directions; n_1, n_2 are refractive indexes of two phases forming interface; i_1, i_2 are incident and reflection angles, respectively.

Equations (2.9) and (2.10) well reflect the two corresponding partial ratios of reflection wave and incident wave, respectively.

Corresponding to these equations, two component ratios of refraction wave and incident wave are as follows:

$$\frac{E_{p2}}{E_{p1}} = \frac{2n_1 \cos i_1}{n_2 \cos i_1 + n_1 \cos i_2} = \frac{2 \sin i_2 \cos i_1}{\sin(i_1 + i_2) \cos(i_1 - i_2)} \tag{2.11}$$

$$\frac{E_{s2}}{E_{s1}} = \frac{2n_1 \cos i_1}{n_1 \cos i_1 + n_2 \cos i_2} = \frac{2 \sin i_2 \cos i_1}{\sin(i_1 + i_2)} \tag{2.12}$$

Fresnel's formula well expresses the electric vector and incident electric vector of refraction and reflection, which are important in optics and electromagnetism. Through Fresnel's formula, physical parameters including reflection and transmission amplitude and intensity under certain incident light angle can be further obtained, which are widely used in an electromagnetic phenomenon. It should be noted that Fresnel was earlier from the time than Maxwell, so the foundation of Fresnel's formula was the early optics, and thus people could also obtain Fresnel's formula by starting from Maxwell's equation and by making use of boundary value relationship of electromagnetic field of medium boundary.

2.2.3.3 Diffraction of light

Light propagates in a homogeneous medium following the straight-line law, which is just a law of approximation, and occurs only when light wavelength is much shorter than an obstacle. In fact, if the obstacle size is comparable to the wavelength or even smaller, obvious deviation might occur during the light propagation, which phenomenon is called as light diffraction. Namely, when meeting obstacles, light diffracts more or less from its original propagation direction, then reflection and refraction on the surface could be observed. Diffraction is endemism of waves, and all waves have diffraction. For example, people can observe wave diffraction of water in a swimming pool or a lake with a blowing breeze.

As the wavelength of visible light is 360–760 nm, quite small as compared to obstacle in a macro-world and common light source being an incoherent surface source, light diffraction is not easily found in our daily life. Only can it be observed clearly when strong light falls on a small particle, such as a pinhole, round screen, slit, filaments, knife edge, straight line, and other obstacles.

As sunlight is a polychromatic light, so we can see a colored diffraction light. In the observation and study of light diffraction, monochromatic light is used with a better effect as compared to polychromatic light (such as sunlight). When light with the same wavelength but different phase propagates after diffraction, both strengthening and weakening of light intensity could be observed, causing light interference. If a black screen is put in front of the propagation direction, alternate dark and bright diffraction figures can be observed (Figure 2.2).

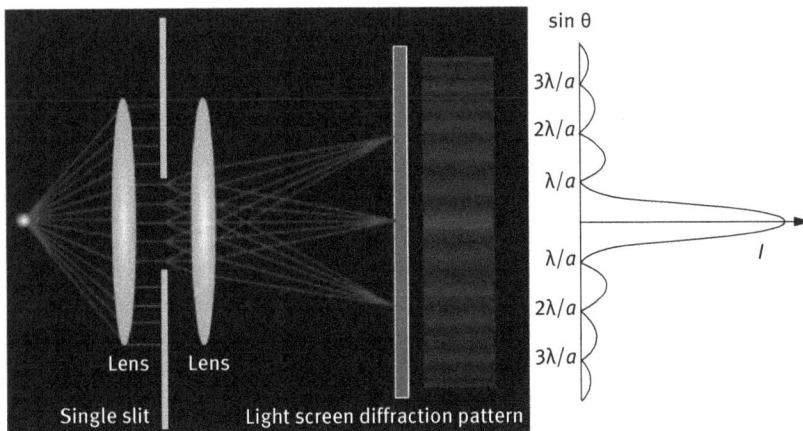

Figure 2.2: Schematic of a single slit diffraction of red light.

From the first discovery of light diffraction by Grimaldi in 1655, there had been a hundred years passed until the fully explanation by Fresnel in nineteenth century. During that time, Dutch famous physicist and astronomer Christiaan Huygens (1629–1695) put forward the view of that "the points wave reached form new wavelets", which could well explain light diffraction, but can not give the parameter of wave amplitude. On this basis, Fresnel put forward the concept of wavelet coherence stack during the propagation the sum of wavelets determines the form of the wave at any subsequent time. So, "Huygens–Fresnel principle" could well explain light diffraction, especially uneven distribution problems of light intensity in a diffraction pattern.

In theory, horizontal line of the aperture ρ has an inverse relation with the diffraction divergence angle $\Delta\theta$ as follows:

$$\rho \approx \frac{\lambda}{\Delta\theta} \tag{2.13}$$

Since Röntgen's finding of X-ray in 1895, scientists put forwards different viewpoints about its essence. Among them, Max Theodor Felix von Laue (1879–1960), Planck's student and a German physicist, came up with one hypothesis according to above diffraction principle: if the distance of mass point is in the same order of magnitude with the X-ray wavelength, it is possible that X-ray can be used for observing light diffraction considering the fact that lattice point (including atom and ions) is a regular three-dimensional structure. Under the guidance of Laue, Ludek Frydrych and other scientists place one parallel wafer between an X-ray source and a photographic plate, and observed a regular spotted groups appearing on the plate, which was later known as the Laue pattern. Based on this, they established X-ray diffraction analysis method and confirmed that the X-ray wavelength is at an order of magnitude of 10^{-8} cm, which is basically consistent with the distance of the atom in a solid. It is that they used this experiment to confirm an electron's undulatory properties, which made great contribution to the later development of physics, Laue won the Nobel Prize in physics in 1914.

2.2.3.4 Interference of light

Interference is the phenomenon that two or more waves with similar properties meet somewhere in the space wherein overlapping occurs to form new waves. For example, by separating one monochromatic light beam with a beam splitter into two beam and then making them meet somewhere in the space, it can be found that the light intensity distribution in their meeting area is not homogeneous, demonstrating alternate dark and bright stripes with their light and shade degree varying with the location in the space. These stripes are interferometric fringes.

Interference is one of the light wave properties, and relative experiments are important evidences to prove and identify the undulatory properties of light in history.

In 1801, 28-year-old British physicist Thomas Young (1773–1829), who loved music, questioned whether the sound is a type of wave just like light, and thus in his own lab he performed a famous "absurd" and "illogic" Young's double-slit interference experiment. He observed bright interference stripes overlapping with two or more light waves in space and some places were always strengthened, whereas other areas were always weak, forming a stable strong-weak distributions.

Interference stripes of Young's double-slit interference experiment are just an unexplainable phenomenon with Newton's corpuscular theory of light, because its experimental facility is the light of one point source propagated to the screen with two slits on one side. As slit distance is rather small and their distances to the light source are almost equal, the two stripes were secondary monochrome point light source with the same phase position and were coherent lights, so interference stripes could be obtained on the screen far from the slits.

Based on his above experimental results, Young did not surrender the authoritative experts of the day, and utilized interference principles for diffraction in 1803 and published a book titled *A Course of Lectures on Natural Philosophy and the Mechanical Arts*, in which he comprehensively described his study of optical theory and experiments, particularly his double-slit interference experiment. His book was a heavy bomb after hundred years of silence in the field of wave theory of light, and his double-slit interference experiment ranked among the five most classic physical experiments. Even though, his theory greatly contracted to a light polarization phenomenon.

In 1818, Fresnel supplemented Huygens principle with light interference principle in one scientific conference held in Paris Academy of Sciences, and came up with Huygens–Fresnel principle, explaining the polarization properties and identifying the that light transverse wave properties of light, which brought light undulation hypothesis to a new stage.

To obtain a coherent source for visible light interference observation, scientists invented a series of optical devices to produce interference light, and eventually the interferometer. By using Michelson interferometer in 1887, American scientists Albert Abraham Michelson (1852–1931) and Edward Morley (1838–1923) completed the famous Michelson–Morley experiment with zero observation results of ether wind. It is for that experiment, Michelson won Nobel Prize in physics in 1907.

Laser, as one high-intensity coherent light source, has been widely applied since the 1960s, in which laser interferometer was widely used in each precision measurement.

2.2.3.5 Polarization of light

Same as interference and diffraction, the polarization of light is also its fundamental property. Research on polarized light is vital for the deep understanding of light.

Rasmus Bartholin (1625–1698), the Danish physicist, was the first to find and study polarized light. In 1669, Bartholin found the existence of double refraction in calcite. In 1690, Huygens found that the two refracted lights from calcite can be separated with another rotary calcite along the direction of light.

During the studies of polarized light, French physicist and military engineer Etienne Louis Malus (1775–1812) made surprising discoveries. On one evening of 1808, when overlooked the Luxembourg Palace at sunset from his house in Paris by using calcite, Malus surprisingly found that the a two-layered widow only looks one layer. On further thinking, he thought that the one-layer image of the two-layered window was owing to the reflection of sunset light, which produce two separated images through rotary calcite as sunlight falls on the window glass. He submitted his experimental results to the French Academy of Sciences in that December, in which he pointed out that the intensity of linearly polarized light with I_0 after passing through an analyzer (without considering absorption) should become as follows:

$$I = I_0 \cos^2\alpha \qquad (2.14)$$

where α is the angle between the incident light in the direction of polarized light vibration and the polarization direction of the polarizer.

Equation (2.14) is the famous Malus law. When $\alpha = 0°$ or $180°$, the transmission light, namely, I_0 is the strongest; $\alpha = 90°$, $I = 0$, transmission light intensity is zero. When α has other values, light intensity is between 0 and I_0. That is why two entirely different images can be seen with Malus rotatory calcite.

2.2.4 Electromagnetic spectrum

Electromagnetic wave has complex components, such as gamma ray, X-ray, ultraviolet, visible, infrared, microwave, and radio waves. Electromagnetic spectrum as shown in Figure 2.3 can be obtained if electromagnetic waves are arranged in sequence of vacuum wavelength increase or frequency decrease.

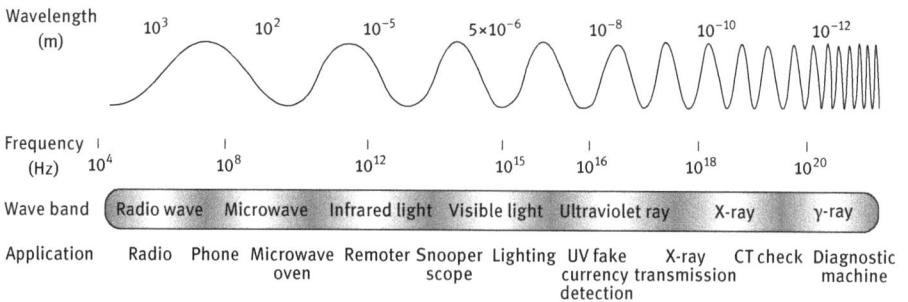

Figure 2.3: Electromagnetic spectrum and its applications.

Range boundaries of an electromagnetic spectrum are gradually varied, which are normally separated by electromagnetic wave generation or measurement methods as shown in Table 2.1.

2.3 Substance radiation

As mentioned earlier, light scattering is a secondary radiation of forced vibrations (Section 1.1.3), and particles can be observed as one type of the new radiators for their scattering properties. Discussion on the radiation properties is quite helpful for our recognition of scattering features of particles.

2.3.1 Black-body radiation

Radiation originating from electromagnetic radiation induced by heat is called as heat radiation. Heat radiation originates from the motion state changes of the micro-particle in the interior of the objects. Any object with temperature higher than 0 K will emit heat radiation to its surrounding space. Under different temperature, objects can emit different electromagnetic waves whose energy distribution in accordance with wavelength is dependant on the temperature.

In 1859, German physicist Gustav Robert Kirchhoff (1824–1887) raised a new theory based on the thermal equilibrium theory, that is, the ratio of the ability to absorb and emit electromagnetic radiation of any object is irrelevant of its property, but a universal function of wavelength and temperature. That is the famous thermal radiation law, or Kirchhoff's law, which states the thermal radiation is in direct proportion to its absorption coefficient. According to Kirchhoff's law, a good absorber is also a good radiator. If one object has a strong absorptive thermal radiation ability, then it also has a strong thermal emissivity ability, and the vice versa. If one object can absorb electromagnetic radiation of any wavelength, then the object would be an absolute black body.

There is only absorption and reflection of an electromagnetic wave when it falls on an opaque object surface:

$$\alpha(\lambda, T) + \rho(\lambda, T) \equiv 1 \qquad (2.15)$$

where $\alpha(\lambda, T)$ and $\rho(\lambda, T)$ represent spectral absorption and spectral reflectance, respectively, which are related to wavelength and temperature. Extreme conditions in equation (2.15) make the absolute black body and absolute white body. For absolute black body, $\alpha(\lambda, T) \equiv 1$, $\rho(\lambda, T) \equiv 0$, while the absolute white body is just the reverse, namely, $\alpha(\lambda, T) \equiv 0$ and $\rho(\lambda, T) \equiv 1$, irrespective of the wavelength and temperature.

Table 2.1: Electromagnetic spectra.

Wave band	Long wave	Intermediate wave	Ultrashort wave (microwave)	Infrared ray	Visible light	Ultraviolet ray	X-ray	Gamma ray
Wavelength	Over 3,000 m	10–3,000 m	1–10 m	0.76–1,000 μm	0.38–0.76 μm	380 nm to 10 nm	10^{-6}–10^{-3} μm	Less than 10^{-6} μm

Ultra far IR	Far infrared	Mid-infrared	Near infrared
15–1,000 μm	6–15 μm	3–6 μm	0.76–3 μm

Red	Orange	Yellow	Green	Indigo	Blue	Purple
0.62–0.76 μm	0.59–0.62 μm	0.56–0.59 μm	0.50–0.56 μm	0.47–0.50 μm	0.43–0.47 μm	0.38–0.43 μm

(1) Sometimes, 0.76–15 μm is seen as near infrared, 15–1,000 μm as far infrared.

In order to investigate the properties of absolute black body, the cavity, as shown in Figure 2.4, was designed for ideal absolute black body experiment. The cavity has a hole and the inner wall made of opaque material for only radiation absorption and reflection. When radiation falls in the cavity of the inner wall through the hole, more than 95% radiation is absorbed, and only a small amount of radiation can be reflected. However, after experiencing n times of reflection, only $(5\%)^n$ radiation can be sent through the hole. Suppose $n > 10$, loss of radiation is only 9.76×10^{-12} %, indicating that the radiation has been fully absorbed, and that the radiation passing through the hole can be neglected. It can be seen from Figure 2.4 that an absolute black body is one ideal model that does not exist in nature, the same as a mass point, a rigid body, and an ideal gas.

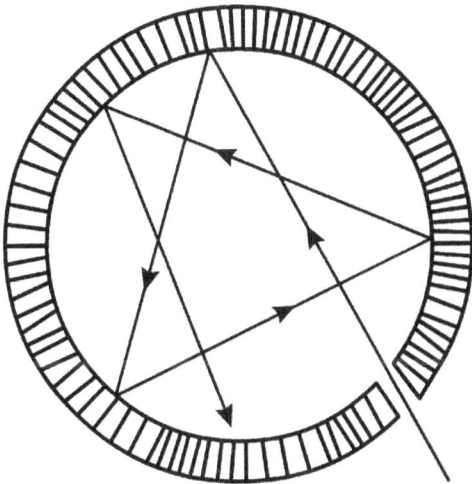

Figure 2.4: Absolute black body.

A breakthrough in quantum theory was first made by Planck in 1900 when he studied the black body radiation. He applied the concept of quantum theory in the black body radiation, and obtained the relationship of flux density (W_λ) with absolute temperature (T) and radiation wavelength (λ):

$$W_\lambda = \frac{2\pi hc^2}{\lambda^5} \cdot \frac{1}{e^{ch/\lambda kT} - 1} \tag{2.16}$$

where W_λ is the radiation flux density (W·cm^{-2}·μm^{-1}), λ is the wavelength (μm), h is the Planck's constant (6.6256×10^{-34} J·s), c is the light speed (3×10^{10} cm·s^{-1}), k is the Boltzmann constant (1.38×10^{-23} J·K^{-1}), and T is the absolute temperature (K).

Figure 2.5 is the black body radiation spectra depicted according to Planck's equation (2.16) at different temperatures. It can be directly observed that heat

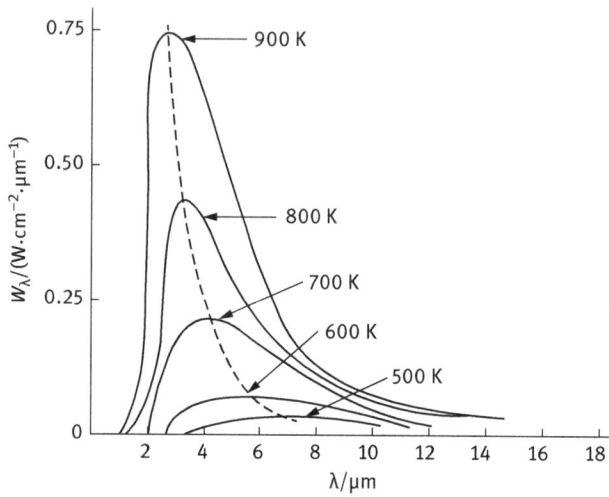

Figure 2.5: Black body radiation spectra at different temperatures.

radiation of low temperature is mainly in the infrared area; at 300 K, the heat radiation of the strongest component is in the infrared region; at 3,000–5,000 K, heat radiation of the strongest component is in the invisible light region.

Black body radiation, besides above properties, has the following three more important features worthy of attention:

(1) Total radiation flux density (W_r) is in direct proportion to the area under the curve, and increases with absolute temperature (T) with fast amplification
(2) The wavelength of the maximum radiation energy density (λ_{max}) moves toward the short wave with the increase of the black body temperature.
(3) The higher the temperature, the larger the spectral radiation flux density of the whole wavelength region, which can be seen from Figure 2.5. That each curve is not intersecting with each other.

2.3.2 Radiation of general objects

Planck believed that the heat radiation of black body only depends on the wavelength and temperature. So far, it has been found that the emitted and absorbed radiations of objects in the nature are lower than the absolute black body under the same conditions. For example, black bituminous coal, which is the most close to the absolute black body, has an absorption coefficient of near to 99%. Only radiation sources such as the sun and other fixed stars are close to the black body. In addition, it has been found that heat radiation of objects depends on their compositions and surface conditions besides the dependance of wavelength and temperature.

Emissivity, ε, is a normally used important parameter for object radiation feature, which is defined as the ratio of spectral radiancy from the thermal radiator to that of black body at the same temperature:

$$\varepsilon = \frac{W'}{W} \qquad (2.17)$$

Emissivity is a number between 0 and 1, which shows how much radiation source is close to a black body. Based on this, actual objects can be divided into selective radiator and gray body according to the changes of the emissivity with the wavelength. Selective radiator has different emissivity value at different wavelength $\varepsilon(\lambda)$, whereas the value of a gray body is constant at any wavelength.

These are the following situations for the objects in nature as compared to the features of absolute black body and absolute white body:
(1) Absolute black body $\varepsilon = \varepsilon(\lambda)=1$
(2) Gray body $\varepsilon = \varepsilon(\lambda)$, but $0<\varepsilon<1$
(3) Selective radiation $\varepsilon = f(\lambda)$
(4) Ideal reflector (absolute white body) $\varepsilon = \varepsilon(\lambda) = 0$.

As emissivity varies with the variations of material, surface roughness, wavelength, and temperature, most of the objects in nature can be seen as gray bodys.

2.3.3 Lambertian radiator

Generally, the radiation of a radiator is not completely identical at different directions, which generally has various directivity. However, if radiation is irrelevant of direction (the angle of intersection between radiation lightness direction and surface normal), the radiator is then called Lambertian radiator or cosine luminous body, whose light intensity can be expressed as follows:

$$L_v = \frac{2\varepsilon_v h}{c^2} \cdot \frac{v^3}{\exp(hv/kT) - 1} \qquad (2.18)$$

where L_v is the light intensity with emission frequency of v, ε_v is the corresponding spectral emissivity.

2.3.4 Solar radiation

Solar radiation characteristics are basically consistent with that of absolute black body (Figure 2.6). That is, radiation spectrum is continuous but has different

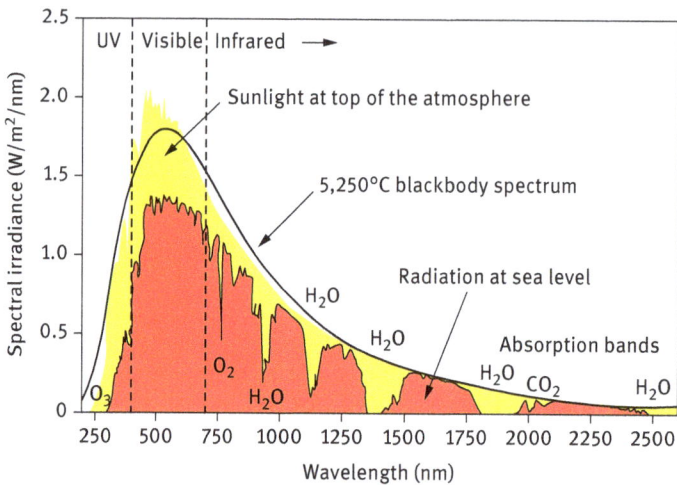

Figure 2.6: Solar radiation distribution curve.

distributions of radiation energy in each wavelength scope, and approximately 99% of the energy is concentrated in the wavelength range of 200–3,000 nm, in which the most concentrated range is from the near UV to mid-infrared, and visible light concentrated is about 38%. In the range of X-ray, γ-ray, far ultra-violet and microwave band, the energy is low with great changes.

When sunlight falls on earth's surface, radiation decreases for the reasons that are related to the wave bands of light. In the visible light band, the absorption of atmospheric molecules is weak, and the main factor to induce the electromagnetic attenuation is molecular scattering. Whereas in the region of ultraviolet, infrared, and microwave, owing to the absorption of oxygen, ozone, water, and carbon dioxide, the main factors for electromagnetic attenuation are molecular absorption besides molecular scattering. For example, oxygen absorption is mainly in the 0.25 and 0.5 cm microband.

As shown in Figure 2.7, ozone absorption is mainly in the ultraviolet region below 0.3 μm, along with the weak absorption at 9.6 and 4.75 μm and relatively weak absorption at 14 μm. The characteristic peaks of the main absorption bands of carbon dioxide are located in the infrared region of 2–2.7, 4.3, 10.0, and 14.4 μm, while the main absorption bands of water vapor are in the far infrared region of 0.70–1.95 μm, 2.5–3.0 μm, 4.9–8.7 μm, and 15 μm to 1 mm and in the microband region of 0.164 and 1.348 cm. Other gases in the air are symmetrical molecules without any polarity and they do not absorb any electromagnetic waves. It should be noted that the atmosphere has a strong absorption for ultraviolet rays, causing remote-sensing image darking. So, ultraviolet band is rarely used in remote

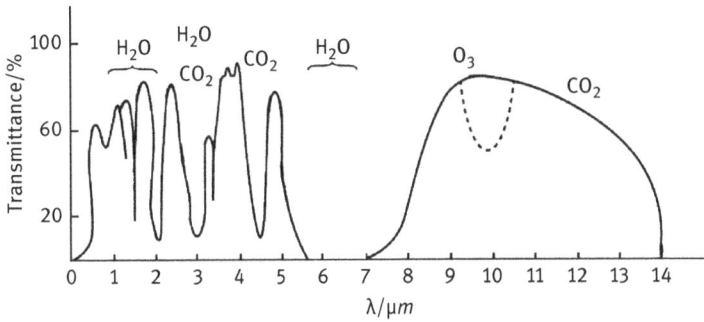

Figure 2.7: Light absorption of the main components of the atmosphere.

sensing. Figure 2.7 shows the atmospheric transmissivity of light with the wavelength less than 15 μm.

When micromagnetic wave of visible light region passes through an inhomogeneous area, scattering occurs and changes the direction of propagation. Scattering properties, including direction, intensity, and wavelength depend on the wavelength of micromagnetic wave the molecule's diameter of the inhomogeneous area, or the aerosol's particle size. But, if the particle size of the region is in the same level of the incident wavelength, Mie scattering occurs; If the particle size of the region is far larger than the incident wavelength, homogeneous scattering occurs; If the uneven level of the medium d is less than one-tenth of the wavelength of the incident micromagnetic wave, Rayleigh scattering occurs. When visible light passes through the atmosphere, its propagation is mostly affected by Rayleigh scattering, while that of infrared light is seldom influenced, and that of microwave is rarely influenced.

Even if for a same object, it shows different properties under the radiation of electromagnetic waves of different bands. In sunny days, for example, Rayleigh scattering occurs when visible light passes through the atmosphere, and the blue scattering light is much stronger than the red light. In cloudy days, for instance again, the floating particles in the air such as water droplets are too large, making the Mie scattering of large particle occur for the visible light of different wavelengths with the same intensity, so cloud appears white and we cannot see the blue sky.

Owing to lower scattering power, the red light of long wavelengths has stronger penetrating power than blue light. Hence, when taking photographs in fog, addition of a red filter to the camera can produce clearer photos. Infrared penetration ability is stronger than visible light, which is used for taking long-distance photographs or for remote sensing; microwaves, owing to having longer wavelengths, much larger particle size, less Rayleigh scattering, and stronger transmission power, can pass through the fog.

2.3.5 Reflection of electromagnetic wave

When an electromagnetic wave falls on the surface, reflection will occur along with mirror reflection, diffuse reflection, and directional reflection according to the different surface properties, as shown in Figure 2.8. The features of these are described below.

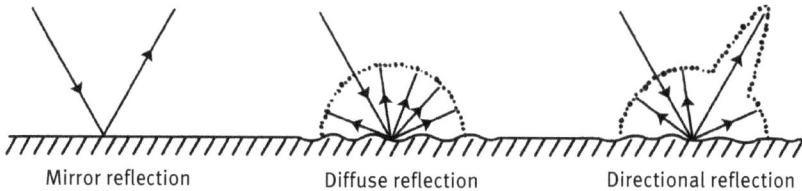

Mirror reflection   Diffuse reflection   Directional reflection

Figure 2.8: Three reflection diagrams of an electromagnetic wave.

2.3.5.1 Mirror reflection

Mirror reflection occurs when the reflection obeys the reflection law. Reflection law is when one light beam passes from one homogeneous medium to another homogeneous medium, the direction of propagation varies on the surface and produces a reflection light. A normal perpendicular to the mirror surface exists on the plane composed by the incident and reflection light beams, and the incident and reflection light beams lie on the both sides of the normal with an equal reflection and incident angle. For an opaque object, the surface reflection energy is the difference of incident total energy and absorbed energy according to energy conservation law. However, real mirror surface is rarely found in nature, and thus sometimes, we suppose that a calm water surface without waving is mirror surface.

2.3.5.2 Diffuse reflection

Diffuse reflection refers to the reflection of different directions when light falls on a rough surface. As the surface is rough, reflection of light irregularly transmits at variable directions owing to the inconsistent normal directions even if the incident light beam is parallel.

    If incident light wavelength (λ) keeps unchanged, but increasing the the surface roughness (h) to the same order of magnitude of wavelengths, then the whole surface can evenly reflect the electromagnetic wave. However, an ideal diffused surface has a light intensity in a direct proportion to the cosine of the direction angle (θ) and has the same brightness. Because of this, snow, white painted wall or rough white paper surface is the ideal diffused surface, and these objects can be named as Lambertian reflector.

2.3.5.3 Directional reflection

Normally, actual objects have a rugged surface, so there might be a strong reflection in one direction, exhibiting directivity, and thus named as the directional reflection. This reflection happens in the case of large surface roughness, and no law could be followed, which in fact is the a combination of a mirror reflection and diffuse reflection.

2.3.6 Light absorption and light scattering in a turbid medium

When light passes through a transparent medium, such as water, no light beam can be seen except that at the incident light direction. If the medium is turbid or filled with dust, however, light path can be seen clearly. For example, we can see the shining dust in front of a car light in a construction site, the reason for that is strong light scattering produced for these large particles.

In these turbid or inhomogeneous medium, if there is a substance with absorptive features, light signal will change greatly. In essence, light absorption and light scattering are two completely different properties, but they are hard to be separated in an actual measurement. When light with intensity of I_0 passes through a turbid system, the transmitted intensity is as follows:

$$I = I_0 e^{-(K+h)l} = I_0 e^{-\alpha l} \tag{2.19}$$

where h is the scattering coefficient, K is the absorption coefficient, and α is the attenuation coefficient. When measuring the absorption properties of a turbid system by using a spectrophotometer, what we actually determined is the attenuation coefficient.

2.4 Rayleigh scattering

2.4.1 Rayleigh scattering theory

2.4.1.1 Rayleigh scattering equation

Rayleigh scattering theory is a simple rule that gives quantitative description of light scattering and particles with classical geometrical physics. Figure 2.9 shows the relation of an incident light, scattering light, and particles with a geometric coordinate system. Scattering particle lies in the original point O of a coordinate system. When one parallel light with 100% Y-axis plane polarization propagates along the

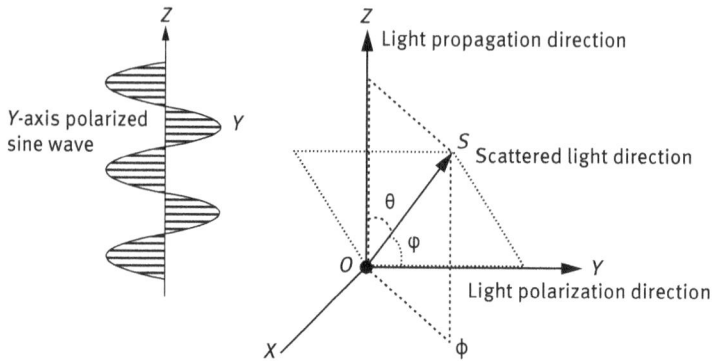

Figure 2.9: Geometric coordinate system of a particle scattering.

positive direction of Z-axis, the as-produced scattering light can emit at any direction. Suppose a light detector is placed at S position (namely, detect the scattering light at S direction), θ is the angle of the detecting scattering light and incident light, and ϕ is the angle of the detecting scattering light of the Y-axis.

Rayleigh scattering is an elastic scattering, which can be viewed as that the electrons transit to a virtual energy level since they absorb the energy of photons, emitting photons with the same frequency and transit back to the original energy level. Rayleigh scattering light is produced mainly from the interaction of an incident light electron and a particle inner electron. As energy level of an inner electron is rather steady, there is no great energy loss in its collision; it can be seen as a hard collision of a photon and an atom.

Suppose all electrons in the scattering particles have the same phase position and frequency as incident light, thus electron and photon have collective oscillation when exposed to an incident light, producing a single large oscillating electron dipole and an instant scattering light. Suppose scattering particles are even with radius (r) and far smaller than an incident wavelength (λ_0), then the theoretical expression of scattering light intensity (I) would be as follows:

$$I = I_0 \cdot \frac{16\pi^4 r^6 n_{med}^4}{d^2 \lambda_0^4} \cdot \left| \frac{m^2 - 1}{m^2 + 2} \right|^2 \cdot \sin^2(\varphi) \tag{2.20}$$

Equation (2.20) is consistent with equation (1.1), where I_0 is the incident monochromatic light intensity, r is the particle diameter, n_{med} is the environment refractive index, d is the particle and tested position distance, λ_0 is the light wavelength in vacuum, Greek fai is the angle of detection direction (S) and the polarization direction of incident light whereas m is the relative refractive index of light at a certain wavelength with an expression equation as follows:

$$m = \frac{n}{n_{\mathrm{med}}} = \frac{n_{\mathrm{rel}} + i n_{\mathrm{im}}}{n_{\mathrm{med}}} \tag{2.21}$$

It should be noted that the refractive index of a particle (n) is normally expressed in a plural form, that is, $n = n_{\mathrm{rel}} + i n_{\mathrm{im}}$, and both n_{rel} and n_{im} represent the real part and imaginary part of index, respectively; $i = \sqrt{-1}$. n_{im} represents light absorption of scattering particle. If the particles are pure and don't absorb light, $n_{\mathrm{im}} = 0$; if the particles could absorb and scatter light, $n_{\mathrm{im}} \neq 0$. The relationship between n, n_{rel} and n_{im} can be shown in Figure 2.10.

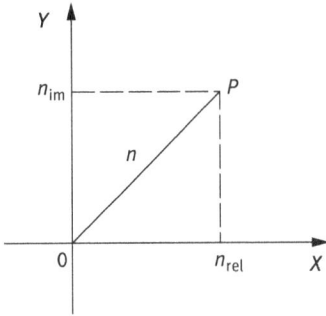

Figure 2.10: Refractive index (n) module is equal to the distance from the original point O to the point of P in a complex plane.

Rayleigh scattering theory only aimed at pure scattering particles (without absorption) at its born early stage. If, however, the refractive index n is expressed with a complex number, it is also suitable for normal scattering particles with absorption as long as $\alpha \ll 1$ (equation (2.1)).

2.4.1.2 Discussion of Rayleigh scattering equation

Next, we will discuss Rayleigh scattering equation (2.20) in three specific situations:
(1) Detector located in the X–Z plane. In the case that the scattering light surface is vertical to the direction of the polarization of the incident light, $\Phi = 90°$. Intensity of all directions in X–Z plane is identical, irrelevant to other parameters. Equation (2.20) can be simplified as follows:

$$I_1 = I_0 \cdot \frac{16\pi^4 r^6 n_m^4}{d^2 \lambda_0^4} \cdot \left| \frac{m^2 - 1}{m^2 + 2} \right|^2 \tag{2.22}$$

(2) Detector located in Y–Z plane. In the case that the scattering light surface is parallel to the incident light polarization direction, $\Phi = 90° - \theta$. The intensity is strongest when the angle of θ is 0° or 180°, while that is zero when the

angle of θ is 90° or 270°, respectively; at other angles, intensity is in direct proportion to $\cos^2\theta$. Equation (2.20) can be simplified as follows:

$$I_2 = I_0 \cdot \frac{16\pi^4 r^6 n_m^4}{d^2 \lambda_0^4} \cdot \left| \frac{m^2-1}{m^2+2} \right|^2 \cdot \cos^2\theta \tag{2.23}$$

(3) The incident propagating in the Z-axis direction without polarization. In the case that the, polarization vector evenly distributes in X–Y space, then the intensity of any angle is as follows:

$$I_{sca} = \frac{I_1 + I_2}{2} = I_0 \cdot \frac{8\pi^4 r^6 n_m^4}{d^2 \lambda_0^4} \cdot \left| \frac{m^2-1}{m^2+2} \right|^2 \cdot (1 + \cos^2\theta) \tag{2.24}$$

2.4.1.3 Deduction of Rayleigh scattering

We can obtain the following deductions from equations (2.20)–(2.24):
(1) The composition of the scattering particles that influences the light intensity entirely depends on the relative refractive index at different wavelengths. That is, the component of the particle decides the scattering light intensity in essence. For a known substance, if r, m, and n, are known, the scattering light intensity at certain detection angle can be estimated. For the scattering light detected by using an ordinary spectrophotometer, the most detected angle is at the right one, and the scattered spectrum (I vs λ_0) can be completely predicted on the basis of the equation.
(2) For a random given wavelength, the scattering intensity is in direct proportion to the particle diameter's sixth power. It can simply explain why the scattering light spectra can be greatly enhanced with an increase in the nanoparticles.
(3) The scattering light intensity gets decreased with an increase in the wavelength quadruplicate, so the longer wavelength has the smaller possibility to get scattered, which can explain why red color is used for alarm.

2.4.2 Scattering cross section

Equation (2.20) expresses the scattering light intensity detected at different angles. Scattering cross section of a scattering particle (C_{sca}) is often used to express the overall scattering ability of a particle. A scattering cross section can be obtained with the following equation by integrating the light scattering of whole space angle according to equation (2.20) (suppose $d = 1$):

$$C_{sca} = \frac{128\pi^5 r^6 n_m^4}{3\lambda_0^4} \cdot \left|\frac{m^2 - 1}{m^2 + 2}\right|^2 \tag{2.25}$$

In the same way, the absorption cross section (C_{abs}) of a scattering particle can be obtained from the integration of equation (2.20):

$$C_{abs} = \frac{8\pi^2 r^3 n_m}{\lambda_0} Im \left(\frac{m^2 - 1}{m^2 + 2}\right) \tag{2.26}$$

where Im means the imaginary part of a complex number expression given in round brackets. If a particle does not absorb, then $n_{im} = 0$, $C_{abs} = 0$. So, after interaction, the total amount of light scattering or absorption can be expressed with an extinction cross section (C_{ext}) defined as follows:

$$C_{ext} = C_{sca} + C_{abs} \tag{2.27}$$

Usually, the absorption value of suspended particles measured with an ultraviolet-visible spectrophotometer is actually the extinction value, including light absorption and scattering. Its molar extinction coefficient is obtained from the following equation:

$$\varepsilon = \frac{NC_{ext}}{2.303 \times 1,000} = 2.63 \times 10^{20} C_{ext} \tag{2.28}$$

The molar extinction coefficient of a suspended particle can help us to estimate particle concentrations on the basis of Lambert–Beer's law.

2.5 Mie scattering theory

2.5.1 Mie scattering equation

A precondition of Rayleigh scattering is the particle size far smaller than the incident light wavelength. In such case, each particle as a scattering point can induce the dipole moment vibration of an electric field, and each electron vibration in a particle has the same phase position with the incident light beam, and does the scattering light beam. When a particle size gets increased to the wavelength level or larger, the electrons in the particle the vibrate at different positions and produce scattering light with different phase. There would be interference for these scattering light with different phase, deviating the intensity and the angle distribution from electronic dipole oscillation.

Mie theory then involves these problems, and treats the scattering light as electric dipoles, four poles, eight poles, or more in particles, and overlapping light of magnetic dipole, quadrupole, or higher magnetic dipoles.

Similar to Rayleigh scattering deduction equation (2.24), with the interaction of an ordinary nonpolarized incident light interacts with particles, scattering light intensity can be seen as addition of both vertical and parallel polarized light.

$$I_{sca} = \frac{I_1 + I_2}{2} = I_0 \cdot \frac{\lambda_0{}^2}{4\pi^2 n_m{}^2 d^2} \cdot \left(\frac{i_1 + i_2}{2}\right) \tag{2.29}$$

where I_1 and I_2 are light intensities of the vertical and parallel polarized light, respectively, whereas i_1 and i_2 are calculated intensity functions according to that the incident light is polarized one of unit irradiance.

In theory, intensity function is the function of a scattering angle, relative refractive index of the surrounding medium and size of the scattering particle, which can be expressed as follows:

$$i_1(\alpha, m, \theta) = |S_1|^2 = \left| \sum_{n=1}^{\infty} \frac{2n+1}{n(n+1)} [a_n \pi_n + b_n \tau_n] \right|^2 \tag{2.30}$$

$$i_2(\alpha, m, \theta) = |S_2|^2 = \left| \sum_{n=1}^{\infty} \frac{2n+1}{n(n+1)} [a_n \tau_n + b_n \pi_n] \right|^2 \tag{2.31}$$

where S_1 and S_2 are complex amplitude functions without dimension, and are, infinite series composed by Bessel function and Legendre function; π_n, τ_n are functions related only to θ, which include the first and second derivative of n-phase Legendre polynomial with an argument of $\cos\theta$, and can be expressed as follows:

$$\pi_n(\cos\theta) = \frac{(2n-1) \cdot \pi_{n-1}}{n-1} \cdot \cos\theta - \frac{n \cdot \pi_{n-2}}{n-1} \tag{2.32}$$

$$\tau_n(\cos\theta) = n\pi_n \cdot \cos\theta - (n+1) \cdot \pi_{n-1} \tag{2.33}$$

where a_n and b_n are complex Mie parameters, fare the functions of m, the particle relative refractive index, and α, the dimensionless parameter, shown as amplitude of n level in an electrical partial wave and n magnetic particle wave as follows:

$$a_n = \frac{m\psi_n(m\alpha)\psi'_n(\alpha) - \psi_n(\alpha)\psi'_n(m\alpha)}{m\psi_n(m\alpha)\xi'_n(\alpha) - m\xi_n(\alpha)\psi'_n(m\alpha)} \tag{2.34}$$

$$b_n = \frac{\psi_n(m\alpha)\psi'_n(\alpha) - m\psi_n(\alpha)\psi'_n(m\alpha)}{\psi_n(m\alpha)\xi'_n(\alpha) - m\xi_n(\alpha)\psi'_n(m\alpha)} \tag{2.35}$$

wherein, $n = 1$ represents the electric dipole and the magnetic dipole, $n = 2$ represents the electric four-rank and magnetic four-rank, and the rest can be done in the same manner:

$$\psi_n(Z) = \left|\frac{Z\pi}{2}\right|^{1/2} \cdot J_{n+1/2}(Z) \tag{2.36}$$

$$\xi_n(Z) = \left|\frac{Z\pi}{2}\right|^{1/2} \cdot H_{n+1/2}(Z) \tag{2.37}$$

where Z is the semi-integer-order Bessel function and second-type Hankel function of α or $m\alpha$ in equations (2.34) and (2.35), $J_{n1/2}$ (Z) and $H_{n1/2}$ (Z); ψ' and ε' are differential quotients of each parameter. When the scattering particles are rather small, $\alpha \ll 1$, for example, then there is only the function of electric dipole, and the electric four-order and magnetic items in the Mie equation can be neglected. In such case, the equation is simplified into Rayleigh scattering equation.

2.5.2 Geometric scattering equation

In Mie scattering, the particle size has a complex relationship with the scattering light intensity, whereas a good optical design can decrease the uncertainty and disclose effective information about the particle, such as its size, shape and even if composition size. In practice, particle size response theoretical curve is not used, but a series of known particles are used to determine its linear response range, and then the size of the actual particle could be determined by using the linear range of the working curve. Even though, the scattering intensity has a more complex relationship with the particle size in the Mie scattering range that other size ranges. When a particle size is far larger than an incident wavelength, a simplified geometric scattering equation can be used as follows:

$$I = I_0 K(n, \theta)d^2 \tag{2.38}$$

where I is the scattering intensity, I_0 is the incident light intensity, K is the function of refractive index (n) and scattering angle (θ), and d is the particle diameter.

It is not difficult to find that Rayleigh scattering equation is simple and applicative, easy to understand, which is much suitable for nonprofessional optical researchers. On the other hand, Mie theory has wide applications and are comparatively accurate.

Even if the simple and applicable equation (2.38), it is not very convenient still. With the development of computer software technology, these calculations can be done through simple software. For example, Mie scattering theory calculation is edited into a software known as the Mie plot [2], through which theoretical scattering light intensity changing rules of various spherical structures at different angles, wavelengths could be obtained through a simple way, making a good foundation for the developments of light scattering technology in analytical chemistry [3, 4].

2.6 Discrete dipole approximation algorithm

Mie scattering theory can be used to calculate the light scattering properties of ideal spheres of any size, but it is unsuitable for the particles of other shapes and mixed components. Discrete dipole approximation (DDA) theory was thus put forward by Purcell and Pennypacker in 1973 [5], and later improved and revised by Draine [6, 7] and gradually developed into a method for the calculations of optical properties of particles of various shapes (such as extinction, absorption, and scattering). In today's rapid development of nanotechnology, DDA has advantages in the calculations of the interactions between light and nanoparticles, and plays a vital role [8, 9].

DDA method treats scattering particle as one cubic made by N polarized points. With the inducing light field, each cubic has a polarization-induced electric field, further affecting adjacent cubics. So, material optical properties can be obtained through self-consistent calculation of the system. As any cubic polarization in a scattering particle is not limited, DDA method can be used to calculate many substances with random shape and composition.

Concretely, DDA theory treats a cubic unit as one dipole, and suppose the dipole polarizability of i unit is α_i located at r_i, then the polarization intensity of interaction between the unit and the local electric field is as follows:

$$P_i = \alpha_i \cdot E_{\text{loc}}(r_i) \tag{2.39}$$

where E_{loc} includes the formed dipole field of incident light field and other dipoles at that point, which can be expressed as follows:

$$E_{\text{loc}}(r_i) = E_{\text{inc}} + E_{\text{dip}} = E_0 \exp(ikr_i) - \sum_{j \neq i} A_{ij} \cdot P_j \tag{2.39a}$$

where E_0 is the amplitude of incident light, E_{inc} is the electric field of the incident wave, and E_{dip} is the formed dipole field of other dipoles at that point. $A_{ij} \cdot P_j$ are the formed electric fields at that point, and k is the incident light wave vector. Then, the interaction matrix has the following forms:

$$A_{ij} \cdot P_j = \frac{\exp(ikr_{ij})}{r_{ij}^3} \left\{ k^2 r_{ij}(r_{ij} \cdot P_j) + \frac{(1 - ikr_{ij})}{r_{ij}^2} \cdot [r_{ij}^2 p_i - 3r_{ij}(r_{ij} \cdot p_j)] \right\} (j \neq i) \qquad (2.40)$$

Combine equations (2.40) and (2.39) into equation (2.40) to obtain

$$A' \cdot P = E \qquad (2.41)$$

where A' is the transport matrix of A. For a cubic with polarizable points with quantity of N, both E and P are $3N$ dimension vectors, so A is $3N \times 3N$ matrix, and thus the particle extinction cross section can be obtained from the following equation:

$$C_{\text{ext}} = \frac{4\pi k}{|E_0|^2} \sum_{j=1}^{N} Im\left(E_{\text{loc},j}^* \cdot P_j\right) \qquad (2.42)$$

DDA can be calculated through DDSCAT software [10, 11]. Through the setting of nanoparticle dipole matrix of different shapes and compositions, and the environment parameters of the particles, the calculations of light absorption and scattering spectrum of any particle can be realized. Now, DDA has been largely used for the theoretical calculation of the optical properties of nanoparticles [12–14]. By comparing the calculated result and the actual nanoparticles spectrum, it can effectively guide the applications of nanoparticles' spectrum properties in each field.

In summary, light scattering is widely known and well investigated with the development of scientific technology, lots of research results are achieved in theory and in application, laying a good foundation for the application of light scattering in analytical science, guiding toward the development of light-scattering analytical chemistry.

References

[1] Bohren C F, Huffman D R. Absorption and scattering of light by small particles. New York: Wiley, 1983.
[2] Mie plot. Available at http://www.philiplaven.com/mieplot.htm. 2018.6.11
[3] Yguerabide J, Yguerabide E E. Light-scattering submicroscopic particles as highly fluorescent analogs and their use as tracer labels in clinical and biological applications. II. Experimental characterization. Anal. Biochem., 1998, 262(2): 157–176.
[4] Yguerabide J, Yguerabide E E. Light-scattering submicroscopic particles as highly fluorescent analogs and their use as tracer labels in clinical and biological applications. I. Theory. Anal. Biochem. 1998, 262(2): 137–156.
[5] Purcell E M, Pennypacker C R. Scattering and absorption of light by nonspherical dielectric grains. Astrophys. J., 1973, 186(2): 705–714.

[6]   Draine B T, Flatau P J.Discrete-dipole approximation for scattering calculations. J. Opt. Soc. Am.
        A 11(4): 1491–1499.
[7]   Draine B T, Flatau P J. Discrete-dipole approximation for periodic targets: Theory and tests.
        J. Opt. Soc. Am. A, 2008, 25(11): 2693–2703.
[8]   Yang W H, Schatz G C, Duyne R P V. Discrete dipole approximation for calculating extinction and
        Raman intensities for small particles with arbitrary shapes. J. Chem. Phys., 1995, 103(3):
        869–875.
[9]   Félidj N, Aubard J, Lévi G. Discrete dipole approximation for ultraviolet-visible extinction
        spectra simulation of silver and gold colloids. J. Chem. Phys., 1999, 111(3): 1195–1208.
[10]   Draine B T, Flatau P J.User guide for the discrete dipole approximation code DDSCAT (Version
        5a10). http://xxx.lanl.gov/abs/astro-ph/0008151v2-2000.
[11]   DDSCAT. Available at http://code.google.com/p/ddscat/. 2016. 9. 22
[12]   Yang Z L, Hu J Q, Li X Y, et al. A DDA calculation on the optical properties of silver nanorods.
        Chinese J. Chem. Phys., 2004, 17: 253–258.
[13]   Guo Y F, Kong F M, Li K. A DDA calculation on the optical properties of metal nanoparticles.
        Opt. Techn., 2012, 38(3): 317–322.
[14]   Kelly K L, Coronado E, Zhao L L, et al. The optical properties of metal nanoparticles: The
        influence of size, shape, and dielectric environment. J. Phys. Chem. B, 2003, 107(16): 668–677.

Jian Ling, Cheng Zhi Huang

3 Detection of light scattering signals

3.1 Overview

Light scattering is a common phenomenon. As shown in Figure 3.1, when the light falls on a particle, the light is scattered. So, the light scattering can be observed from any angle. At this point, the eye really acts as just a detector, which captures and detects the light scattering signal. Therefore, the primary concern of scientists is how to collect and detect light scattering signals and apply them to analytical and practical production.

In analytical chemistry, the simplest method to detect signals is to observe them with naked eyes. A long time ago, the rapid visual analysis method was established using the naked eyes to detect light signal. In 1960s, Tyndall used scattering light for dust detction in air, which is one of the early visual analysis methods. Although the visualization analysis method has a certain disadvantage in accuracy and sensitivity, it is simple, practical and convenient, so it is also an important direction for the analytical chemist to develop new detection technology. Using ordinary equipment for analysis and facilities, analytical chemists have established many methods and technologies for the analysis and detection of optical properties of the substance, resulting in practical applications in different fields of spectrometry. This chapter will focus on the detection of light scattering signals, and introduce some new methods of light scattering analysis and detection technology developed in recent years.

The simplest example of detecting light scatterrng signals in daily life is that we can see a clear beam of light at night when the night sky irradiated with a common flashlight. At this time, the flashlight is the light source, the naked eye is the detector, and the beam is the light scattering signal generated by dust, droplets and aerosols in the air. Similarly, a pretty aqua show is a result of reflection, refraction, and scattering of water drops in air, which is normally performed at night.

3.2 Fluorescence photometer for the detection of light scattering signals

The light scattering properties of materials are presented in the form of spectrum, which is a good way to represent the light scattering characteristics of substances and study the light scattering properties of matter. The research and development of light scattering detection technology has been more than 100 years since Tyndall used

https://doi.org/10.1515/9783110573138-003

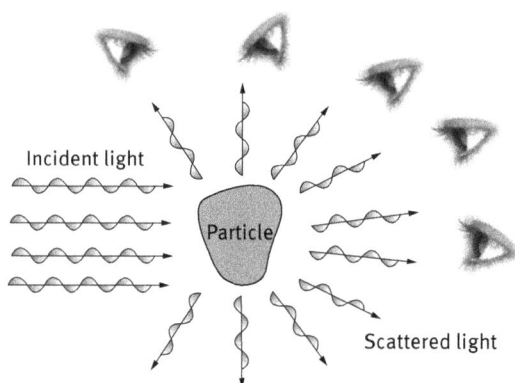

Figure 3.1: The observation of light scattering.

solar energy to determine the purity of the air at the end of 1860s. Although it is difficult to determine that who had first studied and established light scattering spectral technology. The earlier technology of light scattering is the combination of laser source or spectroscopic system and detection system [1]. It is necessary to explain that the application and development of light scattering spectroscopy in various fields are limited by the laser fixed wavelength, the expensive instruments, and the special development of the equipment. So, for rapid development and applications in various fields, fluorophotometer came into existence. This section will mainly introduce the technology of light scattering spectroscopy based on fluorescence spectrophotometer.

3.2.1 Resonance light scattering technique

The researchers who have worked on fluorescence analysis know that the fluorescence spectrophotometer usually uses a xenon lamp or a mercury lamp as a light source, and the composite light emitted through a grating or prism and then passes through a sample cell. The sample in the sample cell absorbs part of the exciting light and then emits a lower energy fluorescence, which is in all directions. However, in order to avoid the interference of exciting light, the fluorescence detection is placed in the angle of 90 to the incident light and enters the photomultiplier to detect signal after light splitting by a monochromator.

However, because of the widespread existence of light scattering phenomenon, the light scattering signal is a very serious interference signal in the detection process of fluorescence signal. In 1993, an American scientist, Robert F. Pasternack, had put forward a resonance light scattering (RLS) technique [2, 3] that uses an ordinary fluorophotometer to detect of light scattering signal. He coupled two monochromators of a fluorophotometer for simultaneous scanning of enhanced RLS signal around the absorption peak after aggregation of chromophore (normally dye molecules), thus to study and determine molecule. And an analytical method for studying and judging molecular aggregation. This signal reflects the functional relationship of a light scattering signal and wavelength, and so it is called the resonance light scattering spectrum. Different from the traditional method, where the absorption band is away from the substance to avoid influence of the absorption, the RLS spectrum involves detecting enhanced scattering signals near the absorption band.

Earlier than Pasternack, Anglister and Steinberg found resonance Rayleigh scattering (RRS) in their study involving properties of dye light scattering in 1980 [1]. But it has not been further utilized and applied to the field of spectroscopic analytical chemistry. One of the important reasons for the extensive research and application of resonance light scattering technology proposed by Pasternack is that this light scattering technique is based on the ordinary fluorescence spectrophotometer, and the synchronous scanning of the excitation wavelength is equal to the emission wavelength to obtain the synchronous luminescence spectrum (Figure 3.2) [2, 3]. An ordinary fluorophotometer is cheap and is widely used in common laboratories. The method is simple, fast and easy to obtain, easier for use, it enables more researchers to learn and use it easily, and has also laid a good foundation for its development.

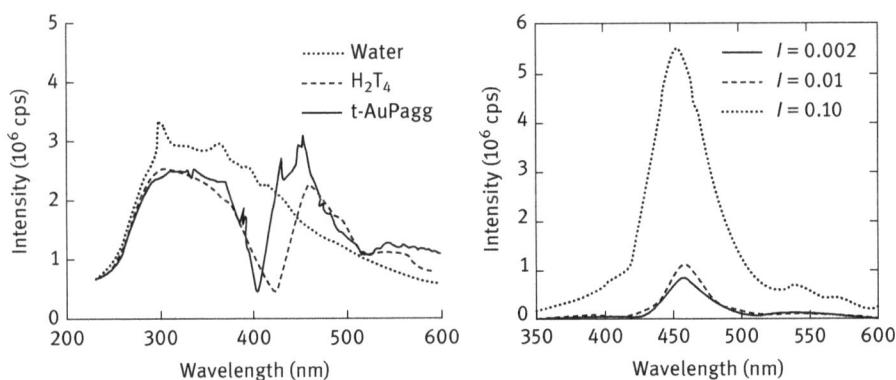

Figure 3.2: The light scattering spectra (published by Pasternack in J. Am. Chem. Soc.). (a) Scattering spectra of different substances. (b) Scattering spectra of 5 mol/L t-H_2Pagg at different ionic strength, where I is the ionic strength.

At the beginning of the resonance light scattering technology establishment proposed by Pasternack, the application mainly focused on the physicochemical research of molecular aggregation state and mechanism. Later, the author found that physicochemical technique can be used to establish highly sensitive analysis method. By using DNA to induce porphyrin aggregation to produce enhanced resonance light scattering signal, a high sensitivity detection of DNA is achieved, with a detection limit of 14 ng/mL [4]; if metal complex, such as Co(II) reagent, were used to detect DNA at the alkaline conditions, its detection limit could be lower than 1.0 ng/mL [4]. Since then, the domestic and foreign counterparts have applied this signal detection method to various fields of analytical chemistry, realizing high sensitivity detection of nucleic acids, proteins, metal ions and small organic molecules. In view of the shortcomings of the light scattering spectroscopy established on the ordinary fluorescence spectrophotometer, many new types of resonance light scattering analysis technique, such as total internal reflection resonance light scattering and resonance light scattering imaging, were established and improved in the experimental methods and data processing [5–7].

The RLS technique is recognized as a simple, fast, sensitive, and reliable spectrometry, and has been applied to different fields in analytical chemistry [7–10]. With the continuous development and applications of the RLS technique in detecting material light scattering signals, the concept of RLS has been widely developed as a new analysis technology. However, most researchers ignore the concept of "resonance" in "resonance light scattering", neglecting the relationship of the scattering signal with the collective absorption band (that is, the incident light wavelength λ_0 makes $m^2 + 2 = 0$, but $n_{im} \neq 0$ in equation 2.20). It is considered that as long as the scattering signal obtained by the synchronous scanning mode of a ordinary fluorescence spectrophotometer and the technology that used for analysis and detection are called as "resonance light scattering technology" [11]. However, the development of resonance light scattering technology in analytical chemistry makes "resonance light scattering" less original, but it can also explain the development potential of this spectral technique and the practicability of the method.

3.2.2 Doubling frequency and half frequency scattering signal

It is well known that a light radiation signal will appear at 2 times (half frequency) and 1/2 (double frequency) at the excitation wavelength when the fluorescence spectrometers are used for measurement. The scattering peak will interfere with the determination of fluorescence spectrum, so people always try to eliminate or avoid it by adding filters or changing the wavelength of measurement. In 1995, a domestic famous scholar, Professor Liu, found that the characteristics and strength of these two kinds of scattered light were not only related to the structure and measurement conditions of the instrument, but also closely related to the properties and concentration of the substance in the solution, and the intensity of scattered light was proportional to the concentration of the substance in the solution under certain conditions. Thus, Professor Liu established a highly sensitive "secondary light scattering (SLS) spectrometry" by combination the intensity of the scattering peak to the properties, state and concentration of the material, which was successfully applied for detecting mercury and selenium particles in water [12, 13]. Later, when Professor Liu studied the spectral properties of the ion associates formed by rhodamine dye and anions, he found that the "resonance luminescence spectrum" was obtained by synchronous fluorescence scanning mode by fixing the excitation and emission at the same wavelength [14]. This novel resonance luminescence can be used for highly sensitive detection of selenium, cadmium, and mercury ions [15, 16]. In his subsequent studies, he found that the "resonance luminescence" was actually the resonance Rayleigh scattering spectrum (RSS) of the ionic association formed in the solution [17–20].

Resonance luminescence proposed by Liu and RLS proposed by Pasternack have no difference in essence. All of them are light scattering signals produced by molecule aggregates or bulk mass molecules under a monochromatic irradiation, and were discovered in different fields and test systems, and thus the different names. This can be confirmed by the three-dimensional spectral scanning with a fluorophotometer.

A three-dimensional luorescence spectrum originated in 1980s. It is a graphical representation of the function of the excitation wavelength and the emission wavelength by the fluorescence intensity, also known as the total luminescence spectrum or contour spectrum. As the three-dimensional fluorescence spectrum concept is not quite clear, including fluorescence and other scattering signals, this book considers the concept of total luminescence spectrum for discussion. From the total luminescence spectrum, we can clearly obtain the fluorescence intensity information when the excitation wavelength and the emission wavelength change. Except for the strong resonance Rayleigh scattering (RRS) signal when $\lambda_{ex} = \lambda_{em}$, there are other light scattering signals (as shown in Figure 3.3). These signals include secondary light scattering (SLS, $2\lambda_{ex} = \lambda_{em}$), anti-secondary light scattering (ASLS, $\lambda_{ex} = 2\lambda_{em}$) and Raman scattering. For fluorospectrophotometer without a three-dimensional spectral function, we can manually choose $2\lambda_{ex} = \lambda_{em}$, $\lambda_{ex} = 2\lambda_{em}$ to obtain SLS and anti-SLS spectra.

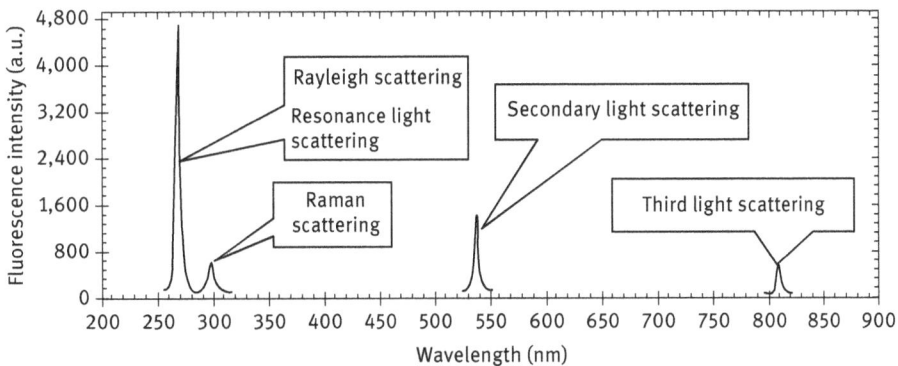

Figure 3.3: Total luminescence spectrum for light signal distribution.

In 1999, Huang et al. [6] used fluorospectrophotometer to measure the total luminescence spectrum of water (Figure 3.4). A stronger scattering signal appears at $\lambda_{ex} = \lambda_{em}$, while there is a weak secondary scattering signal and a weak antisecondary

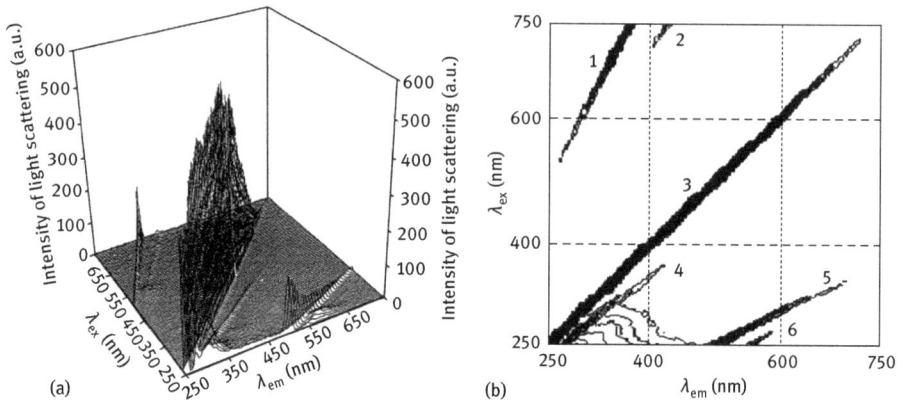

Figure 3.4: Total luminescence spectra of distilled water: (a) total luminescence spectrum and (b) contour spectra, where 1. ASLS, 2. anti-second Raman scattering, 3. RLS, 4. Raman scattering, 5. SLS, and 6. secondary Raman scattering.

Table 3.1: The linear relationship between light scattering wavelengths and excitation wavelengths ($\lambda_{ex} > 250$ nm).

Line	Linear regression equation	Relative coefficient	Affiliation	Intensity order[a]
1	$\lambda_{ASLS} = 0.0 + 0.50\lambda_{ex}$	0.9999	Anti-secondary light scattering (ASLS)	3
2	$\lambda^R_{ASLS} = -52.0 + 0.64\lambda_{ex}$	0.9984	Anti-secondary Raman scattering (Raman$_{ASLS}$)	6
3	$\lambda_{RLS} = 0.0 + 1.00\lambda_{ex}$	0.9999	Resonance light scattering (RLS)	1
4	$\lambda^R_{RLS} = -49.0 + 1.28\lambda_{ex}$	0.9998	Raman scattering	4
5	$\lambda_{SLS} = 0.0 + 2.00\lambda_{ex}$	0.9999	Secondary light scattering (SLS)	2
6	$\lambda^R_{SLS} = -171.7 + 2.86\lambda_{ex}$	0.9983	Secondary Raman scattering (Raman$_{SLS}$)	5

[a]Scattering intensity order: 1>2>3>4>5>6.

scattering signal at $2\lambda_{ex} = \lambda_{em}$, $\lambda_{ex} = 2\lambda_{em}$. There is a quite weak Raman scattering signal near $\lambda_{ex} = \lambda_{em}$. Relationships of these signals are listed in Table 3.1. It can be clearly seen from Figure 3.4 that secondary scattering signal and antisecondary scattering signal are weaker than the resonance scattering signal.

3.2.3 Effects of fluorescence spectroscopic instruments on light scattering spectra

The light scattering spectroscopy, based on the ordinary fluorospectrophotometer, makes it easy to detect the light scattering spectrum, which is the foundation for the rapid development of light scattering spectroscopy. But there are still many problems in terms of spectroscopic properties, which need more discussion based on the equipment's condition and target being tested.

3.2.3.1 The possible fluorescence signal included in light scattering signal

The choice of the excitation wavelength by the fluorescence spectrophotometer is based on the diffraction of the grating, which is chosen by turning the width of the grating and the slit. Therefore, the fluorescence spectrum of the spectrofluorometer is not completely the scattered light when the excitation wavelength is equal to the emission wavelength. Suppose that the slit width is 5 nm, a scanning with $\lambda_{ex} = \lambda_{em}$ actually results in a spectrum of $\lambda_{ex} \pm 2.5$ nm with a light excitation at $\lambda_{ex} \pm 2.5$ nm. So, if there is fluorescence emission around the wavelength (Stokes shift is rather little), it is likely to include $\lambda_{em} + 2.5$ nm fluorescence or other signals excited at $\lambda_{ex} - 2.5$ nm wavelength. This situation is possibly exist in the fluorophores with relatively little Stokes shift, such as anthraquinones and new type of nanometer fluorescence probes, including quantum dots and carbon dots.

3.2.3.2 Half-frequency and double frequency light scattering signals are derived from grating

The spectral signals obtained from the ordinary spectrofluorometer are affected by the structure of the instrument itself. For example, in a fluorescence spectrum, there is a secondary and antisecondary scattering peak at twice and half of the excitation light wavelength. The so-called secondary scattering does not result in a real peak but is led by the diffraction in beam splitting system, which can be explained by the interference phenomenon of diffraction light (Refer to Figure 2.2).

The light source of an ordinary fluorospectrophotometer is mainly a continuous white light source, and the spectra are obtained for excitation and emission wavelengths with a grating beam splitting system, based on the grating equation:

$$(a+b) \sin \phi = k\lambda \tag{3.1}$$

where a and b are the slit width and slit distance of the grating, respectively, this is, the grating parameters. ϕ is the diffraction angle; k is the diffraction level, $k = 0, \pm 1, \pm 2, \pm 3, \ldots$; and λ is the wavelength of the light.

Under certain grating parameters of a and b as mentioned in equation 3.1, the same level diffraction angle at different wavelengths is related to wavelength. Thus, when the white light passes through the grating, the central bright stripes are still white light mixed with each color light, but bright stripes at both sides are colored bands with a symmetrical arrangement from purple to red, which is the diffraction spectrum. There is a certain spectral overlapping among diffraction bands at different levels. For example, the first order diffraction stripe of the green light at 500 nm overlaps with the secondary diffraction stripe at 250 nm, and the first order diffraction stripe at 1000 nm overlaps with the secondary diffraction at 500 nm.

The ordinary fluorescence spectrophotometer uses the grating principle to divide the light, and the corresponding position in the diffraction spectrum is used to obtain a certain beam band and for further spectrometry. For example, suppose that we set 500 nm for a full-wavelength fluorescence scanning, three clear peaks of 250, 500, and 1,000 nm can be obtained in the spectrum. It is because that when the excitation wavelength is set at 500 nm, the light that passes through the slit would produce the secondary diffraction at 250 nm and primary diffracted at 500 nm, and so 250 and 500 nm scattering peaks could be detected, while the peak at 1,000 nm is actually the secondary diffraction light at 500 nm in the emission beam-splitting system.

In addition, the third level diffraction light will overlap with the primary diffraction light. For example, the third order diffraction light at 250 nm can be detected at 750 nm, and the three order diffracted light at 500 nm can be detected at 1,500 nm; this is the reason for scattering peaks at 1/3, 3/2, 3 times of the excitation wavelength. As spectroscopic detection of fluorescence spectrometer is limited to a certain range, a photomultiplier has different sensitivities at different wavelengths, and the diffraction series increase gradually attenuates the intensity of the light; and thus these peaks are not necessarily present.

3.2.3.3 Distinct light scattering spectra are obtained with different equipments

The light scattering signals and spectra detected by the fluorescent spectrophotometer are also related to the design of the ordinary fluorescence spectrophotometric system, so that the obtained spectrum of distinct instruments can be completely different. The light source, beam-splitting system, and optical system design of fluorescence spectrofluorometer may be quite different, resulting in an obvious difference in a spectrum. Different models of fluorophotometer with the same brand, for example, Hitachi F-2500, F-4500, and F-7000, are the same in the main structure but there are differences in the optical path and some of the parts, so that the scattering spectra of the same solution are different. Another example is that the fluorescence spectrophotometer produced by Perkin Elmer company uses pulse as a light source, which obtain a quite different light scattering spectrum from that of other brands. These differences are mainly reflected in the details of the weak scattering spectrum part and the ultraviolet spectral part, but some typical scattering peaks of some materials, such as the metal nanoparticles with strong plasma resonance absorption or the porphyrin molecule aggregates with strong Soret absorption band, will not be too great irrespective of the kind of fluorescence spectrophotometer, as the signal is strong and thus its nuances are often ignored. For example, 30 nm gold nanoparticles have plasma resonance absorption at about 530 nm, and the plasma resonance light scattering peak is at about 560 nm regardless of the fluorospectrophotometer.

Although there are some problems with regard to the spectrum and equipment, in terms of analytical chemistry, the relationship between signal and substance concentration and nature can be established, an analytical method can be developed. Therefore, when it is difficult to analyze the optical signal itself, we should first consider building a relationship between the intensity of light scattering and the substance concentration, thus to establish the light scattering analysis method.

3.2.4 Total internal reflection-light scattering detection technology

The light scattering signal based on an ordinary fluorescence spectrophotometer can build highly sensitive resonance scattering methods with poor selectivity. Thus, the author combines the total internal reflection (TIR) with the resonance light scattering analysis technique at the liquid-liquid interface. On the one hand, the targets are separated and the selectivity is improved. On the other hand, the targets are enriched to the interface to improve the analysis sensitivity.

We know that when the light spreads from an optically denser medium to a thinner medium with an incident angle larger than the total reflection, the incident light does not refract through the surface but occors a total reflection. As most light of the total reflection is reflected, and only a small part of it passes through the surface to form an invisible wave or evanescent wave, propagating along the surface of thinner medium (Figure 3.5). The frequency of the evanescent wave is the same as that of the incident light, and its intensity decreases exponentially with the vertical distance from the interface. The intensity attenuation is in accordance with the following rule:

$$I_{(Z)} = I_0 e^{-\frac{z}{d}} \tag{3.2}$$

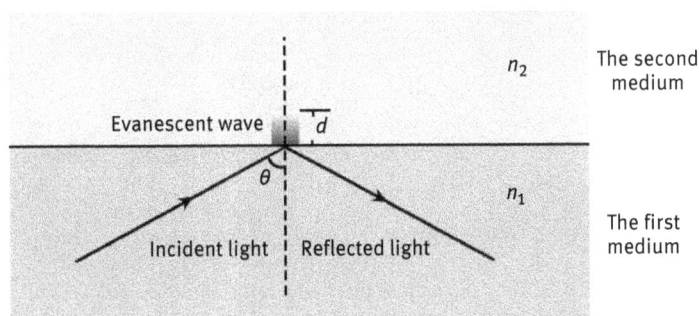

Figure 3.5: The generation and propagation of the evanescent wave.

where $I_{(Z)}$ is the light intensity with Z distance from the interface, and I_0 is the light intensity at the critical surface.

Its penetration distance (d) conforms to the following equation:

$$d = \frac{\lambda}{2\pi n_2} \frac{1}{\sqrt{n_1^2 \sin \theta^2 - n_2^2}} \tag{3.3}$$

where n_1 and n_2 are the refractive indices of the optically denser medium and thinner medium, respectively; θ is the TIR angle; and d is the propagation distance of the evanescent wave in the thinner medium. Normally, the penetration of the visible light is less than 200 nm. Thus, the evanescent wave could excite a substance within a 200-nm thin layer range on an adsorption interface but can not excit substances below 200 nm. Therefore, the incompatible two phases are used for the total internal reflection light scattering analysis, which can detect the light scattering signals of the adsorbed material on the interface, and eliminate the scattering interference of the material in the solution below the interface. The authors used TIR principles to detect a chemical reaction at the interface or extracted or adsorbed the substance to the interface for testing and analysis (Figure 3.6), to establish a series of methods based on total internal reflection -resonance light scattering (TIR-RLS) analysis [21–23]. The method presents the following advantages:

(1) The material under test can be selectively adsorbed at the liquid–liquid interface while other substances are kept in the solution to improve the selectivity of the method.

(2) The material under test gathers at the liquid–liquid interface, and so detection selectivity of the analysis method is improved.

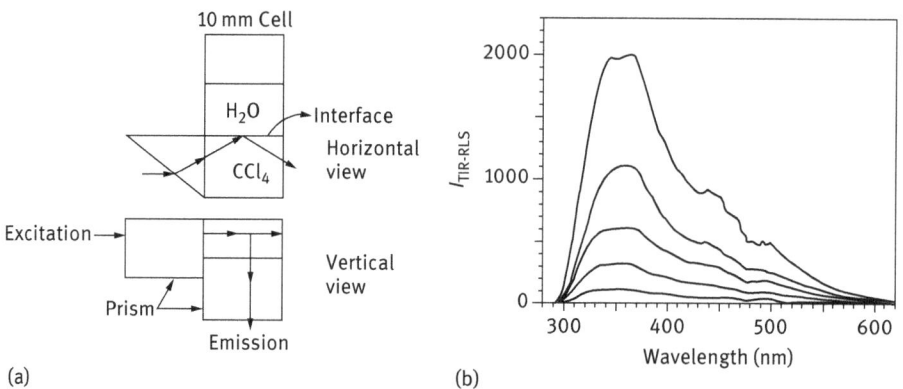

(a)   (b)

Figure 3.6: TIR-RLS measurement: [21] (a) The schematic diagram of a total internal reflection light scattering measurement device (b) a liquid–liquid interface light scattering spectra.

(3) Because of the special properties of the interface, more objects research can be realized, such as the adsorption mechanism of the material at the gas-liquid interface and the adsorption state of the material at the interface, and so on.

It can be seen that the combination of total internal reflection and resonance light scattering helps to overcome the drawbacks of traditional light scattering technique and can be expanded in more research field.

When the light falls on an optically thinner medium with an angle larger than the need for total reflection, although most energy of the incident wave is reflected back to the dense medium, there is still an evanescent wave in the thinner medium with a penetration depth at the level of the wavelength of the incident light interface. After the interface layer adsorbs molecules or particles, it is no longer an ideal smooth surface, but an optical surface with a little roughness. The light scattering should be the part side of the scatterer (molecules or particles) exposed to the incident light, and the area of a particle or molecule is related to the concentration of the adsorbed substance on the interface. Therefore, the light scattering on the interface consists of two parts: the first part is the scattering of an incident light connected to the optically denser medium in the interface layer; the second part is the light scattering of the evanescent wave propagating in the interface layer.

In summary, when the incident light arrive at the liquid-liquid interface with the angle greater than the critical angle, the light scattering signal of substance that is adsorbed or enriched on the interface can build new, highly sensitive and selective light scattering analysis methods based on the total internal reflection characteristics. In order to produce a significantly enhanced scattering signal at the interface, the following two points should be satisfied: firstly, the assembly should have both hydrophobic group and hydrophilic group, and secondly, the assembly should have a certain particle size to form a large scattering cross section.

3.2.5 Light scattering polarization spectroscopy

Tyndall found polarization of part scattering light in his observation of turbid liquid. Because of that, senior photography enthusiasts often use a polarizer to get beautiful images. As shown in Figure 3.7, because partial white light is filtered by polarizing mirror, the blue polarized light can pass through the sky, thus the sky appears bluer, and the white clouds are darker for a better effect.

In fact, the particles in a medium, such as dust, raindrops, or molecules, their light scattering are all anisotropic. As described in Chapter 1 (Section 1.1.3), light scattering is secondary radiation, so the uneven nature of each scattering particle is like an uneven radiator luminous source to produce inhomogeneous scattering in all directions (see Section 2.3.3; Chapter 2). Of course, the deeper reason for the anisotropy of light scattering is that the periodicity and density of the particles or

Figure 3.7: The visual effect with a polarizer improvement (from Internet). Left: the imaging was obtained without polarizer; Right: the imaging was obtained with polarizer.

molecules along different directions are not exactly the same, so that all or part of the physical and chemical properties of the object show the direction selectivity. At this time, if the incident light is also polarized, that is to say, the incident electric field vector has direction selectivity, then it shows result unlike that radiated by the natural light.

When linearly polarized light is irradiated in a gas or liquid, partial polarized light can be obtained laterally, so that the linearly polarized light is depolarized

$$p = \left| \frac{I_y - I_x}{I_y + I_x} \right| \tag{3.4}$$

where I_x and I_y are the light intensity along the direction of x and y axis, respectively, and P is the degree of polarization of a partial light in z-direction.

To demonstrate the degree of side polarization, the degree of depolarization concept needs to be introduced:

$$\Delta = 1 - P \tag{3.5}$$

where Δ is the degree of depolarization, related to the properties of the particle; for example, the degree of depolarization of hydrogen, nitrogen and CO_2 are 1%, 4% and 7%, respectively. The degree of depolarization of CS_2 can reach as high as 14%. It can be found that the measurement of the degree of depolarization can be used to determine particle anisotropy and symmetry property.

The light scattering technique based on the ordinary fluorescence spectrophotometer is used to characterize the spectral characteristics of molecular assembly and aggregation, it can not obtain the spectral information about the shape and spatial orientation of aggregates. The propagation of polarized light depends on the factors such as particle size, particle spatial orientation and composition, and spatial distribution. Research of the variation of polarization scattering light in space, direction, coherence, and polarization, is helpful to understand scatterer's shape and spatial orientation and also provide meaningful insight for the development of new biomedical optics and new applications.

The polarizer is installed on the light source and the detection port of the ordinary fluorescence spectrophotometer, which can help to obtain the scattering spectrum with polarized light by the scanning of the scattering spectrum and the adjustment of the polarization angle. By analyzing the shape, intensity and light polarization angle of the scattering spectrum, and combining the basic information of the scatterer, it helps to sum up the relationship between the polarization spectrum of light scattering and the morphology of scattered particles, the spatial orientation, the composition and distribution, realizing the monitoring and analysis of the shape, space orientation and composition of the unknown scatterers. In the actual study, it is found that the polarization spectra of different forms of bacteria such as *Escherichia coli* and *staphylococcus aureus*, have a clear distinction in a spectrum, and the polarization spectra of bacteria with similar morphology and size have some regularity [24].

3.2.6 The signal detection of backward light scattering

The theoretical basis of backscatter analysis is that the intensity of light scattering is closely related to the scattering angle. On the basis of the Mie scattering theory, the light intensity after an interaction with a particle depends on the detection angle. For small particles, forward scattering and direct scattering have greater intensity, while for larger particles, backward scattering is stronger (Figure 3.8). Thus, backward scattering technique is especially suitable for larger particles.

Only the right angle scattering signal can be obtained by an ordinary fluorescence spectrophotometer measurement. The light scattering signal at this angle is weaker than other directions, and so it is necessary to use the scattering intensity

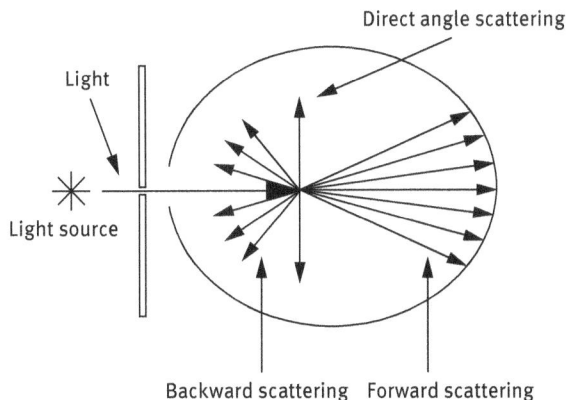

Figure 3.8: The relationship of the scattering signal distribution and scattering angle.

signal at different angles for different targets, especially for larger bioaggregates and large-size nanoparticles, the backward detection will have better signal response.

The detection of a backward light scattering signal can be realized only with a small improvement of fluorospectrophotometer's excitation and emitting light path. For example, by using the optical fiber to change the light path can detect 180° angle backward light scattering signal [25]. In the sample cell of the spectrofluorometer, prism is used to change the angle of excitation and emission, and the backscattering signal detection can also be realized [26]. The choice of facility improvement and detection angle can be carried out based on actual sample's scattering features (Figure 3.9).

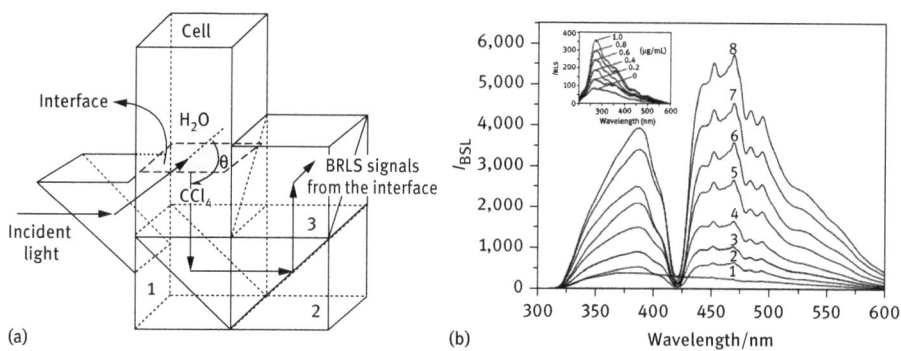

Figure 3.9: The signal detection of backward light scattering. [26] (a) Backward light scattering signal detection device; (1)–(3) are total reflection prism. (b) Backward light scattering spectra: 1. ctDNA; 2. TPP-Al(III); 3. TPP-ctDNA; 4–8. scattering spectra of TPP-Al(III)-ctDNA on H_2O/CCl_4 interface; the insert shows resonance light scattering of Al(III)-ctDNA in liquid phase.

3.2.7 Detection of light scattering signal in solid phase sample

It is of great significance to study the chemical reactions and interactions that occur at the solid-liquid interface. The study of the light scattering properties of solid surface materials is an important way to study the solid-liquid interface reaction, and the solid phase analysis has its unique advantages.

Fig. 3.10 is a schematic diagram of detecting biological macromolecules enriched by solid phase surface using an ordinary fluorescence spectrophotometer. In order to realize the light scattering detection on the surface of the glass, it is necessary to test and optimize the glass slide, the angle of the incident light and the slide, and the detection angle, so as to avoid the influence of the reflection and refraction of the incident light on the surface of the slide.

After refitting the sample cell of fluorescence spectrophotometer, it can be directly used for signal detection of glass-loaded samples. For example, signal difference before and after glass surface modification with antigen-antibody can be used to establish immunoassay based on solid phase light scattering [27].

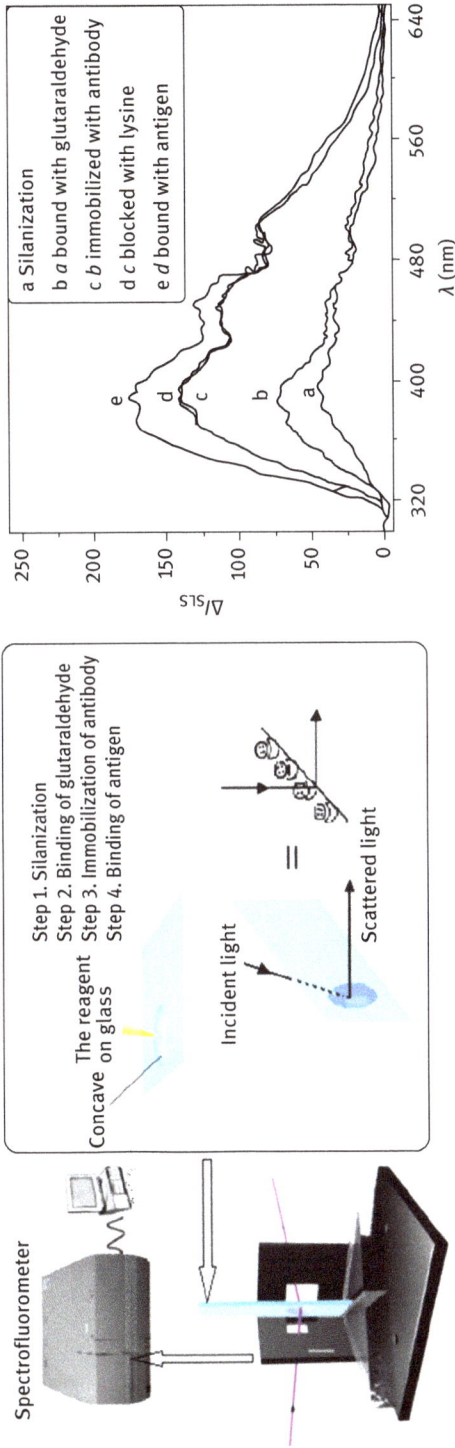

Figure 3.10: Solid phase immunity device and protein immune light scattering spectra [27].

3.3 Light scattering microscopic image analysis

In Section 3.2, the light scattering signal scanned with an ordinary fluorescence spectrophotometer was introduced. Actually, there are many ways to detect scattered light. For example, light scattering signals form microscope is one of very important methods. Using optical microscopy, the light scattering signals of particles or molecules are displayed in the form of images. By analyzing the light scattering images, it can be used for quantitative or qualitative analysis of materials, thus producing light scattering microscopic imaging (iLSM) technology. The establishment of a light scattering microscopic image analysis exhibits particle optical properties and morphology features, which are different from fluorescence. Especially its combination with optical microscopic image analysis has resulted in light scattering representation of microworld, biological tissue, and cytoskeleton shape features, and so it is very important.

3.3.1 Ordinary microscopic imaging

Light scattering imaging for particles or molecules is based on microscopic systems, and an ordinary optical microscope provides various optical observation modes. When the resolution of the dark field imaging does not meet the requirement, the self-made scattered light excitation device can be combined with ordinary microscope. For example, Japanese scientists improved facilities and clearly observed light scattering of a single metal nanoparticle [28]. In 2003, the author used laser as a source to observe an organic collective light scattering signal of biomolecular aggregates in the solution (Figure 3.11). He established a "particle counting" quantitative analysis method based on the scattering image according to the relationship between aggregate quantity and biomacromolecule concentration [29]. The images shown in Figure 3.11 are very

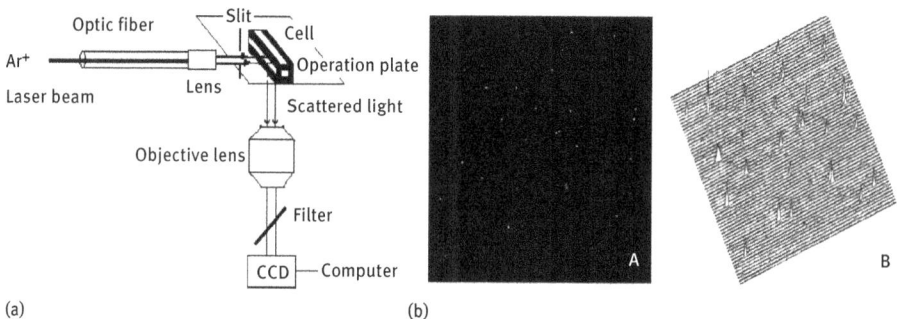

Figure 3.11: Laser scattering imaging analysis. (a) Laser scattering imaging facility and (b) light scattering image of tetrsulfonate porphyrin aggregates, where A is the surface imaging figure and B is stereo imaging [29].

sensitive for protein analysis, but the background is very high. To improve the sensitivity, the background signal needs further decrease.

3.3.2 Dark-field microscopic imaging

Dark-field microscopic imaging (iDFM) is a major means to characterize the scattering imaging of materials. Figure 3.12 shows advanced iDFM system. Compared with the self-built microscopic scattering imaging system shown in Figure 3.11, it significantly reduces background with a higher resolution, which is able to observe light scattering imaging of a single particle and molecular aggregates more conveniently.

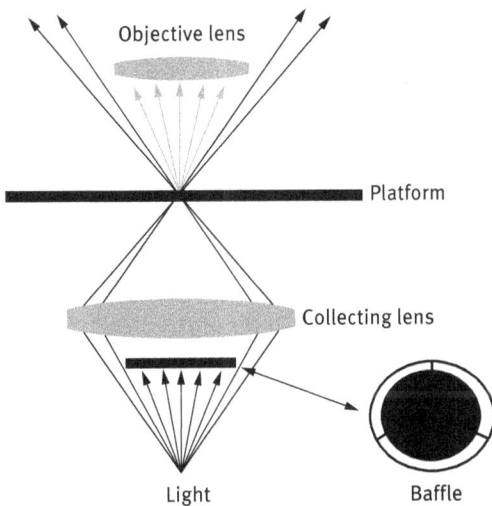

Figure 3.12: The principle of an inverted dark-field focus imaging.

To obtain a lower scattering background, the dark-field microscope uses a dark-field condenser as a key core component combined with high-power objective lens suitable for a dark-field observation. In a commercial dark-field microscope, high numerical aperture dark-field condenser and 100 times dark-field objective lens have appeared. The difference between the dark-field condenser and the common bright field condenser is that there is a baffle in the center of a dark-field condenser, blocking the light from passing through center. Thus, the light could go only through its of baffle edge and fall on the sample at a certain angle. The objective lens of the above sample matched with the dark-field condenser and could only observe the scattered light from the direction of the incident light (forward scattered light), so that the particles can be seen with the aid of scattered light.

The characteristic of dark field light scattering imaging is that light scattering imaging of single nanoparticles is easily observed, but it does not mean that the dark field microscope breaks through the "Abbe limit" of optical microscopy.[*] For example, the light scattering imaging of a single nanoparticle observed by a dark field microscope is just a specific performance of the light scattering properties of the nanoparticles rather than the true morphology of the nanoparticles. That is to say, for a 40 nm silver nanoparticle and 60 nm gold nanoparticle, smaller than the theoretical resolution of an optical microscope, their size cannot be directly determined by the size of the light spot in the dark-field scattering imaging . In morphology, 40 nm silver nanoparticles are smaller than 60 nm gold nanoparticles, but as 40 nm silver nanoparticles have stronger light scattering signals, thus the observation of a single 40 nm silver nanoparticles under a microscope is more possible to be larger than 60 nm gold nanoparticles [30]. Although the single nanoparticles observed by the dark field optical microscope are not the morphology of nanoparticles themselves, their light scattering signals appear in the form of spot. However, it is easy to capture the existence of individual particles by light scattering signal and establish a single particle based spectrometry. It needs to be demonstrated that the scattering light distribution of single spot needs to be further studied with the combination of high-resolution technique, which may provide more valuable information.

The dark field microscopic imaging system mostly uses a tungsten lamp or mercury lamp as the source of light with a feature of continuous spectra and white light. If the observed particle can scatter light at a certain wavelength, the color of particle light scattering can be clearly seen even without a filter. Although dark field observation is simple and convenient, there are also uncertainties in optical signals. For example, for fluorescent and scattering nanoparticles, both fluorescence and scattering signals can be observed simultaneously in the dark field observation mode. Therefore, before dark field scattering imaging, it is necessary to identify other optical signals such as particles without fluorescence, so as to ensure the authenticity of scattering imaging.

3.3.3 TIR imaging

TIR imaging ($_i$TIR) principles have been widely applied to single molecular fluorescence imaging analysis and cell membrane fluorescence imaging analysis, which have gained remarkable achievements. The advantage of total internal reflection

---

[*] Abbe limit: found by a German physicist Ernst Karl Abbe (1840–1905) in 1770s, who contributed to the microscopic theory. Visible light cannot limitlessly focus beam with its undulatory property diffraction, Abbe equation $f = 0.61/NP$, where f is resolution, NP is numerical aperture. On the basis of this, he determined a microscopic resolving limit for the visible light band. He believed that the minimum diameter focused by the visible light is one-third of the wavelength, no lower than 200 nm. The Abbe limit is a principle that must be followed in the design of an optical instrument. Objects need to be enlarged by 1–2,000 times to reach 0.2–0.4 mm to be seen by human eyes (clearly).

imaging lies in the selective excitation of the evanescent wave generated by the total reflection at the interface, which avoids the optical signal interference of the material outside the interface. The introduction of total internal reflection technology to light scattering imaging can improve the quality of light scattering imaging and expand the application fields of light scattering imaging. The author used an external laser to excite liquid-liquid interface aggregates in total internal reflection mode, and realizes the light scattering imaging of the biological macromolecule aggregates at the liquid interface [31]. The Ren's reseach group reassembled the dichroic mirror of the total internal reflection fluorescence microscope, and realized the scattering imaging of the single metal nanoparticles in the solution, and established the analytical methods of counting the particles number in the solution [32].

3.3.4 Single particle light scattering spectroscopy

In Sections 3.1 and 3.2, an ordinary fluorescence spectrophotometer was used to detect optical signal and establish a resonance scattering method to determine the average light scattering property of the total solution or a certain substance, and can not characterize the light scattering of a single particle in the microscopic field. In Section 3.3.3, the dark field light scattering microscopy can only obtain the light scattering imaging of the microscopic material, but can not obtain accurate or precise light scattering spectral information. In order to realize the optical microscopic imaging of the sample and the spectral characterization of the observed region, it is necessary to combine the spectral system with the microscope to obtain the microscopic spectral information of the particles and make up the problems of the microscopic system and the spectral system. This combination has led to a new research field, namely single particle light scattering spectroscopy analysis technology.

Figure 3.13 shows a system that assembles the optical microscopy and the dark field spectral systems, which consists of three parts: a dark field microscope, a true color charge coupled device (CCD) imager and a microscopic spectral system. The light scattering information of objects is first obtained through dark field microscopy. After passing through the objective lens, the light scattering signals are divided into two images by binary mirror, which are transported to the true color CCD imaging system and the microspectral system by two imaging connectors. The whole scattering image of the target can be observed after the capture with the true color CCD. The microscopic spectral system can be used to characterize the spectral properties of a specific region, and to capture and obtain linear spectra by spectral CCD. Because the microspectrometer can scan quite small area observed by the microscope, the light scattering spectral information of single particles in these regions can be acquired.

The single particle light scattering spectroscopy plays an important role in the study of the microscopic light scattering properties of single particles, multiple

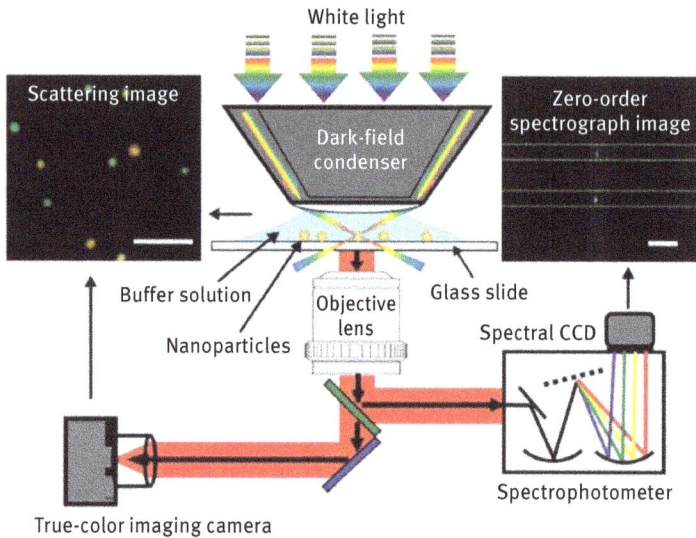

Figure 3.13: Single-particle light scattering spectrometry system [33].

particles and particle assemblies, and the establishment of highly sensitive optical analysis methods. In recent years, popular studies include the following: single metal nanoparticles light scattering spectral properties, the plasma coupling spectra of metal nanostructures, local surface plasma resonance spectroscopy of single metal nanoparticle, the energy transfer of single metal nanoparticles, and so on.

3.4 Dynamic light scattering

Different from light scattering analysis foundation discussed earlier, dynamic light scattering (DLS) is a technology for particle size, state, and other message analysis based on real-time monitoring and statistics of the total light scattering intensity of all scattered particles varies with time. The theoretical basis of dynamic light scattering analysis is the light scattering signal generated by Brown motion of molecules or particles in the medium under laser irradiation. As the Brownian movement is related to the particle size, medium viscosity, temperature, and so on, the dynamic fluctuation of signal intensity can be used to analyze particle or molecule size, molecule weight, dispersion, uniformity, and so on.

The dynamic light scattering signal fluctuation of nanoparticles is directly related to the particle size. If the particles are aggregated or assembled, the size of formed aggregates or assemblies can be considered as the particle size, which will linearly increase with the particles aggregation degree detected by the dynamic light scattering instrument. Therefore, the quantitative analysis method based on the change of particle size can be established by connecting the dynamic light scattering technique with the particle size change and the factors that affect these changes.

Based on the principle of DNA induced gold nanoparticles aggregation, the dynamic light scatterometer was used to detect the linear relationship between the particle size of gold nanoparticles and the concentration of DNA, a quantitative method for quantitative analysis of biological macromolecules based on dynamic light scattering was established by Qun's group [34]. Similar principles can be applied to immunoassay to examine specific protein, such as cancer biomarker protein [35].

References

[1] Anglister J, Steinberg I Z. Resonance Rayleigh scattering of cyanine dyes in solution. J. Chem. Phys., 1983, 78(9): 5358–5368.

[2] Pasternack R F, Bustamante C, Collings P J, et al. Porphyrin assemblies on DNA as studied by a resonance light-scattering technique. J. Am. Chem. Soc., 1993, 115(13): 5393–5399.

[3] Pasternack R F, Collings P J. Resonance light scattering: A new technique for studying chromophore aggregation. Science, 1995, 269(5226): 935–939.

[4] Huang C Z, Li K A, Tong S Y. Determination of nucleic acids by a resonance light-scattering technique with-tetrakis[4-(trimethylammoniumyl)phenyl] porphine. Anal Chem., 1996, 68(13): 2259–2263.

[5] Huang C Z, Li K A, Tong S Y. Determination of nanograms of nucleic acids by their enhancement effect on the resonance light scattering of the cobalt(II)/4-[(5-chloro-2-pyridyl) azo]-1, 3-dia-minobenzene complex. Anal. Chem., 1997, 69(3): 514–520.

[6] Huang C Z, Li Y F, Mao J G, et al. Determination of protein concentration by enhancement of the preresonance light-scattering of-tetrakis(5-sulfothienyl)porphine. Analyst, 1998, 123(6): 1401–1406.

[7] Huang C Z, Li Y F. Resonance light scattering technique used for biochemical and pharmaceutical analysis. Anal. Chim. Acta, 2003, 500(1–2): 105–117.

[8] Huang C Z, Lu W, Li Y F. Analytical applications of resonance light-scattering signals totally reflected at liquid/liquid interface. Rev. Anal. Chem., 2002, 21(4): 267–278.

[9] Ling J, Huang C Z, Li Y F, et al. Recent developments of the resonance light scattering technique: Technical evolution, new probes and applications. Appl. Spectrosc. Rev., 2007, 42(2): 177–201.

[10] Lu W, Band B S F, Yu Y, et al. Resonance light scattering and derived technique in analytical chemistry: Past, present, and future. Microchim. Acta, 2007, 158(1–2): 29–58.

[11] Lu W, Shang J. A. Resonance light-scattering (RLS)serving for various quantitative events since 1995: A comment proposed towards how to apprehend well the meaning of RLS and its corresponding guiding role. Spectrochim. Acta A, 2009, 74(1): 285–291.

[12] Liu S P, Liu Z F, Li M. Analytical application of double scattering spectra of ion-association complex(II). Acta Chim. Sin., 1995, 53(12): 1153–1184.

[13] Liu S P, Liu Z F, Li M. Double scattering spectra of mercury(II)-thiocyanate-rhodamine dye systems and their analytical applications. Acta Chim. Sin. 1995, 53(12): 1185–1192.

[14] Liu S, Liu Z. Studies on the resonant luminescence spectra of rhodamine dyes and their ion-association complexes. Spectrochim. Acta A, 1995, 51(9): 1497–1500.

[15] Liu S P, Liu Z F. Spectra of resonant luminescence and double scattering of selenium (IV)-iodide-crystal violet system. 1996, 17(8): 1213–1215.

[16]   Liu S P, Liu Z F. Spectra of resonant luminescence and double scattering of mercury(II)-thiocyanate-victoria blue 4R system and their analytical application. Chem. J. Chinese U., 1996, 17(6): 887–892.

[17]   Liu S, Luo H, Li N, et al. Resonance Rayleigh scattering study of the interaction of heparin with some basic diphenyl naphthylmethane dyes. Anal. Chem., 2001, 73(16): 3907–3914.

[18]   Wei X Q, Liu Z F, Liu S P. Resonance Rayleigh scattering spectra of tetracycline antibiotic-Cu(II)-titan yellow systems and their applications in analytical chemistry. Anal. Bioanal. Chem., 2006, 385(6): 1039–1044.

[19]   Liu S P, Chen S, Liu Z F, et al. Resonance Rayleigh scattering spectra of interaction of sodium carboxymethylcellulose with cationic acridine dyes and their analytical applications. Anal. Chim. Acta, 2005, 535(6): 169–175.

[20]   Liu S P, Yang Z, Liu Z F, et al. Resonance Rayleigh scattering study on the interaction of gold nanoparticles with berberine hydrochloride and its analytical application. Anal. Chim. Acta, 2006, 572(2): 283–289.

[21]   Feng P, Shu W Q, Huang C Z, et al. Total internal reflected resonance light scattering determination of chlortetracycline in body fluid with the complex cation of chlortetracycline-europium-trioctyl phosphine oxide at the water/tetrachloromethane interface. Anal. Chem., 2001, 73(17): 4307–4312.

[22]   Huang C Z, Li Y F, Feng P. Determination of proteins with-tetrakis (4-sulfophenyl) porphine by measuring the enhanced resonance light scattering at the air/liquid interface. Anal. Chim. Acta, 2001, 443(1): 73–80.

[23]   Huang C Z, Feng P, Li Y F, et al. Adsorption of penicillin-berberine ion associates at a water/tetrachloromethane interface and determination of penicillin based on total internal-reflected resonance light scattering measurements. Anal. Chim. Acta, 2005, 538(1): 337–343.

[24]   Huang C Z, Chen S F. Quantitation and differentiation of bioparticles based on the measurements of light-scattering signals with a common spectrofluorometer. J. Phys. Chem. B, 2008, 112(37): 11785–11793.

[25]   Tan K J, Huang C Z, Huang Y M. Determination of lead in environmental water by a backward light scattering technique. Talanta., 2006, 70(1): 116–121.

[26]   Wang Y H, Guo H P, Tan K J, et al. Backscattering light detection of nucleic acids with tetraphenylporphyrin-Al(III)-nucleic acids at liquid/liquid interface. Anal. Chim. Acta, 2004, 521 (1): 109–115.

[27]   Zhao H W, Huang C Z, Li Y F. A novel optical immunosensing system based on measuring surface enhanced light scattering signals of solid supports. Anal. Chim. Acta, 2006, 564(2): 166–172.

[28]   Yguerabide J, Yguerabide E E. Light-scattering submicroscopic particles as highly fluorescent analogs and their use as tracer labels in clinical and biological applications. II. Experimental characterization. Anal. Biochem., 1998, 262(2): 157–176.

[29]   Huang C Z, Liu Y, Wang Y H, et al. Resonance light scattering imaging detection of proteins with α, β, γ, δ-tetrakis(p-sulfophenyl)porphyrin. Assnal. Biochem., 2003, 321(2): 236–243.

[30]   Ling J, Li Y F, Huang C Z. Visual sandwich immunoassay system on the basis of plasmon resonance scattering signals of silver nanoparticles. Anal. Chem., 2009, 81(4): 1707–1714.

[31]   Ling J, Huang C Z, Li Y F. Directly light scattering imaging of the aggregations of biopolymer bound chromium (III) hydrolytic oligomers in aqueous phase and liquid/liquid interface. Anal. Chim. Acta, 2006, 567(2): 143–151.

[32]   Xie C, Xu F, Huang X, et al. Single gold nanoparticles counter: an ultrasensitive detection platform for one-step homogeneous immunoassays and DNA hybridization assays. J. Am. Chem. Soc., 2009, 131(35): 12763–12770.

[33] Liu G L, Long Y-T, Choi Y, et al. Quantized plasmon quenching dips nanospectroscopy via plasmon resonance energy transfer. Nat. Methods, 2007, 4(12): 1015–1017.
[34] Dai Q, Liu X, Coutts J, et al. A one-step highly sensitive method for DNA detection using dynamic light scattering. J. Am. Chem. Soc., 2008, 130(26): 8138–8139.
[35] Liu X, Dai Q, Austin L, et al. A one-step homogeneous immunoassay for cancer biomarker detection using gold nanoparticle probes coupled with dynamic light scattering. J. Am. Chem. Soc., 2008, 130(9): 2780–2782.

Jian Ling, Cheng Zhi Huang

4 Resonance light scattering spectroscopy

In the history of light scattering analysis, two main aspects of application should be mentioned here. One is light scattering analysis for atmosphere, water, and environmental particle detection on the basis of Tyndall effect; the other aspect is the particle size characterization with dynamic light scattering technology and Raman light scattering spectroscopy developed after the widely using of laser.

Light scattering technology has been widely used and it provides a great deal of information for research. Inelastic light scattering spectrum (Raman spectroscopy) has a good selectivity and low signal level, and for this, a laser must be used. In terms of analytical method establishment and analysis technology development, light scattering signal application is still in its initial phase, and further development is needed.

4.1 Introduction

As above-mentioned, every medium can scatter light. For systems with no absorbing material, its scattering signal intensity is only related to a medium refractive index, particle size, intensity, testing angle, and distance. However, if a system has substances with a characteristic absorption band, when the incident light wavelength is near to the absorption band, the scattered light would be greatly enhanced and a new scattering feature would appear, that is, resonance Rayleigh scattering enhancement (RRSE). Its appearance is further explained the theoretical basis of Rayleigh scattering.

American scholar Pasternack [1] obtained the dependence relation of a light scattering signal and wavelength by simultaneous scanning excitation and emission monochromator on a common fluorescence spectrophotometer (see Section 3.2.1). They developed a resonance light scattering (RLS) method and applied it for DNA J-type stacking aggregate of porphyrin, trans-bi(N-picoline-4) 2-diphenyl porphyrin, and its Cu^{2+} ramification, as well as the interaction between a new synthetic metal binder and calf thymus DNA. This is the first connection of light scattering with the aggregation process of small organic molecule.

Although Pasternack was the first to build the RLS technology, which later developed and was applied widely, at that time, RLS was only used as a method for substance interaction without being used in the field of analytical chemistry. By 1996, the author of this book found that the RLS signal produced by super spiral assembly of nucleic acid and meso-4 (p-aminophenyl) porphyrin has a linear relation with nucleic acid concentration in a certain range, and built a highly sensitive quantitative method for nucleic acids [2, 3]

https://doi.org/10.1515/9783110573138-004

4.2 Theoretical basis of resonance light scattering analysis

4.2.1 Rayleigh scattering signal intensity and ratio

Based on the Miller macroscopic wave theory, molecular scattering is due to the fluctuation of refractive index (m), whereas refractive index can be divided into real and imaginary parts, that is:

$$m = n - ik \tag{4.1}$$

where m is the complex refractive index, reflecting light spread in an area, whereas n is the medium average refractive index, and k is the absorptivity. Equations (4.1) and (2.21) are actually consistent. It means the refractive index of a solution is related to the incident light wavelength and the molar absorption coefficient near the absorption bands, and the relationship is also related to the absorption of the molecule in a overall wavelength range.

The average refractive index in equation (4.1) can be expressed in Kronig–Kramers equation as follows [4]:

$$n = n_0 + \frac{2.303 c \lambda_0^2}{2\pi^2} \int_0^\infty \frac{\varepsilon(\lambda)}{\lambda_0^2 - \lambda^2} d\lambda \tag{4.2}$$

where n_0 is solvent refractive index; c is the mole concentration of the solution; λ_0 is the characteristic absorption wavelength of the molecule; λ is the random incident light wavelength on the molecular absorption band; $\varepsilon(\lambda)$ is the molar absorption coefficient of the molecule.

Different from n, k is only related to molecular transitions of absorption band, and its relation with the molar absorption coefficient $\varepsilon(\lambda)$ is

$$k = \frac{2.303 \varepsilon(\lambda) c \lambda^2}{4\pi} \tag{4.3}$$

Based on this definition, Rayleigh ratio is introduced to represent a light scattering feature system. At an angle of $90°$ with the incident light, the tested ratio is as follows: [5]

$$R_{90} = \frac{4,000\pi^2 n^2 c}{\lambda_0^4 N_A} \left[\left(\frac{\partial n}{\partial c}\right)^2 + \left(\frac{\partial k}{\partial c}\right)^2 \right] C_v \tag{4.4}$$

where N_A is the Avogadro constant; $\partial n/\partial c$ and $\partial k/\partial c$ are refractive index increase of real and imaginary parts for 1.0 mol/L solution; C_v is the Cabannes factor, representing a light scattering increase. Combine n and k to obtain the following:

$$R_{90} = \frac{(2.303)^2 1,000cn}{N_A} \left\{ \left[\frac{1}{\pi} \int_0^\infty \frac{\varepsilon(\lambda)d\lambda}{\lambda_0^2 - \lambda^2} \right] + \frac{\varepsilon^2(\lambda_0)}{4\lambda_0^2} \right\} C_v \qquad (4.5)$$

It can be acquired from equation (4.5) that
(1) If all conditions are certain, then R_{90} is in direct proportion to the substance concentration (c)
(2) If incident light wavelength (λ) is away from molecular absorption band (λ_0), then $\partial k/\partial c = 0$, the acquired spectrum is mainly dependent on the real part of the refractive index. If $d \leq 0.05\lambda_0$, then the intensity follows Rayleigh scattering law, where the intensity is inversely proportional to the wavelength of the biquadrate [5]:

$$I = \frac{24\pi^3 N v^2}{\lambda^4} \left(\frac{n_1^2 - n_0^2}{n_1^2 + 2n_2^2} \right)^2 I_0 \qquad (4.6)$$

where I_0 is the incident light intensity, λ is the wavelength of the incident light or scattered light, N is the particle number in unit volume, v is the volume of a single particle, and n_1 and n_0 represent the refractive rate of the scattering phase and the medium, respectively.
(3) If incident light wavelength (λ) is near the molecular absorption band (λ_0), then both the real part and the imaginative part of the refractive index contribute a lot to light scattering. Especially, when the absorption band is strong, the resonance Rayleigh scattering is enhanced, which is related to electron transition in the band. It needs to be demonstrated that even resonance Rayleigh scattering is enhanced, its intensity is still 1–2 in the order of the magnitude that is lower than the solvent intensity, so that the laser is required for being the strong light source and higher concentration of the organic dye is required for producing detectable scattering signal [4]. While for molecular aggregation, particle size is larger. Strong light scattering enhancement can be detected by a common fluorescence spectrophotometer as the scattering intensity depends on the particle size (Figure 4.1). It is noticeable that as a molecule absorbs incident light and scattered light, low scattering intensity is acquired at the molecular absorption maximum value.

As shown in equation (4.3), the relation of ε and k depends on the imaginary part of the refractive index that has a similar shape with molecular absorption spectrum; contrary to the Kronig–Kramers integration in equation (4.2), imaginary part of the

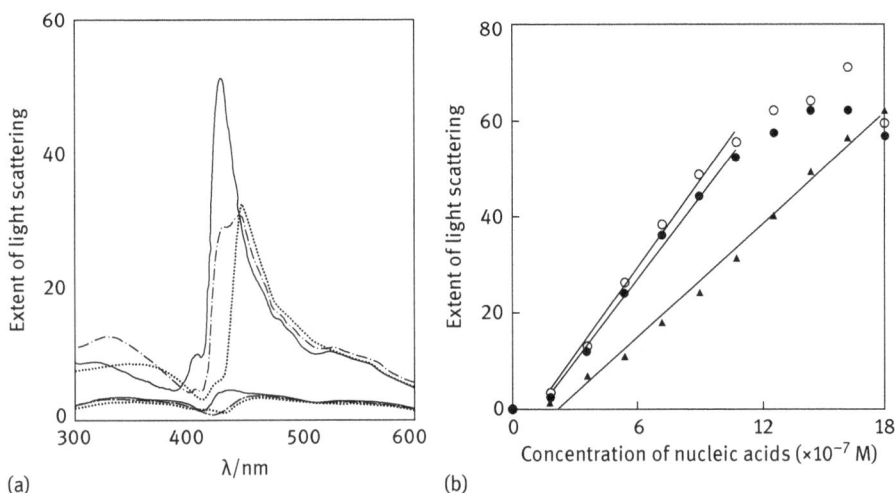

Figure 4.1: TAPP and nucleic acid interaction induced the enhancement of light scattering spectra and DNA resonance light scattering measurement [2]. Symbols: (-) pH 7.48; (- . -), pH 3.79; (...) pH 2.02. Upper, DNA + TAPP; lower, TAPP. (a) Resonance light scattering before and after ctDNA interaction with TAPP under different acidity. (b) Resonance light scattering measurement for quantitative detection of ctDNA, fsDNA, and yeast RNA standard curve.

refractive index is related to high-energy and low-energy area of the band, so in a high-energy area, the imaginary part has a positive contribution and the real part has a negative contribution and thus the two parts cancel each other out; whereas in a low-energy area, both are positive. In a normal RLS spectrum, the strongest signal is usually from the low-energy area in resonance absorption, whereas strictly, the pre-resonance light-scattering spectrum can be obtained, as shown in Figure 4.2 [6].

It is notable that equation (4.5) considers molecule scattering mainly from the resonance light scattering enhancement. Molecular scattering signal is normally weak and hardly obtained accurately. So, in an actual mechanical process, a larger excitation and emission monochromator slit (normally > 5 nm) was used for light scattering measurement. The obtained resonant light scattering spectrum often has a dynamic scattering component; meanwhile, it may not accord with the condition of $d \leq 0.05\lambda_0$. When particle concentration is high, the resonance light scattering intensity has a larger fluctuation, which can be proven by a decrease in the slit. When the slit width is smaller than 5 nm, the spectral feature has a large variation because it is closer to the real signal, meanwhile, the smaller slit decreases the incident light intensity and then weakens the scattered light intensity.

So, the measured scattering signal by an ordinary fluorophotometer is not a pure resonance Rayleigh scattering signal, the maximum signal may not be obtained at $\Delta\lambda = 0$ nm [5], this can be contributed to existence of complex components. For example, when particles are larger (does not accord with condition of $d \leq 0.05\lambda_0$), the obtained spectrum composes not only Rayleigh scattering signal, but also Mie scattering.

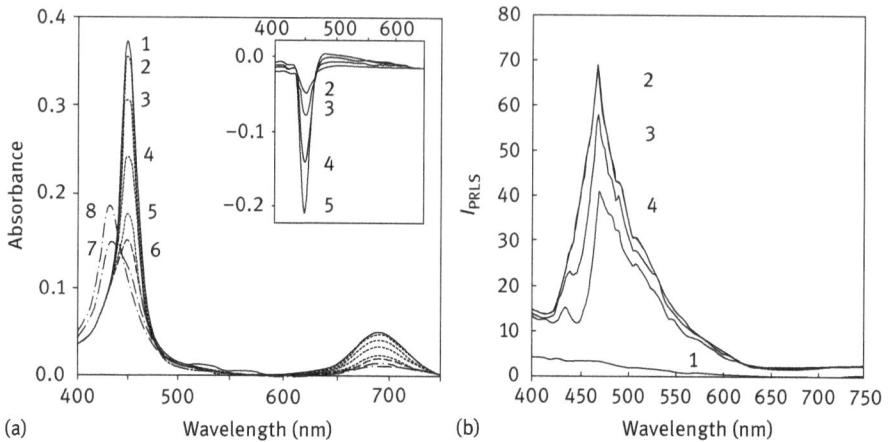

Figure 4.2: Preresonance light-scattering spectra enhancement of α,β,γ,δ-4(5-fluoro(5-sulfothiophene) porphyrin [T-(5-ST)P] induced by protein [6]. (a) The absorption spectra of the aggregation of T(5-ST)P in the presence of BSA. T-(5-ST)P, 1.2×10^{-6} mol/L; BSA (1–7)0.0, 1.0, 2.0, 3.0, 4.0, 5.0, 10.0, and 20.0 μg/mL; pH 1.81. (b) pre-resonance light-scattering-enhanced spectra at various T-(5-ST)P concentration. BSA, 2.0 μg/mL; T-(5-ST)P (1–4), 0.0×10^{-6}, 0.8×10^{-6}, 1.2×10^{-6}, and 2.0×10^{-6} mol/L.

4.2.2 Measurement of resonance light scattering

It can be known from equation (4.4) that enhanced resonance Rayleigh scattering signal is normally found in a low-energy zone (right side or long-wavelength area); rather it can correspond with its molecular absorption band. For this, the resonance light scattering spectrum can be obtained by synchronously scanning excitation and emission monochromator on an ordinary fluorescence spectrophotometer. The specific steps for operation of a spectrophotometer are choosing a suitable excitation and emission slit on a spectrophotometer, and synchronous scanning at the same excitation and emission wavelength ($\Delta\lambda = 0$) [2, 3].

4.2.3 Quantitative method of resonance light scattering

Currently, there is no authoritative quantitative equation of resonance light scattering. We can understand how to use light scattering signal for quantitative analysis form two aspects. One is, the method of obtaining resonance light scattering spectrum, that is the synchronous spectrum, and the other is the theoretical deduction of Rayleigh equation. In comparison, the former one is simple but does not refer to the principle of light scattering, the latter one is complicated but is the essence.

4.2.3.1 Synchronous scanning spectra method

Resonance light scattering spectrum obtained on a common fluorescence spectro-photometer in essence is a synchronized spectrum, it cannot entirely reflect the real light scattering property of a system. If the elastic light scattering is the light scattered by a particle, which is the same wavelength to incident light, the scattering particle is actually a new emitter, which is able to emit light with the same wavelength of the excited light. So the resonance light scattering signal can be expressed in a simply synchronous luminescence signal [2, 3]. Based on the synchronous luminescence following equations can be derived:

$$I_{SL} = KcbE_{ex}(\lambda_{ex})E_{ex}(\lambda_{ex} + \Delta\lambda) \tag{4.7}$$

$$I_{SL} = KcbE_{em}(\lambda_{em})E_{em}(\lambda_{em} - \Delta\lambda) \tag{4.7a}$$

where E_{ex} is the excitation function at a given excited light wavelength (λ_{ex} = $\lambda_{em} - \Delta\lambda$); E_{em} is the emission function ($\lambda_{ex} = \lambda_{em} + \Delta\lambda$) at a corresponding excited light wavelength ($\lambda_{ex} = \lambda_{em} - \Delta\lambda$), K is a constant related to instrument condition, such as intensity, slit width, instrument stability, and so on; b is the path length of the light. When $\Delta\lambda = 0$, the following resonance light scattering intensity is obtained:

$$I_{RLS} = KcbE_{ex}(\lambda_{ex})E_{em}(\lambda_{ex}) \tag{4.8}$$

$$I_{RLS} = KcbE_{ex}(\lambda_{em})E_{em}(\lambda_{em}) \tag{4.8a}$$

When instrument condition is fixed, then the resonance light scattering intensity is in direct proportion to the particle concentration, which can be used for particle quantitative determination.

4.2.3.2 Rayleigh scattering method

Equation (1.1) or (2.20) shows a light scattering signal of one particle. For a system with N particles of the same component and shape, approximately 20 times smaller than an incident wavelength, then the overall scattered light intensity is as follows:

$$I = I_0 \cdot \frac{16VN\pi^4 r^6 n_{med}^4}{d^2 \lambda_0^4} \cdot \left|\frac{m^2 - 1}{m^2 + 2}\right|^2 \cdot \sin^2(\varphi) \tag{4.9}$$

where V is the solution volume. If the Avogadro constant is introduced as Rayleigh ratio (equation (4.4)), then the intensity has the following relation with the particle concentration:

$$I = I_0 \cdot \frac{16VN\pi^4 r^6 n_{med}^4}{N_A d^2 \lambda_0^4} \cdot \left|\frac{m^2-1}{m^2+2}\right|^2 \cdot \sin^2(\varphi) \qquad (4.9a)$$

That is,

$$I = I_0 \cdot \frac{16c_P\pi^4 r^6 n_{med}^4}{d^2 \lambda_0^4} \cdot \left|\frac{m^2-1}{m^2+2}\right|^2 \cdot \sin^2(\varphi) \qquad (4.9b)$$

where $c_p = VN/N_A$. As refractive index of a particle (m) is identical to that of a medium environment (m), and the particle has the same anisotropy value (r), with certain detection angle (φ) and distance (d), then equation (4.9b) can be expressed as the linear relationship between the intensity and the concentration:

$$I = I_0 k c_P \qquad (4.10)$$

Equation (4.10) is the quantitative basis of analysis, where k is a constant, which is related to the following factors:

$$k = \frac{16\pi^4 r^6 n_{med}^4}{d^2 \lambda_0^4} \cdot \left|\frac{m^2-1}{m^2+2}\right|^2 \cdot \sin^2(\varphi) \qquad (4.10a)$$

It needs to be demonstrated that c_p in equation (4.10) is the particle concentration, which is obtained by the conversion of particle number in a unit volume and Avogadro constant.

4.3 Factors affecting the sensitivity of resonance light scattering spectroscopy

4.3.1 Instrument factor

In equations (4.8) and (4.9), the scattered light intensity is directly proportional to the incident light intensity I_0. When we use a fluorospectrophotometer for a signal acquiring, where there is a voltage fluctuation or environmental state variation, then the fluctuation in the exciting light intensity will occur. So, in the measurement of signals, a voltage stabilizer is needed for a steady signal output.

The obtained spectrum of the same model of instrument has its own character-istic peak. Taking a spectrum from RF-540 photometer as an example, it is observed that as long as a medium has no strong absorption at 470 nm, there is one RLS peak around 470 nm, which is mainly caused by large emission intensity of a xenon lamp.

Different instrument have different light output with a difference of intensity and wavelength. In addition, monochromator has its own resolution for light splitting and detector (photomultiplier) has different sensitivity to light.

Different instruments have different characteristic peaks. Figure 4.3 is RLS spectra obtained from F-2500 and F-4500 fluorospectrophotometers. It can be observed that the same sample presents great differences in the RLS spectra, even under identical testing conditions (voltage and slit is kept the same). RLS spectrum of F-2500 fluorospectrophotometer displays a scattering peak at 380 and 570 nm, and the peak obtained at 380 nm is stronger than the latter; RLS spectrum of F-4500 fluorospectrophotometer also has two peaks at 380 and 570 nm, and the 380 nm peak shows a lower intensity than the latter signals from both the instruments present various peak shapes and intensity. For example, the signal at 380 nm is nearly 600, which is obtained from F-2500, whereas the value is only 150 obtained from F-4500. For the signal at 570 nm, the signal from F-2500 is much stronger than that from F-4500.

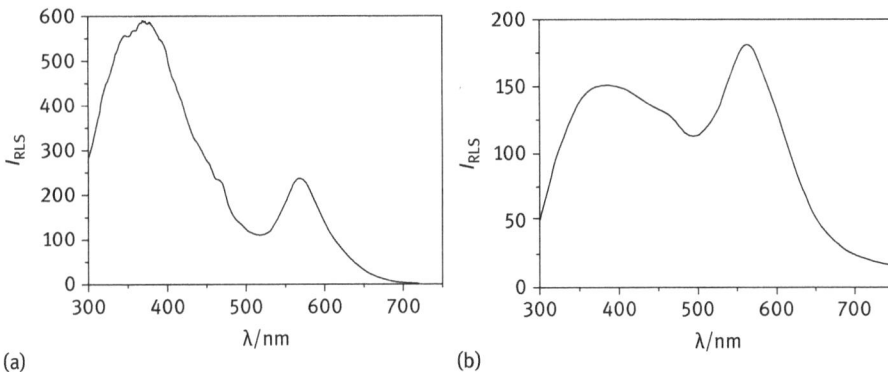

Figure 4.3: Light-scattering spectra of a gold nanoparticle solution tested by different fluorospec-trophotometers. (a) F-2500 fluorospectrophotometer (b) F-4500 fluorospectrophotometer.
Note: Both the fluorospectrophotometers work at 400 V voltage and an exciting and emission monochromator slit of 5 nm.

Figure 4.4 is the light scattering spectra of H_2TAPP and Heparin (HP) complex measured by F-2500 fluorospectrophotometer using LED and Xe lamp [7]. As shown in Figure 4.4 (a), when a blue LED is used as an external light source for a complex, then two spectra with a characteristic peak at 459 nm are acquired for both H_2TAPP and H_2TAPP-HP

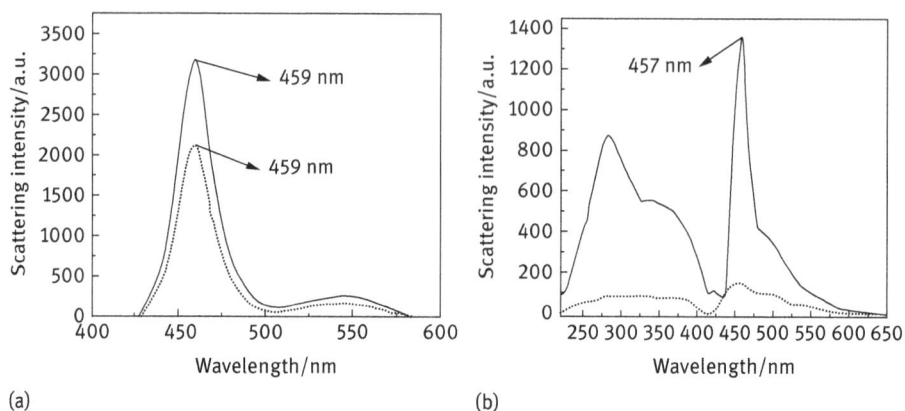

(a)   (b)

Figure 4.4: The influence of a light source on the resonance scattering spectra.

Note: An external blue LED (a) and Xe lamp excitation (b) light scattering spectra of $H_2TAPP-HP$ complex on F-2500 photometer: H_2TAPP, 6.36×10^{-6} mol/L; HP, 4.0 µg/mL. Full line: HP + H_2TAPP; imaginary line: H_2TAPP [7].

complex, whereas the scattering intensity of H_2TAPP has a weaker scattering intensity than its complex. When a F-2500 equipped a Xe lamp is directly used for measurement, then UV-visible region bands can be obtained. Besides obtaining a characteristic peak at 457 nm, there is scattering signal in the UV region. $H_2TAPP-HP$ complex has a 420 nm scattering peak that is caused by the quenching of scattering light by the strong Soret absorption of H_2TAPP at 412 and 430 nm.

In addition, both external battery voltage and current influences the light intensity, the increase of which brings a strong incident light intensity, and vice versa. An external battery voltage and a current fluctuation directly interfere with the repeatability of the test. If the voltage varies randomly in a test, then the output light intensity also changes randomly, producing a random scattered light intensity and measuring error.

4.3.2 Incident light wavelength

The incident light wavelength depends on the substance absorption with the following two situations:

(1) If the incident light wavelength is away from the molecular absorption band, then $\partial k/\partial c = 0$. Based on equation (4.6), the scattered light intensity is inversely proportional to the biquadrate of an incident light wavelength (λ); the shorter the wavelength, the stronger will be the scattering.

(2) If an incident light wavelength is near the molecular absorption band, then $\partial k/\partial c \neq 0$. Based on equation (4.5), the Rayleigh ratio is closely related to the molar absorption coefficient (ε); the higher the coefficient, the stronger will

be the scattering. The scattered light intensity is not inversely proportional to the biquadrate of wavelength of an incident light (λ); the feature of a spectrum is directly influenced by an absorption spectrum. Molecules with different absorption spectrum have various features, producing characteristic scattering peaks.

It needs to be demonstrated that as there is an inner filter effect in the absorption of a molecule that changes fluorescence spectra, if there is a decreasing factor in the system, as the medium absorbs excitation and scattered light, then the collected signal is weakened and the scattered light intensity is reduced. For reducing the influence of molecular absorption on establishing a light scattering method, water is usually added for dilution, which will minimize the scattering test influence by the molecular absorption. However, it is not a universal solution, for example, when the amount of testing component is little but has strong absorption, the large amounts of dilution will "lose" substances and an error will occur. Meanwhile, dilution changes the substance structure, solvation, association degree, and so on.

4.3.3 Particle size and concentration

As per equation (4.6), the scattered light intensity is in direct proportion to the particle number (N) and volume (v), which is the fundamental basis of quantitative analysis of a light scattering signal. In a homogeneous system, aggregates are formed with an increase in the particle volume and decrease in the particle quantity, and in which the light scattering intensity of the system is enhanced in the aggregation process. This is clearer in nanoparticle aggregates, where the changes in volume and number can be confirmed by a dynamic light scattering and imaging technique. However, when the formed aggregates is too large or is placed too long, then the sediment is likely to happen with a decreased signal, then a lower light scattering intensity would be observed when the concentration of target exceeds the linear range.

4.3.4 Refractive index of scattering particles and the surrounding medium

As given in equation (4.6), the scattering particle and the medium have a refractive index difference; the larger the difference, the stronger is the scattering. In a homogeneous system, scattering happens by the fluctuation of a medium refractive index that causes by the of insolubility of the scattering particle phase and the medium phase. If the medium is water and the particle is large enough, then a hydrophobic interface is formed, so that the particle would reflect an incident light. If an organic solvent such as methanol or acetone is added, then the signal will decrease significantly, because organic solvent breaks the hydrophobic interface of particle and media, forming a new medium refractive index fluctuation system [8].

4.4 Calibration of a resonance light scattering spectrum

A light-scattering spectrum is related to the molecular absorption and the condition of the instrument. As described earlier, an inner filter effect in the absorption of a molecule reduces the scattered RLS signal; when instrument condition changes, the shape of the RLS spectrum also changes. So, similar to calibration of fluorescence spectra, it is also necessary to calibrate the inner filter effect in the absorption of a molecule to improve its sensitivity, as well as the instrument's influence, thus separating the scattering component from the detected spectrum to obtain a real RLS spectrum.

To reduce error in different instruments, the authors of the book [9] designed an absorption optical cell equipment by changing the light path of an incident light in a fluorescence spectrophotometer and realized the calibration of RLS spectrum on the same fluorescence spectrophotometer.

4.4.1 Calibration principle

As shown in Figure 4.5, a absorption device is used to change the light path of the fluorospectrophotometer. Perform synchronous scanning for light source and sample at same excitation and emission wavelength to obtain an intensity spectrum $I_0(\lambda)$ of a light source and sample transmission spectrum $I_1(\lambda)$ and use performance function of a photometer to obtain its rate. When an incident light with intensity of I_0 passes through an absorption cell, its intensity is reduced with a increase in the absorption and scattering of the cell and solvent. For one detection, the same cell and solvent

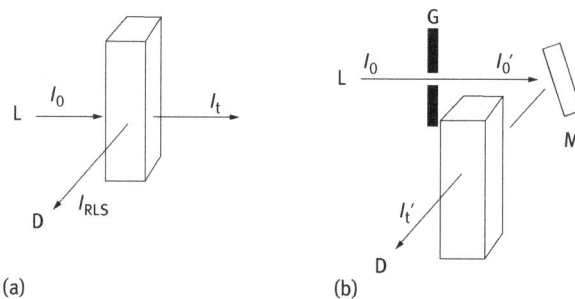

Figure 4.5: Two examples of spectral conversion. G is the steel gauze and M is the plane mirror.

can be used. In a theoretical detection, cell and solvent effect can be neglected. Based on Bill's law:

$$\log \frac{I_x(\lambda)}{I_0(\lambda)} = -\varepsilon(\lambda)xc \qquad (4.11)$$

Or

$$\log \frac{I_t(\lambda)}{I_0(\lambda)} = -\varepsilon(\lambda)bc \qquad (4.12)$$

where $I_x(\lambda)$ is the light transmission intensity and $I_0(\lambda)$ is intensity of the incident light that passes through x distance in an absorption cell. $I_t(\lambda)$ is the intensity after passing through the overall absorption cell; $\varepsilon(\lambda)$ is the molar absorption coefficient; b is the light path of the cell, and c is the sample concentration; combining equations (4.11) and (4.12) following equation is obtained:

$$I_t(\lambda) = I_0(\lambda)T^{x/h} \qquad (4.13)$$

where T is the light transmittance at a wavelength of λ, $T = I_t(\lambda)/I_0(\lambda)$.

RLS intensity can be obtained from the integration of infinitesimal volume $dxdydz$. When the intensity of an incident light is $I_x(\lambda)$, its RLS intensity can be expressed as follows:

$$d(I_x(\lambda)) = \beta I_x(\lambda)dxdyZ_0 \qquad (4.14)$$

where β is the coefficient of proportion, dx and dy are partial differential on X and Y directions. Z_0 is the solution height that can be expressed by the light beam in the Z direction, which is a constant decided by the property of the instrument.

In Y direction, as the detection direction is vertical to the excitation light beam, the detected RLS intensity ($dxdydz$) is lower because of molecular absorption. The final detected RLS intensity of the infinitesimal volume is as follows:

$$d\left(I_{app}(\lambda)\right) = d(I_x(\lambda))T^{y/h} \qquad (4.15)$$

where $I_{app}(\lambda)$ is the apparent RLS intensity that can be directly detected, y is the distance from infinitesimal volume to the inner surface of the absorption cell in the Y direction. Substituting equations (4.13) and (4.14) into equation (4.15) the following equation is obtained:

$$d\left(I_{app}(\lambda)\right) = \beta I_0(\lambda)Z_0 T^{x/h}dxT^{y/h}dy \qquad (4.16)$$

If there is no absorption, then $T = 1$, equation (4.16) would become

$$d(I_{corr}(\lambda)) = \beta I_0(\lambda) Z_0 dx dy \tag{4.17}$$

where $I_{app}(\lambda)$ is the apparent RLS intensity after calibration. Integrate equations (4.16) and (4.17) to obtain the following:

$$I_{app}(\lambda) = \beta I_0(\lambda) Z_0 \int_0^{x_0} T^{x/h} dx \int_0^{y_0} T^{y/h} dy = \beta I_0 Z_0 B^2 \frac{\left(1 - T^{x_0/h}\right)\left(1 - T^{y_0/h}\right)}{(-InT)^2} \tag{4.18}$$

$$I_{corr}(\lambda) = \beta I_0(\lambda) Z_0 \int_0^{X_0} \int_0^{Y_0} dx dy = \beta I_0(\lambda) X_0 Y_0 Z_0 \tag{4.19}$$

If detected with an absorption cell of 1.0 cm, there is $X_0 = Y_0 = b = 1$ cm, so

$$I_{corr}(\lambda) = \beta I_0 Z_0 \tag{4.20}$$

$$I_{app}(\lambda) = I_{corr}(\lambda)\left(\frac{1 - [I_t(\lambda)/I_0(\lambda)]}{2.303\log[I_0(\lambda)/I_t(\lambda)]}\right)^2 \tag{4.21}$$

If we define $C(\lambda) = \left(\frac{2.303\log[I_0(\lambda)/I_t(\lambda)]}{1 - [I_t(\lambda)/I_0(\lambda)]}\right)^2$, equation (4.21) can be expressed as follows:

$$I_{corr}(\lambda) = C(\lambda) I_{app}(\lambda) \tag{4.22}$$

where $C(\lambda)$ is the calibration factor, $I_0(\lambda)/I_t(\lambda)$ is obtained from the performance function. Based on equation (4.21), use $I_0(\lambda)/I_t(\lambda)$ and apparent absorbance ($\log[I_0(\lambda)/I_t(\lambda)]$) to calibrate the RLS intensity, which are the results of common function of the absorption and scattering of a molecule.

4.4.2 Separation of scattering components in a apparent absorption spectrum

When an incident light with intensity of $I_0(\lambda)$ passes through a system, the intensity reduces to $I_t(\lambda)$ because of absorption and scattering of the medium. The relation can be expressed as follows:

$$I_t(\lambda) = I_0(\lambda) e^{-2.303[\alpha_a(\lambda) + \alpha_s(\lambda)]bc} = e^{-2.303\alpha(\lambda)bc} \tag{4.23}$$

where $\alpha(\lambda)$ is the apparent absorption factor; $\alpha_a(\lambda)$ is the absorption factor; $\alpha_s(\lambda)$ is the scattering factor; b is the thickness of the absorption cell; c is the sample concentration; $\alpha(\lambda) = \alpha_a(\lambda) + \alpha_s(\lambda)$. So, in a apparent absorption spectrum, the relationship of absorption and scattering component is as follows:

$$A_{\text{app}}(\lambda) = A(\lambda) + S(\lambda) \qquad (4.24)$$

where $A_{\text{app}}(\lambda)$ is the apparent absorbance; $A(\lambda)$ is the absorbance; $S(\lambda)$ is the scattering component of the apparent absorbance spectrum, expressed as $S(\lambda) = \alpha_s(\lambda)bc$.

RLS signal intensity from an common spectrophotometer is reduced because of the apparent absorption; after calibration, the RLS intensity can be shown as the following:

$$I_{\text{corr}}(\lambda) = I_{\text{app}}(\lambda) \left(\frac{2.303 \log[I_0(\lambda)/I_t(\lambda)]}{1 - [I_t(\lambda)/I_0(\lambda)]} \right)^2 \qquad (4.25)$$

If we define $\Delta I(\lambda) = I_0(\lambda) - I_t(\lambda)$
The equation will become

$$\Delta I(\lambda) = I_0(\lambda) \left(2.303\alpha(\lambda)bc - \frac{1}{2}(2.303\alpha(\lambda)bc)^2 + \frac{1}{3!}(2.303\alpha(\lambda)bc)^3 - \cdots \right) \qquad (4.26)$$

If $\alpha(\lambda)bc \le 0.3$, the relative error is less than 10% when neglect the items later the second item in a trace analysis, so equation (4.26) can be simplified into the following:

$$\Delta I(\lambda) = I_0(\lambda) \left(2.303\alpha(\lambda)bc - \frac{1}{2}(2.303\alpha(\lambda)bc)^2 \right) \qquad (4.27)$$

In a non-absorption spherical particle scattering system, $\alpha(\lambda) = \alpha_s(\lambda)$, $A_{\text{app}}(\lambda) = S(\lambda)$, the incident light intensity is lowered mainly by scattering, so the intensity has a quadratic function with $S(\lambda)$. For this system, only scattering contributes to $\Delta I(\lambda)$. So, after calibration, a relation is obtained as follows:

$$I_{\text{corr}}^{\text{sphere}}(\lambda) = I_0(\lambda) \left(2.303 \log \left(I_0(\lambda)/I_1(\lambda) \right)_{\text{sphere}} - \frac{1}{2} \left(2.303 \log \left(I_0(\lambda)/I_1(\lambda) \right)_{\text{sphere}} \right)^2 \right) \qquad (4.28)$$

For an absorption spherical particle scattering system, if only the scattering component contribution to $I(\lambda)$ is considered, following equation is obtained:

$$I_{\text{corr}}^{\text{sphere}}(\lambda) = I_0(\lambda) \left(2.303S(\lambda)_{\text{sphere}} - \frac{1}{2} \left(2.303S(\lambda)_{\text{sphere}} \right)^2 \right) \qquad (4.29)$$

$I_{corr}^{sphere}(\lambda)$ is the RLS intensity after calibration. Quadratic function of fitted curves of I_{RLS} and $\log(I_0/I_t)$ at different wavelengths is quite different, which is mainly caused by the difference of $I_0(\lambda)$. To diminish $I_0(\lambda)$ influence on the RLS spectrum, divide equation (4.29) with equation (4.28) to obtain the following:

$$\frac{S(\lambda)_{sample} - 1.152 S(\lambda)_{sample}^2}{\log\left(I_0(\lambda)/I_t(\lambda)\right)_{sphere} - 1.152\left(\log\left(I_0(\lambda)/I_t(\lambda)\right)_{sphere}\right)^2} = \frac{I_{corr}^{sample}}{I_{corr}^{sphere}} \qquad (4.30)$$

Acquire $S(\lambda)_{sample}$ from equation (4.30).

4.4.3 Spectral calibration application

Taking Rhodamine 6G (Rh6G) resonance light scattering spectrum at pH 7.0 as an example [9]. The instrument's characteristic factor and molecular absorption factor are introduced to discuss the influence of the system's molecular absorption and sensitivity of detector on the RLS spectrum, thus the spectral calibration equation is proposed forward. As shown in Figure 4.6, in an absorption system, the scattering light intensity decreases near the absorption band, leading to a spectral distortion (Figure 4.6(a)). As absorption and scattering follow different rules, there is no linear relationship between the signal and the Rh6G concentration. After calibration (Figure 4.6 (b)), the influence of the light source of the instrument and the variation of the detection system for an RLS spectrum

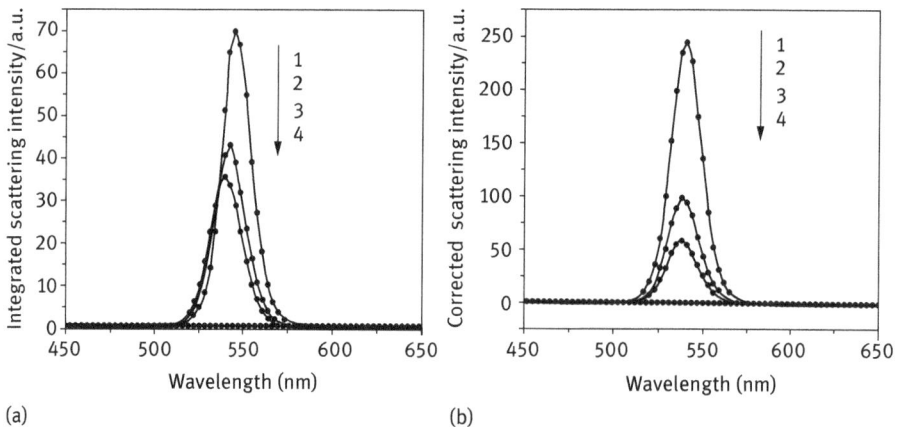

Figure 4.6: Rh6G scattering light spectral calibration. (a) Rh6G RLS spectra and (b) calibrated Rh6G RLS spectra. 1–4 represent Rh6G concentration of ($\times 10^{-5}$ mol/L) 1.6, 0.8, 0.4, and 0 respectively.

feature as well as molecular absorption is also eliminated, thus improving the sensitivity of the test.

4.5 Resonance light scattering ratio analysis

4.5.1 Limitations of resonance light scattering analysis

With further development of RLS in analytical application, following disadvantages of RLS methods were found on the basis of large numbers of experiments data:
(1) Resonance light scattering spectrum based on the bulk solution reaction system similar to other spectral methods that also has complex interference factors. The present of lots of unquantifiable factors, such as the incident light intensity, sample concentration, pH of the solution, ion strength, temperature, and medium polarity, limit the applications of unstable analytical reagents in the RLS spectrometry.
(2) Resonance light scattering technology based on a single wavelength measurement is largely affected by conditions of the instrument, battery voltage, and other external factors that produce large systematic errors.
(3) The obtained spectrum is a apparent scattering signal, which is unable to exhibit dynamic light scattering changes of particles in a system.

To compensate for its disadvantages and reducing its errors, recently, some new data processes and improvement in some experimental technologies appeared; among which, the dual-wavelength resonance light scattering ratiometry is performed by the combination of dynamic light scattering technology and imaging technology.

4.5.2 Dual-wavelength resonance light-scattering ratiometry

Proportional analysis is the most important method in a financial analysis with many advantages; it has wider application prospects that are introduced in analytical chemistry, especially for the sensor analysis in photochemistry. In analytical chemistry, the using of ratiometry can eliminate the affects of reagent concentration, light scattering intensity, instrument condition and stability of excitation light source on the fluctuation of the result, and also can avoid the uncontrollable factors on the operation of an instrument.

Under these conditions that are aimed toward the disadvantages of RLS, it is quite necessary to establish a more accurate dual-wavelength resonance light-scattering ratiometry and to develop simple, low-cost, and feasible facilities, further expanding the applicability of the RLS technology.

4.5.2.1 Wavelength selection

According to Lakowicz's wavelength selection principle [10], dividing the absorption spectrum of post-reaction to that of pre-reaction, a ratio absorption spectrum can be obtained. Two wavelengths that at greater absorption ratio difference are selected as the two wavelengths for RSL ratiometry measurement. Taking the example of aggregation of Janus Green B (JGB) on heparin (Hep) to illustrate this principle [11].

In an alkaline medium, the RLS characteristic peak of Hep is at 275 nm (Figure 4.7 (a)), for JGB characteristic peak is at 325 nm with one shoulder peak at 280 nm. It is observed that a separated JGB and Hep shows a weak RLS signal, whereas their mixture has an enhanced RLS signal with a characteristic peak of 314 nm and a few shoulder peaks at 350–600 nm, that confirming their interaction. JGB has two absorption bands at 285 and 493 nm (Figure 4.7(b)); when it interacts with Hep, obvious hypochromic effect can be observed. Dividing the absorption spctrum after its interaction with the blank JGB absorption spectrum, a ratio absorption spectrum can be obtained. Based on this, we can observe in the range of 220–600 nm that the maximum difference in absorption ratio is at the 285 and 345 nm. Now, we choose the two equations of I_{285}/I_{345} and A_{285}/A_{345} to discuss the molecular absorption and the RLS spectrum characteristic of JGB and Hep.

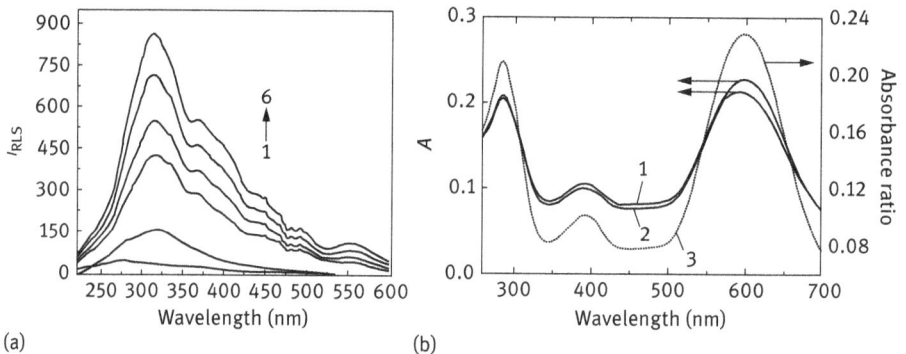

Figure 4.7: Wavelength selection of RLS ratiometry. Scattering spectra (a) and absorption spectra (b) of the aggregation of JGB with Hep. pH 7.96; JGB, 1.0×10^{-5} mol/L (except for 1 in (a)); (a) Hep in (a, 1–6): 500, 0.0, 200, 400, 600, and 500 ng/mL; Hep in (b), 1,000 ng/mL [11].

We can observe from the effect of medium pH on I_{285}/I_{345} and A_{285}/A_{345} ratio (Figure 4.8 (a)), with pH variation that the ratiometry is more steady as compared to a single wavelength-enhanced RLS signal. Constant I_{285}/I_{345} value is in the pH range of 2.0–9.9, whereas for A_{285}/A_{345} pH is 5.0–8.0 and single wavelength-enhanced RLS intensity is only kept steady in the pH range of 7.0–9.9.

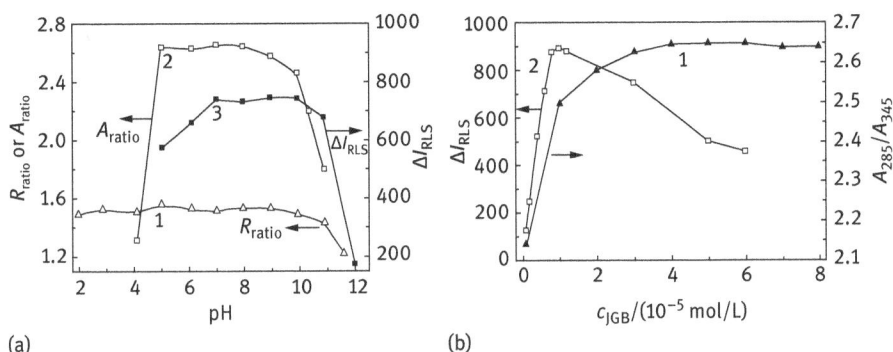

Figure 4.8: Influence in the variation of medium condition on the scattering ratio signal. (a) Influence of pH on the double-wavelength RLS ratio (Curve 1); absorption ratio (Curve 2); and single wavelength RLS (Curve 3) (b) Influence of JGB concentration on absorption ratio (Curve 11) and single wavelength RLS (Curve 2).

However, JGB concentration could obviously affect the RLS ratio, as the change in the concentration of the dye brings an obvious shift in the maximum scattering peak. As shown in Figure 4.8(b), RLS ratio has a little increase of value with an almost unchangeable scattering intensity with a JGB concentration of 0.8×10^{-5}–1.2×10^{-5} mol/L; when JGB concentration exceeds this range, the RLS intensity decreases as well as the ratio of RLS. Corresponding to this, the absorbance ratio increases with the increase in JGB concentration in the solution; it is steady until JGB concentration increases to approximately 3.0×10^{-5} mol/L. It is obvious that we can solve some practical problems with a difference between the RLS ratio and the absorbance ratio.

4.5.2.2 Ratio analysis method

It is found that the ratio value R has a linear relationship with the logarithm of substance concentration; it can be expressed as follows:

$$R = a - b \log c \tag{4.31}$$

where R is the ratio of scattering intensity of two selected wavelengths, a and b represent two constants, c is the concentration of the substance. A simple equation for the ratio of scattering and concentration of the substance is determined, which can be used for analysis.

The ratio can also be used to analyze the degree of completion of a reaction. Under an experimental condition, with the addition of a substance, there is a linear relationship between the ratio and the concentration of the substance according to the equation (4.31). When the concentration reaches a certain value, the scattering rate is almost kept constant, demonstrating that the reaction is almost finished. So ratiometry can not only be used for the analysis of a sample but it can also be used in the study of reaction kinetics. Compared with a single-wavelength measurement in RLS technology, it can show the completion degree of a reaction more accurately. As the ratio value R is proportional to $\lg c$, ratiometry can increase the resolution at a low concentration, with an advantage of high resolution in a low concentration analysis [12].

In addition, dual-wavelength ratiometry can not be used in all resonance light scattering systems. Based on the principle of wavelength selection, in an RLS system, the RLS probe must have UV-vis absorption and the analytical target should have no or negligible absorption at the selected wavelength. When a RLS probe has no UV-vis absorption or a analytical target has a rather strong absorption, then the dual-wavelength RLS ratiometry is also possible. RLS ratiometry overcomes the shortcomings of too many interferences in the single-wavelength RLS measurement, by acquiring a relationship between a simple functional ratio and the target concentration at two specific wavelengths. RLS ratiometry can be used for accurate characterization of the completion degree of a reaction and also for measuring of the related constant in chemical kinetics.

4.5.3 Scattering-fluorescence ratio method

Spectrofluorimetry has an important application in the modern analytical science. On one hand, light scattering signal always coexists as an interference signal in a fluorescence spectrum and decreases the sensitivity of fluorescence analysis. On the other hand, fluorescence and scattering signal is more sensitive than others in the method of analytical signal transfer. Fluorescence and scattering analysis are based on single wavelength data gathering. The data from single wavelength is often influenced by the instrument's response, concentration, and environment. Although these influences can be partly solved by the dual-wavelength light scattering or by the fluorescence method, it was not perfect. It promotes us to use scattering and fluorescence signal for analysis at the same time.

As scattering and fluorescence signal can be obtained on a three-dimensional spectrum by a fluorescence spectrophotometer. By using the interaction of water-soluble CdS-quantum dots (QDs) and organic molecules (aminoglycoside antibiotics) as a model, the authors of the book established a new fluorescence-scattering ratiometry based on the enhancement of fluorescence signal and the quench of the scattering signal (Figure 4.9). They discussed the principle of excitation wavelengths selection and deduced theoretical derivative equation of the method.

Choosing a suitable excitation wavelength to obtain a stronger scattering and fluorescence signal is quite important for the method. To minimize the overlapping of interference of noise and spectrum, an ideal reagent should have a strong signal in

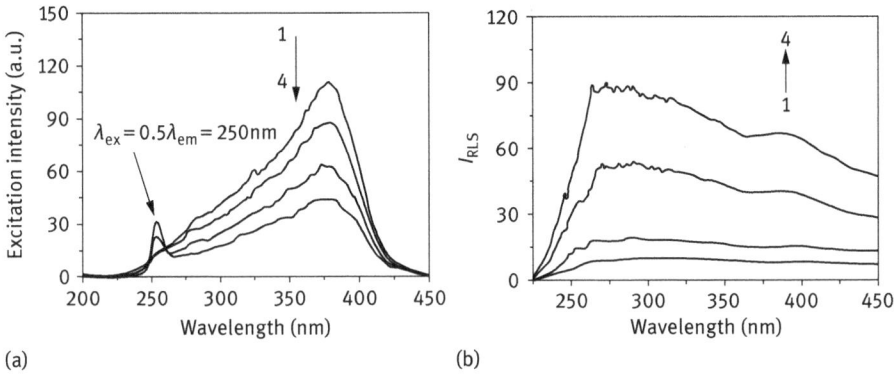

Figure 4.9: Scattering and fluorescence spectra of quantum dots. Excitation spectra (a) and scattering spectra (b) of CdS–QDs under the existence of different concentration of tobramycin (TOB) [13]. CdS–QDs concentration: 4×10^{-5} mol/L; TOB concentration: (1) 0 µmol/L; (2) 0.21 µmol/L; (3) 0.43 µmol/L; (4) 0.64 µmol/L; 1–4 means the interaction of QDs with different concentration of TOB.

the excitation and emission wavelength with great Stokes shift. Compared with other organic molecules, quantum dots are nanoparticles, which is easy to aggregate, rather than being steady. The Stokes shift of quantum dots are great that the scattering and fluorescence peak almost has no overlap. Thus, the quantum dots are ideal probe ideal for ratiometry. As shown in Figure 4.9, choose 376 nm as the excitation wavelength of CdS–QDs probe. If the fluorescence quenching of the fluorescent substance accords to Stern–Volmer equation, whether static or dynamic quenching, it can be expressed as follows:

$$\frac{F_0}{F} - 1 = K_{sv}c \tag{4.32}$$

where F_0 and F are fluorescence intensities with and without analytical target, c is the target concentration, and K_{sv} is the Stern–Volmer quenching constant. An enhanced scattering intensity (ΔI) at an excitation wavelength and concentration of the analyst follows the following equation:

$$\Delta I = I - I_0 = kc \tag{4.33}$$

where I_0 and I are light scattering intensities with and without the target, respectively, c is the target concentration, and k is the slope.

If we define ratio, $R = \frac{(F_0 - F) \cdot \Delta I}{F}$, it can be expressed as follows:

$$R = kK_{sv}c^2 \tag{4.34}$$

K_{sv} is the Stern–Volmer quenching constant, c is the target concentration, and k is the slope

So,

$$\lg R = \lg(kK_{sv}) + 2\lg c \tag{4.35}$$

Theoretically, the linear relationship between $\lg R$ and $\lg c$ has a slope of 2, whereas, in fact, equations (4.34) and (4.35) always have deviations, the slope is not 2. Equation (4.35) can be further written as follows:

$$\lg R = b + a \lg c \tag{4.36}$$

where a, b are constants.

4.6 Flow injection coupled resonance light scattering technology

Resonance light scattering method based on an common fluorescence spectrophotometer actually has some flaws in its application, such as low selectivity and bad repeatability. The modifications on a spectrophotometer and its combination with other methods are the ways to solve these problems. This section introduced an example of improved RLS technology that modification the component of a spectrophotometer for the combination of another technology.

Flow injection analysis (FIA) is a continuous sample injection method put forward by J. Ruzicka and E.H. Hansen in 1975. The idea ensures high repeatability in the reagent mixture and the reaction time, realizes effective analysis online and avoids the analytical error of manual operation. Combination of FIA and other techniques greatly promoted the automation and miniaturization of an analytical method, so it is one of the important techniques in the disposal of a flowing reagent (current-carrying) and in the analysis of the flowing quantitative sample.

Because of this, the combination of the flow injection technology with light scattering analysis technology can effectively avoid errors in the manual injection and operation, thus improving the repeatability of the results. Many researchers have successfully applied this technology for detecting proteins, sugars, and nucleic acids. The experimental results with the use of flow injection are obviously better than manual operation. The repeatability of the results is enhanced with the relative standard error obviously reduced with a more simpler operation (Figure 4.10) [7].

The combination of a flow injection technology with light scattering analysis technology will promote the application of light scattering technology in the field of automatic analysis and microfluidic chip analysis.

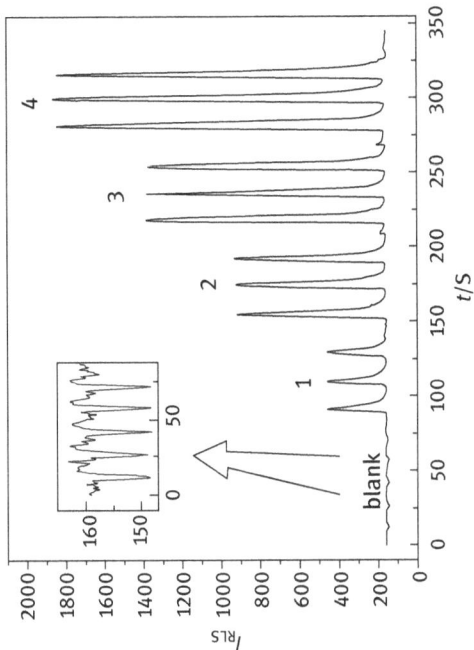

(b)

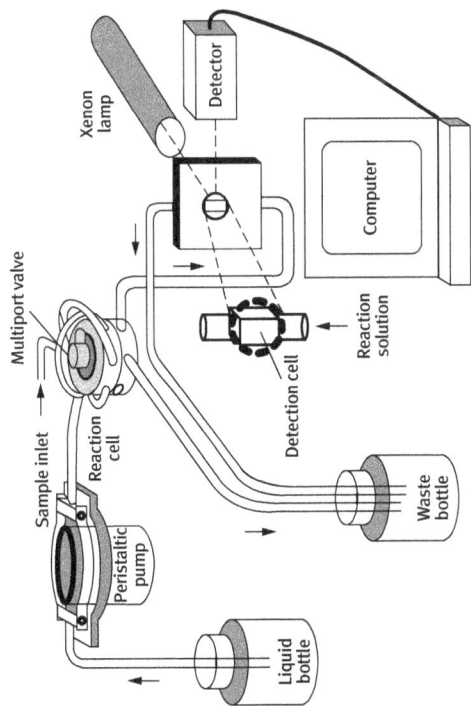

(a)

Figure 4.10: Combination of a flow injection-resonance light scattering. (a) Structural diagram of the instrument (b) Analytical result of an practical sample [7].

References

[1]   Pasternack R F, Bustamante C, Collings P. J, et al. Porphyrin assemblies on DNA as studied by a resonance light-scattering technique. J. Am. Chem. Soc., 1993, 115(13): 5393–5399.

[2]   Huang, C. Z., Li, K. A., Tong, S. Y. Determination of nucleic acids by a resonance light-scattering technique with $\alpha,\beta,\gamma,\delta$-tetrakis[4-(trimethylammoniumyl)phenyl]porphine. Anal. Chem., 1996, 68(13): 2259–2263.

[3]   Huang C Z, Li K A, Tong S Y. Determination of nanograms of nucleic acids by their enhancement effect on the resonance light scattering of the cobalt(II)/4-[(5-chloro-2-pyridyl)azo]-1,3-diaminobenzene complex. Anal. Chem., 1997, 69(3): 514–520.

[4]   Anglister J, Steinberg I Z. Measurement of the depolarization ratio of Rayleigh scattering at absorption bands. J. Chem. Phys., 1981, 74(2): 784–791.

[5]   Wei Yongju. Study of protein molecules absorption and scattering spectrum probe. Beijing: Doctoral dissertation of Peking University. 1997.

[6]   Huang C Z, Li Y F, Mao J G, et al. Determination of protein concentration by enhancement of the preresonance light-scattering of $\alpha,\beta,\gamma,\delta$ -tetrakis(5-sulfothienyl)porphine. Analyst, 1998, 123(6): 1401–1406.

[7]   Pang Xiaobing. Application study of resonance light scattering technology in biochemical drugs analysis. Chongqing: Southwest Normal University master's thesis, 2004, 41.

[8]   Wilson J D, Cottrell W J, Foster T H. Index-of-refraction-dependent subcellular light scattering observed with organelle-specific dyes. J. Biomed. Opt., 2007, 12(1): 014010.

[9]   Huang Chengzhi, Li Yuanfang, Feng Ping. Calibration of resonance light scattering spectrum. Anal. Chem., 2001, 29(7): 832–835.

[10] Lakowicz J R. In Topic in Fluorescence Spectroscopy. New York: Plenum Press, 1994: Vol. 1V, P3.

[11] Long Y J, Li Y F, Huang C Z. A wide dynamic range detection of biopolymer medicines with resonance light scattering and absorption ratiometry. Anal. Chim. Acta, 2005, 552(1): 175–181.

[12] Huang C Z, Pang X B, Li Y F, et al. A resonance light scattering ratiometry applied for binding study of organic small molecules with biopolymer. Talanta, 2006, 69(1): 180–186.

[13] Liao Q G, Li Y F, Huang C Z. A light scattering and fluorescence emission coupled ratiometry using the interaction of functional CdS quantum dots with aminoglycoside antibiotics as a model system. Talanta, 2007, 71(2): 567–572.

Yun Fei Long, Jian Ling, Qie Gen Liao, Cheng Zhi Huang

5 Light scattering spectral probes of organic small molecule

5.1 Types of light scattering spectral probes

5.1.1 Probe

5.1.1.1 Concept of a probe

A probe is a commonly used term in spectrometry, but as of now there has been no authoritative definition. Generally, it refers to a small device for detecting the molecular size (even up to nanosize), which is usually contact with analyze and test.

In electronic test field, a probe is defined as a contact medium for electrical test, which is a high-level precision electronic hardware component. Of course, the element must be small with accurate targeting and quick response performance; the former can guarantee precise recognition of substance to be tested, whereas the latter could give out real-time, rapid response consistent of component under test.

The probe is quite different from sensor in size, structure, working principle, and so on, but it is also closely related. Besides, in essence, the probe is quite different from the analytical reagent. The probe emphasizes more about targeting and microcell characteristics, while the analytical reagent emphasizes the ability to give the signal construction analysis method. Sometimes, analytical reagent, probe, and sensor are confused to name probe reagent or sensor probe.

There are many types of probes, which can be classified based on their property, application, working principle, and so on. Analytical chemistry often uses molecular probe, nanoprobe; inorganic probe, organic probe, biological probe, luminescence probe, nonluminescence probe, and so on.

5.1.1.2 Probe application

An ideal probe has four fundamental modes, giving out sample batch test resulting from small area (nanolevel or smaller) to deep, large area:

Point analysis: This is a qualitative analysis or a quantitative analysis of a selected point, and perform quantitative analysis for its concluded element.

Line analysis: This is a continuous analysis of a function or a linear direction analysis for a pointed subject, obtaining one-dimensional distribution information of a layer component.

https://doi.org/10.1515/9783110573138-005

Surface analysis: This is used for the distribution of concentration of an element under test in a chosen microzone, to acquire a two-dimensional surface distribution figure.

Three-dimensional analysis: This is used for the elemental distribution of overall individual, resulting in a three-dimensional spatial scanning image.

5.1.1.3 Dimension of analysis method

It needs to be demonstrated that the above four working methods exclude time factor and multicomponent factor, only considering of space dimension, which result in zero-dimensional, one-dimensional, two-dimensional, and three-dimensional messages. But actually, objective world is changing all the time, so real-time distribution with time is required for complete description of a component under test. Of course, if multicomponent analysis is involved, dimensions should have new changes. For example, including time, all the above-mentioned analysis should add one more dimension. Besides, if further consideration of a one-component internal property is required, then its dimension will increase, that is, analysis of a single component at one point is of zero dimension. If conformational change is considered, then the time and space factor should be added with greatly increasing dimensions.

5.1.2 Types of light scattering spectral probes

Working principles of light scattering probes include real-time scattering response signal of an object in its given area. Hence, molecules, ions, aggregates or nanoparticles can all be classified as light scattering probes.

Many small molecules can dissociate into water solution with some charges; with electrostatic interaction, it forms new aggregates or ionic associates targeting with opposite charges, such as protein, nucleic acid, and so on, greatly increasing particle number in system and strong light scattering enhancement. Generally, chemists use this for analysis and measurement of a series of target objects. There are lots of this type probe. According to their structural properties of these common probes, they can be divided into two main types: organic small molecules and inorganic ions. Organic small molecules mainly include organic positive-ion small-molecule probe, organic negative-ion small-molecule probe, and coordination organic small-molecule probe. Inorganic probes mainly include metal ion probes, metal complex probes, and inorganic negative ion probes.

As nanoparticles can usually give out Rayleigh scattering and Tyndall scattering optical signals, and these signals can provide sensitive, rapid, and characteristic response for the surrounding environment, nanoparticles are often a very good scattering probe. There are many types of nanoparticles: organic, inorganic, metal, and complex. As there is a localized surface plasmon resonance feature, light scattering spectral probes of the noble metal nanoparticle have wide applications. It is notable that in recent years, a study on semiconductor nanoprobe is quite popular. Quantum dot and carbon dot have luminescent properties with good application, but their mechanism is not clear; thus related exciton light scattering of nanoparticles in these quantum confinements is worth further discussion.

5.2 Inorganic ion probes

5.2.1 Metal ion probes

Under appropriate conditions, metal ions produce insoluble compound with measured ion. With stabilizers or dispersants, compound disperses in solution, enhancing resonance light scattering (RLS). For example, in the determination of Cl^-, strong light scattering signals can be generated from AgCl precipitates by interaction with $AgNO_3$ [1]. In salted food $K_4[Fe(CN)_6]_3$ measurement, the signal of insoluble Prussian blue $Fe_4[Fe(CN)_6]_3 \bullet x\ H_2O$ fine crystals can be detected [2].

A characteristic RLS signal obtained after interaction of metal ion and deoxyribonucleic acid (DNA) can be used to detect DNA. Yang Chuanxiao [3] found that, in acidic medium of pH 2.21, Al^{3+} has electrostatic interaction with DNA, producing RLS spectrum with a characteristic peak at 291 nm. Reasons for light scattering enhancement are combination of Al^{3+} with phosphate group on DNA molecule surface, and after DNA thermal denaturation, Al^{3+} combines with DNA basic group, lowering the signal.

5.2.2 Metal complex ion probe

A metal complex ion serves as an integrated recognition unit with good application prospect in interaction probe of an organic, an inorganic or a biological system. On the one hand, it adjusts the molecular structure of organic molecules, enhancing its diversity and recognition specificity; on the other hand, it produces a new electro-optic activity.

For example, as Co-Co reagent (Co(II)/5-Cl-PADAB) complex has RLS enhancement with stacking on nucleic acid, this can be applied for highly sensitive measurement of nucleic acid [4]; electrostatic interaction of Fe^{3+} and DNA at pH 2.20 BR buffer and formation of association enhance the RLS of a system [5]; with nonionic surfactant Triton X-100, a small amount of protein could dramatically increase the signal of (DBHPF)-Mo(IV) complex [6]; at pH 5.0–7.0, bovine serum albumin (BSA) and human serum albumin (HSA) could enhance the RLS of bromine chloride replacement arsenazo-A1^{3+} metal complex [7].

Interaction of rare earth complex with DNA also brings light scattering enhancement, such as Ru(bpy)-(dppx)$^{2+}$-SDS (bpy = 2,2′-dipyridyl, dppx = 7,8-dimethyl–pyridine phenazine), which has a strong RLS signal. It decreases with the addition of DNA [8, 9]. Lanthanum mixed content LaL$_1$L$_2$•3H$_2$O (L$_1$ is phthalic acid and L$_2$ is phenanthroline) has long-distance self-assembly of enhanced RLS peak at 470 nm [10]. In Tris-HCl buffer at pH 7.8, DNA could dramatically enhance the intensity of Terbium–salicylic acid–phthalate ternary complex with a characteristic scattering peak at 351 nm [11].

5.2.3 Inorganic negative ion probes

Under suitable conditions, some simple inorganic negative ions form insoluble compound with target ion. With stabilizers or dispersants, insoluble compound disperses in solution, enhancing its RLS. Li Xiaoyan added K$_2$Cr$_2$O$_7$ into water with Pb^{2+}, producing insoluble PbCrO$_4$, realizing Pb measurement in water sample with PbCrO$_4$ suspension signal [12].

It needs to be demonstrated that an inorganic negative ion probe, especially those inorganic ions with larger volume, is easily polarized, possibly with special optical activity on electric field of opposite charges, producing new RLS features.

5.3 Types and architectural features of organic small-molecule probes

Organic small-molecule probes are most widely used probes because these kinds of molecules have larger chromophores with certain water solubility and charges in certain acidity, so easily polarized. These molecules are also sensitive to environment with diversified properties; therefore, they have found quite wide applications.

Based on their charges and features, they mainly include organic small-molecule positive ions, organic small-molecule negative ions, and cooperative organic small molecules. Among which, cooperative organic small molecules are formed by molecules with some cooperative reagents. When there is a component with opposite charges in solution, each probe could form new complexes with these components, such as bioassemblies or ion associates, further enhancing RLS of the system.

5.3.1 Organic small-molecule positive ion probes

Organic small-molecule positive ion probes mainly include organic positive ion dye, and cationic surfactant. As they usually have positive charges, they could interact with the substances with negative charges through electrostatic interaction. For example, nucleic acid has lots of negative charges; hence, the probe is easily stacked on its molecular surface through electrostatic attraction. It is as if biomacromolecules aggregate probe molecules, increasing surface concentration far larger than the dissociative concentration, which is similar to aggregate production, increasing scattering particle size. This kind of aggregation sometimes brings absorption spectral changes, but anyway, a strongly enhanced light scattering signal can always be observed [13].

Reported probes include positive ions such as porphyrin derivatives, meso-4 (*p*-trimethyl amide) porphyrin (TAPP) [4], and quinone-imine-type dyes such as safranine T (ST) [14], neutral red (NR) [15], and phenosafranin (PS) [16]. Figure 5.1 shows the part of quinone-imine-type dye. This dye causes light scattering enhancement with the interaction of a nucleic acid. Methylene blue (MB) [17], Nair blue (NB) sulfate [18, 19], and brilliant cresyl blue (BCB) [20] interact with the nucleic acid to change the refractive index of the system, dramatically increasing the intensity of RLS. These probes have two common features: first, they have positive charges with formation of quaternary ammonium salt, and second, they have parallel hexatomic ring structures with heteroatoms such as N, O, and S.

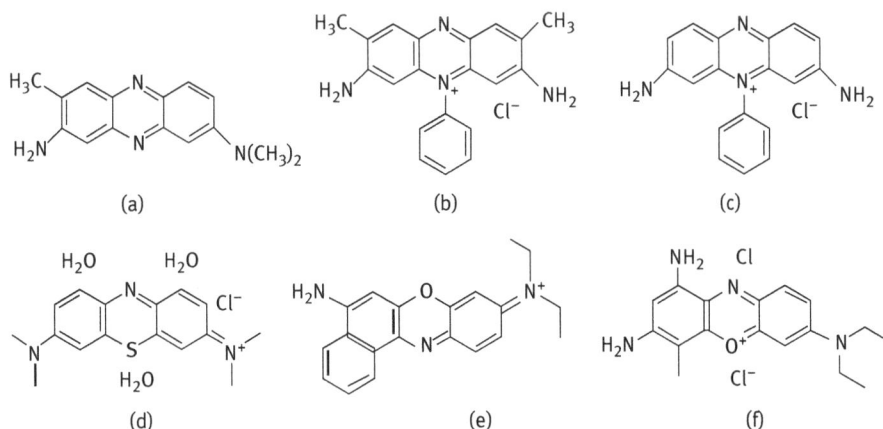

Figure 5.1: Molecular structure of a part of quinone-imine-type probes: NR (a), ST (b), PS (c), MB (d), NB (e), and BCB (f).

Some basic triphenylmethane dyes are pararosaniline chloride (PRL) [21], ethyl violet (EV) [22], crystal violet (CV) [23], methyl violet (MV) [24], and rosaniline (RL) [25]. On the one hand, triphenylmethane structure has positive charges to combine with nucleic acid; on the other hand, it forms larger complex with combination as molecules possess large conjugated structure (see Figure 5.2), dramatically enhancing RLS signal, and producing new RLS spectrum with increase of DNA concentration (Figure 5.3).

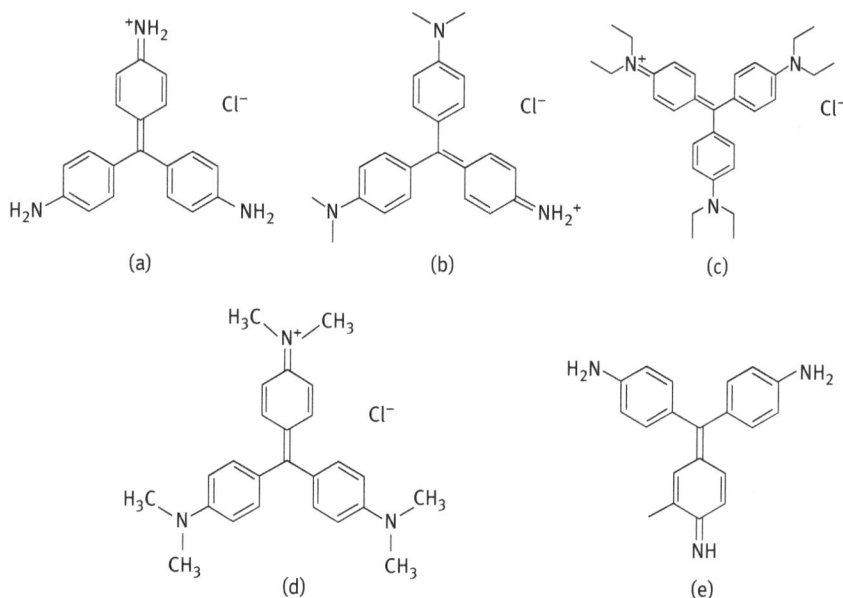

Figure 5.2: A molecular structure of a part of triphenylmethane positive ion probes: PRL (a), MV (b), EV (c), CV (d), and RL (e).

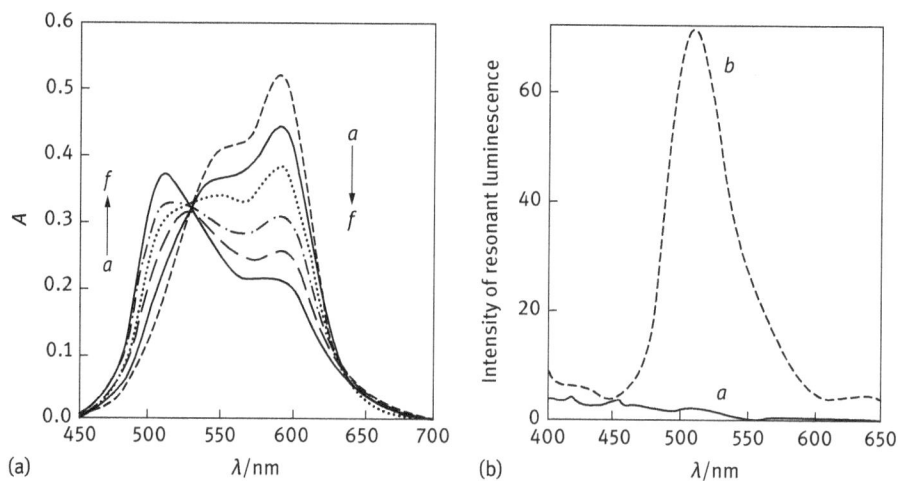

Figure 5.3: Absorption (A) and scattering (B) spectra of EV and DNA interaction. (A) EV (7.5 × 10^{-6} mol/L) interaction with DNA of different concentrations; DNA concentration (a–f): 0, 1.0 × 10^{-6} mol/L, 2.0 × 10^{-6} mol/L, 3.0 × 10^{-6} mol/L, 4.0 × 10^{-6} mol/L, 5.0 × 10^{-6} mol/L. (B) a is EV and b is EV+ DNA.

An RLS study of rhodamine dyes such as rhodamine B (RB) [26], b-tyl-rhodamine B [27], and rhodamine 6G [28] interacting with nucleic acid has been reported. Figure 5.4 shows some rhodamine dye molecular structures. With the interaction of nucleic acid, rhodamine dye formed a positive ion dye–nucleic acid complex with larger particles, showing a strong scattering signal. This spectrum is closely related to its absorption with fluorescence quenching.

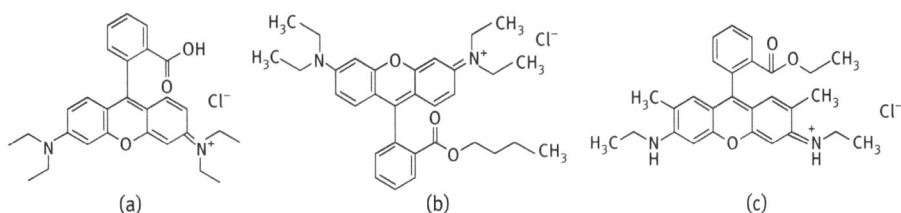

Figure 5.4: Molecular structure of rhodamine B (a), btyl-rhodamine B (b), and rhodamine 6G (c).

5.3.2 Organic small-molecule negative ion probes

Organic small-molecule negative ion probes mainly include organic negative ion dye and anionic surfactant. The RLS signal changes with anion interacting with a measure of molecules that can realize assay determination of different types of substances. As of now, more studies focus on organic anion dye application in protein quantitative analysis and inorganic ion detection.

In current assays, organic small-molecule probes for protein scattering measurement mainly include polyphenols [7, 29–37], azo type [38–44], porphyrin [45, 46], phthalocyanines [47–50] as well as some new reagents with a few reports [51–59]. These dyes are most widely used probes in protein detection, which is mainly used for BSA, HSA, and human immunoglobulin. Organic small-molecule dye probes have common features as described below:

(1) It contains easily dissociated group that brings charge to probe, as hydroxyl, carboxyl, sulfonate and so on. For example, water-soluble porphyrin reagent 4(5-thiophenyl sulfonate) porphyrin and 4(4-sulfonic acid phenyl) porphyrin (TPPS$_4$) molecules have four sulfo groups and tetracarboxyl phenyl porphyrin (TCPP) molecule has four carboxyl groups; some azo dyes have multiple hydroxide group and sulfo groups, as flavonoid reagents' parent is hydroxy-flavone with five hydroxide groups in molecule; sulfated or carboxylated phthalocyanine dyes also have four sulfo groups and carboxyl groups.

(2) Their chromophores are able to produce spectral absorption. For organic small molecules with chromophores, after aggregation, light absorbance weakens,

producing RLS effect near the absorption peak, when RLS peak can be obtained through scanning.

(3) Other organic small molecules like drugs can be used for protein detection as probes (such as ampicillin sodium and rifamycin sodium) [60, 61] can also be used to produce enhanced signal with protein surface stacking. Besides, there are some composite new reagents with specific groups, which are also used for protein detection [62].

In terms of principle and result, the following protein methods are with similar results. Normally, the detection range of protein is 0.01–10 µg/mL, and tested proteins include BSA, HSA, and immunoglobulin.

5.3.3. Cooperative organic small molecules

This kind of probe mainly uses the third party as medium, such as surfactant, to change system charges and compositions, and structure, so that an organic small-molecule probe has quite wide application range.

With the existence of positive ion surfactant, through bridge function of surfactant charge and hydrophobic grouping, DNA interacts with positive ion dye for RLS enhancement. For example, basic triphenylmethane dyes such as Brilliant Green (BG) and malachite green (MG) show no obvious intensity change with the combination of nucleic acid, while with the existence of positive ion surfactant hexadecyl trimethyl ammonium bromide (CTAB), RLS intensity is greatly enhanced for the DNA system with BG [63] and MG [64].

A positive ion surfactant not only combines DNA with positive ion dye but also combines DNA with negative ion dye, including CTAB combines negative ion dye, resorcinol yellow (RY) [65] with DNA, and the existence of cetyl pyridinium bromide interacts m-cresol purple [66] with DNA, greatly enhancing the RLS intensity of a system, that is to say, surfactant has sensitization for RLS intensity of DNA and dye. Other reports include: with the existence of CTAB, DNA has RLS enhancement for organic dye, methylene green [67]; negative ion dye, xylene cyanol FF, has RLS enhancement for positive ion surfactant CTAB and nucleic acids (fsDNA, ctDNA, and yRNA) [68]; with the existence of sodium dodecyl sulfate (SDS), trace nucleic acid has RLS enhancement for benzidine-lauryl and SDS [69].

In terms of protein analysis, the same strategies can be used with positive ion dye as probe and establish protein analysis method with cooperative interaction of surfactant. For example, in sulfuric acid, trace amounts of protein can enhance the weak RLS signal of pyrrole-red Y-SDS system [70]; with Triton X-100, interaction of protein and Titan yellow greatly enhances the system RLS [71].

Meanwhile, there were reports about a new method for sulfonamides with RLS at 468 nm of nitrite-8-hydroxyquinoline diazotization-coupled reaction products [72]; phorate and sodium tungstate can interact to form heteropolyacid in acidic solution, which can interact with basic triphenylmethane dye RB, producing a characteristic RLS enhancement at 606 nm, this result can be used to measure phorate [73]; tin could be measured by RLS intensity decrease in chlorophenol red-HSA-SDS system [74].

5.3.4 Other organic small-molecule probes

Under certain conditions, the measured ions can be reduced to elementary substance, sol or nanoparticle by reducing agent, and dispersed in solution, which could be applied to establish new analytical methods by measuring the change of resonance light scattering intensity. For example, reducing Se(IV) with VC and dispersing in solution in nanoform, the elementary substance Se could be well used for VC measurement. A study found that, in neutral medium, amide ligand [N,N'-double(2-aminophenyl) diamide oxalate, NAPO] could selectively complex with cupric ion, leading to strong enhancement of RLS signal [75], thus, the detection of cupric ion could be achieved.

5.4 Aggregation of organic small-molecule probes

5.4.1 Formation of molecule aggregates

When molecules or ions aggregate in a system, large particles are obtained, which can be directly observed through atomic force microscope (AFM), transmission electron microscope (TEM) or scanning electron microscope (SEM). As an RLS signal is related to a particle, large particles lead to an enhanced light scattering signal. For colored aggregates, there is a new scattering peak corresponding to a color absorption peak. Of course, this may not be completely identical with scattering wavelength and absorption wavelength, which is likely to appear before and after the absorption zone. For this reason that absorption brings not only changes of refractive index imaginary part and enhancement, but also reduces incident light intensity and inner filter effect.

So, when an enhanced RLS signal and a new scattering peak appear, it is suggested that aggregate is formed. Different from that, a system with no absorption or only inconspicuous absorption change, the increase in scattering light intensity is more important. Hence, through RLS analysis, aggregate formation can be determined, and its formation process can be observed through dynamic observation of time and signal relation, as well as best condition and mechanism exploration of fast, rapid aggregate formation to determine factors influencing its formation.

5.4.2 Aggregation of conjugated structure molecules

Substituted porphyrin is a conjugated molecule, which is more studied. As the porphyrin molecule has a large ring with four pyrrole rings, there is one dimethyl bridging in 2 and 5 positions of each ring, and 4 substituted phenyl can be bound in 5, 10, 15, and 20 positions, forming a tetra-substituted phenyl porphyrin (Figure 5.5). Its conjugated structure is alternating single and double bonds in the ring, with 18π electrons and aromaticity.

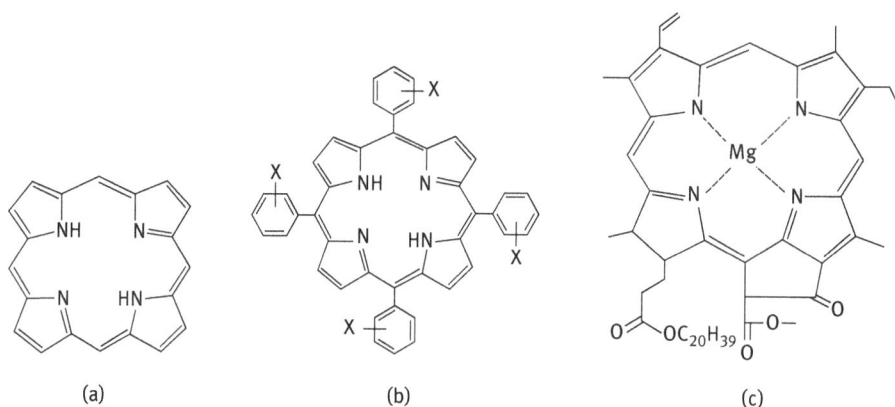

(a) (b) (c)

Figure 5.5: The porphyrin molecular structures. (a) a porphyrin ring; (b) substituted porphyrin, where X can be different functional groups, such as sulfonate, hydroxyl, carboxyl, amino groups, with positive, negative, or no charge; and (c) chlorophyll a.

With electrostatic attraction and hydrophobic interaction, negative ion porphyrin forms a molecule aggregate to improve degeneracy of a surface oscillator. With high acidity, high ion strength, and large porphyrin concentration, protonized negative ion porphyrins such as 3(4-sulfo phenyl) porphyrin (H_2TPPS_3) and 4(4-sulfo phenyl) porphyrin (H_2TPPS_4) have a J-type aggregate, producing a 490 nm aggregate absorption peak. Accordingly, the J-type aggregate scattering peak is also at 490 nm [76]. Besides, 2(4-sulfo phenyl) porphyrin (t-H_2TPPS_2) produces 480 and 450 nm J-type aggregate absorption peaks in acidic and neutral conditions. The J-type aggregate absorption peak of beta-8-bromine-4(4-sulfo phenyl) porphyrin at 521 nm has good correspondence with the strongest RLS peak [76]. Different from that, nonionic porphyrin 4(4-hydroxy phenyl) porphyrin (H_2THPP) exists as aggregative state in water, with strong RLS signals at 440–480 nm, but organic solvent as ethyl alcohol depolymerizes THPP, reducing the RLS signal [77].

In biology, chlorophyll [Figure 5.5(c)] aggregate is a determinant of chloroplast photochemical and photochemical behavior. The same as porphyrin, chlorophyll aggregate affects the ring structure. Pasternack applied RLS technology to find that chlorophyll has a strong RLS enhancement signal in Soret band at 469 nm, with corresponding RLS in Q_y band at 699 nm. Combined with absorption and circular dichromatic spectra, they believe that its aggregate includes three steps: 1. dimer nucleation; 2. formation of macropolymer with secondary structural tendency; and 3. spiral aggregate formation [78].

It needs to be demonstrated that organic small molecules with macroconjugated system such as porphyrin aggregate is usually related to exciton. Authors believe that, with laser system, the exciton message in the aggregate process can be captured through a light scattering signal.

5.4.3 Aggregate of a surfactant

With an amphipathic structure, a surfactant molecule has one fixed property, when it associates with micelle in solution. When micelle concentration reaches a critical micelle concentration (CMC), as it changes a lot, it is sure to change the light scattering signal.

For example, at room temperature, the light scattering signal intensity of sodium dodecyl benzene sulfonate (SDBS) solution increases with SDBS concentration, and the maximum RLS intensity is obtained at SDBS CMC. The peaks are observed at 330 and 396 nm. Where 396 nm RLS intensity has an S-type relation curve with SDBS concentration, the intersection point of two tangent lines at the curve rising point is SDBS CMC; the result is basically the same as that tested in fluorescent pyrene probes and electrical conductivity [79].

For example, cholate is a kind of biosurfactant, with the parent structure of one steroid ring skeleton (Figure 5.6) with a good hydrophobic property. Its tail chain is normally of carboxyl group (such as sodium glycolic acid, sodium cholate, and deoxycholic acid) or sulfonic acid base (e.g., sodium sulfonate), dissociating into –COO– or $-SO_3^{2-}$ in solution. As the tail chain is of negative ions with negative

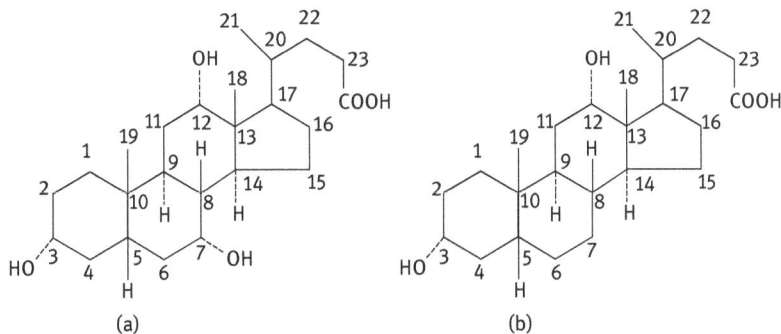

Figure 5.6: The molecular structures of a cholate (a) and a deoxycholic acid (b).

charges, it has strong hydrophily, easily forming a hydrophobic convex surface and hydrophilic concave, the molecule has amphipathy. In strong acidic medium, as cholate hydrophilic groups protonize, greatly reducing its hydrophilia, the molecule becomes a hydrophobic organic molecule, losing its surface activity. With its property, molecules aggregate with Van der Waals force, hydrogen bonding, and hydrophobic repulsive force.

In view of a cholate molecular structure, there are two possible aggregate methods: one is that cholate forms a hydrogen bond head–head combined dimer through hydrogen bond in hydroxyl or water bridge, and the second is carboxyl or sulfonate at cholate end is easy to protonize in acidic medium, and the protonized carboxyl aggregates monomer through the hydrogen function. Through TEM characterization, average particle size of the aggregate product in acidic medium is around 30 nm, while the simple micelle size of a cholate is around 1.0 nm and the secondary micelle size of cholate is around 30 nm.

The aggregate is produced by protonation of a molecule in a strong acidic medium. Different from the formation of a surfactant dimer or a tetrasome, it enlarges the volume of a molecule and further improves hydrophobicity of a molecule for hydrogen interaction among molecules, likely to form liquid–solid interface in aggregate molecule and hydrone. The interface is another important factor for scattering enhancement [80].

A Gemini surfactant is a double-headed and double-end surfactant connected by hydrophilic head groups of two identical surfactant monomers (Figure 5.7). Different from two hydrophilic molecules of single chain, a Gemini amphiphilic molecule is connected by one group through a chemical bond at its polar head through two single-chain hydrophilic molecules. A Gemini amphiphilic molecule has diversified structures and properties with changes of structural factors, such as chemical structure, length, stiffness and flexibility, hydrophilic hydrophobicity, and hydrocarbon chain length.

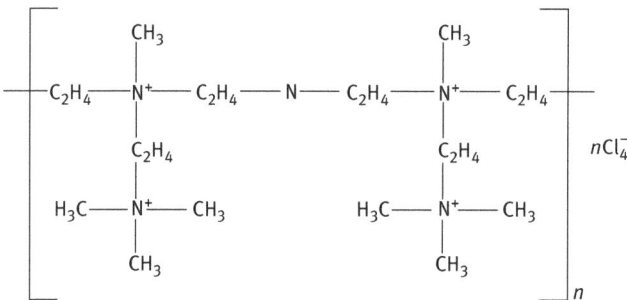

Figure 5.7: A Gemini surfactant.

Two phosphate ester quaternary ammonium salt (PQAS) is a Gemini surfactant with amphoteric polar groups showing good performance and low toxicity. In PQAS amphiphilic molecular structure, two heads are connected by chemical bonds of groups, closely related to two single-head amphiphilic molecules. This structure increases hydrophobic effect of hydrocarbon chain on the one hand, increasing aggregate micelle trend of escaped hydrophobic groups from water; on the other hand, limited by chemical bonds, electrostatic repulsion of polar heads are widely weakened. So, great improvement of Gemini amphiphilic molecular surface activity has given it specific properties, such as singular aggregation form causing super high surface adsorption capability and strong aggregate formation ability (micelle is the most common one). A micelle with a certain diameter is obtained by aggregation. It can produce a strong RLS signal based on Rayleigh scattering theory. Researches found that [81, 82], under certain conditions, the Gemini surfactant PQAS serves as an RLS probe, to put forward new RLS method for protein and polysaccharide measurement based on controllable RLS weakening.

5.4.4 Association of different small molecules or ions

Ionic associates or complexes formed by electrostatic force, hydrophobic force, and charge transfer (CT) among small molecules also have a strong RLS. Based on the chemical structure, we can analyze their functional groups and possible charges, and choose small molecules with opposite charges to achieve analysis. Take an example of trypan red and kanamycin interaction to demonstrate the function [83].

In acidic medium (pH 3.12), single trypan red and kanamycin have weak RLS, while they interact and form ionic associates, enhancing RLS dramatically and producing new RLS spectrum with three new scattering peaks. The maximum scattering wavelength is 400 nm, while 479 and 535 nm are two peaks with lower intensity.

The binding ratio of trypan red to kanamycin in ion association complex was determined by isomolar continuous change method, which was 1:1. To further confirm their binding site, the approximate calculation of their ground-state charge distribution with quantum chemistry AM1 method was performed. Figure 5.8 shows that in five sulfo

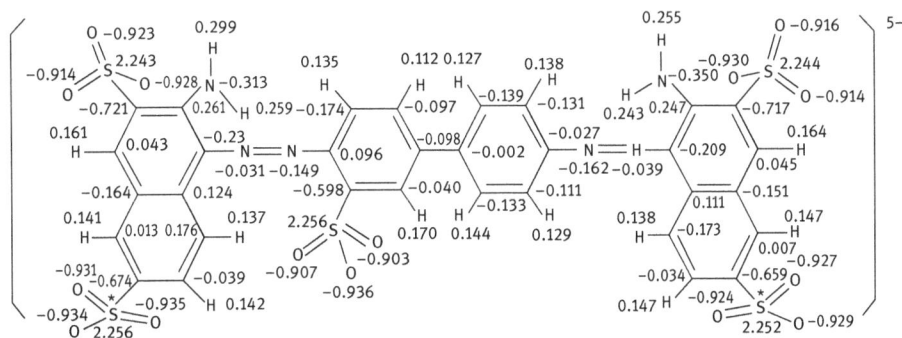

Figure 5.8: The charge pattern of trypan red.

groups of trypan red, negative charge distribution is not uniform, SO_3^- negative charge density on two naphthyl is higher, expressed with an asterisk. Figure 5.9 shows that in the same kanamycin molecule, two amino groups have higher negative charge density, easily protonized with positive charges in acidic medium, expressed with asterisks. The possible structure of 1:1 ionic associate by trypan red and kanamycin is shown in Figure 5.10. Trypan red is an acidic bisazo dye, in faintly acidic or neutral medium. It exists with negative ion state with sodium ion dissociation on many sulfo groups. At that time, aminoglycoside antibiotic sulfate dissociates into protonized positive ion and acid radicals. As trypan red and kanamycin are organic ions with opposite charges, they have stronger electrostatic attraction with each other. Besides, trypan red has two hydrophobic naphthyl and one biphenyl, while kanamycin has strong hydrophobicity with cyclol matrix except for protonized amino group. Through electrostatic attraction and hydrophobic force, when they get close to a certain distance, there is likely to be CT.

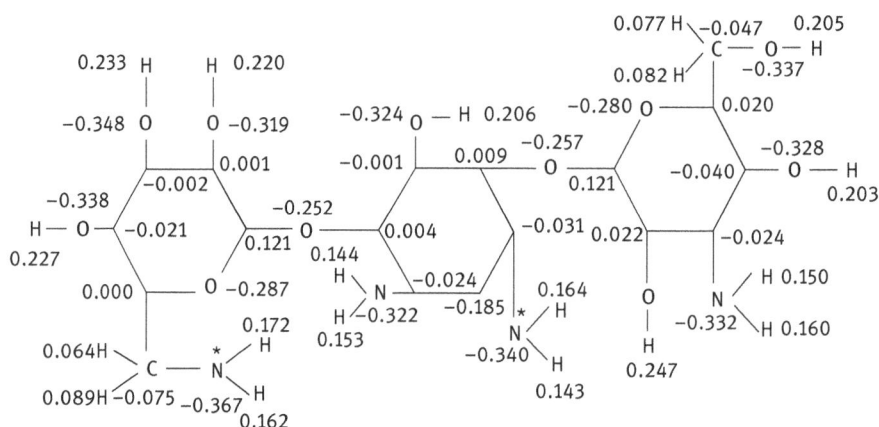

Figure 5.9: The charge pattern of kanamycin.

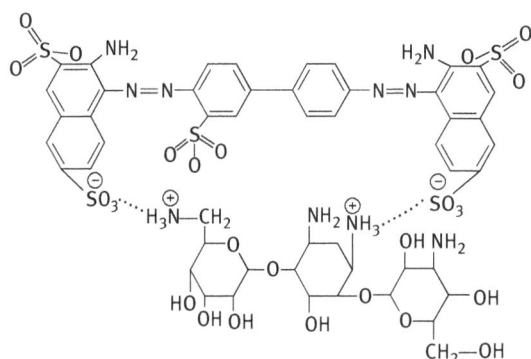

Figure 5.10: An ionic associate diagram of 1:1 trypan red and kanamycin.

Quantum chemistry AM1 method describes trypan red and kanamycin interaction in terms of charge distribution. Trypan red total discharges from -5.0 before reaction to -3.92 after reaction, while protonized kanamycin changes from 2.0 to 0.92, demonstrating CT. Absorption spectrum can also show the CT, as shown in Figure 5.11, the kanamycin itself is colorless and only displays absorption band near 200 nm in the ultraviolet region, while trypan red has 505 nm maximum absorption, with ionic associate formation, λ_{max} has red shift to 570 nm, $\Delta\lambda = 65$ nm, suggesting the formation of CT complex.

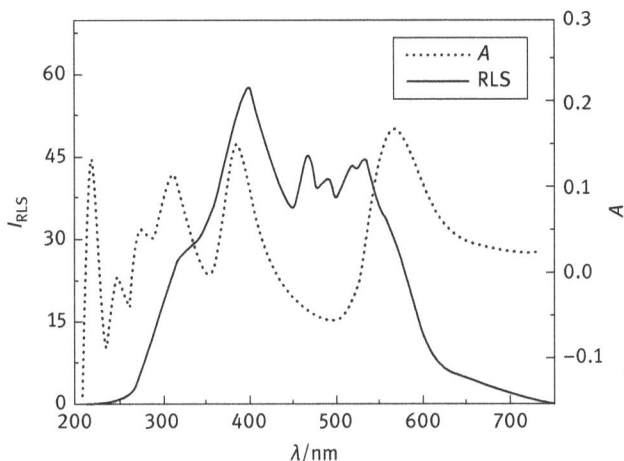

Figure 5.11: Correspondence of molecule absorption and scattering in RLS enhancement.

In a dye-conjugated system, electronic cloud distribution changes, reducing excitation energy ΔE, and absorption band has a red shift; hence, besides electrostatic attraction and hydrophobic force, CT cannot be neglected. RLS spectrum is closely related to absorption spectrum. It can be observed from the comparison between RLS and absorption spectra that RLS is located in its absorption band, which is a necessary condition for the enhancement of RLS, and the RLS peak here is also related to the absorption spectra. The reason for resonance scattering enhancement is its imaginary part of aggregate refractive index. By measuring with Shimadzu RF-540-type fluorospectrophotometer, there is stronger scattering at 400 and 470 nm. Scattering peaks at 400 and 535 nm are close to the corresponding absorption peaks at 396 and 570 nm. It can be considered that the 400 nm scattering peak is strong because of its double influence by the absorption spectrum and equipment factor, while RLS signal at 553 nm is mainly influenced by the absorption spectrum, irrelevant with the equipment factor. It can be seen as a characteristic peak of ionic associate, while the peak at 470 nm is a scattering peak mainly produced because of the equipment factor.

When a small-molecule probe and an analyte have the same kind of charge, or that one of them barely has no charge, the second probe is usually used as a bridge to form a complex with double inverse charge. For example, in NRF(Norfloxacin)-Pd(II)-eosin Y system [84], Pd(II) and NRF form a 1:1 chelate cation $[Pd(NRF)]^{2+}$, when pH is 4–5, eosin Y mainly exists with dianion form, so that they could form a triple ionic compound through electrostatic attraction and hydrophobic force [Pd(NRF)]•L, as shown in Figure 5.12.

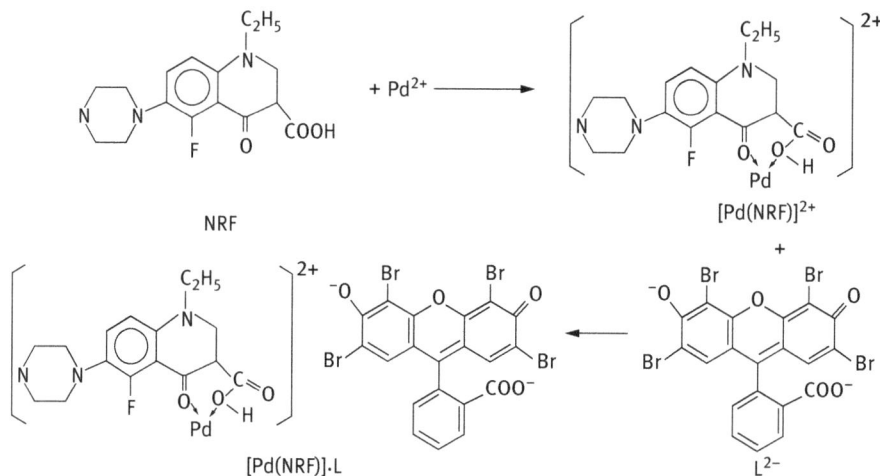

Figure 5.12: The diagram of the formation of NRF-Pd(II)-eosin Y ionic associate.

Rayleigh scattering of a triple ionic compound is in its absorption band, while scattering peak is well corresponding to the absorption spectrum. The RLS spectrum scattering peaks at 324, 368, and 550 nm are around their absorption peaks at 296, 337, and 517 nm, while the scattering intensity dramatically increases in 324–368 nm. NRF has a stronger fluorescence at 444 nm, while it forms NRF-Pd(II) cation with Pd (II) complexing, there is fluorescence quenching with unchanged λ_{em}. Eosin Y also has stronger fluorescence, and its maximum excitation wavelength (λ_{ex}) and maximum emission wavelength (λ_{em}) lie at 518 and 538.5 nm, while there is fluorescence quenching for NRF-Pd(II) cation with eosin negative ion, and λ_{ex} and λ_{em} are 517 and 537 nm, respectively.

5.5 Interaction of an organic small-molecule probe and a biomacromolecule

5.5.1 Scatchard plotting

Composition of small molecules or ions can be obtained through fixation of one component first, then change other components continuously, and test scattering signal and concentration relation for one preliminary data, then analyze molecule charges, possible weak interactions, as hydrogen bond, π–π conjugate action etc., then confirm rationality of constituent ratio in theory.

Through Scatchard plotting, constant and incorporation ratio of organic small-molecule probes such as dye, stack, or aggregate on biomacromolecules can be determined . Compared with organic small-molecule probe self-aggregate, there are more probes combined on one biomacromolecule, and a long-range assembly is possible on it. As long-range assembly is much larger than a small-molecule aggregate, stronger light scattering enhancement signal appears.

Scatchard plotting was put forward in 1949 by George Scatchard (1892–1973) from MIT, USA. It is a method to determine small molecules or ions (such as drug) and macromolecules (such as protein and receptor) combination with plotting. It shows the ratio of coupled receptor and noncoupled receptor changes with receptor concentration, expressed in the following Scatchard equation:

$$\frac{m}{c} = nK_a - mK_a \qquad (5.1)$$

where m is the rate of combined ligand concentration and total combination sites, c is the free ligand concentration, n represents the combination sites of each macromolecule template, and K_a is the combination constant. A plot with m/c versus m obtains $-K_a$ from the slope, and the intercept is nK_a, besides distinguishing the combination ability of two sites from a straight line and the curve.

5.5.2 Interaction of organic small molecules and nucleic acids

Nucleic acid, protein, and sugar are important macromolecules that play an important role in daily life; hence, studies about organic small molecules and nucleic acid interaction are significant for drug treatment mechanism study, rapid and simple drug in vitro screening, guidance of design and synthesis of new drugs, and disease diagnosis, prevention, and rational drug use.

5.5.2.1 Binding method

The binding site of organic small molecules and nucleic acid may be base, phosphate backbone, and pentose, or on nucleic acid molecule surface; hence, there are three main interactions: the first one is covalent binding, the second is noncovalent binding, which is reversible, and the third is shear binding, which is irreversible.

(1) Covalent binding: Covalent binding is mainly a chemical bond breakage and new bond formation of small molecules and bases, such as nucleophilic reaction and

electrophilic substitution reaction. The result is unwinding of nucleic acid double strands and bending, changes of conformation, molecule changes from structure, conformation, function, and so on.

A study of the function is mainly about new drug development and drug function mechanism discussion. For example, at present, alkylation of anticancer drug mechanism inhibits DNA copy or RNA synthesis with DNA base alkylation, reaching antidrug function.

(2) Noncovalent binding: As a nucleic acid has a phosphate group, ribose and bases, negative charges and hydrophobic group, as well as spatial structures, noncovalent binding is a complex one; there are electrostatic interaction, groove binding, intercalative interaction, long-range assembly, and external binding. The binding method involves weak interaction such as hydrogen bond, ionic bond, Van der Waals' force, and hydrophobic bond. Influencing factors include medium acidity, ionic strength, molar ratio of small molecules, and nucleic acid. Obviously, when conditions change, these binding methods are likely to change, and these functions are reversible.

The essence of an electrostatic interaction is mainly the reaction of a double-helix structure framework and a negative-charge phosphate group, resulting in new ion bond formation. The interaction is not selective, which can be determined based on the influence of ionic strength on a tested system.

The double-helix structure of a nucleic acid has a major groove and a minor groove area different in electric field, spatial effect, hydrogen bond, hydration, and so on. Groove binding is the direct interaction of a molecule and a major groove or a minor groove base edge. As small molecules have minor volume, with the precondition of electrostatic attraction, it binds to the minor groove with the hydrophobic effect. The mechanism of Hoechst 33255, 4',6-diamidino-2-phenylindole is studied as a classic representation.

An intercalative interaction is one layer deeper than the groove binding. It mainly occurs in small molecules with a rigid plane structure, such as aromatic ring. These planar constructions have flat structures, easily implanted onto double-helix structure bases. With implantation, aromatic nucleus and base have a $\pi-\pi$ stacking, a dipole–dipole interaction, and a hydrophobic effect. The result is direct improvement of melting point and viscosity changes, and the most classic intercalator is ethidium bromide (EB).

The long-range assembly is put forward by Pasternack in the study of porphyrin and DNA. He believes that in the interaction of organic cationic molecule, such as porphyrin, and DNA, if the molar ratio is higher and medium ion strength is low, then porphyrin would stack externally on the DNA surface until the charges surrounding the DNA are in balance with solvent medium charges. The stacking structure obviously has large-scale refractive index inhomogeneity with a strong RLS signal [79].

(3) Shear binding: Metalloporphyrin complexes, rare earth ion, transition metal, and complexes interact with nucleic acid and bound to certain nucleotide sequence of a

nucleic acid chain, leading to a specific fracture. Molecules with a shear function have an activity similar to mimic enzyme, and sometimes it is called nucleic acid enzyme.

5.5.2.2 A classic binding mechanism

(1) Interaction of a flat cationic organic small molecule: It needs to be demonstrated that the interaction of small molecules and nucleic acid has one or more functions as mentioned earlier, but not all of them could enhance scattering. For example, cationic dye such as Janus Green B (JGB) is an azo dye and a quinone amide dye with positive charges, and ideal organic small molecules for nucleic acid interaction. When it has higher mole ratio with nucleic acid and medium ion strength is low, it induces the formation of a nucleic acid superhelical structure [85]. The superhelical structure is formed by the arrangement or stacking on the molecular surface. Analyze the result as Scatchard figure (Figure 5.13), and each base on calf thymus DNA (ctDNA) can be bounded to 4.7 JGB molecules, while fish sperm DNA (fsDNA) base can bind 4.2 JGB molecules, showing that the binding of JGB and nucleic acid is of a completely stacking type.

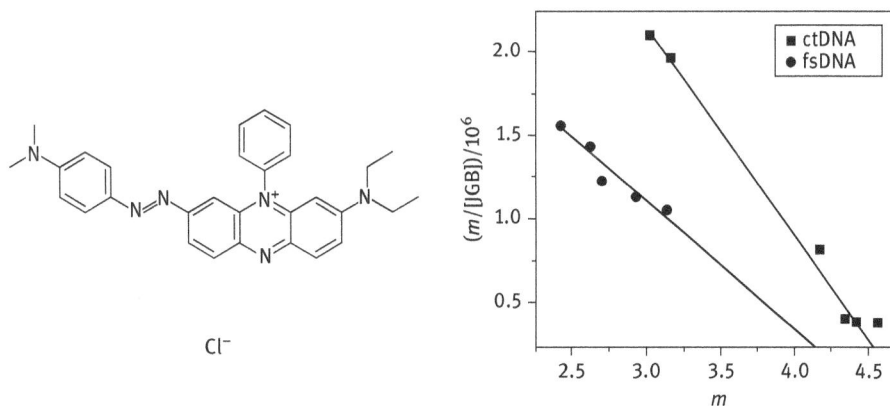

Figure 5.13: A Scatchard figure of JGB molecular structure and its binding with calf thymus DNA and fish sperm DNA. pH 6.37; ionic strength 0.006, fitted curve of ctDNA and fsDNA binding is $m/[JGB] = 5.93 \times 10^6 - 1.25 \times 10^6$ m ($r = -0.9939$), $m/[JGB] = 3.63 \times 10^6 - 8.56 \times 10^5$ m ($r = -0.9495$). JGB–ctDNA binding of n and K values is 4.7 and 1.25×10^6 L/mol, while JGB-fsDNA binding of n and K values is 4.2 and 8.56×10^5 L/mol. All data are obtained at 20°C, 416.0 nm.

EB embedding experiment confirms that under low ion strength and high JGB/nucleic acid mole ratio conditions, JGB and DNA embedding binding is not involved. JGB interaction with nucleic acid is on molecular surface, which is long-distance assembly of organic molecules on nucleic acid surface. Experiment shows that long-distance assembly is highly dependent on their molar ratio (m). When $m > 1.67$, the scattering signal increases with the increase of nucleic acid concentration, and the main function is long-distance assembly; when $m < 1.67$, the scattering signal decreases with nucleic acid concentration, compared with that of $m > 1.67$. It reduces a bit, showing changes of long-distance assembly; further, when $m < 0.22$, RLS signal is quite weak. There is obvious subtractive effect and blue shift phenomenon, demonstrating intercalation of JGB and nucleic acid bases. When $0.22 < m < 1.67$, there is a balance between intercalation and long-distance assembly.

Water-soluble porphyrin reagent, such as TAPP, is the first dye used for the light scattering analysis of nucleic acid. As TAPP is a big-ring conjugate molecule with positive charges, its binding with DNA is first about electrostatic interaction of porphyrin substituent group positive charges and nucleic acid skeleton phosphate group, and then consider nonelectrostatic force of porphyrin big-ring conjugate system and nucleic acid [86]. As TAPP and protonized TAPP stack on nucleic acid surface, there is a long-distance assembly and formation of superhelical structure with nucleic acid helical structure. In these structures, porphyrin molecules stacking on acid surface have high linear density, causing electron excitation delocalization, which exactly produced RLS enhancement [87].

Nucleic acid increases TAPP RLS signal, and the maximum RLS enhancement is when $m = 0.85$. The mole ratio corresponds to the low ionic strength TAPP-induced nucleic acid superhelical structure formation, described by Feil [86]. As TAPP has +4 charge, it is easy to have electrostatic interaction with nucleic acid molecule and polarize it, forming a superhelical structure. At high m value and low ion intensity, there is more possibility for a superhelical structure. Similar to that, Gibbs thinks that porphyrin can have a spiral arrangement on the nucleic acid surface; when light interacts with it, photon produces resonance in each porphyrin unit. It can be seen that the RLS enhancement is caused by a superhelical structure formation.

In addition, as TAPP positive charges were neutralized by negative charges of nucleic acid, there is possibly TAPP molecules $\pi–\pi$ stacking existing on nucleic acid molecule surface in the form of a polymer. It needs to say that the binding does not involve porphyrin embed and nucleic acid structure major and minor groove, but adjusted by nucleic acid negative ions; the formation can only occur with porphyrin saturation on surface.

Before the formation of a superhelical structure, TAPP molecule stacks on nucleic acid surface through "face" and "flat" method, experiencing one turn from TAPP molecule fluorescence quenching to new fluorescence complex. At the turning point, $m = 0.25$, equal to binding of one porphyrin with two bases. When $m = 0.85$, there is a strongest RLS enhancement signal, equal to binding of one base with 1.7 TAPP molecules [88].

Organic small molecules stacking on nucleic acid surface are basically cationic organic dye with a rigid plane or porphyrin; they embed, aggregate, or assemble in long-distance on a molecule to produce strong RLS spectrum. In fact, some colorless cationic surfactant with positive charges could bind multiple negative charges, to form supramolecular compound, causing strong RLS [89]. In these cationic surfactants, aryl and large-molecule-weight scattering signal enhancement is the most obvious, which is probably because of electrostatic attraction, hydrophobic force, π–π stacking function, caused by collaborative resonance of aromatic base, and nucleic acid base absorption.

It can be seen that electrostatic attraction plays a vital role in binding of two or more substances. For example, negative ionic dye and nucleic acid have the same negative charges and electrostatic repulsive force so that dye would not have weak interactions as embedding, aggregation, or long-distance assembly on nucleic acid, and there is no enhanced RLS. While with cationic surfactant such as CTAB, negative ionic dye forms a large aggregate on the DNA surface through CTAB bridging, enhancing the RLS system. System's maximal absorption reduces and there is barely red shift or blue shift in the absorption curve. The phenomenon shows that in the interaction of organic small molecules and nucleic acid, electrostatic attraction is the main function, and embedding, aggregation, or long-distance assembly occur only after electrostatic attraction is satisfied.

(2) Interaction of cationic small-molecule drugs and DNA: As drugs that can interact with DNA in vitro could interact with DNA in vivo to some extent, that is to say, interaction between drugs and DNA has consistency [90], so in vitro study of drug and DNA has directive significance for drug selection.

Currently, interaction of small organic molecule drugs with nucleic acid usually adopts ultraviolet spectroscopy and fluorescence method. In fact, more information can be obtained with the RLS technique in drug and pure DNA in vitro function, drug and DNA conformational change, and so on, which are beneficial for activity screening with the RLS technique. For example, with parameters of scattering signal enhancement degree, mechanism, and binding constants, the measurement of RLS spectrum with small-molecule anticancer drugs such as mitoxantrone, daunorubicin, pirarubicin, and adriamycin amycin with ctDNA induction to simply and rapidly screen drug with best cure effect [76].

Consider another example: dequalinium chloride [91] and berberine hydrochloride [92] bind on nucleic acid molecule surface through electrostatic force and hydrophobic action, thus RLS is strongly enhanced. While in acidic conditions, pirarubicin hydrochloride interacts with nucleic acid, enhancing RLS and a new RLS spectrum is obtained [93].

5.5.3 Interaction of organic small molecules and protein

5.5.3.1 Light scattering of protein solution

As an amphipathic molecule, protein has hydrophilic amino and carboxyl groups and hydrophobic alkyl group. It exhibits many similar properties to surfactant; hence, in theory, protein could form a micelle with a characteristic CMC.

For example, when bovine serum albumin (BSA) and human serum albumin (HSA) are in low concentration, the molecule is dispersed in small numbers, and the scattering intensity is weak. With the increase in concentration, the protein numbers in solution keeps on increasing as well as the scattering monomer quantity and the RLS intensity. When the concentration further increases as amphipathic molecule, its structure is similar to the surfactant, and interaction of molecules and chances of collision increased greatly. There is aggregate and micelle formation (equal to the pre-micelle phase of a surfactant), reducing free protein monomer quantity in solution. As the micelle was rather small in number in the first stage, the RLS signal is weak, and the scattering intensity is determined by the monomer quantity; hence, there is reduction in the intensity of RLS solution.

The above process can be understood as the required protein monomer quantity is constant; and the saturation concentration of monomer quantity is CMC. With more protein concentration, the formed micelle number increases dramatically, leading to a dramatic increase in RLS intensity. So, inflection point appeared after RLS decrease, which can be seen as CMC value exhibition of the phenomenon [94].

5.5.3.2 Interaction mechanism of organic small molecules and protein

As an amphoteric matter, when pH of the medium is lower than the isoelectric point (pI) of protein, protein has a positive charge, and when pH > pI, then it has a negative charge. Isoelectric point of each protein is different, so organic small-molecule probes of different molecular structures have different functions with protein at various acidities, at least involving electrostatic interaction and hydrophobic interaction. Let's take an example of interaction between Cal-Red and HSA [95].

As Cal-red is an organic acid (pK_{a1} = 2.7, pK_{a2} = 3.8, pK_{a3} = 9.4, pK_{a4} = 14.4), it exists in different negative ion form in solution with changes of acidity in environment (Figure 5.14). Contrary to that, pI value of HSA is 4.7–4.9, which is able to exist in solution in negative ion or positive ion form in solution.

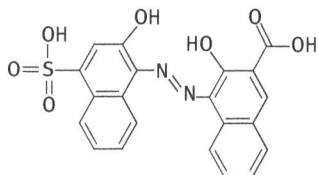

Figure 5.14: The structure of a Cal-Red molecule.

When pH > pI, both Cal-Red and HSA have negative charges, and electrostatic repulsive force makes them unable to bind. Protein loses its activity in strong acidic or alkaline medium, but at pH 4.1, the system has higher RLS value, showing a strong binding, that is to say Cal-Red could bind with HSA to produce one complex. Scatchard figure shows that at room temperature, the maximum binding value of

Cal-Red with HSA is 215, the number is great, and aggregate enlarges clearly with the increase in Cal-Red absorption and molar extinction coefficient, which is helpful for scattering signal enhancement.

From thermodynamic parameters of Cal-Red and HSA, with temperature increase, binding constant and maximum binding number decreases, showing that their binding is weakened. Binding reaction of free energy changes $\Delta G^{\circ} < 0$ shows that binding is a spontaneous exothermic process. Entropy change $\Delta S^{\circ} > 0$ and enthalpy change $\Delta H^{\circ} < 0$ show that the binding has a main function of electrostatic interaction, but the hydrophobic effect is not excluded. Cal-Red and HSA molecular simulation shows that Leu219 in HSA molecule and Phe211 are close to Cal-Red molecule, which is strong evidence for hydrorepellent force in their binding, resulting in multiple weak intermolecular forces such as hydrophobic force, electrostatic force, and hydrogen bond force.

Besides, when $pH < pI_0$, protein has a positive charge, and anionic surfactant stacks on protein with principal interaction of electrostatic force and hydrophobic effect. It shows more obvious interactions in anionic surfactant with aryl group, and anionic surfactant not only works as a stacking small molecule itself, but also enhances the interaction of small molecules and other protein as sensitizer.

5.5.4 Interaction of organic small molecules and saccharides

As saccharide molecules seldom have characteristic functional groups, research about their life mechanism still has a lot to do. Generally, more studies focus on repeated long-chain polysaccharide by a disaccharide unit, one of which is hexosamine (glucosamine or galactosamine), also known as glycosaminoglycan with another unit of uronic acid.

Glycosaminoglycan includes hyaluronic acid, 4-chondroitin sulfate, 6-chondroitin sulfate, dermatan sulfate, heparan sulfate, heparin, and keratan sulfate. These glycosaminoglycans bound conjugately with sulfuric acid or carboxylic acid, as they have many negative charges on surface with strong repulsive force in each unit, as well as linear structure. As the molecular chain is not easy to curve and cross-link, it is named as a linear anionic polyelectrolyte with an enhanced RLS signal. Let's take an example of heparin and alkaline diphenylnaphthyl methane for further discussion [96].

Heparin is named as it was found in the liver, a widely existing natural anticoagulant substance in animals, and is widely applied in clinic. Its molecular structure is composed of mucopolysaccharide sulfate by glucosamine, L-aldoglycoside, N-acetylglucosamine and D-glucuronic acid (Figure 5.15), with average molecule weight of 15 kDa and strong acidity. As heparin has 3 -O-SO_3H, 2 -$NHSO_3H$, and 2 -COOH in each four unit, where $-O$-SO_3H and -$NHSO_3H$ have strong acidity, when $pH \geq 3.0$, related protons would dissociate completely; while -COOH is a weak acidic group, pK_a value of D-glucuronic acid is 3.6, it does not dissociate completely until the pH is above 5.0.

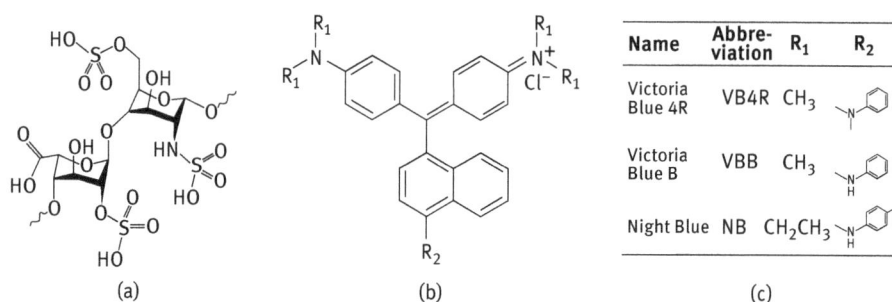

Name	Abbre-viation	R_1	R_2
Victoria Blue 4R	VB4R	CH_3	
Victoria Blue B	VBB	CH_3	
Night Blue	NB	CH_2CH_3	

(a)   (b)   (c)

Figure 5.15: (a) Heparin and (b, c) alkaline diphenylnaphthyl methane dye molecule structures.

As protonated groups in heparin is a weak acid, when acidity is rather high (such as 2 mol/L HCl), it barely exists in weak acid, and –COOH does not only dissociate, but also –O-SO_3H and –NHSO_3H exists in molecular state with no charge, losing electrostatic attraction of positive ion, so that interaction of heparin and alkaline diphenylnaphthyl methane hardly occurs. Under high acidity, positive ion would become a double-charge positive ion with increased surface charge density and decreased hydrophobicity, unsuitable for stacking.

While in alkaline medium, a positive ionic dye loses positive charge, reducing stacking function. As heparin has seven binding sites of each tetrose unit (five -SO_3H and two -COOH), each molecule of heparin has 42 monosaccharide; hence, in theory, each molecule of heparin has 73.5 binding sites with dye. Actually, binding sites can be calculated through Scatchard equation (5.1).

Interaction of positive ion and heparin is also influenced by their molar ratio. Normally, there is aggregation possibility with positive ion concentration of 10^{-3}–10^{-6} mol/L to form dimer or polymer with a dynamic balance of dye monomer, dimer, and polymer. So, positive ions stacking on heparin molecules are not single form. When dye concentration is rather low with monomer form as main existing form, their binding product has strong RLS, when dye concentration is further improved, RLS intensity decreases, suggesting that the existence of dimer or polymer is not beneficial for binding interaction. It is possibly because of competition between positive ion, heparin binding, and positive ion aggregate, or it is probably because of large combination of aggregate with heparin, or formation of sediment, which reduced the scattering signal.

Under near neutral condition, heparin mostly exists in multivalence anion state with molecular skeleton of hydrophobic glycosyl connected by oxygen bridge, and its charges are from -SO_3H group, 2 -NHSO_3H groups, and 2 –COO$^-$ groups. When hydrophobic polyaryl maternal cation organic small molecules get close to heparin, ionic associate is formed through electrostatic attraction of opposite charges and hydrophobic effect between aryl maternal and heparin skeleton. Dye molecule is trapped on heparin surface through positive and negative electric fields, the RLS signal is greatly enhanced.

Although heparin is a macromolecular compound, it has no conjugated system itself. There is no molecular absorption in near UV-visible region, so there is no obvious RLS. When it binds with dye cation, there is strong light absorption because of aggregate of multiple dye chromophore on molecule surface, and there is a large molar absorption coefficient. Based on previous light scattering theory, with the increase in molar absorption coefficient, light scattering intensity enhances with increase in the imaginary part of refractive index.

5.6 Interaction of metal complex molecule probes and biomacromolecules

Besides the interaction of organic small molecules and biomacromolecule, there are interactions of metal complex molecule probes and biomacromolecules. A study of that has an important reference value for metal drug development, and structure and functional assignment of biomacromolecules.

5.6.1 Interaction with nucleic acid

As shown in Figure 5.16(a), [Pt(terpy)Me]Cl is a platinum-trialpyridine complex, with aggregate trend in phosphate buffer medium, able to form extended aggregate on nucleic acid surface, which is related to the molar ratio of platinum-trialpyridine complex/nucleic acid. For a certain amount of platinum-trialpyridine complex, if nucleic acid concentration is rather low, molar ratio is pretty high, and easy to form self-aggregate and extended aggregate balance, producing enhanced RLS signal along with the increase in nucleic acid concentration. If nucleic acid concentration is rather high, complex has embedded binding with nucleic acid, RLS signal decreases with the increase of nucleic acid concentration [97]. Hence, as a difunctional intercalating agent [Pt(terpy)Me]Cl and nucleic acid interaction, maximum RLS enhancement signal actually marks the qualitative change process from surface stacking to embedded binding.

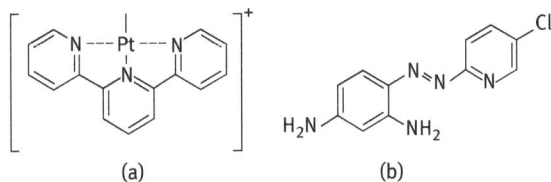

(a)   (b)

Figure 5.16: (a) [Pt(terpy)Me]Cl and (b) Co reagent molecular structure.

The study on light scattering features with metal cobalt complex, and biomacro-molecule interaction was rather early. Molecular absorption spectroscopy shows that, in alkaline medium, Co (II)/cobalt complex [Figure 5.16(b)] reacts with thermotropic nucleic acid, each nucleotide residue binds two binary complex molecules, forming stacking of molecule on single-stranded nucleic acid. The stacking of Co(II)/cobalt complex was obtained on nucleic acid with RLS technique, which strongly increased RLS signal at 547 nm, with a corresponding sharp peak with ternary complex absorption [98].

Interaction of metalloporphyrin and nucleic acid is quite different from that of pure porphyrin with nucleic acid. For example, interaction of tetraphenyl metalloporphyrin is sensitive to pH condition, which is quite different from TAPP [91], showing two different mechanisms. Red shift and hypochromic effect of absorption spectrum prove that metalloporphyrin binds nucleic acid in the insert mode, while TAPP binds nucleic acid with supramolecular self-assembly. Compared with water-soluble porphyrin, the metal ion of metalloporphyrin coordinates with the nitrogen atom on porphyrin ring, greatly improving the plane rigid structure of a molecule. Besides, the metal ion has the axial coordination capability, which is able to combine bases of nucleic acid.

Anionic surfactant such as sodium dodecyl sulfate (SDS) exists with premicelle aggregate, inducing positive complex ion $[Ru(bpy)_2(dppx)]^{2+}$ to aggregate on surface with a strong RLS signal; with DNA addition, RLS decreased greatly [8]. As $[Ru(bpy)_2(dppx)]^{2+}$ has stronger intercalative effects with DNA, that is to say, $[Ru(bpy)_2(dppx)]^{2+}$ embeds onto DNA double-helix structure through ligand of dppx, separating $[Ru(bpy)_2(dppx)]^{2+}$ from SDS and thereby reducing RLS.

RLS enhancement can be produced even without chromophore and long carbon chain hydrophilic or hydrophobic simple metal ions. For example, in acidic medium (pH 2.21), Al^{3+} can bind DNA surface phosphate group through electrostatic interaction to produce larger particle and RLS enhancement signal with a characteristic peak at 291 nm [3].

5.6.2 Interaction with protein

Phosphotungstic acid mainly exists in $[W_7O_{24}]^{6-}$ in pH 4.1 solution [99]; at this time, pH is lower than the BSA isoelectric point, so that BSA peptide residue exists positive ion. After electrostatic interaction, BSA side-chain hydrophobic group coordinates with phosphotungstic acid with Van der Waals force, aggregating molecules around BSA molecule surface to form large-scale complex, suspending in water solution. Although phosphotungstic acid has no chromophore, larger scale could also enhance the scattering signal. When the anionic surfactant SDS is added in solution, which is formed micelle promotes their coordination, especially in CMC, thus system scattering is further enhanced.

5.7 Factors affecting organic small-molecule probe aggregation

Organic small-molecule probe and macromolecules mostly interact on the basis of electrostatic interaction, involving weak interactions such as hydrogen bond and Van der Waals force. These functions highly depend on environmental conditions such as acidity, ion strength and temperature, and it is necessary to discuss related conditions to obtain maximum light scattering signal, to help for sensitivity improvement of analysis and measurement of biomacromolecule with probe.

5.7.1 Acidity influence

In an aggregate system with the module of biomacromolecules, based on its structure, charges of them can be predicted at various acidity conditions. Nucleic acid with phosphate skeleton has negative charges, and heparin, chondroitin sulfate and other glycosaminoglycan with sulfate skeleton have strong negative charges, while charges of protein are related to their isoelectric point (pI_0). When the pH is lower than pI_0, protein is protonated with positive charges. In the aggregate system, selective light scattering probes mainly need negative charges, the charges closely related to acidity. Some probes even have opposite charges under different acidity conditions.

Some light scattering probes are formed by two or more molecules or ions, where one of which could be charge-free or with the same charges. Similar to aggregate of ion associate, in organic small molecules and biomacromolecule interaction system, the pH plays an important role. In some self-aggregate system, by adjusting pH to make

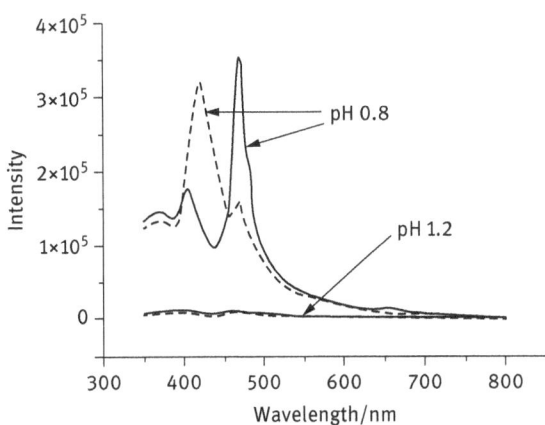

Figure 5.17: An RLS spectra of TCPP aggregate in different acidity. RLS spectra of nitric acid solution (full line) and aqueous hydrochloric acid solution (imaginary line). They are almost coincident at pH 1.2.

molecules charge-free, molecules aggregate with hydrogen bond, Van der Waals force and hydrophobic force. While most dyes assemble in high concentration or biomacromolecule, their charges aggregate as well, especially for porphyrin, as H_2TCPP^{2+} [100] aggregates with interplanar conjugate action. Acidity influence can be further determined through reagent adding sequence, adjusting adding sequence and measure immediately after mixing, solution that has buffer solution in the last addtion has the minimum changes, and obvious electrostatic attraction can be determined in aggregation.

5.7.2 Ion strength influence

As the ionic strength increases, the scattering signal decreases in the aggregates with electrostatic attraction, indicating that the aggregates are gradually depolymerizing. Ionic strength adjustment usually adopts sodium chloride solution to reduce or influence the interaction between the probe molecule and the aggregate, as coordination, sediment. High concentration of chloride and sodium ions forms ion atmosphere around aggregate positive and negative charges, effectively blocking the original coulombic interaction. The shielding effect can use the same amount of 1 mol/L NaCl and 1 mol/L $MgCl_2$ to detect the signal changes. If $MgCl_2$ induces more signal reduction, then the scattering signal change is from the increase in ionic strength. Different ions induce different aggregate states, which is also obvious in the aggregate study of porphyrin, such as H_2TCPP^{2+} [101].

5.7.3 Influence of organic solvent

Currently, most homogeneous systems are aqueous solution. When there is aggregate in a system with particles larger than water, or unbonded small molecules or ions, these aggregates do not easily precipitate or become turbid, forming interface in water medium, which shows hydrophobicity of aggregate. When an organic solvent soluble with water is added, such as acetonitrile and methanol, solvent breaks the microenvironment of an aggregate. When more organic solvent is added, aggregate in water solution depending on the hydrophobic effect would gradually dissolve. In the formation process, there is electrostatic attraction, coordination function and weak interactions as Van der Waals force, $\pi - \pi$ conjugation, which can be weakened by an organic solvent. Besides, organic solvent changes the system refractive index, aggregate light scattering signal is likely reduced with changes of system refractive index even without depolymerization, which can be confirmed by polymer beads light scattering changes by environment refractive index.

5.7.4 Influence of temperature and time

In formation, an aggregate is influenced by electrostatic attraction, coordination function and Van der Waals force. When system temperature rises, molecule's moving speed accelerates, and these interactions may decrease, reducing aggregate scattering signal to different degrees.

Aggregate formation is a dynamic equilibrium process, influenced by temperature and time. It is easy to decide accurate time of aggregate completion with changing curve of scattering signal with time. After mixing up of aggregate, previously formed aggregate is likely to form larger aggregates; then there is sediment in the end, reducing scattering signal.

5.8 Interface behavior of organic small-molecule probes

Organic small-molecule probes exhibit inconsistent behavior on interface with that in water phase. For example, aggregates with electrostatic interaction expose hydrophobic group to the outside, while most aggregates are not easily adsorbed on to the interface, as a hydrophobic function is one of the most important functions of aggregates. To adsorb onto the interface, the hydrophobic function is likely to disappear while aggregates would depolymerize. When an amphiphilic substance adsorbs onto the interface as a surfactant, its oleophylic functional groups (as alkyl groups) extend to organic phase, while a hydrophilic groups with charges extend in water phase, a single surfactant has a weak scattering signal on the interface.

Cationic surfactant (as CTAB) would easily adsorb anionic compounds, such as DNA [100], protein [41] in water phase on the interface, though they may aggregate on the interface with enhanced scattering signal, but the enhancement is not that obvious. This is probably because though particles magnify with the binding of a surfactant and these biomacromolecules, based on Rayleigh scattering theory, scattering is related to absorption, and a surfactant and these biomacromolecules have weak absorption ability. Besides, after binding, active binding sites still exist. So, if another sensitized assembly with strong absorption of cationic compounds is added, such as acridine orange [101] and tribromoarsenazo [41] dye, it not only neutralizes extra charges, bringing closer combination, but also increases assembly size and improves assembly molar absorption coefficient, which is helpful for scattering signal enhancement.

Some organic small molecules have charges and hydrophobic functional groups, showing amphipathy, even with a surfactant function partially. For example, under experimental conditions, fluorescein [102] has complete ionization with negative charges, while N in berberine hydrochloride has positive charges, which can be combined with electrostatic interaction and hydrophobic interaction. The combination does not cause obvious scattering light enhancement in water phase, which means the combination is rather weak. However, as fluorescein shows amphipathy and can be adsorbed on to phase interface, after combination, its water solubility reduces and lipid solubility increases, so that fluorescein berberine hydrochloride complex is easily adsorbed onto H_2O/CH_2Cl_2 interface. The strong adsorption makes interface concentration far higher than that in phase, enhancing an interface light scattering signal. As amphipathy fluorescein has stronger absorption itself, there is no need to add a third sensibilization reagent as a surfactant.

Properties of soluble molecules of two insoluble phases are at opposite poles; if one substance can connect molecules of two phases, interface assembly is possible. The substance is usually metallic ions, such as Eu(III) [103] and Al(III) [104]. For example, Eu (III) forms a complex with chlortetracycline (CTC) in water, while its coordination number is not saturated, which is occupied by water molecules. At this time, if there is a synergistic ligand as trioctyl-phosphine oxide (TOPO), water molecules will be replaced by synergistic ligand to satisfy central ion Eu(III) coordination number. As TOPO, CTC–Eu(III) has lipophilic and hydrophilic properties and the interface has amphipathy, so Eu(III)–CTC distributes in water phase and interface while Eu(III)–TOPO distributes in dichloromethane phase and interface. These scattering light signals are actually weak, so the interface formed "CTC–Eu(III)–TOPO complex" has a quite strong light scattering signal.

A metal composite probe does not only adsorb organic small molecules as aureomycin on the interface, but also adsorbs biomacromolecules for the enhanced light scattering signal. For example, in water medium with pH 3.0, binding of Al(III) and DNA produces an enhanced light scattering signal. When carbon tetrachloride dissolved with tetraphenyl porphyrin (TPP) mixes with aqueous phase dissolved with Al(III) and DNA, they can form a ternary complex, which is lipophilic and hydrophilic and able to adsorb onto the interface with an enhanced RLS signal.

As a liquid–liquid interface has a thickness at wavelength level, substances with amphipathy are adsorbed in interface of the range. As common inorganic substances, amino acid and saccharides cannot reach the interface, the total amount of interference material allows the established RLS method. An interface light scattering method has a sensitivity of RLS method and good selectivity, and can be directly used for trace-level protein and drug measurement in body fluid such as urine.

References

[1]   Li S W, Su X D, Wang X H, et al. Determination of chloride ion by resonance light scattering method. Chem. Res. Appl., 2004, 16(4): 535–536.
[2]   Xie L S, Li Y L, Liu Z F, et al. Resonance Rayleigh scatting method for the determination of potassium ferrocyanide in salinizde foods. Chinese J. Anal. Chem., 2006, 34(10): 1511.

[3]   Yang C X, Li Y F, Feng P, et al. Study on the interaction of alaminum ion with deoxyribonucleic acid by resonance light scattering measurement. Chinese J. Anal. Chem., 2002,30(4): 473–477.

[4]   Huang C Z, Li K A, Tong S Y. Determination of nanograms of nucleic acids by their enhancement effect on the resonance light scattering of the cobalt(II)/4-[(5-chloro-2-pyridyl)azo]-1,3-diaminobenzene complex. Anal. Chem., 1997, 69(3): 514–520.

[5]   Tao Y L, Ao D G W. Resonance Rayleigh scattering study of the interaction of iron(III) with deoxyribonucleic acid. J. Instr. Anal., 2007, 26(6): 876–879.

[6]   Guo Z X, Shen H X. A highly sensitive assay for protein using resonance light-scattering technique with dibromohydroxyphenylfluorone-molybdenum (VI) complex. Spectrochim. Acta A, 1999, 55(14): 2919–2925.

[7]   Jia R P, Dong L J, Li Q F, et al. A highly sensitive assay for protein with dibromochloro-arsenazo-Al^{3+} using resonance light scattering technique and its application. Talanta, 2002, 57(4): 693–700.

[8]   Huang J P, Mei P, He Z K. Quenching effect of DNA on resonance light scattering of $Ru(bpy)_2$ $(dppx)^{2+}$-SDS system and its application. Chinese J. Appl. Chem., 2010, 27(7): 849–854.

[9]   Huang J P, Mei P, Hu S Q, etc. Study on resonance light scattering spectrum of $Ru(bpy)_3^{2+}$ with DNA and its analytical application. Chinese J. Anal. Lab. 2008, 27(8): 36–39.

[10]   Xiao Z B, Cai Z X, Li L, et al. Long range assembly of mixed-complexes on the molecular surfaces of nucleic acid. J. Anal. Sci. 2003, 19(4): 317–320.

[11]   Yu J H, Yang H X, Yang S Y, et al. Resonance light scattering method for the determination of DNA by terbium salicylic acid and phenanthroline ternary complex. J. Xiangtan Nor. Univ. (Sci.), 2009, 31(3): 53–55.

[12]   Li X Y, Li S W, Wang X H, etc. Determination of Pb^{2+} by resonance light-scattering turbidimetry method. J. Southwest Univ. Natl. (Nat. Sci. Ed.), 2005, 31(2): 192–194.

[13]   Ma C Q, Liu Y, Li K A, et al. Rayleigh light scattering and its application in biochemical analysis. Chinese Sci. Bull., 1999, 44(7): 682–690.

[14]   Huang C Z, Li K A, Tong S Y. Determination of nucleic acids at nanogram levels with safranine T by a resonance light-scattering technique. Anal. Chim. Acta, 1998, 375: 89–97.

[15]   Wang Y T, Zhao F L, Li K A, et al. Molecular spectroscopic study of DNA binding with neutral red and application to assay of nucleic acids. Anal. Chim. Acta, 1999, 396: 75–81.

[16]   Damaola J L L, Huang C Z. Resonance light-scattering study on the interactions of phenosafranine with deoxyribonucleic acid and the light-scattering determination of trace deoxyribonucleic acid. Chinese J. Anal. Chem., 1999, 27(10): 1204–1207.

[17]   Huang C Z, Li Y F, Deng S Y, et al. Assembly of methylene blue on nucleic acid template as studied by resonance light-scattering technique and determination of nucleic acids of nanogram. Bull. Chem. Soc. JPN, 1999, 72(7): 1501–1508.

[18]   Huang C Z, Li Y F, Pu Q H. Interactions of Nile blue sulphate with nucleic acids as studied by resonance light-scattering measurements and determination of nucleic acids at nanogram levels. Anal. Lett., 1999, 32(12): 2395–2415.

[19]   Huang C Z, Li Y F, Hu X L, et al. Three-dimensional spectra of the long-range assembly of Nile Blue sulfate on the molecular surface of DNA and determination of DNA by light-scattering. Anal. Chem. Acta, 1999, 395: 187–197.

[20]   Liu C, Chen X M. Resonance light scattering method for the determination of deoxyribonucleic acid with brilliant crystal blue. Chinese J. Anal. Chem., 2001, 29(6): 685–688.

[21]   Xiang H Y, Chen X M, Zhou D W. Determination of deoxyribonucleic acid with pararosaniline by a resonance light-scattering method. Spectrosc. Spectr. Anal., 2002, 22(6): 1051–1053.

[22]   Li T J, Shen H X. Studies on the resonant luminescence spectra of ethyl violet with deoxyribonucleic acid and its analytical application. Chemical J. Chinese U., 1998, 19(10): 1570–1573.

[23]   Liu C, Hu H, Chen X M. Determination of deoxyribonucleic acid with rosaniline by a resonance light scattering method. Chinese J. Anal. Lab., 2001, 20(2): 30–33.

[24] Liu Y, Ma C Q, Li K A, et al. Rayleigh light scattering study on the reaction of nucleic acids and methyl violet. Anal. Biochem., 1999, 268: 187–192.

[25] Liu C, Chen X M, Xiang H Y, etc. Determination of deoxyribonucleic acid with rosaniline by a resonance light scattering method. Spectrosc. Spectr. Anal., 2001, 21(5): 697–700.

[26] Liu Y, Ma C Q, Li K A, et al. Simple and sensitive assay for nucleic acids by use of Rayleigh light scattering technique with rhodamine B. Anal. Chim. Acta, 1999, 379(1–2): 39–44.

[27] Yu Y, Huang F D. Resonance scattering spectra of nucleic acid-butyl rhodamine B system. Chinese J. Anal. Chem., 2002, 30(10): 1234–1236.

[28] Wu H L, Li W Y, He X W, et al. Study on rhodamine 6G as a new probe for the determination of nucleic acids by resonance light scattering. J. Anal. Sci. 2002, 18(2): 94–99.

[29] Wang Y T, Zhao F L, Li K A, et al. A novel protein assay with an Azo dye using Rayleigh light scattering technique. Anal. Lett., 2000, 33: 221–235.

[30] Jia R, Dong L, Li Q, et al. Study of the interaction of protein and dibromomethyl-Arsenazo- Al(III) by Rayleigh light-scattering technique. Anal. Chim. Acta, 2001, 442: 249–256.

[31] Li Q, Dong L, Jia R, et al. Rayleigh light scattering enhancing reaction of dibromochloroarsenazo and proteins system and its analytical application. Spectrosc. Lett., 2001, 34: 407–417.

[32] Yang C X, Li Y F, Huang C Z. Determination of total protein content in human serum samples with fast red VR by resonance light scattering technique. Anal. Lett., 2002, 35: 1945–1957.

[33] Yang C X, Li Y F, Huang C Z. Determination of proteins with fast red VR by a corrected resonance light-scattering technique. Anal. Sci., 2003, 19: 211–215.

[34] Li Y F, Huang C Z, Li M. A resonance light-scattering determination of proteins with Fast Green FCF. Anal. Sci., 2002, 18: 177–181.

[35] Huang C Z, Yang C X, Li Y F. Determination of proteins with Ponceau G by compensating for the molecular absorption decreased resonance light scattering signals. Anal. Lett., 2003, 36: 1557–1571.

[36] Yao G, Li K A, Tong S Y. A study on the reaction of protein with Arsenazo III by Rayleigh light scattering technique. Anal. Lett., 1998, 31: 1689–1701.

[37] Wu X, Sun S, Guo C, et al. Resonance light scattering technique for the determination of proteins with Congo Red and Triton X-100. Luminescence, 2006, 21: 56–61.

[38] Feng P, Hu X L, Huang C Z. Determination of proteins at nanogram levels with their enhancement effects of resonance light-scattering on quercetin. Anal. Lett., 1999, 32: 1323–1338.

[39] Wu H L, Li W Y, He X W. Study of the reactions of proteins with Epsilon Blue by resonance light scattering technique and its analytical application. Chin. J. Anal. Chem., 2003, 31: 989–991.

[40] Tan K J, Li Y F, Huang C Z. Flow injection resonance light scattering detection of proteins of nanogram. Luminescence, 2005, 20: 176–180.

[41] Dong L, Chen X, Hu Z. Total internal reflected resonance light scattering determination of protein in human blood serum at water/tetrachloromethane interface with Arsenazo-TB and Cetyltrimethylammonium bromide. Talanta, 2006, 71: 555–560.

[42] Dong L J, Li Y, Zhang Y H, et al. Determination of proteins in human serum by combination of flow injection sampling with resonance light scattering technique. Microchim. Acta, 2007, 159: 49–55.

[43] Dong L J, Li Y, Zhang Y H, et al. A flow injection sampling resonance light scattering system for total protein determination in human serum. Spectrochim. Acta A., 2007, 66: 1317–1322.

[44] Li Y, Dong L J, Zhang Y H, et al. Determination of proteins by flow injection analysis coupled with the Rayleigh light scattering technique. Talanta, 2007, 71: 109–114.

[45] Huang C Z, Li Y F, Feng P. Determination of proteins with-tetrakis(4-sulfophenyl)porphine by measuring the enhanced resonance light scattering at the air/liquid interface. Anal. Chim. Acta, 2001, 443: 73–80.

[46] Huang C Z, Liu Y, Wang Y H, et al. Resonance light scattering imaging detection of proteins with-tetrakis(p-sulfophenyl)porphyrin. Anal. Biochem., 2003, 321: 236–243.
[47] Chen X L, Li D H, Zhu Q Z, et al. Determination of proteins at nanogram levels by a resonance light-scattering technique with tetra-substituted sulphonated aluminum phthalocyanine. Talanta, 2001, 53: 1205–1210.
[48] Wang L, Li Y X, Zhu C Q, et al. Determination of proteins based on their resonance light scattering enhancement effect on manganese- tetrasulfonatophthalocyanine. Microchim. Acta, 2003, 143: 275–279.
[49] Zhong H, Wang K, Chen H Y. Protein analysis with tetra-substituted sulfonated cobalt phtha-locyanine by the technique of Rayleigh light scattering. Anal. Biochem., 2004, 330: 37–42.
[50] Tang N L, Peng J Y. Determination of proteins with tetracarboxy manganese(II) phthalocyanine by resonance light scattering technique. Spectrochim. Acta A, 2008, 71: 1246–1249.
[51] Guo Z X, Shen H X. Sensitive and simple determination of protein by resonance Rayleigh scattering with 4-azochromotropic acid phenylfluorone. Anal. Chim. Acta, 2000, 408: 177–182.
[52] Li Q, Zhang H, Xue C, et al. Determination of proteins at nanogram levels based on their enhancement effects of Rayleigh light scattering on dibromomethylchlorophosphonazo. Spectrochim. Acta A, 2000, 56: 2465–2470.
[53] Li Y F, Huang C Z, Huang X H, et al. Determination of proteins based on their resonance light scattering enhancement effect on Alcian Blue 8GX. Anal. Sci., 2000, 16: 1249–1254.
[54] Feng S, Pan Z, Fan J. Determination of trace proteins with pyronine Y and SDS by resonance light scattering. Anal. Bioanal. Chem., 2005, 383: 255–260.
[55] Zhan G Q, Zhang L X, Li C Y. Resonance light scattering spectral method for the determination of serum albumin with the interaction of neutral red-sodium dodecyl sulfonate. Colloids Surf. B, 2009, 71: 84–87.
[56] Guo C, Wu X, Xu W, et al. Resonance light-scattering enhancement effect of the protein-Y [3+]-TTA-SLS system and its analytical application. Luminescence, 2008, 23: 404–409.
[57] Cai C, Chen X. Determination of nanograms of proteins based on the amplified resonance light scattering signals of Tichromine. Spectrochim. Acta A, 2010, 75: 1057–1060.
[58] Wu L H, Mu D, Gao D J, et al. Determination of protein by resonance light scattering technique using dithiothreitol-sodium dodecylbenzene sulphonate as probe. Spectrochim. Acta A, 2009, 72: 178–181.
[59] Wang W P, Tang J H, Peng X H, et al. A laser light scattering study of the interaction between human serum albumin and ampicillin sodium. Sci. China Ser. B, 2006, 49: 332–337.
[60] Yang J D, Cao T W, Liu Z F, et al. Effects of the interaction of rifamycin SV with serum albumins on the resonance Rayleigh scattering spectra and their analytical application. Chin. J. Chem., 2008, 26: 893–897.
[61] Li Y F, Shen X W, Huang C Z. A coupled reagent of o-phthalaldehyde and sulfanilic acid for protein detection based on the measurements of light scattering signals with a common spectrofluorometer. Talanta, 2008, 75: 104–1045.
[62] Zhang S H, Fan Y S, Feng S, et al. Microdetermination of proteins by resonance light scattering technique based on aggregation of ferric nanoparticles. Spectrochim. Acta A, 2009, 72: 748–752.
[63] Huang X H, Shu W Q, Li Y F, et al. The sensitizing effect of cetyltrimethylammonium bromide onthe resonance light-scattering enhancement between the interactionof deoxyribonucleic acids and brilliant green. Chinese J. Anal. Chem., 2001, 29(3): 271–275.
[64] Feng S L, Liu X P, Fan J. Determination of deoxyribonucleic acids based on the enhanced effect of resonance light scattering of malachite green sensitized by cetyltrimethylammonium bromide. Spectrosc. Spectr. Anal., 2004, 24(5): 606–609.

[65] Chen Y J, Yang J H, Wu X, et al. Study on the RLS system of resorcinol yellow, CTAB and nucleic acids and its application. Chinese J. Spectrosc. Lab., 2002, 19(3): 299–302.

[66] Wang Y B, Wu X, Yang J H, et al. Determination of DNA with m-cresol purple and DNA systems by resonance light scattering technique. J. Shandong Univ. (Sci. Ed.), 2002, 37(6): 537–539.

[67] Feng N C, Lin F, Liu L J, etc. Determination of deoxyribonucleic acid by methylene green and cetyltrimethylammonium bromide system with a resonance light scattering method. Chinese J. Health Lab. Technol., 2005, 15(2): 172–173.

[68] Ouyang Z Z, Bai S, Du Z H. Resonance light scattering spectra of the xylene cyanal FF-cetyltrimethyl ammonium bromide-nucleic acid system and its analytical application. Chinese J. Spectrosc. Lab. 2010, 27(2): 643–647.

[69] Pan Z H, Chen L H, Hao C J. The spectroscopic study of proteins-pyronine Y-SDS system by resonance light scattering. Chem. Ind. Times, 2007, 21(12): 27–29.

[70] Rao Z H, Xiong Y P, Ren L P, etc. Determination of deoxyribonuceic acids by resonance light scattering technique with benzidine in the presence of the sensitizing sodium dodecyl sulfate. Spectrosc. Spectr. Anal., 2004, 24(19): 1035–1038.

[71] Feng S L, Wang J, Fan J. Microdetermination of proteins with titan yellow by the resonance light scattering method. Spectrosc. Spectr. Anal., 2004, 6(6): 927–929.

[72] Liang L N, Ge G H, Dong C Z. Determination of sulfonamides using resonance light scattering method. Chinese J. Health Lab. Technol., 2008, 18(8): 1561–1562.

[73] Zhang G S, Dong X Z, Wang A F, et al. Study on determination of phorate residue with sodium tungstate-rhodamine B-P system by resonance light scattering technique. J. Instr. Anal., 2008, 27, 10(10): 1062–1066.

[74] Li S D, Meng J H, Li M Q. Determination of trace Sn^{2+} in environmental water by resonance light-scattering. Chinese J. Spectrosc. Lab., 2006, 23(5): 981–983.

[75] Long Y F, Huang C Z, He R X, et al. Selectively light scattering spectrometric detection of copper (II)based on a new synthesized oxamide ligand. Anal. Chim. Acta, 2008, 624(1): 128–132.

[76] Chen Z G, Song T H, Wang S B, et al. Screening DNA-targeted anticancer drug in vitro based on the drug-conjugated DNA by resonance light scattering technique. Biosens. Bioelectron., 2010, 25(8): 1947–1952.

[77] Pasternack R F, Schaefer K F, Hambright P. Resonance light-scattering studies of porphyrin diacid aggregates. Inorg. Chem., 1994, 33(9): 2062–2065.

[78] Huang C Z. Studies on the reaction and application of bioactive substances and spectroscopic probes – Application of nucleic acid spectral probes and resonance light scattering techniques in the analysis of biological macromolecules. Beijing: Doctoral dissertation of Peking University, 1996.

[79] de Paula J C, Robblee J H, Pasternack R F. Aggregation of chlorophyll a probed by resonance light scattering spectroscopy. Biophys. J., 1995, 68(1): 335–341.

[80] Nan H J, Liu Z F, Liu S P. Resonance Rayleigh scattering spectra of self-aggregation for bile salts in acidic media and their analytical application. Acta Chim. Sin., 2006, 64(12): 1253–1259.

[81] Chen Z G, Liu G L, Chen M H, et al. Determination of nanograms of proteins based on decreased resonance light scattering of zwitterionic Gemini surfactant. Anal. Biochem., 2009, 384(2): 337–342.

[82] Chen Z G, Liu G L, Chen M H, et al. High-sensitivity detection of polysaccharide using phosphodiesters quaternary ammonium salt as probe by decreased resonance light scattering, Talanta, 2009, 79(2): 171–176.

[83] Hu X L, Liu S P, Luo H Q. Resonance Rayleigh scattering spectra of interaction of aminoglycoside antibiotics with trypan red and their analytical applications. Acta Chim. Sin., 2003, 61(8): 1287–1293.

[84] Yang Z P, Liu Z F, Hu X L, etc. Resonance Rayleigh scattering spectra of some fluoroquinolones antibiotics-Pd(II)-eosin Y systems and their analytical application. Chinese J. Appl. Chem., 2007, 24(3): 261–267.

[85] Huang C Z, Li Y F, Huang X H. Interactions of Janus Green B with double stranded DNA and the determination of DNA based on the measurement of enhanced resonance light scattering. Analyst, 2000, 125(7): 1267–1272.

[86] Carvlin M J, Fiel R J. Intercalative and nonintercalative binding of large cationic porphyrin ligands to calf thymus DNA. Nucl. Acids Res., 1983, 11(17): 6121–6139.

[87] Huang C Z, Li K A, Tong S Y. A resonance light-scattering analysis of the suprahelical helixes of nucleic acids induced by 5, 10, l5, 20-tetrakis[4-(trimethylammonio) phenyl] porphine. Bull. Chem. Soc. Jpn, 1997, 70(8): 1843–l849.

[88] Huang C Z, Li K A, Tong S Y. Spectroscopic studies on the interactions of water-soluble porphyrins with nucleic acids. Chem. J. Chinese U., 1997, 18(4): 525–529.

[89] Liu S P, Hu X L, Luo H Q, etc. Resonance Rayleigh scattering spectral properties and application of cation surfactants and nucleic acid reaction, China Sci. Ser. B-Chem., 2002, 32(1): 18–26.

[90] Li Z P, Li K A, Tong S Y. Determination of nucleic acids in acidic medium by enhanced light scattering of large particles. Talanta, 2000, 51(1): 63–70.

[91] Chen Z G, Ding W F, Liang Y Z, et al. Determination of nucleic acid based on their resonance light scattering enhancement effect on metalloporphyrin derivatives. Microchim. Acta, 2005, 150(1): 85–93.

[92] Liu C, Chen X T, Li S Q, etc. Determination of deoxyribonucleic acid with berberine by a resonance light scattering method. Chinese J. Anal. Chem., 2002, 30(10): 1218–1221.

[93] Liu S P, Wang F, Liu Z F, etc. Study on the interaction between epirubicin hydrochloride and nucleic acid and its analytical application by resonance Rayleigh scattering spectra. Acta Chim. Sin., 2007, 65(10): 962–970.

[94] Zhao H W, Huang C Z, Li Y F. Validation of micelle formation of proteins and determination of their critical micelle concentrations by measuring synchronous light scattering signals using a common spectrofluorometer. Chem. Lett., 2006, 35(4): 418–419.

[95] Dong L J, Chen X G, Hu Z D. Study of the effect of Cal-Red on the secondary structure of human serum albumin by spectroscopic techniques, J. Mol. Struct., 2007, 846(1–3): 112–118.

[96] Liu S P, Luo H Q, Li N B, et al. Resonance Rayleigh scattering study of the interaction of heparin with some basic diPhenylnaphthylmethane dyes. Anal. Chem., 2001, 73(16): 3907–3914.

[97] Arena G, Soolaro L M, Pasternack R F, et al. Synthesis, characterization, and interaction with DNA of the novel metallointercalator cationic complex (2, 2′: 6′, 2″-terpyridine)methylplatinum (II). Inorg. Chem., 1995, 34(11): 2994–3002.

[98] Huang C Z, Li K A, Tong S Y. Determination of nanograms of nucleic acids by their enhancement effect on the resonance light scattering of the cobalt(II)/4-[(5-chloro-2-pyridyl) azo]-1,3-dia-minobenzene complex. Anal. Chem., 1997, 69(3): 514–520.

[99] Chen Z G, Ding W F, Ren W F, et al. A simple and sensitive assay of nucleic acids based on the enhanced resonance light scattering of zwitterionics. Anal. Chim. Acta, 2005, 550(1–2): 204–209.

[100] Doan S C, Shanmugham S, Aston D E, et al. Counterion dependent dye aggregates: Nanorods and nanorings of tetra(p-carboxyphenyl) porphyrin. J. Am. Chem. Soc., 2005, 127(16): 5885–5892.

[101] Huang C Z, Lu W, Li Y F. Total internal reflected resonance light scattering detection of DNA at water/tetrachloromethane interface with acridine orange and cetyltrimethylammonium bromide. Anal. Chim. Acta, 2003, 494(1–2): 11–19.

[102] Pang X B, Huang C Z. A selective and sensitive assay of berberine using total internal reflected resonance light scattering technique with fluorescein at the water/1,2-dichloroethane interface. J. Pharm. Biomed. Anal., 2004, 35(1): 185–191.

[103] Feng P, Shu W Q, Huang C Z, et al. Total internal reflected resonance light scattering determination of chlortetracycline in body fluid with the complex cation of chlortetracycline-europium-trioctyl phosphine oxide at the water/tetrachloromethane interface. Anal. Chem., 2001, 73(17): 4307–4312.

[104] Wang Y H, Guo H P, Tan K J, et al. Backscattering light detection of nucleic acids with tetraphenylporphyrin-Al(III)-nucleic acids at liquid/liquid interface. Anal. Chim. Acta, 2004, 521(1): 109–115.

Yun Fei Long, Jian Wang, Jian Ling

6 Light scattering nanospectral probes

The sizes of nanoparticles are in the range of Rayleigh scattering and Tyndall scattering, which are good light scattering spectral probes. They can be divided into two types: (1) metallic nanometer probe and (2) nonmetallic nanometer probe. There are many detection methods and analytical techniques that are established on the basis of their light scattering properties with good development trend. So, wider studies are performed in the preparation of unique optical metallic and nonmetallic probes, some innovative methods were also mentioned, including preparation of probes with a solid-phase method, liquid-phase method, and gas-phase method, with advantages of a smooth surface, controllable particle size, highly resistant to aggregation, and excellent thermal stability, further promoting the development of light scattering spectrometry.

6.1 Spectral probe of nonmetallic nanometer light scattering

Nonmetallic nano-light scattering spectral probes mainly refer to nano-sized particles formed by the nonmetallic elementary substance, or one or more metallic/nonmetallic element, or one nonmetallic element (C, O, N, and so on, usually it is O), which are mostly metallic oxides and nonmetallic oxides.

Nonmetallic nanometer particles have good light scattering features. As described in equation (2.20), the differences in the refractive index of nanoparticles and the surrounding medium is the fundamental reason for an enhanced light scattering signal, which can be further confirmed by the RLS spectrum of a liquid-phase inorganic nanoparticle system [1–4]. Some nonmetallic nanoparticles, such as selenium, mercury iodide, lead iodide, and iron oxide have characteristic resonance scattering peaks with a characteristic absorption in the visible region, which are incomparable for small organic molecules and complex probes. Especially for semi-conductor nanocrystals, as they have plasma resonance absorption and scattering characteristics in the visible and near infrared region, the observed nanoparticle color is related to resonance scattering; and they have found a wide application in the analytical chemistry of light scattering. So, light scattering is the fourth important factor for the color of a particle system after light transmission, light absorption, and fluorescence [5].

More studies focus on nonmetallic nanoparticles, including quantum dots, carbon cluster nanoparticles (including carbon nanoparticles and carbon dots

https://doi.org/10.1515/9783110573138-006

[CDs]), sulfur cluster nanoparticles (including sulfur nanoparticles, selenium nanoparticles, and sulfide nanoparticles), oxide nanoparticles, composite non-metallic nanoparticles, and so on. However, these semiconductor nanoparticles have neither been widely used in the area of the analysis of light scattering spectra nor they have shown their specific advantages. The hidden exciton excitation, emission, and scattering in these nanoparticles are worth for further study, especially exciton resonance light enhancement or possible RLS enhancement of exciton and electron, magnetic vibrators, phonons, and local domains of plasmon coupling, which is a promising field of study in analytical chemistry.

6.1.1 Quantum dots

Quantum dots are semiconductor nanoparticles composed of II–VI group or III–V group elements with sizes between 1 and 10 nm, which can be called semiconductor nanocrystals. As exciton of an electron and a hole passes through quantum confinement, the previously continuous energy band structure becomes a discrete-level structure with molecular properties that are able to emit fluorescence after excitation with a narrow, symmetric, and color-tunable emission spectrum. Quantum dots have excellent fluorescence characteristics, high photochemical stability, long fluorescent lifetime, which can act as an ideal fluorescence probe.

Professor He Z. K. found in a RLS spectral study of CdSe quantum dot and lysozyme interaction that CdSe quantum dot and lysozyme have weak light scattering signals in the range of 225–800 nm. Whereas when they coexist, the signal is apparently enhanced with a characteristic scattering at 294 nm. Through a circular dichromatic spectrum, transmission electron microscopy and a fluorescent lifetime test, it was found that their combination not only changed the lysozyme configuration, but also changed the CdSe quantum dot dispersive state and the fluorescent lifetime [6].

It needs to be demonstrated that photoluminescent CDs and quantum dots have photoluminescence ability, whereas in a light scattering spectroscopic analysis, photoluminescence signal is likely to interfere with the scattering signal. However, it is possible to use a light scattering signal and photoluminescence signal to obtain a multifunctional or a multimode probe. For example, through electrostatic bonding of quantum dots and aminoglycoside antibiotics, enhancement in the scattering at 376 nm is caused, whereas there is fluorescence quenching at 500 nm; based on this, the fluorescence/scattering ratio method (Figure 6.1) [7] is established, which is a selective and sensitive method.

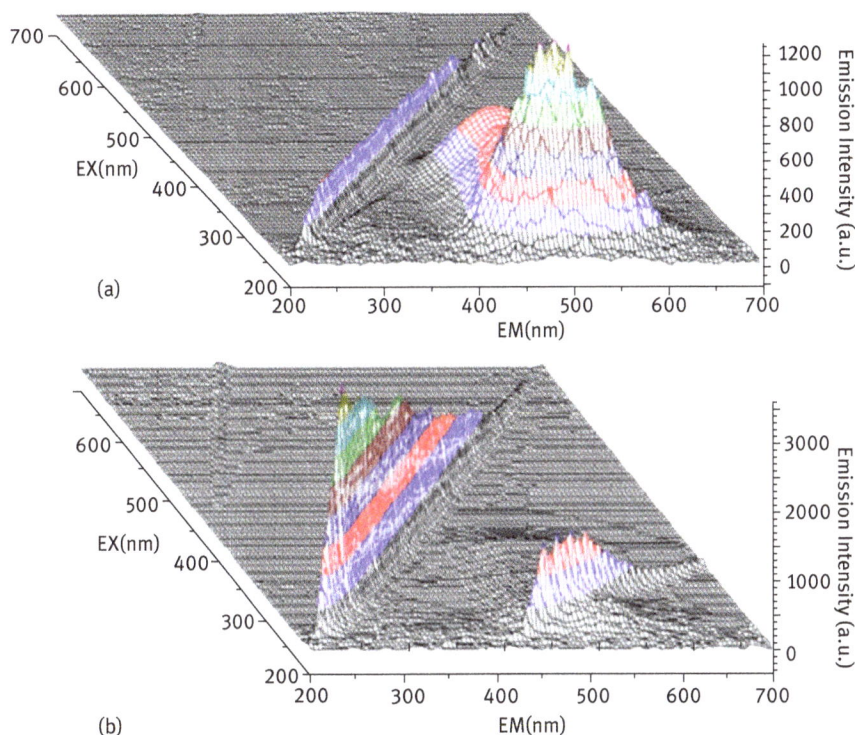

Figure 6.1: Three-dimensional spectrometry of aminoside antibiotic with quantum dots [7]. (a) Three-dimensional spectrometry of quantum dots. (b)Three-dimensional spectrometry of the mixture of quantum dots and aminoside antibiotic.

6.1.2 Carbon nanoparticles

6.1.2.1 Luminescent carbon dots

In 2006, the Sun Research Group successfully synthesized Luminescent carbon nanoparticles with a polymer shell, and named it as CDs [8]. Usually, these carbon nanoparticles with photoluminescence have chemical inertness, better optical properties and quantum size effect, low toxicity, good biocompatibility, and anti-bleaching properties [8, 9]. These properties have good application prospects, which have been already applied in biosensing and bioimaging analysis [9, 10]. The yield of some CDs can reach over 90% with good biocompatibility, and the color can cover a broad spectrum (Figure 6.2) [11, 12], adjusting a surface oxidation degree to obtain dots of different colors [13], so its application in the biological applications of fluorescence property can be compared with that of quantum dots.

CDs' absorption mainly distribute in the UV to visible region, most of which have no characteristic peaks, whereas RLS properties with their aggregates have successfully

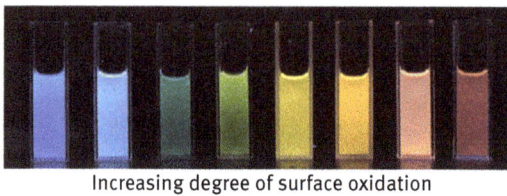

Increasing degree of surface oxidation

Figure 6.2: Fluorescence changes in carbon dots with surface oxidation degree [13]. (refer to the colored postscript picture).

realized accurate, rapid, and sensitive measurement of lysozyme [14], as well as a good probe for DNA specific structure [15].

It needs special attention that as CDs have very good fluorescent luminescence performance, and Stokes shift is not much great, and there is a possibility for light scattering signal interfered by a fluorescent signal.

6.1.2.2 Nonluminous carbon nanoparticles

Although some carbon nanoparticles, such as carbon nanodiamond, fullerene, carbon nanotube, graphene, and spherical carbon nanoparticles do not have photoluminescence with specific structure and peculiar physical and chemical properties, such as light absorption and electrical and mechanical properties, they have become hot topics in the domestic and foreign school physics, chemistry, materials, and other fields with promising application. Their special light scattering properties have caused a wide attention in the scientific society. These features might be related to carbon semiconductor properties. The exciton-coupled light scattering enhancement is worth of further discussion.

Studies have found that by increasing its surface electronegativity and repulsive force, single-stranded DNA twines on the surface of carbon nanotubes (CNTs). There is no aggregate at certain concentration of sodium chloride (NaCl); after centrifugation at a certain rotary speed, CNTs' concentration in the upper liquid is high with strong scattering. While with the target DNA, double helical structure of phosphate has electronegativity and CNTs' surface electrostatic repulsion force makes the DNA unable to adsorb on the surface of CNTs. With a charge shielding effect, NaCl at the same concentration brings aggregation of CNTs; the upper liquid has weak scattering intensity after centrifugation. With the difference in the scattering intensity, one simple, effective, and label-free DNA analysis method can be established (Figure 6.3). This method can selectively distinguish a single-base mismatch and double-base mismatch, with an important application in PCR product detection and single nucleotide polymorphisms of a diseased gene [16].

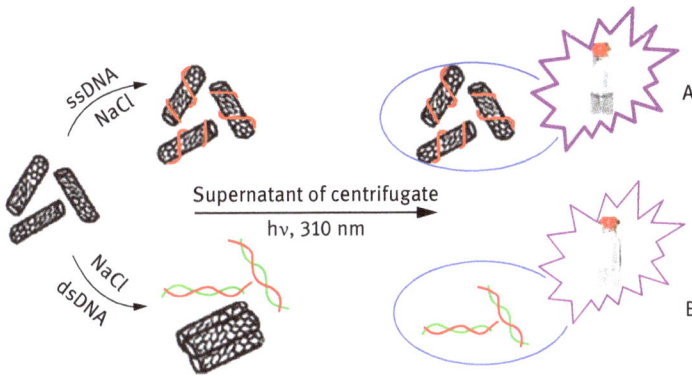

Figure 6.3: Light scattering analysis of DNA based on the adsorption of CNTs [16].

Based on equation (4.10), the RLS intensity function can be written as follows:

$$I_{RLS} = I_0 k c_p \tag{6.1}$$

It can be known that under certain conditions and nanoparticle concentration(c_p), as k can be seen as an experimental constant, RLS intensity at one wavelength (I_{RLS}) is in direct proportion to the equivalent wavelength of intensity of the incident light I_0 (known as emission intensity of light source). That is, the higher the emission intensity I_0, the greater is the resonance scattering light intensity of I_{RLS}.

In an emission spectrum of fluorospectrophotometer with a xenon lamp and semiconductor laser, xenon lamp has the strongest intensity at 470 nm; even without molecular absorption, carbon nanoparticles have stronger light scattering signal at 470 nm. If a semiconductor laser with an emission wavelength at 650 nm is used as a light source, then carbon nanoparticles have the strongest intensity at 650 nm. Usually, a xenon lamp and a semiconductor laser can produce light of other wavelengths, so carbon nanoparticles have stronger characteristic scattering peaks at 400, 470, 510, and 650 nm. Recently, scattering signals of carbon nanoparticles have been used to establish a series of analytical methods.

For example, Jiang Z. L. used an inker bought from the market made in the city of Huizhou to obtain 24 g/L carbon nanoparticles through several grindings on an inkstone [17]. The carbon nanoparticle suspension has 5 RLS peaks at 400 nm, 470 nm, 510 nm, 800 nm, and 940 nm. Liquid-phase nanoparticles with carbon mass concentration lower than 0.6 g/L have similar RLS outline, whereas there is no RLS with concentration over 800 g/L. 800 nm and 940 nm peaks are half-frequency light scattering signal of 400 nm and 470 nm peaks [18]. Arabic gum and gelatin are good stabilizers and dispersants of carbon nanoparticles, their aqueous solution has a RLS peak at 470 nm, which is rather weak. With a mass concentration of 3 g/L, they do not influence the measurement of RLS spectrum of 0.2 g/L carbon nanoparticles. When mass concentration is in the range of 0.45–4.5 g/L, $I_{400\ nm}$ and $I_{470\ nm}$ have a linear relation with the concentration. Therefore, a new method can be built to evaluate carbon black ink quality.

Similar to regular carbon nanoparticles, CNTs also have absorption that reduces with the increase in wavelength from the UV-visible region to the near infrared region without characteristic absorption peak, so there is no characteristic RLS spectrum. With different emission characteristics of light source, a characteristic scattering signal that changes with light source can also be obtained, which can be applied for the detection of some targets.

6.1.3 Sulfur cluster nanoparticles

6.1.3.1 Sulfur nanoparticles

Sulfur nanoparticles have features of small-sized particles, that is, good uniformity and dispersion, large specific surface area, and so on, which are widely used in fields of chemical engineering, rubber, agriculture, medicine, dye, chemicals, chemical fiber, paper-making, and so on. So, preparation of sulfur nanoparticle and its application in analytical chemistry is one of the major fields of research. For example, Guo Y.M. et al. [19] used sulfur powder as a material and formic acid as a precipitant, polyethylene glycol-400 as a dispersant, they synthesized 50–80 nm spherical monoclonal sulfide nanoparticles at normal temperature, normal pressure, and liquid-phase conditions.

For example, with a xenon lamp, sulfur nanoparticles have a strong Rayleigh scattering light, which increases with the concentration of the sulfur mass [20], affected by solvents, such as ethanol, acetone, and organic dyes, (such as bromophenol blue, bromomethyl violet, crystal violet, and bright green) especially, light scattering intensity reduces with the increase in nonradiative absorption of organic dyes.

6.1.3.2 Sulfide nanoparticles

Metallic sulfide nanoparticles have quantum-size effect, small-size effect, surface effect, and macroscopic quantum tunnel effect, as well as good light, electricity, magnetic, and lubrication and catalysis properties with special composition and structure. These properties make sulfide nanoparticles to have good applications.

For example, in-situ polymerization of acrylic acid onto nanoPbS initiated by ultrasonic radiation and potassium persulfate ($K_2S_2O_8$) resulted in a large number of carboxyl groups covering the surface of nano-particles, which not only increased the water-solubility and stability of the nanoparticles, but also endowed them good biocompatibility. The result was that when protein interacted with the nanoparticles, it showed very good resonance light scattering enhancement [21].

In an alkaline medium, a RLS spectrum of Cu–sodium sulfide-hexadecyl trimethyl ammonium bromide system is used to realize Cu^{2+} measurement in 0.25–3.50 g/mL [22]; in a micro ammonia medium, $Ag(NH_3)_2$ and excessive Na_2S is used to form Ag_2S colloidal particles [23]. As Ag^+ concentration is low, the produced Ag_2S colloidal particle is difficult for coagulation and sedimentation, then Ag_2S colloidal particle in a solution further adsorbs ions to form Ag_2S/S^{2-} inorganic nanoparticles modified by a homologous ion functionalization. When it encounters a cationic surfactant cetyltrimethylammonium bromide (CTAB) with positive charges, an ion associate is formed, enhancing the RLS intensity. Similarly, at room temperature, CuS nanoparticles (CuS–NP) [24] formed by Cu^{2+} and S^{2-} in water with sodium citrate as a stabilizer can be used for detecting trace amounts of mercury in a light scattering test in water. Inorganic nano sol CuS/Cu^{2+}, prepared with ultrasound, can bind with γ-globulin for sensibilization of RLS [23]. Preparation of stable inorganic nano sol ZnS/Zn^{2+} with ultrasound can bind with γ-globulin for the enhancement of RLS [25].

At pH 5.12, NaAc–HAc medium of L-cysteine covered ZnS nanoparticles have RLS peaks at 308 nm [26], which is enhanced by protein interaction. While in Tris-HCl buffer, the water-phase synthetic L-cysteine is covered with ZnS nanoparticles to form large nanoparticles with intermolecular forces, with two RLS peaks at 304.5 nm and 373.8 nm [27] (Figure 6.4).

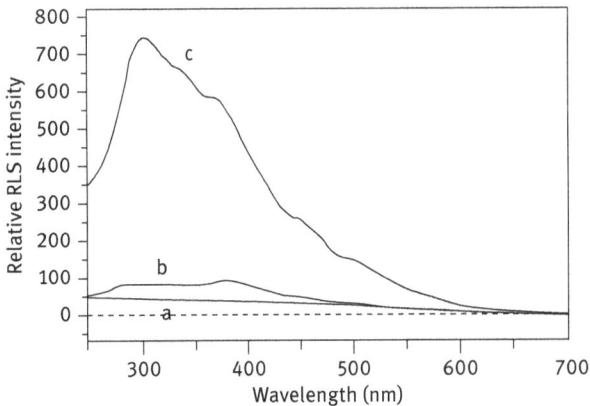

Figure 6.4: ZnS nanoparticles and nucleic acid interaction induced scattering spectra [27]. (a) ctDNA; (b) ZnS; and (c) ZnS+ctDNA.

Although the following studies have shown that sulfide nanoparticles have exhibited excellent light scattering features with their applications in organic, inorganic, and biochemical analysis, these features are not specific, many of which did not show resonance features. The relationships between absorption and scattering as well as between scattering spectrum and particles are not clear; therefore, further study is needed.

6.1.3.3 Se nanoparticles

Se is one of the important essential trace minerals in life, participating in many physiological activities, as active center of each selenium-containing protein or sigma subunit, plays a vital role in the healthy growth of humans, animals, and plants. Compared with other forms of Se, nanoSe is a red elementary substance with biological activity and with advantages of high biological activity, high safety, high immunoregulation capability, and high inoxidizability and low toxicity.

Jiang Z. L. reseach group found that reduced Se(IV) with vitamin C in a hydrochloric acidic medium can produce steady Se nanoparticles [28]. Although the prepared Se nanoparticle has no obvious characteristic absorption peak in the visible light range, it has 4 RLS peaks at 340, 400, 470, and 520 nm, which are influenced by a light source. They prepared AuSe with $NaBH_4$ reduction method, and used a single-stranded nucleic acid aptamer to modify AuSe nanoalloy for Apt–AuSe. At pH6.8, where buffer and NaCl exist, Apt–AuSe is dispersed with a low scattering intensity. In the presence of Hg^{2+}, it can form steady $T–Hg^{2+}–T$ double-stranded mismatch compound with T–T mismatch in a nucleic acid aptamer sequence, releasing and aggregating to form larger particles with AuSe nanoalloy, producing 590 nm localized surface plasmon of the enhanced RLS peak [29].

6.1.4 Oxide nanoparticles

There were many studies on oxide nanoparticles, such as TiO_2, SiO_2, ZnO, Al_2O_3, ZrO_2, CeO_2, and Fe_2O_3, with their applications in solar cells and bioelectric sensors. In a liquid-phase medium, $CoFe_2O_4$ nanoparticles have 5 RLS peaks at 400, 470, 510, 800, and 940 nm [30], where 800 nm and 940 nm peaks are half frequency (or secondary) of RLS peaks of 400 nm and 470 nm, which are similar to the outline of $CoFe_2O_4$ system with a concentration lower than 8.4×10^{-6} mol/L. For a system with higher concentration, the spectrum is related to its concentration.

To analyze a liquid-phase $CoFe_2O_4$ nanoparticle of a RLS spectrum, Jiang Z. L. et al. defined the white particle, quasi-white particle, black particle, and gray particle in scattering process, and also studied the RLS spectra of them. They believed that white particle refers to the liquid phase that can absorb all optical radiations and radiate out rapidly (or with scattering efficiency at different wavelengths), producing secondary radiations called the white particle RLS; the absorption is resonance absorption. RLS spectrum of many diluted nanoparticle solution is similar to that of white particles, and can be thought as a quasi-white particle scattering. Black particle refers to the liquid-phase particle that can absorb all optical radiations and radiate nothing. In fact, a real black particle does not exists, almost an ideal black particle can be obtained through reducing the light resource intensity and improving the particle concentration (refer to Section 2.3.2). That is, black particle produces no resonance light scattering, whereas a gray particle is a liquid particle between the following two particles, its scattering is called resonance light scattering of a gray particle [30]. Gray particle scattering efficiency changes with different wavelengths.

$$f(\lambda) = \frac{E(\lambda)}{I(\lambda)} \tag{6.2}$$

where $E(\lambda)$ is the scattering light intensity at λ, $I(\lambda)$ is the exciting light intensity. Resonance scattering spectrum is decided by the light source and the detector and system absorption.

In essence, involved systems in a RLS analysis are mostly gray particle system, so the method has selectivity of (molecule) electronic absorption spectrum and sensitivity of molecular emission spectra (such as fluorescence), which can improve the sensitivity through excitation of light intensity.[30]

Based on this, it was found that the solution-phase synthesis of manganese dioxide nanoparticles at 12.5 nm, by $KMnO_4$ and $NH_3 \cdot H_2O$ in a microwave and high pressure, has a resonance nonlinear scattering phenomenon. When excitation wavelength is 210 nm (maximum absorption), there is no scattering in the region of 200–1,000 nm, because stronger molecular absorption at 210 nm has transferred into energy of other forms. When excitation wavelength is 340 nm, where an exciting MnO_2 nanoparticle is characteristic of absorption, there is resonance Rayleigh scattering peaks in the region of 200–1,000 nm, (λ_{RRS}=340 nm) and half frequency scattering peak ($\lambda_{1/2r}$=2λ_{RRS}=680 nm). When excitation wavelength is 470 nm (λ_{maxRRS}= 470 nm), resonance scattering peaks and half frequency scattering peaks can also be observed, which are larger than that at 340 nm excitation [31].

For excitation wavelength of 680 nm (2×340 nm) and 940 nm (2×470 nm), resonance scattering peaks and double-frequency scattering peaks can also be observed at 200–1,000 nm. A double-frequency scattering peak at 470 nm is larger than 340 nm because 470 nm is at the strongest emission position [30], so at the resonance Rayleigh scattering peak (470 nm) and double-frequency scattering peak, frequency division peaks are stronger than others. Liquid-phase MnO_2 nanoparticle in a nonlinear scattering phenomenon can also be explained on the basis of Fourier transform theory of light and nanoparticle interaction [32].

Li Y. T. et al. found that a combination of 6 nm titanium oxide nanocrystalline and L-tyrosine has susceptibility for RLS. Based on this, the RLS method for a solution of L-tyrosine with TiO_2 nanocrystalline sol as a probe is established [33].

6.1.5 Halide nanoparticles

In certain concentrations, with Hg^{2+} and X^- rate of 1:2, $[HgClX_2]_n$, $[HgBr_2]_n$, and $[HgI_2]_n$ particles with average size less than 4, 9, and 70 nm can be formed [34]. As $[HgClX_2]_n$ and $[HgBr_2]_n$ have no absorbance in the visible light region, their scattering light is away from the absorption band; there is only a real part of scattering signal, which is irrelevant to absorption, and there is no light scattering enhancement signal of an imaginary part; that is, intensity mainly depends on the particle size and its concentration. Its scattering peak can depend on equipment factors to produce a 470 nm scattering peak. $[HgI_2]_n$ has a certain absorption band in the visible light region, and its scattering peak is in the absorption band, which can produce a stronger resonance Rayleigh scattering; thus, dramatically enhancing the scattering intensity by nearly 60 and 24 times as compared to $[HgClX_2]_n$ and $[HgBr_2]_n$ signals at the same concentration.

Light scattering signal of $AgNO_3$ solution is weak, whereas RLS signal greatly enhances with the interaction with KX (X=Cl$^-$, Br$^-$, I$^-$) [35], where AgCl and AgBr nanoparticles have 4 resonance scattering peaks at 330, 400, 470, and 680 nm. The strongest peak is at 330 nm; the 680 nm peak can be observed as a half-frequency peak of 330 nm peak. While AgI nanoparticle has 5 resonance scattering peaks at 340, 400, 437, 470, and 680 nm, the strongest peak is at 437 nm, which corresponds to the resonance absorption peak of AgI at 425 nm. The 680 nm peak can be seen as a half-frequency peak of the 340 nm peak. The 470 nm peak is related to a light source and a detector. The occurrence of these peaks is related to the silver halides' nanoparticle refractive index differences with that of the surrounding medium.

6.1.6 Other nonmetallic nanoparticles

Under an ultrasound irradiation, hydrogen peroxide has an oxidative degradation on polyvinyl alcohol (PVA) with Zn^{2+} to form nanosize diketone-type polymer microsphere of Zn–PVK [36]. The RLS study of Zn–PVK properties has found that its RLS signal increases dramatically in the formation of its associates.

Ultrasonic synthesis of nanoparticles is a good method. Under ultrasound irradiation, use V_{water}: $V_{acetone}$=1:1 as solvent, potassium peroxodisulfate as an initator, and perform a soap-free emulsion polymerization with styrene and acrylic acid that can form a polystyrene–acrylic acid nano-sol with an average parameter of approximately 9 nm [37]. The infrared spectral representation shows that the synthesized nano-sol surface is rich in free carboxyl groups that can be used as active groups to capture biomacromolecules as proteins.

6.2 Complex nonmetallic nanoparticles

6.2.1 Carbon nano/complex nanoparticles

Complex nanoparticles have advantages of different materials and have various applications. For example, Long Y. F. et al. [38] have found that CNTs disposed with a mixed acid oxidation can form steady nano-complex with a single-stranded DNA (Figure 6.5). The complex is well-dispersive and steady in an aqueous solution with stronger RLS property. Due to the formation of specific T-Hg^{2+}–T structure between thymine and mercury ions, different concentrations of mercury ions were added to CNTs/PolyT, and the specific T-Hg^{2+}–T double-stranded structure was formed by trapping mercury ions in the polyT, which made it dissociate from CNTs. The RLS intensity (I_{RLS}) of CNTs decreases due to the aggregation of CNTs in the presence of electrolytes and their easy centrifugation and subsidence [38].

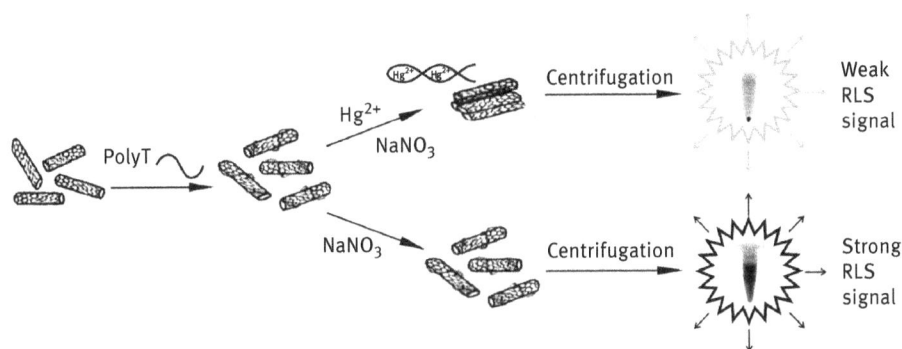

Figure 6.5: An interaction diagram of PolyT, CNTs, CNTs/PolyT, and Hg^{2+} [38].

6.2.2 Carbon nano/metal complex nanoparticles

Ye L. Y. et al. have used CNTs and $AgNO_3$ as materials, $NaBH_4$ as a reductant to obtain CNTs/Ag nano-complex in water [39], and studied its ultraviolet-visible absorption spectrum, RLS spectrum, and its morphological features. Results show that a complex has a stronger UV absorption peak and a single RLS spectrum peak as compared to Ag sol without CNTs'. After a complete reaction of CNTs/Ag with different concentrations of standard hydrogen peroxide solution, it was found that with the increase in concentration of hydrogen peroxide, CNTs/Ag complex system has a regular decrease in UV absorption and RLS intensity.

Detection of light scattering for DNA hybridization assay based on separation properties of magnetic bead and strong signal of CNTs is a simple, selective, sensitive, and repetitive [40] probe construction path as shown in Figure 6.6. A sequenced DNA probe P1 is fixed on a bead surface, and another sequenced DNA probe P2 is modified on a multiwall carbon nanopipe surface. P1-bead and P2-nanotube modified nano-materials can exist steadily for 6 months at 4°C, showing good performance after 17 repeats uses. The hybrid-modified nano-material with a target DNA T1 to measure the upper layer light intensity after a magnetic separation is found to have a detectable target DNA as low as 1 nmol/L with good specificity. In a real sample analysis, three target DNA in PCR amplification were detected.

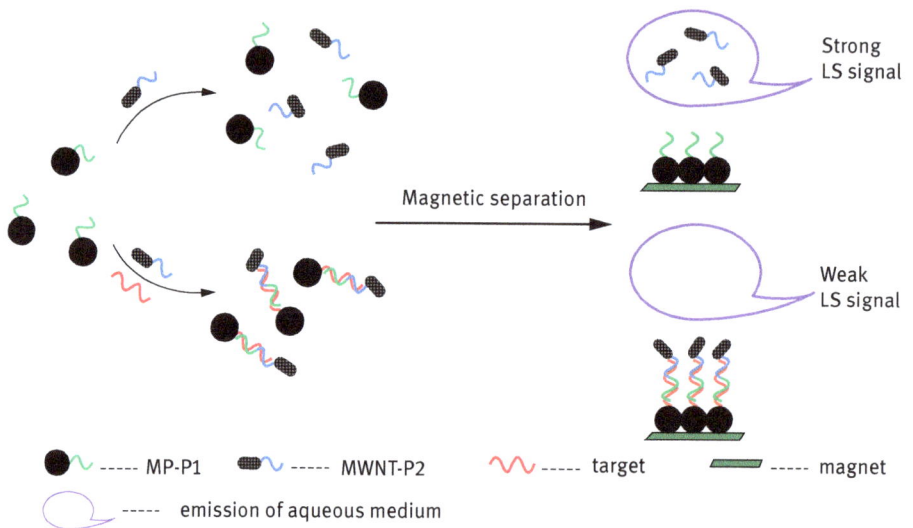

Figure 6.6: DNA analysis with carbon nanotube–magnetic nanoparticles [40].

6.2.3 TiO$_2$/SiO$_2$ nano-complex

Dong C. Z. et al. used mixed, stirred, and aged SiO$_2$ and TiO$_2$ sol with Si:Ti=1:1 mole ratio to obtain a homogeneous, and transparent TiO$_2$–SiO$_2$ nano-sol [41]. It was found that TiO$_2$–SiO$_2$ nano-sol has a secondary scattering intensity lower than that of TiO$_2$ nano-sol with no change in its wavelength. Addition of small amount of RNA can obviously enhance its scattering light intensity.

6.3 Metallic nanoparticle probe

6.3.1 Plasma resonance of noble metal nanoparticles

6.3.1.1 Surface plasma resonance

Surface plasma (SP) refers to an electron dilatational wave that spreads along a metal surface, produced by the interaction of a photon and a free electron that exists on the metal surface in light. If light falls from an optically denser medium to an optically thinner medium with a direction greater than a critical angle, there would be a total internal reflection with a vertical angle to interface on an optically thinner medium. With increase in the depth, there would be an evanescent wave that reduces with exponential form.

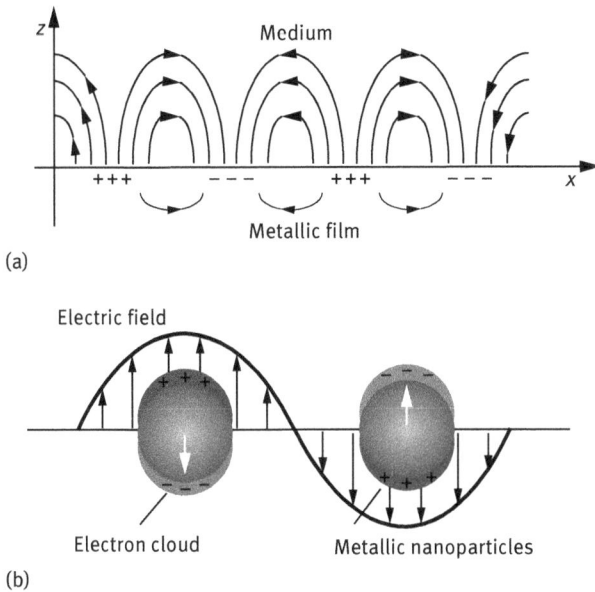

(a)

(b)

Figure 6.7: A surface plasmon resonance. (a) Surface plasmon resonance on a metallic thin film and (b) localized surface plasmon resonance on a metal particle.

If there are dozens of nanometers of thin metallic membrane in two medium interface (Figure 6.7 [a]), then the total-reflection-produced evanescent wave polarized component would enter the membrane and interact with the free electron, which is an exciting surface plasmon resonance (SPR) wave that spreads along the metal surface. Under a specific incident angle, the evanescent wave component that parallels to the metal (electric medium) is likely to have an identical wave vector (or frequency) of surface plasmons, then the two modes will couple strongly; that is, there is a plasma resonance at the interface of the metal membrane and the substance, leading to part absorbance of the incident light energy by the wave, in which energy is transferred and the total reflection light intensity is reduced. This is called SPR [42].

Noble metal nanoparticles produce special light scattering properties from local SP resonance (LSPR) effect in a light interaction [43, 44]. Flowing electrons of metal nanoparticles have a collective oscillation with electromagnetic waves and form a SP with the interaction of an incident light, there would be resonance if the oscillation frequency of these electrons are the same as that of electromagnetic waves [45]. As metal nanoparticle size is far smaller than incident light wavelength($d \ll \lambda_0$),its SP resonance is localized on the nanoparticle surface, bringing out the special optical properties of a nanoparticle. To distinguish from SP resonance on metal membrane surface, this is named as localized surface plasmon resonance (LSPR) [46]. (As shown in Figure 6.7 [b]).

6.3.1.2 Surface plasma resonance scattering

Metallic nanoparticles have different materials, shapes, and sizes, whose LSPR frequency changes with these. Figure 6.8 shows the plasmon resonance properties of a conductor, semiconductor, and so on [47]. This brings a change in the SPR absorption and light scattering wavelength, which is further shown as a colorful scattered light similar to a fluorescence. For example, a 40 nm silver particle is blue, whereas a 40 nm gold particle is green. It also shows other colors with an increase in the particle size. 78, 118, and 140 nm gold nanoparticles are yellow, orange, and red, respectively (Figure 6.9) [43].

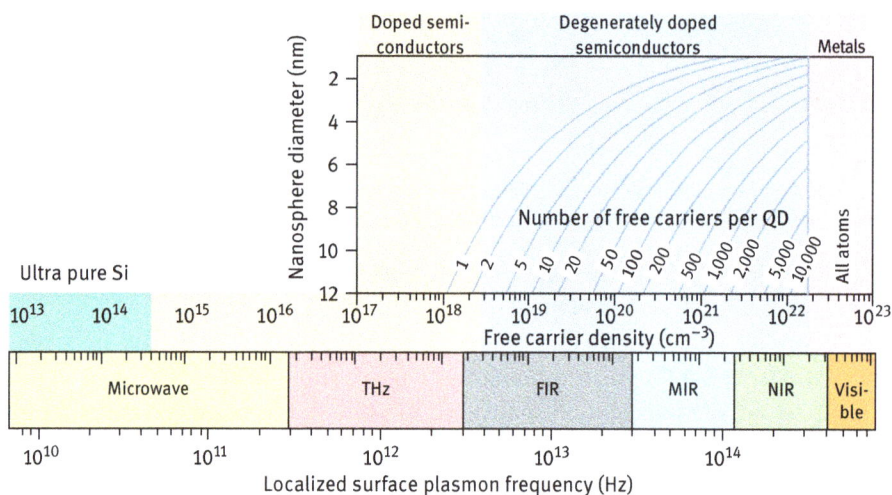

Figure 6.8: Plasma resonance of different substances.

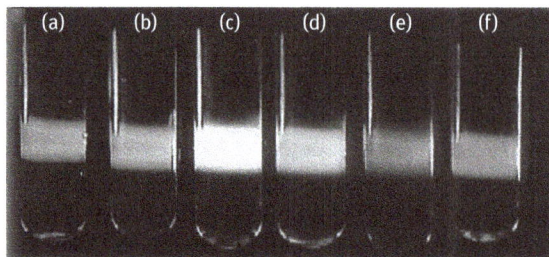

Figure 6.9: Scattered light of nanoparticles with white light irradiation [43]. (See colorful photos at the back of the book) Left to right: (a) 40 nm silver nanoparticles; (b–e) 40 nm, 78 nm, 118 nm, 140 nm gold nanoparticles; and (f) fluorescein.

Based on metallic nanoparticles, scattered light has the following features:

(1) Scattered light wavelength range can be adjusted through choice of materials, sizes, and shapes, which is able to adjust from visible light region to the near infrared or even infrared region with current technology. Scattered light of noble metals, such as Au and Ag, and transition metal, such as Cu are sensitive to the changes of the surrounding medium, and lot of plasmon resonance absorption and highly sensitive analytical scattering methods have been established using these changes.

(2) Metallic nanoparticles have good stability. That is, its scattered light does not have photobleaching in the long-time illumination, so it has a steady light scattering and anti-photobleaching properties.

(3) The light scattering of metal particles has high quantum yield, high luminous efficiency and visible light scattering from a single particle.

(4) Most current metal nanoparticles are noble metal particles of Au, Ag, and so on, with good biocompatibility.

From twentieth century to the present, metal nanoparticles have always been popular in a nanoparticle study. With the development of nano-science and technology, there have been nano-materials of different shapes (including spheres, rod, flower shape, flake, and cage), adding new elements for its development. Their different structures and materials decide their property differences and special plasma resonance absorption and scattering properties. So, we begin with the synthesis of these nanoparticle probes, and then introduce their absorption and scattering properties and basic principles for a light scattering analysis.

6.3.2 Gold nanoparticles

6.3.2.1 Gold nanospheres

With features of a convenient preparation, steady property, good biocompatibility, easy modification, and so on, gold nanospheres (AuNSs) are one of the most-studied metal nanoparticles [47, 48]. It is of worth illustration that AuNSs are symmetric, one kind of a non-anisotropic nanomaterial. The absorption spectrum only has one peak near 520 nm, and it gradually red shifts toward a longer wavelength with increase in AuNSs size. After AuNSs aggregate with polarization and coupling of adjacent nanoparticles, their absorption spectrum gradually red shifts toward a longer wavelength or there is a new absorption peak at a longer wavelength. At present, the spectrophotometric analysis method based on the change of surface plasmon absorption properties has been well developed, and the application of plasma resonance scattering method has been gradually expanded.

(1) AuNSs synthesis: With the development of nanotechnology, there are diversified nanoparticle preparation methods. In terms of methods on AuNSs synthesis,

there are wet chemical method, photochemical method, electrochemical method, microwave method, ultrasonic method, and so on. In general, extra surfactants are not added in the preparation process, because some compounds such as chitosan, cyclodextrin, ionic liquids, citrate, and so on, are used as reductants and protective agents, so that nanomaterials present good stability.

Wet chemical synthesis uses reductant to reduce $HAuCl_4$ to obtain AuNSs in mild conditions. Sodium citrate reduction is the mostly used method [49]. Citrate acid group serves as a reductant, protective, and simplifying steps with controlled size. However, the method has the disadvantage of weak salt tolerance for AuNSs, which is easy to aggregate.

As $NaBH_4$ has a stronger reducibility, it prepares small-size AuNSs (<10 nm) [50], which are able to produce photoluminescence. The produced AuNSs have a poor stability. The freshly prepared nanoparticles are light brown with barely any clear peak in ultraviolet–visible spectrophotometer. After 20 days, the solution becomes red with an obvious absorption peak at 520 nm, suggesting an increase in the particle size.

Reducibility of ascorbic acid is weaker than $NaBH_4$, the reaction is more controllable with more homogeneous AuNSs and it is stable for more than half a year [51].

Photochemical synthesis prepares AuNSs with uniform particle size and better stability [52]. Factors such as as acidity, $HAuCl_4$ concentration, and radiation time can influence their size, quantity, and shape. The disadvantage of this method is that the reaction time is longer.

Electrochemical synthesis directly uses electron as a reductant to reduce $HAuCl_4$ [53], which controls the nanoparticle size through the use of current and voltage with faster response speed. Usually, a surfactant is used to keep monodispersity of AuNSs.

Microwave synthesis method has the advantage of being fast [54], whereas nanoparticles are not uniform enough, which present spherical nanoparticles and triangular nanosheets and nanorods. In addition, the power of the microwave apparatus influences the radiation time. Low power and longer time are helpful for the complete reduction of $HAuCl_4$.

An ultrasound method can be used to prepare AuNSs with uniform size [55]. However, synthesized AuNSs have bad stability and they are easy to form sediment.

There are many synthesis methods for AuNSs. Actually, AuNSs have become commercial, which they are able to satisfy application requirements for a laboratory. Currently, in AuNSs performance modification field, researchers pay more attention to the development of its good performance, adapting AuNSs for requirements of biochemical sensing, and so on, and in serving the society and production.

(2) Plasma resonance absorption of AuNSs. As shown in Figure 6.10 (a), AuNSs have only one absorption peak approximately near 520 nm. With increase in the particle size, absorption wavelength has a red shift. Plasma resonance absorption peak of 25 nm of AuNSs is at 520 nm, whereas, when AuNSs aggregate with the polarization and coupling effect of the adjacent nanoparticle electrons, their

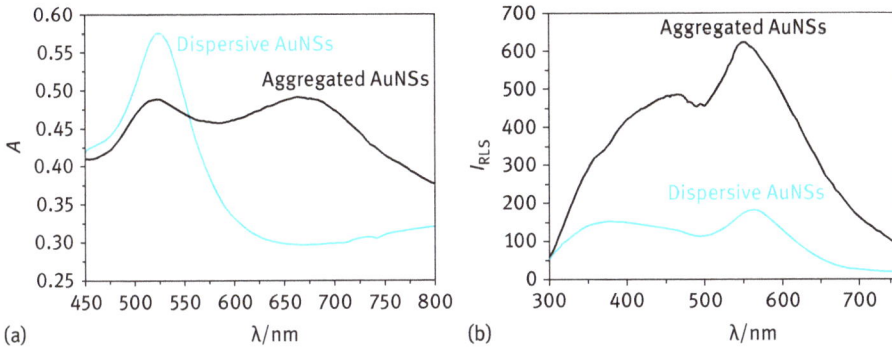

Figure 6.10: Plasma resonance spectra of AuNSs. (a) Plasma resonance absorption spectra. (b) Resonance light scattering spectra.

absorption spectrum has a red shift toward a longer wavelength, or there is a new absorption peak at the longer wavelength [56].

(3) Plasma resonance light scattering of AuNSs: Corresponding to the plasma resonance absorption spectrum, AuNSs display scattering peaks around 550 nm (Figure 6.10[b]). On one hand, the scattering signal intensity is enhanced with an increase in the nanoparticle size (or aggregate) [56, 57]; on the other hand, scattering signal is enhanced with the increased particle number [58]. While resonance light scattering usually shows an enhanced intensity and no obvious shift in the wave crest position. AuNSs play a more important role as a probe to achieve better analysis, preparation, and surface modification, which will gain popularity in future research in the field of biochemistry.

6.3.2.2 Gold nanorods

Gold nanorod (AuNRs) is a one-dimensional gold nanostructure. Different from AuNSs, AuNRs have two adjustable plasma resonance absorption peaks in the transverse direction and the longitudinal direction, so AuNRs solution color is highly dependent on its particle aspect ratio, so that red, blue, and purple solution can be obtained through adjusting aspect ratio [59–61]. Plasma resonance absorption peak of AuNRs in the longitudinal direction can be adjusted to the infrared region, whereas the light energy absorption can transfer into heat energy, which is widely applied in photothermal therapy [62–64]. In addition, as AuNRs' shape owns anisotropy, it is able to realize the diversified assembly method [65–70].

(1) Gold nanorod synthesis: Synthetic method of AuNRs is an advanced method, including the chemical reduction method [71], photochemical method [72], electrochemical process [73], high temperature synthetic method [74], mesoporous templating method [75], and so on. These methods can be made to prepare ideal gold nanorods

effectively and controllably. For example, chemical synthesis adopts a good seed as a crystal nucleus to obtain gold nanorods with controllable aspect ratio with gradual growth and good reproducibility [71]; photochemical synthesis produces gold nanorods with good uniformity, whereas long-time illuminance is needed [72]; electrochemical process produces gold nanorods with good uniformity, controllable aspect ratio, and complex reaction mechanisms [73]. However, most of these synthetic methods need CTAB as a soft module [76]. As CTAB has certain biotoxicity [77], surface functionalization is needed in the later application [78], bringing out challenges for biological applications [77].

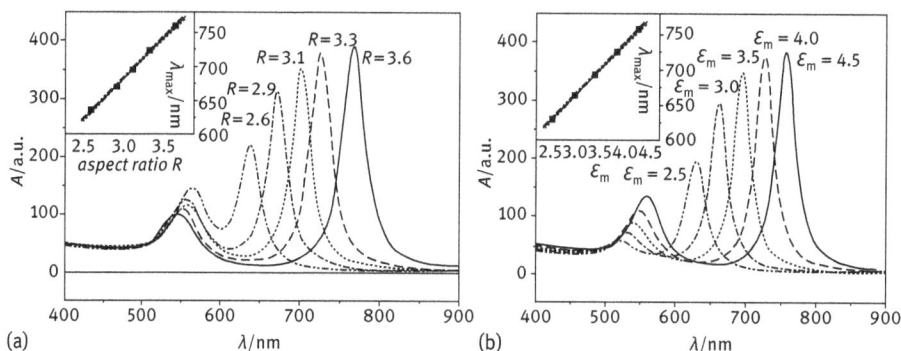

Figure 6.11: Changes in the calibrated absorption of AuNRs with aspect ratio and dielectric constant [79].

(2) Plasma resonance absorption of gold nanorods: As shown in Figure 6.11, AuNRs have 2 absorption peaks at 520 nm and at longer wavelength, corresponding to peaks of transverse direction and longitudinal direction. With increase in the aspect ratio, transverse direction absorption wavelength nonlinearly moves toward the direction of the short wavelength, whereas the longitudinal direction absorption wavelength moves toward the direction of the longer wavelength. When the ratio is 1, the transverse direction and the longitudinal direction absorption wavelength is overlapped around 530 nm. El-Sayed and his group [79] deduced the relationship of aspect ratio with absorption wavelength on an experimental basis, as shown in equation (6.3)

$$\lambda_{max} = (53.71R - 42.29)\varepsilon_m + 495.14 \qquad (6.3)$$

where λ_{max} is the longitudinal mode absorption wavelength, ε_m is the dielectric constant, and R is the aspect ratio of nanorods. While Pileni group [80] gave a similar equation based on an experiment on DDA simulation

$$\lambda_{max} = (52.95AR - 41.68)\varepsilon_0 + 466.38 \qquad (6.3a)$$

where λ_{max} is the longitudinal mode absorption wavelength, ε_0 is the dielectric constant, and AR is the aspect ratio of nanorods.

Equation (6.3) shows that AuNRs' longitudinal wavelength moves linearly to infrared; with an increase in the aspect ratio, the shift degree is closely related to the environment dielectric constant (Figure 6.11). The dependency lays foundation for a new application on the expansion of adjusting AuNRs' absorption position. For example, when assembled with AuNRs, their shape does not change, whereas the assembled wavelength shifts with the surrounding environmental changes, including medium dielectric constant and electron coupling [81].

As AuNRs' longitudinal wavelength is adjustable with a large section area, they show specific optical properties and the application is related to their absorption properties. For example, different assemblies of AuNRs can be judged through the shift of absorption spectrum [65–70]. As nanorods present 2 absorption peaks, which can quench different fluorescence molecules at the same time [61], the absorbed infrared light can be transferred into heat energy for photothermal therapy [62–64].

(3) The local plasma resonance light scattering of gold nanorods has features of good optical absorption. AuNRs' local plasma resonance light scattering properties have been a popular research topic in recent years [75, 76].

Combined with the theoretical simulation, Zhu et al. [82] have found that similar to plasma absorption, the resonance light scattering with 2 characteristic peaks correspond to the transverse and longitudinal resonance led by AuNRs anisotropy. With increase in the aspect ratio, the transverse resonance peak has a blue shift and reduction in the scattering intensity, whereas a longitudinal resonance peak has a red shift with an increasing scattering value. When the ratio is 1, the transverse resonance peak and the longitudinal resonance peak have overlap at 450 nm (Figure 6.12).

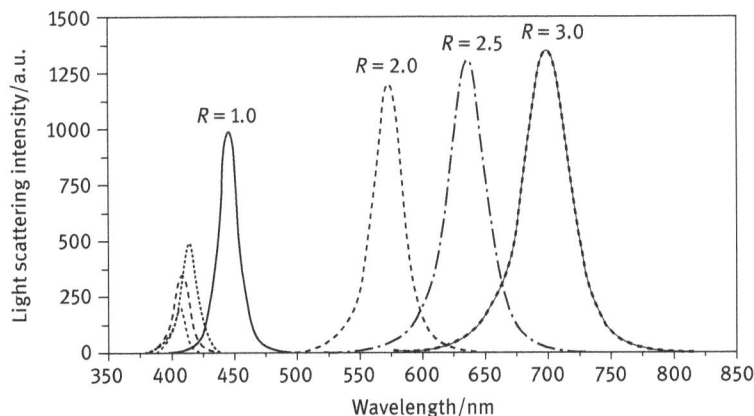

Figure 6.12: Resonance light scattering spectra of gold nanorods [82].

As described earlier, because AuNRs assembly has special longitudinal plasma resonance absorption features, it is natural to think of the local plasma resonance light scattering characteristics of assemblies. Figure 6.13 shows single AuNRs scattering spectrum and the spectrum after coupling of two AuNRs, showing local plasma resonant light scattering properties of the following aspects [83, 84]:

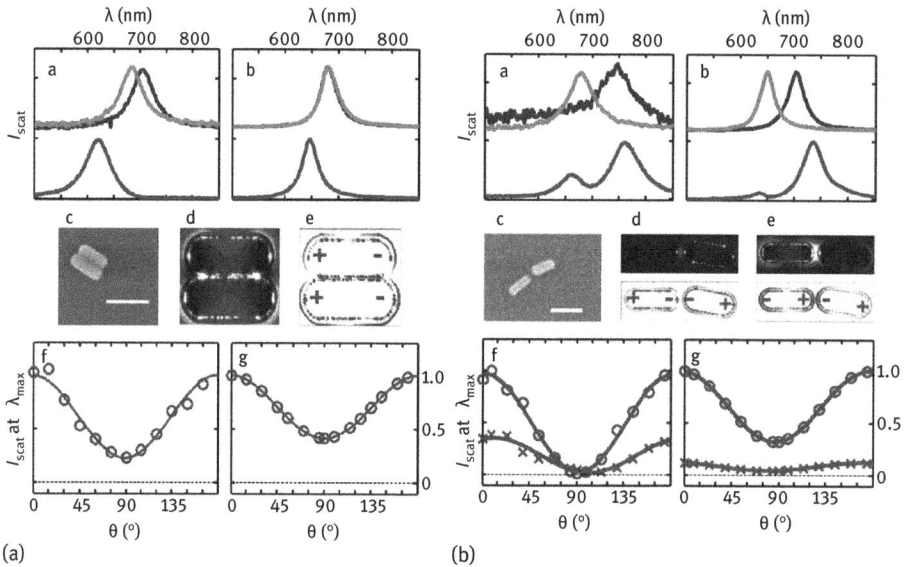

Figure 6.13: Plasma resonance scattering spectra of gold nanorod assembly. (a) side-by-side (b) head-to-head AuNRs dark-field scattering spectra. a. Nonpolarized dark field scattering spectra; b. integrated dark field scattering spectra; c. SEM of gold nano-dimer; d. FDTD calculated gold nanorods electric field enhancement e. FDTD calculated gold nanorods charge distribution; f. tested gold nanorods scattering peak position changes with the polarized angle; and g. simulated gold nanorods scattering peak position changes with polarized angle [83]

(1) There is an obvious blue shift for a scattering peak when two AuNRs are arranged in the side-by-side mode.

(2) There is an obvious red shift for scattering peak when two AuNRs are arranged in head-to-head mode.

(3) If the polarizer is added to change the angle of incident light, the intensity would also change. When an incident light is parallel to AuNRs, scattering light is the strongest; when they are vertical, scattering light is the weakest. Thus shape and the relative arrangement will directly decide AuNRs scattering spectrum.

(4) AuNRs distance, relative angle, solvent dielectric constant, and substrate property will also influence the scattering spectrum.

The above four features have been confirmed by an experiment and Finite-Difference Time-Domain (FDTD)-simulated longitudinal scattering spectrum in theory. The result is related to an electron coupling, for the reason that side-by-side assembly expands the electron transverse resonance, but head-to-head arrangement prolongs the electron longitudinal resonance.

(4) Nanorods serve as a dark-field imaging probe: As AuNRs have good light scattering signal in the near infrared region, which usually work as a dark field imaging probe [85]. As shown in Figure 6.14, under the dark field microscope, AuNRs modified with anti-epidermal growth factor antibodies were randomly dispersed in normal cells, but could be specifically bound to the surface of cancer cells by antigen-antibody recognition, and aggregated on the cancer cells, showing orange-red scattered light, and providing more obvious cells outline than normal cells. Besides, compared with bright field imaging, because of the dark background of dark field imaging the background interference is diminished; thus, cancer cells are more easily observed by providing a more obvious contrast effect [86].

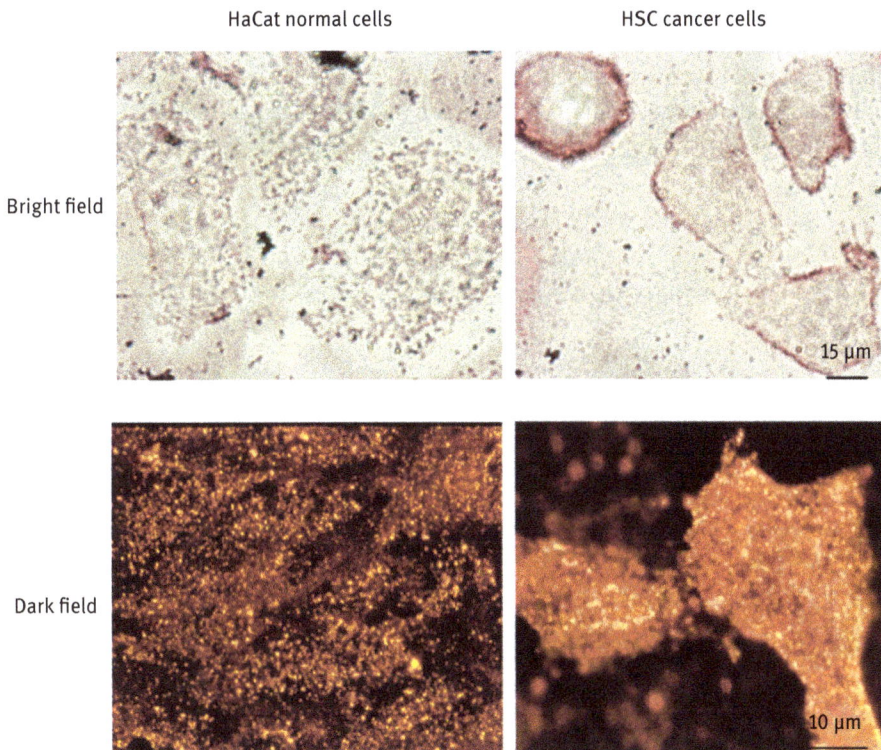

Figure 6.14: AuNRs scattering signal is applied in the dark field imaging.

Currently, AuNRs are widely applied in analysis and detection, cell imaging, disease diagnosis and treatment, as well as surface-enhanced Raman scattering (SERS) and metal-enhanced fluorescence and catalysis, that is, Fano resonance. The use of nanorods as scattering light prove and further apply to biomedical analysis and cancer cell therapy, which have an important prospect. In addition, as nanorods have plasma resonance absorption and scattering characteristics in longitudinal and transverse directions, light scattering polarization probe was built on these properties, even for a five-dimensional spectral probe (three-dimensional plus time, and probe dimension) [87].

6.3.2.3 Gold nanocages

Gold nanocages (AuNCs) are a hollow-core nanomaterial [88], with complex synthesis. It is normally obtained through an electrochemical substitution reaction rather than direct synthesis (Figure 6.15). The method is to transfer single-dispersed solid cubic silver nanoparticle into hollow-core gold nanobox particles in a chloroaurate solution; atfer adding chloroaurate solution, eight angles of gold nanobox particles with round holes forms nanocage particles. The side length of the cages could be as small as 30 nm and wall thickness can be as small as 3 nm. As gold solution has different valence states, different cage structures can be obtained. The use of $AuCl_4^-$ solution to obtain a hollow-core cubic box structure with equation (6.4a) and the use of $AuCl_2^-$ solution to obtain hollow-core cages structure with equation (6.4b) are as follows:

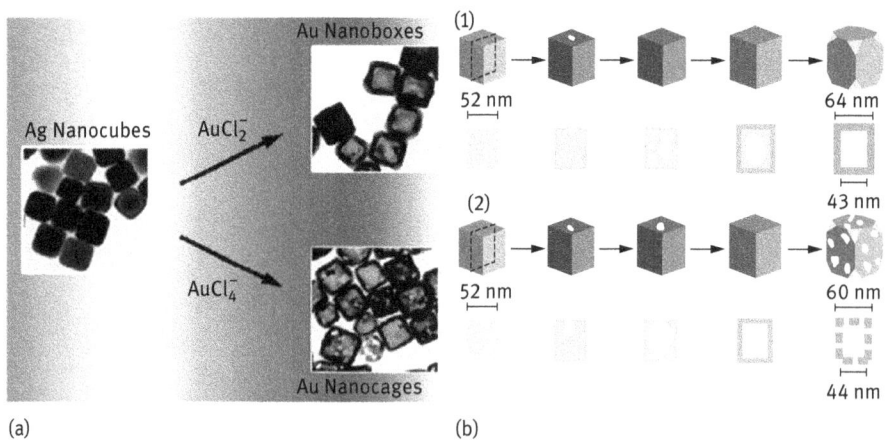

$$3Ag(s) + AuCl_4^-(aq) \rightarrow Au(s) + 3AgCl(s) + Cl^-(aq) \qquad (6.4a)$$

$$Ag(s) + AuCl_2^-(aq) \rightarrow Au(s) + AgCl(s) + Cl^-(aq) \qquad (6.4b)$$

(a)   (b)

Figure 6.15: Principles of AuNCs synthesis. (a) Reaction diagram of silver nano-cubic and two different reaction precursors ($AuCl_2^-$ and $AuCl_4^-$) (b) Mechanism of (1) $AuCl_2^-$ and (2) $AuCl_4^-$

Experiment and theoretical calculations show that absorption wavelength of these AuNCs can be adjusted to near 800 nm, with absorption section far larger than the scattering section (As in equation [2.25]), and five order magnitudes higher than traditional dyes (such as indocyanine green, ICG). It shows that AuNCs have good light absorbency with wide prospect in the field of photothermal therapy [89].

So far, plasma resonance scattering properties of AuNCs have not been fully developed with only a few related theories, but their cage structure provides a good carrier for loading and delivering of drug molecule and Raman's signal molecules [90].

6.3.3 Silver nanoparticles

6.3.3.1 Shape and local plasma resonance light scattering of silver nanoparticles

Silver nanoparticles have different forms, exhibiting different light scattering properties [91]. The study found that the scattering light of a silver nanosphere is blue, light of a silver nanocubic is cyan, light of a silver nano-trigonal bipyramid is yellow, and light of a silver nanorod is red. So, the study of scattering of different-form silver nanoparticles is quite attractive.

Silver nanospheres are similar in structure to gold nanospheres, but they differ in properties because of their different materials. With a larger scattering section area, silver nanoparticles have a stronger plasma resonance absorption and scattering. Compared to gold nanoparticles of the same diameter, their absorption and scattering is at a shorter wavelength (near 400 nm). So, scattering light has different colors, as described earlier; 40 nm silver nanoparticle is blue, whereas a 40 nm gold nanoparticle is green (Figure 6.16).

Silver nanoparticles have different colors, indicating that they are different in size and shape, which are similar to other noble metal nanoparticles. According to Mie theory, small-sized metal sphere particles have only one SP absorption peak, whereas particles with optical anisotropy or larger-sized metallic nanoparticle have 2 or 3 absorption peaks with different forms. For larger metallic nanoparticles, as plasma resonant dipole excitation or four level, and even higher excitation, there will be an extra-wide absorption peak in the UV-visible wavelength [92].

As silver nanoparticles have special optical properties, they are widely used in fields of analytical chemistry. Especially with their local SP resonance characteristic, it has good application in SP resonance spectroscopy, surface-enhanced Raman scattering, colorimetry detection, light scattering detection, cell imaging, and so on. Among which, chromaticity detection based on color changes is a simple and rapid method and does not need large instruments, thus becoming a promising detection method. For example, silver nanosphere resonance light scattering signal

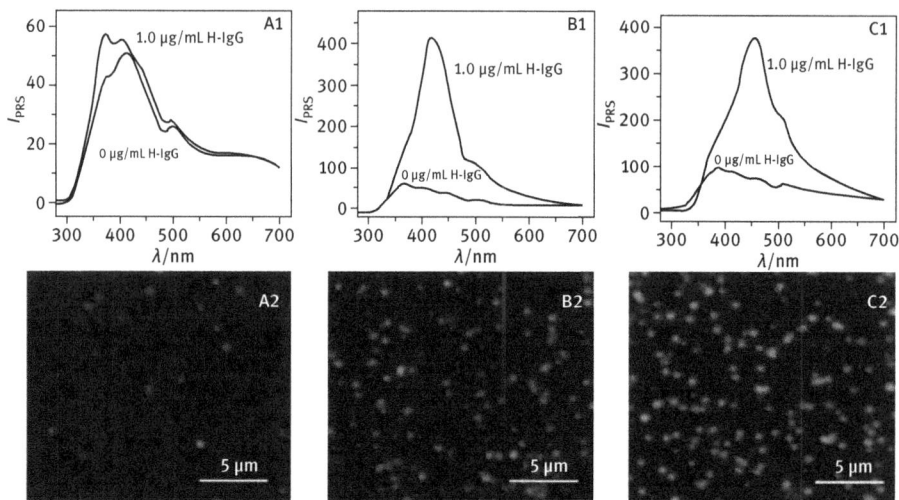

Figure 6.16: Light scattering spectra and light scattering imaging of silver nanoparticles. Glass scattering spectra (above) and corresponding dark-field imaging (below) of different size of silver nanoparticles. AgNPs size (nm): (A) 35.4; (B) 58.0; (C) 66.0.

based on nanoparticle aggregates [93] or production enhancement [94, 95] can realize drug analysis and food safety control.

6.3.3.2 Silver nanospheres

Scattering section of a silver nanosphere (equation [2.25]) is larger than a gold nanosphere with stronger light scattering signal, but as it is more difficult to synthesize, its application is limited. On the one hand, oxidizing ability of silver ion is weaker than that of gold; on the other hand, two reductants [96] need to be added in the chemical reduction method or adding NaOH for fine control [97]. Even so, the as-prepared silver nanoparticle is not uniform enough, producing a scattering light of different colors. In recent years, the study staff used reductant substances in fruits and natural proteins as a surfactant or a protective agent to prepare silver nanoparticles with good performance [98].

Silver nanospheres can be prepared through light radiation [99] and a microwave-assisted method. It is worth demonstrating that as silver ion is quite sensitive to light, sunlight can reduce it to the elementary substance silver, which is stable in a wide range of acidity range [100]. However, it is time-consuming, usually consumes more than 10 hours. Microwave-assisted method prepares nanosphere rapidly with the disadvantage of uneven particle size [101].

In general, different synthesis methods have their own merits and demerits, but excellent properties, such as stability and uniformity are a constant aim for researchers.

6.3.3.3 Silver nanorods

Relatively, the aspect ratio and shape is more uncontrollable for silver nanorods, and the study of which is not as wide as gold nanorods and is mostly on their synthesis. Their synthesis method includes chemical reduction method [102], photochemical method, electrochemical method, ultrasonic method, and so on. Currently, most of the methods need surfactants, such as PVP as a template [103].

Similar to gold nanorods, silver nanorods have two plasma resonance absorption peaks in transverse and longitudinal directions, but their scattering section is much larger than that of gold nanorods [104].

6.3.3.4 Silver nanoprism

American scholar Mirkin et al. first reported that the triangular silver nanoprism prepared in solution under illumination had obvious multipolar resonance absorption peak [105]. The plasma resonance band of the triangular silver nanocolumn was ascertained by DDA calculation and experimental observation. It was clear that a single nanoparticle had two special quadrupole plasma resonance. Different from Rayleigh scattering of blue light sphere of silver particle, these silver nanoprisms scatter red light (Figure 6.17). Meanwhile, they synthesized silver nanoprisms of different sizes through a simple thermal reduction method, manually adjusting their LSPR peak position. It is found that with the increase of the edge length, the increase of the tip angle and the decrease of thickness will redshift the LSPR peak.

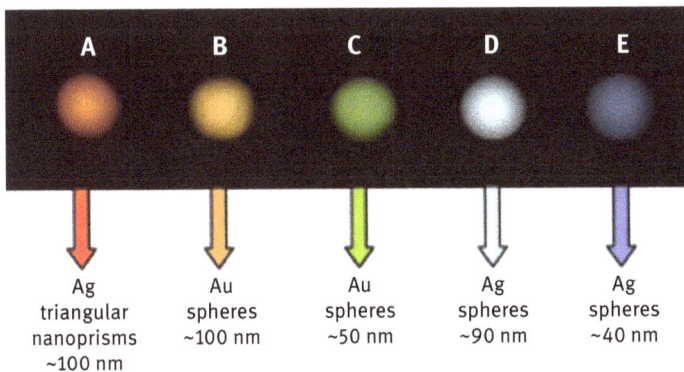

Figure 6.17: Different scattering light of gold and silver nanoparticle (refer to the colored picture at the *end of the book*).

He R. et al. used polymer as a stabilizer to prepare silver nanoprism with special optical properties in an organic solvent medium with a microwave-assisted method [106]. They found that the formation of a regular triangle (or angle-lacked triangle) nanoprism is gradually transferred from 10 nm sphere particle along with a reaction. As shown in Figure 6.18, in an early stage, there is only a small spherical nanoparticle or a nano-cluster; after approximately 12 minutes, there are large particles with regular geometrical shapes; when the reaction ends, small spherical nanoparticles decrease with lots of nanoparticles with regular geometrical shapes, most of which are triangles or angle-lacked triangles, and each silver nanoparticle is a single crystal [106]. Corresponding to this process, UV-visible absorption peak changes from a single plasma resonance absorption peak to the coexistence of multipole absorption peaks with color changes.

Besides the following synthesis methods, preparation methods also include heat transfer method, microemulsion method, and reduction method and so on. For example, Mirkin group put forward using PVP and sodium citrate as a protectant to reduce

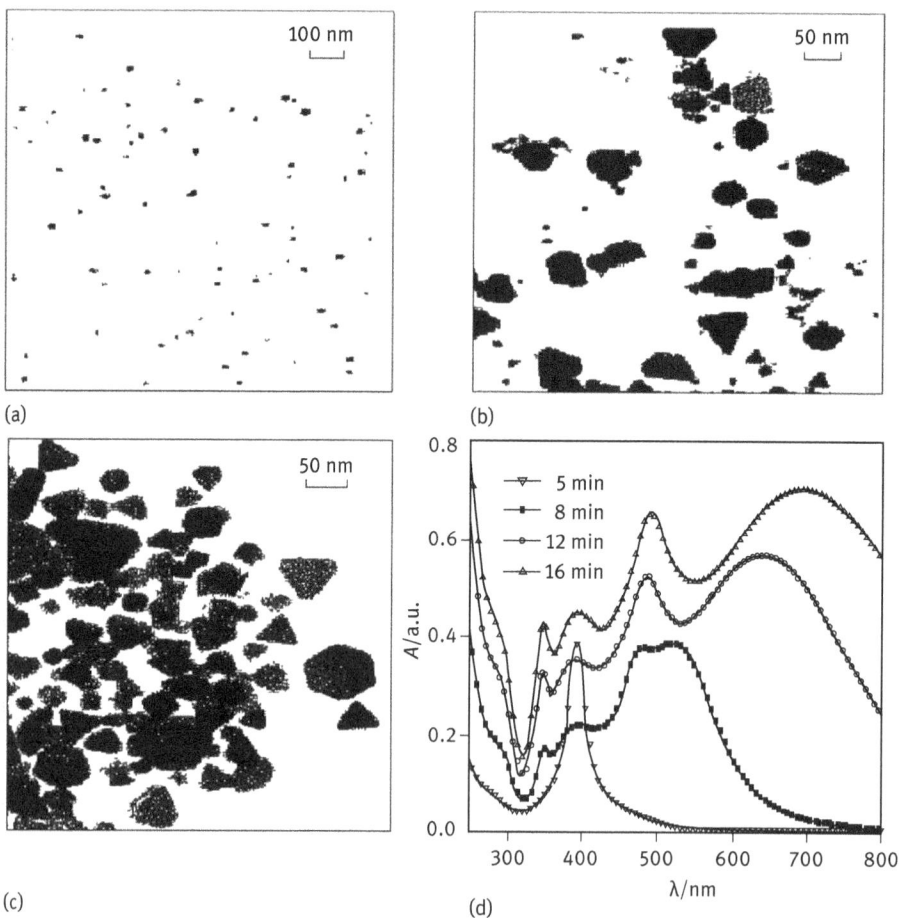

Figure 6.18: The growth process of silver nanoprisms (a–c) and the corresponding absorption spectra (d) [106].

AgNO$_3$ with sodium borohydride at room temperature to obtain silver nanotriangle with good monodispersity, which are able to store for months under dark [107]. With different amounts of sodium borohydride, pink to green products can be obtained. What is more important is that this method can precisely adjust the thickness of the silver nanoprism, which is better than previous optical or thermal methods.

6.3.3.5 Other silver nanoparticles

In terms of nanomaterial forms, besides the previously mentioned sphere, rod, and cage, there are shell, star shape, branch, flower, piece, trigonal bipyramid, and so on. Of course, their differences in the material, size, and shape decide differences of their properties. Figure 6.19 shows SEM and corresponding dark-field light scattering imaging of silver nanocrystalline twins, single crystal nanosheets and twin crystal decanedron. It can be seen that silver nanoparticles of different shapes and structures could be yellow, yellow–green, and red [108].

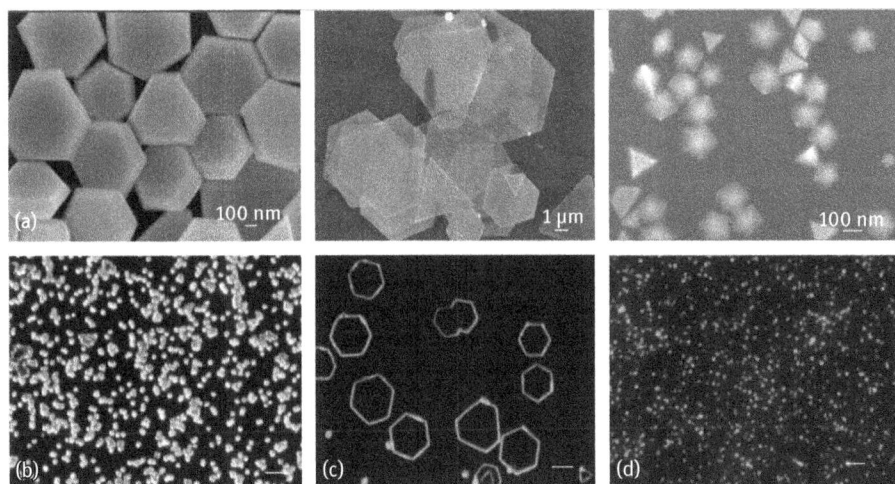

Figure 6.19: Dark-field scattering image of silver nanoparticles [108]. Above (a): SEM; Below: Dark-field scattering images, from left to right: (b) nanocrystalline twins (c) single crystal nanosheets, and (d) twin crystal decanedron.

It needs special demonstration that monocrystalline nanocrystal has larger size with scattering light on the edge and transparent shape in the middle. Further study is needed for its mechanism. Nanocrystals of three shapes exhibit excellent near-infrared absorption and dark-field scattering properties, which can be used as a cell imaging probe for cancer diagnosis and therapy. By far, there are few reports on this, which are mainly closely related to its bad structural stability. It has great application potentials in biomolecules and inorganic ion-sensing detection field; however, the potential has not released yet.

6.3.4 Other metallic nanoparticles and its light scattering

In the last 20 years, with booming development of nanotechnology, there are more and more nanomaterials. On the one hand, nanomaterials of different materials are gradually developed; on the other hand, nanoparticles of the same materials but different forms are gradually prepared.

In terms of materials, besides common gold and silver nanomaterials, there are also Pt, Cu, Ni, Co, Fe, alloy Ag/Au, Ag/Hg, Pd/Pt, $Cd_xZn_{1-x}S$, NiCo–Pt, Au–PtAg, and other metallic or alloy nanoparticles, as well as oxide and composite oxide nanoparticles, such as Fe_3O_4, TiO_2, and ZnO and $MnFe_2O_4$, CdS, CdSe, CdTe, and other semiconductor nanomaterials have good luminescent property, which are always classified as quantum dots.

6.4 Metallic nanoparticles aggregate and light scattering assembly

When two metallic nanoparticles are close to each other, their distance would influence the localized surface resonance plasma of a single nanoparticle and there might even be a localized resonance plasmon coupling. At this time, their plasma resonance scattering or absorption spectrum maximum wavelength shifts. For example, a single 40 nm silver nanoparticle plasma resonance scattering peak is around 420 nm, when two silver nanoparticles get close and become dimerization particles, their wavelength has a red shift to approximately 500 nm. At this time, the silver nanoparticle scattering turns from blue to green; 50 nm gold nanoparticle has a localized resonance plasmon coupling, as the peak has a red shift from 530 nm to 580 nm, and the scattering light changes from green to orange. Resonance plasmon coupling caused by distance reduction of two or more metallic nanoparticles widely exist in the aggregation and assembly process of particles.

Dispersion and aggregate/assembly is a common state of nanoparticles. Usually, dispersed nanoparticles have a weaker plasma resonance light scattering signal, whereas aggregated or assembled nanoparticles have a stronger signal. Aggregate and assembly are spatial arrangements for large number of nanoparticles. It can be simply understood that a disordered array is thought as an aggregation, whereas an ordered array is thought as an assembly. Usually, isotropic nanomaterials as zero-dimensional spatial structure nanoparticles cannot perform specific spot modification with homogeneity of active sites, then they normally do not have a spatial array of aggregation; while one-dimensional or multidimensional anisotropic nanomaterials can perform controllable assembly, such as head-to-head array for AuNRs, [66, 67] side-by-side assembly forms [68], and so on.

6.4.1 Light scattering of nanoparticles aggregate

A zero-dimensional spatial structure nanoparticle (as sphere) normally has spatial array of aggregation; based on differences of nanoparticle functions, aggregation can be divided into cross-linking aggregation and non-cross-linking aggregation.

In 1996, Mirkin and his colleagues first studied AuNPs cross-linking aggregates [109]. They divided the probe DNA into two segments, which were modified on the surface of AuNPs and then mixed with the target DNA. Different AuNPs were linked through the target DNA. Because of the change of the distance between the gold nanoparticles, the plasmon resonance coupling was produced, so that the solution color changed from red to blue, and the target DNA molecule was analyzed. This way of linking AuNPs to each other through molecular linkage is cross-linking (Figure 6.20 [a]).

In 2003, Maada et al. reported non-cross-linking aggregation of AuNPs [110]. When the target DNA is completely hybridized with the probe DNA, the double stranded DNA can not protect the AuNPs, so that the AuNPs aggregate at a certain ionic strength. In the aggregate, nanoparticles have non-cross-linking aggregation because of their low electrostatic repulsion rather than molecular connection (Figure 6.20 [b]).

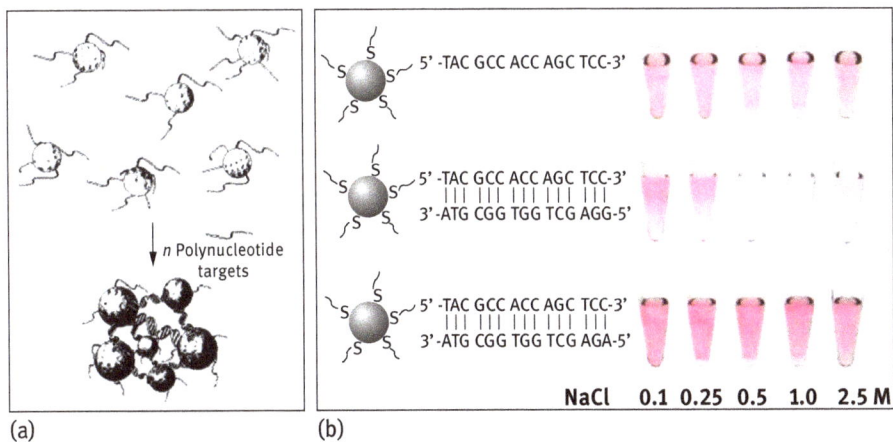

(a)   (b)

Figure 6.20: Aggregation modes of nanoparticles. (a) Cross-linking aggregation [109] and (b) non-cross-linking aggregation [110].

In the study of interaction of nucleotides and metal ions, T–Hg^{2+}–T mismatch structure is easily formed between Hg^{2+} and thymine base. T–Hg^{2+}–T mismatch structure is meaningful, and it is applied in the molecular assembly field. For example, gold nanoparticles are used as probe to measure Hg^{2+} in a non-cross-linking aggregation (Figure 6.21) [111], and to detect melamine in a cross-linking aggregation mode (Figure 6.22) [112].

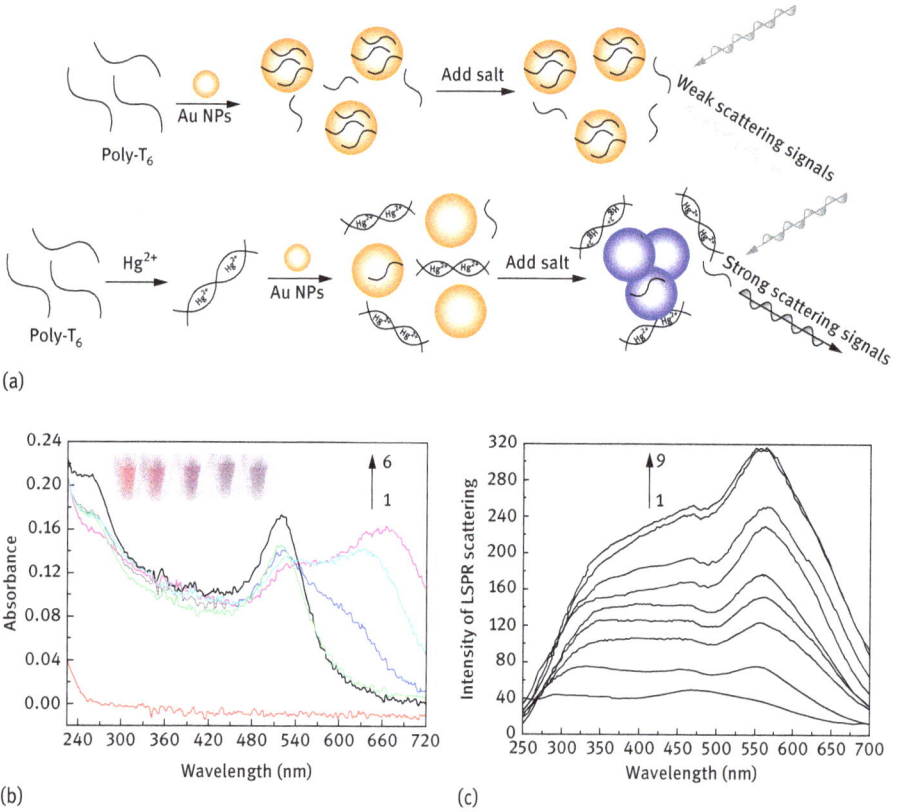

(a)

(b)   (c)

Figure 6.21: The detection of Hg^{2+} using T–Hg^{2+}–T structure with T base-modified gold nanosphere as light scattering probe [111] (a) Mechanism; (b) system absorption spectra, curves 1–6 mean Hg^{2+} concentration(mol/L) is 1.0, 0, 0.04, 0.1, 0.3, and 0.6, where curve 1 is without gold nanospheres, curve 2–6 are with gold nanospheres; (c) system localized resonance plasmon spectra, curve 1–9 are Hg^{2+}concentrations (mol/L) 1.0, 0, 0.04, 0.1, 0.2, 0.3, 0.4, 0, and 0.6, where curve 1 only has Hg^{2+}, curve 2 only has Poly-T_6 gold nanospheres, curves 3–7 and 9 have Hg^{2+} and gold nanosphere, curve 8 has T_6 and A_6 gold nanocomplex

Non-cross-linking aggregation (Figure 6.21) mainly occurs because ssDNA is quite flexible and easily adsorbed on the surface of gold nanosphere to keep dispersed even in higher ionic strength [71]. So, when the surface of gold nanosphere is adsorbed with 6 thymine base length single-stranded T_6, it is still in a monodispersed

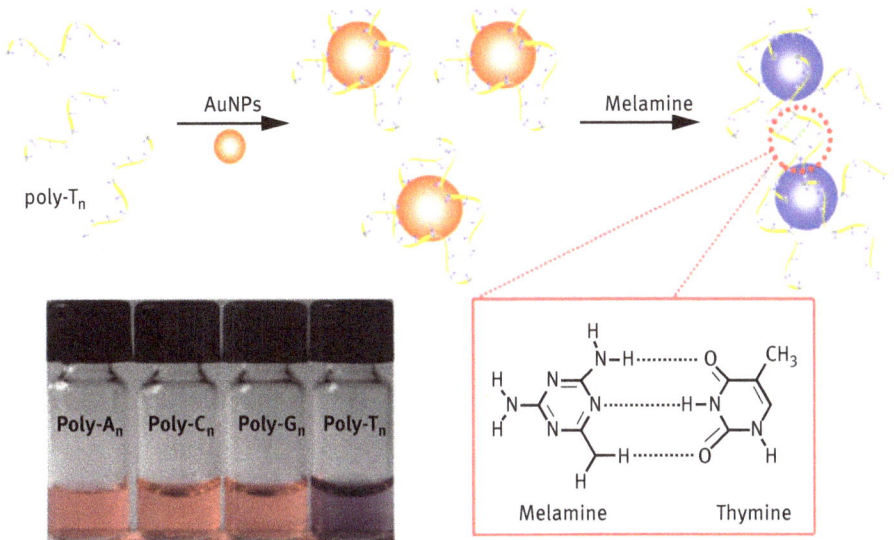

Figure 6.22: Use of T base modified gold nanosphere to detect melamine [112].

state with a weaker plasma resonance scattering intensity. While in the prescence of Hg^{2+}, T base forms $T–Hg^{2+}–T$ double-stranded mismatch structure with Hg^{2+}, the double-stranded DNA structure is rigid, which is unable to adsorb on the surface of the gold nanosphere, so there is non-cross-linking aggregation, enhancing the plasma resonance light scattering [112].

Similarly, the T base single-stranded DNA protection of gold nanosphere can be used for light scattering detection of melamine in milk (Figure 6.22) [113]. In this detection, the T base sequence is required to be longer; the shorter T base detection result is not as good as that of a longer T base sequence. This method is based on the cross-linking aggregation; longer T base sequence is partially adsorbed on the surface of gold nanosphere, whereas the other part works with melamine through hydrogen bonds. When melamine exists, different T bases on the surface interact with the melamine to shorten their distance, inducing an aggregate and enhanced light scattering signal. Based on similar mechanisms, Professor Lu L. H. and his group [114] used 3-(4,5-dimethyl-2-thiazolyl)-2,5-diphenyl-2-H-tetrazolium bromide to test melamine in infant formula milk powder.

Chen and his reseach group studied the relation of melamine, Hg^{2+}, and gold nanosphere, finding their non-implicit logical relationships [115]. As shown in Figure 6.23, non-implicit logical gate working process includes melamine and mercury ion as input, and output is definited by the solution color. Define "lack" and "existence" input as "0" and "1", respectively. In output, "1" is the blue solution with aggregated gold nanoparticles and "0" is the red solution with dispersed gold nanoparticles. At (0,0), a 13 nm nanoparticle is stabilized by the T base DNA, the solution is red, the output is "0" at (0,1) state, only Hg^{2+} cannot induce change of the solution, so the

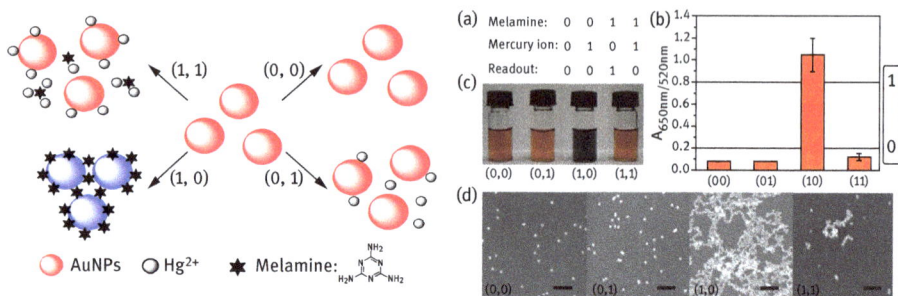

Figure 6.23: Logic gate with T base-modified gold nanosphere to detect Hg^{2+} and melamine [115]. (a) Melamine/Hg^{2+}–gold nanoparticle logical gate input and output; (b) Logic gate output by $A_{650\,nm/520\,nm}$ form; (c) Corresponding color changes of logic gate; (d) Corresponding SEM imaging.

output is "0" at (1,0) state, melamine is added to induce gold nanoparticle aggregates with color change from red to blue, thus the output is "1". When mercury ions and melamine is added into a gold nanoparticle solution at the same time, that is, in (1, 1) state, there is no dramatic change color of the solution, because melamine has to combine with Hg^{2+}, losing its ability to aggregate gold nanoparticles, thus the output is "0". So, there is only one positive output with melamine (color change of nanoparticle solution from red to blue), this is a non-implicit logic gate.

6.4.2 Light scattering assembly

Multidimensional nanomaterials with their special structures have different active sites, and a controllable ordered array can be performed. For example, as AuNRs have a special one-dimensional structure and anisotropy, there are multiple assembly methods, including head-to-head, [66–69] side-by-side, [65, 70] and so on. As mentioned earlier, different assembly causes longitudinal local SP resonance absorption to produce different shifts. Whatever the assembly method is, the resonance light scattering signal is enhanced. As shown in Figure 6.24, HIV virus sequence analysis method is established on the basis of gold nanorods.

AuNRs have a weak light scattering intensity. Without the target DNA, the probe DNA is adsorbed onto the surface of gold nanorod. Through the electrostatic interaction of phosphate groups and CTAB positive charges, DNA adsorbed gold nanorods are in a dispersive state with the electrostatic repulsion. With target DNA, it adsorbs gold nanorods with the same mechanism, and the complementary base pairing forms a double helical structure. With a double-stranded DNA as bridge, gold nanorods aggregate, thus greatly enhancing their SP resonance scattering signal, and achieving HIV sequence detection [116]. The method has good recognition ability for a single base mismatch, which is important for the study of a single nucleotide polymorphisms, which can be further applied for a HIV gene piece test and recognition in a PCR product, with potential application value for HIV clinical diagnosis.

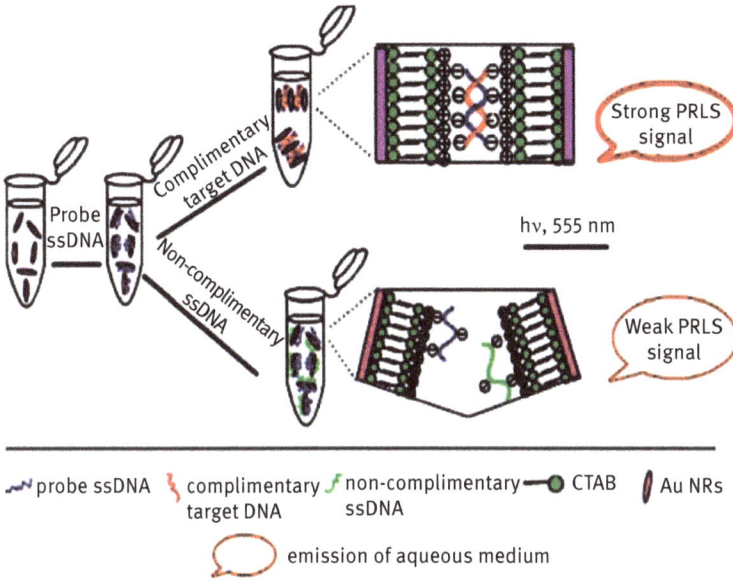

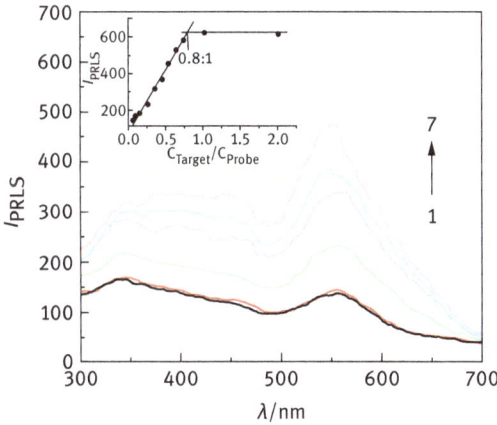

Figure 6.24: Gold nanorod assembly is used to detect HIV sequence [116]. Curve 1, gold nanorods; curve 2, gold nanorods and single-stranded DNA; curves 3–7, gold nanorods and a single-stranded DNA as well as a target DNA (with the concentration of 3.33, 5.0, 6.67, 8.33, and 11.67 nmol/L).

When in the presence of adenosine, aptamers change from single-stranded structure to rigid G-quadruplex structure, then phosphates are exposed to the outside and electronegativity is enhanced. The electrostatic interaction with CTAB-coated gold nanorods induces the assembly of gold nanorods in side-by-side mode, resulting in enhanced light scattering signal [70].

Besides, gold nanorod assembly is widely applied in immunoassay. Neuron-specific enolase is a biomarker in a small-cell lung cancer and neuroblastoma, since neuron-specific enolase (NSE) is very high in patients; while when therapy is effective, NSE

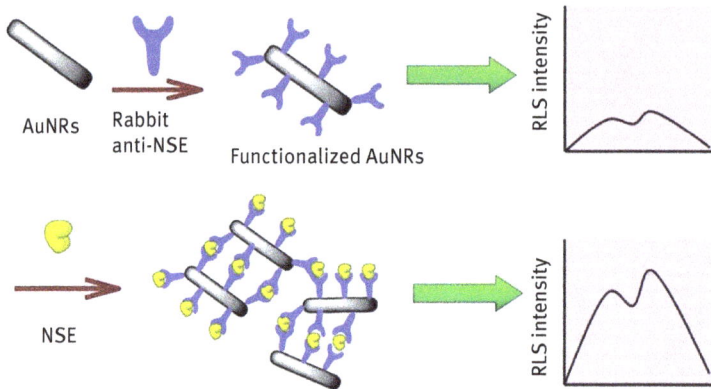

Figure 6.25: Use of gold nanorod assembly to detect neuron-specific enolase [117].

concentration gradually falls to a normal level and the serum NSE concentration rises when the sickness comes back, so NSE can be used for an early stage diagnosis, therapy, and relapse prediction, which is a good clinical index in the observation and monitoring. Based on Figure 6.25, anti-NSE antibody-modified gold nanorod is dispersed with a lower scattering signal. When in the presence of NSE through immune reaction of antigen–antibody, there are assembly forms of side-by-side and head-to-head, enhancing the scattering signal [117].

Similar to this, side-by-side and head-to-head assembly of gold nanorods enhances light scattering signal, realizing selective detection of cancer antigen 125 (Figure 6.26) [118].

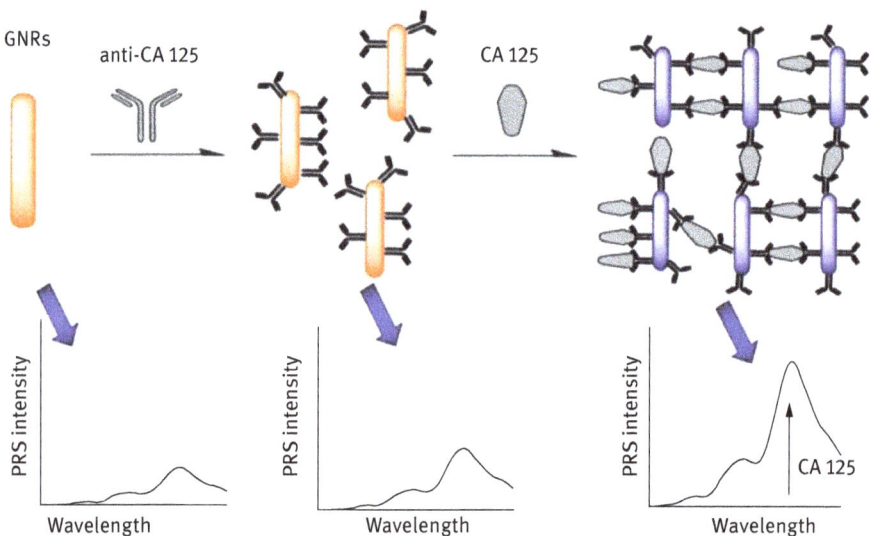

Figure 6.26: Assembly detection of cancer antigen 125 with gold nanorods [118].

References

[1]   Jiang Z L, Feng Z-W, Li F, et al. Resonance scattering spectroscopy of gold nanoparticle. Sci. China B, 2001, 44(2): 175–181.

[2]   Liu S P, Jiang Z L, Kong L, et al. Absorption and Rayleigh scattering and resonance Rayleigh scattering spectra of $[HgX_2]_n$ nanoparticles. Sci. China B, 2002, 45(6): 616–624.

[3]   Jiang Z L, Zhong F X, Li S T, *etc*. Study on the resonance scattering spectra of green silver nanoparticles. Acta Chim. Sin., 2001, 59(3): 438–441.

[4]   Jiang Z L, Li u Q Y, Liu S P. Resonance scattering spectral analysis of chlorides based on the formation of $(AgCl)_n(Ag)_s$ nanoparticle. Spectroichim. Acta A, 2002, 58(12): 2759–2764.

[5]   Berns R. Principle of color technology. Li X M, Ma R, Chen L R, etc. Beijing: Chemical Industry Press, 2002: 10.

[6]   Qie G L, Yuan F L, Cheng Z H. A light scattering and fluorescence emission coupled ratiometry using the interaction of functional CdS quantum dots with aminoglycoside antibiotics as a model system. Talanta, 2007, 71, 567–572.

[7]   Huang S, Xiao Q, He Z K, *etc*. Resonance light scattering method for the determinaiton of lysozyme using CdSe quantum dots as probe. Chem. J. Chinese U., 2009, 30(10): 1951–1955.

[8]   Sun Y P, Zhou B, Lin Y, et al. Quantum-sized carbon dots for bright and colorful photolumi-nescence. J. Am. Chem. Soc., 2006, 128(24): 7756–7757.

[9]   Gao M X, Liu C F, Wu Z L, et al. A surfactant-assisted redox hydrothermal route to prepare highly photoluminescent carbon quantum dots with aggregation-induced emission enhancement properties. Chem Commun., 2013, 49(73): 8015–8017.

[10]   Zhao H X, Liu L Q, Liu ZD, et al. Highly selective detection of phosphate in very complicated matrixes with an off–on fluorescent probe of europium-adjusted carbon dots. Chem. Commun., 2011, 47(9): 2604–2606.

[11]   Wu Z L, Zhang P, Gao M X, et al. One-pot hydrothermal synthesis of highly luminescent nitrogen-doped amphoteric carbon dots for bioimaging from Bombyx mori silk-natural pro-teins. J. Mater. Chem. B., 2013, 1(22): 2868–2873.

[12]   Yuan Y H, Li R, Wang Q, et al. Germanium-doped carbon dots as a new type of sensitive and selective fluorescent probe for visualizing the dynamic invasions of mercury(II) ions into cancer cells. Nanoscale, 2015, 7(40): 16841–16847.

[13]   Ding H, Yu S B, Wei J S, et al. Full-color light-emitting carbon dots with a surface-state-controlled luminescence mechanism. ACS Nano, 2016, 10(1): 484–491.

[14]   Liu L Q, Li Y F. Carbon dots used for the detection of lysozyme with resonance light scattering signals. J. Southw. Univ. (Nat. Sci. Ed.), 2010, 32(11): 25–29.

[15]   Dong W, Wang Y, Song Y T, et al. Carbon dots used for the detection of nucleic acids with resonance light scattering signals. Chem. Reagent, 2013, 35(2): 147–149.

[16]   Zhang L, Huang C Z, Li Y F, et al. Label-free detection of sequence-specific DNA with multiwalled carbon nanotubes and their light scattering signals. J. Phys. Chem. B, 2008, 112(23): 7120–7122.

[17]   Jiang Z L, Liu S P, Liu Q Y. Resonance scattering spectroscopy of fluid carbon naoparticle. Chinese J. Appl. Chem., 2002, 17(1): 22–25.

[18]   Zhong F X, Jiang Z L, Li T S, *etc*. Photochemical preparation of green silver and yellow silver colloid and their resonance scattering spectroscopic study. Photogr. Sci. Photochem., 2001, 19(1): 13–18.

[19]   Guo Y M, Deng Y H, Zhao J Z, et al. Synthesis and characterization of sulfur nanoparticles by liquid phase precipitation method. Acta Chim. Sin., 2005, 63(4); 337–340.

[20]   Jiang Z L, Jiang H L, Liu F Z, *etc*. Effect of dyes on resonance scattering spectrum of S ultrafine particles. Chinese J. Appl. Chem. 2003, 20(4): 351–354.

[21] Chen H Q, Ma J, Diao X L, et al. Determination of proteins with PbS-PAA composite nanoparticle by resonance light scattering technique. J. Instr. Anal., 2005, 24(6): 56–58.

[22] Zhong M H. Resonance light scattering method for the determination of copper with the sulfide-cetyltrimethyl ammonium bromide system. Chinese J. Health Lab. Technol., 2005, 15(11): 1286–1287.

[23] Zhong M H, Huang J S, Lin Y R. Resonance light scattering method for the determination of silver with the sulfide -cetyltrimethyl ammonium bromide system. J. Hanshan Norm. Univ. (Nat. Sci. Ed.), 2007, 28(6): 60–63.

[24] Long Y F, He Q, Chen S. Synthesis of copper sulfide nanoparticles and its application to the determination of mercury ion. Chinese J. Appl. Chem., 2010, 27(11): 1318–1321.

[25] Xu F G, Wang L, Wang L Y, et al. Preparation and application of ZnS-Zn^{2+} nanoparticles as a resonance light scattering probe. J. Anal. Sci., 2005, 21(3): 265–267.

[26] Zhang P, Zhuo S J, Chen J L, et al. Application of L-cysteine-capped nano-ZnS as a new fluorescence probe for the determination of proteins by resonance light scattering technique. Chinese J. Anal. Lab., 2005, 24(3): 86–90.

[27] Li Y, Chen J, Zhuo S, et al. Application of L-cysteine-capped ZnS nanoparticles in the determination of nucleic acids using the resonance light scattering method. Microchim. Acta, 2004, 146: 13–19.

[28] Cheng Y Y, Zhang B M, Jiang Z L, et al. Resonance scattering spectral properties of Se nanoparticles and its application. J. Guangxi Normal Univ. (Nat. Sci. Ed.), 2005, 23(3): 56–60.

[29] Jiang Z L, Zhang Y, Tan H M, et al. Resonance scattering detection of trace Hg^{2+} using aptamer modified AuSe nanoalloy. Spectros. Spectr. Anal., 2011, 31(5): 1371–1374.

[30] Jiang Z L. Fraction frequency scattering spectra for Au clusters in liquid. Acta Photon. Sin., 2001, 30(4): 460–464.

[31] Chen S, Jiang Z L, Liu S P. Synthesis and resonance Rayleigh scattering spectra properties of nanometer $(MnO_2)_n$ colloid particle. J. Southwest Normal Univ. (Sci. Ed.), 2002, 27(1): 73–77.

[32] Li Y F, Huang C Z, Hu X L. The principles and the applications of resonance light scattering technique in the research of biochemistry and bioanalysis. Chinese J. Anal. Chem., 1998, 26(12): 1508–1515.

[33] Li Y T, Wang Y Q, Wei Y J, et al. Determination of L-tyrosine by TiO_2 nanocrystals as a resonance light scattering spectroscopy probe. J. Sichuan Univ. (Nat. Sci. Ed.), 2008, 45(1): 110–114.

[34] Liu S P, Jiang Z L, Kong L, et al. Absorption spectra, Rayleigh scattering and resonance Rayleigh scattering spectra of $[HgX_2]_n$ nanoparticles. China Sci. Ser. B, 2002, 32(6): 554–560.

[35] Jiang Z L, Zhai H Y, Zhang B M, et al. Interface fluorescence and resonance scattering spectral properties of AgX nanoparticles in liquid phase. Acta Chim. Sin., 2004, 62(14): 1272–1276.

[36] Zhou Y Y, She S K, Zhang L, et al. The Resonance Light Scattering Spectra from the Interaction of Zn-PVK Nanoparticles with Nucleic Acids and Its Analytical Application. Chinese J. Spectrosc. Lab., 2005, 22(1): 25–29.

[37] Chen H Q, Wang L Y, Li L, et al. Preparation and resonance light scattering property of poly (stryene-acrylic acid) nanoparticles. Chinese J. Appl. Chem., 2004, 21(2): 197–199.

[38] Ye L Y, Chen S, Yue M, et al. Spectroscopic studies on the interaction between carbon nanotubes/polythymine oligonucleotide composites and mercury ions. Chinese J. Light Scatt., 2011, 23(1): 46–51.

[39] Ye L Y. Spectroscopic analysis of functionalized carbon Nanocomposites. Xiangtan: Master's thesis of Hunan University of Science and Technology, 2011.

[40] Hu P, Huang C Z, Li Y F, et al. Magnetic particle-based sandwich sensor with DNA-modified carbon nanotubes as recognition elements for detection of DNA hybridization. Anal. Chem., 2008, 80(5): 1819–1823.

[41] Dong C Z, Liang L N, Zhang S J. TiO$_2$-SiO$_2$ nanoparticles synthesis and determination of ribonu-cleic acid by second-order scattering method. Chinese J. Anal. Chem., 2009, 38(2): 279–282.

[42] Homola J, Yee S S, Gauglitz G. Surface plasmon resonance sensors: Review. Sensor. Actuat. B Chem., 1999, 54: 3–15.

[43] Yguerabide J, Yguerabide E E. Light-scattering submicroscopic particles as highly fluorescent analogs and their use as tracer labels in clinical and biological applications I. Theory. Anal. Biochem., 1998, 262: 137–156.

[44] Yguerabide J, Yguerabide E. Resonance light scattering particles as ultrasensitive labels for detection of analytes in a wide range of applications. J. Cell. Biochem., 2001, 37: 71–81.

[45] Su K-H, Wei Q-H, Zhang X, et al. Interparticle coupling effects on plasmon resonances of nanogold particles. Nano Lett., 2003, 3: 1087–1090.

[46] Willets K A, Van Duyne R P. Localized surface plasmon resonance spectroscopy and sensing. Annu. Rev. Phys. Chem., 2007, 58: 267–297.

[47] Luther J M, Jain P K, Ewers T, et al. Localized surface plasmon resonances arising from free carriers in doped quantum dots. Nat. Mater., 2011, 10(5): 361–366.

[48] Murphy C J, Gole A M, Stone J W, et al. Gold nanoparticles in biology: beyond toxicity to cellular imaging. Acc. Chem. Res., 2008, 41: 1721–1730.

[49] Creighton J A, Blatchford C G, Albrecht M G. Plasma resonance enhancement of Raman scattering by pyridine adsorbed on silver or gold sol particles of size comparable to the excitation wavelength. J. Chem. Soc., Faraday Trans., 1979, 75: 790–798.

[50] Philip D. Synthesis and spectroscopic characterization of gold nanoparticles. Spectrochim. Acta A, 2008, 71: 80–85.

[51] Lee J H, Kamada K, Enomoto N, et al. Morphology-selective synthesis of polyhedral gold nanoparticles: What factors control the size and morphology of gold nanoparticles in a wet-chemical process. J. Colloid Interf. Sci., 2007, 316: 887–892.

[52] Uechi I, Yamada S. Photochemical and analytical applications of gold nanoparticles and nanor-ods utilizing surface plasmon resonance. Anal. Bioanal. Chem., 2008, 391(17): 2411–2421.

[53] Guo S, Wang E. Synthesis and electrochemical applications of gold nanoparticles. Anal. Chim. Acta, 2007, 598: 181–192.

[54] Bhuvanasree S R, Harini D, Rajaram A, et al. Rapid synthesis of gold nanoparticles with Cissus quadrangularis extract using microwave irradiation. Spectrochim. Acta A, 2013, 106: 190–196.

[55] Okitsu K, Mizukoshi Y, Yamamoto T A, et al. Sonochemical synthesis of gold nanoparticles on chitosan. Mater. Lett., 2007, 61: 3429–3431.

[56] Du B A, Li Z P, Liu C H. One-step homogeneous detection of DNA hybridization with gold nanoparticle probes by using a linear light-scattering technique. Angew. Chem. Int. Ed., 2006, 45: 8022–8025.

[57] Shen X W, Huang C Z, Li Y F. Localized surface plasmon resonance sensing detection of glucose in the serum samples of diabetes sufferers based on the redox reaction of chlorauric acid. Talanta, 2007, 72: 1432–1437.

[58] Ling J, Huang C Z, Li Y F, et al. Light-scattering signals from nanoparticles in biochemical assay, pharmaceutical analysis and biological imaging. TrAC-Trend. Anal. Chem., 2009, 28: 447–453.

[59] Huang X H, Neretina S, El-Sayed M A. Gold nanorods: From synthesis and properties to biological and biomedical applications. Adv. Mater., 2009, 21: 4880–4910.

[60] Wang J, Zhang H Z, Li R S, et al. Localized surface plasmon resonance of gold nanorods and assemblies in the view of biomedical analysis. TrAC-Trend. Anal. Chem., 2016, 80: 429–443.

[61] Zeng Q, Zhang Y, Liu X, et al. Multiple homogeneous immunoassays based on a quantum dots-gold nanorods FRET nanoplatform. Chem. Commun., 2012, 48: 1781–1783.

[62] Wang J, Sefah K, Altman M B, et al. Aptamer-conjugated nanorods for targeted photothermal therapy of prostate cancer stem cells. Chem. Asian J., 2013, 8: 2417–2422.

[63] Wang J, You M, Zhu G, et al. Photosensitizer-gold nanorod composite for targeted multimodal therapy. Small, 2013, 9: 3678–3684.

[64] Wang J, Zhu G, You M, et al. Assembly of aptamer switch probes and photosensitizer on gold nanorods for targeted photothermal and photodynamic cancer therapy. ACS Nano, 2012, 6: 5070–5077.

[65] Wang L, Zhu Y, Xu L, et al. Side-by-side and end-to-end gold nanorod assemblies for environmental toxin sensing. Angew. Chem. Int. Ed., 2010, 49: 5472–5475.

[66] Zhen S J, Huang C Z, Wang J, et al. End-to-end assembly of gold nanorods on the basis of aptamer-protein recognition. J. Phys. Chem. C, 2009, 113: 21543–21547.

[67] Wang Y, Li Y F, Wang J, et al. End-to-end assembly of gold nanorods by means of oligonucleo- tide- mercury (II) molecular recognition. Chem. Commun., 2010, 46: 1332–1334.

[68] Wang J, Zhang P, Li C M, et al. A highly selective and colorimetric assay of lysine by molecular-driven gold nanorods assembly. Biosens. Bioelectron., 2012, 34: 197–201.

[69] Zhang L, Chen H, Wang J, et al. Tetrakis (4-sulfonatophenyl) porphyrin-directed assembly of gold nanocrystals: Tailoring the plasmon coupling through controllable gap distances. Small, 2010, 6: 2001–2009.

[70] Wang J, Zhang P, Li J Y, et al. Adenosine-aptamer recognition-induced assembly of gold nanorods and a highly sensitive plasmon resonance coupling assay of adenosine in the brain of model SD rat. Analyst, 2010, 135: 2826–2831.

[71] Jana N R, Gearheart L, Murphy C J. Wet chemical synthesis of high aspect ratio cylindrical gold nanorods. J. Phys. Chem. B, 2001, 105: 4065–4067.

[72] Kim F, Song J H, Yang P. Photochemical synthesis of gold nanorods. J. Am. Chem. Soc., 2002, 124: 14316–14317.

[73] Yu Y-Y, Chang S-S, Lee C-L, et al. Gold nanorods: Electrochemical synthesis and optical properties. J. Phys. Chem. B, 1997, 101: 6661–6664.

[74] Zijlstra P, Bullen C, Chon J W M, et al. High-temperature seedless synthesis of gold nanorods. J. Phys. Chem. B, 2006, 110: 19315–19318.

[75] Gao C, Zhang Q, Lu Z, et al. Templated synthesis of metal nanorods in silica nanotubes. J. Am. Chem. Soc., 2011, 133: 19706–19709.

[76] Huang Y F, Lin Y W, Lin Z H, et al. Aptamer-modified gold nanoparticles for targeting breast cancer cells through light scattering. J. Nanopart. Res., 2009, 11: 775–783.

[77] Alkilany A M, Nagaria P K, Hexel C R, et al. Cellular uptake and cytotoxicity of gold nanorods: Molecular origin of cytotoxicity and surface effects. Small, 2009, 5: 701–708.

[78] Wang C, Ma Z, Wang T, et al. Synthesis, assembly, and biofunctionalization of silica-coated gold nanorods for colorimetric biosensing. Adv. Funct. Mater., 2006, 16: 1673–1678.

[79] Link S, Mohamed M B, El-Sayed M A. Simulation of the optical absorption spectra of gold nanorods as a function of their aspect ratio and the effect of the medium dielectric constant. J. Phys. Chem. B, 1999, 103: 3073–3077.

[80] Brioude A, Jiang X C, Pileni M P. Optical properties of gold nanorods: DDA simulations supported by experiments. J. Phys. Chem. B, 2005, 109: 13138–13142.

[81] Liu M, Guyot-Sionnest P. Synthesis and optical characterization of Au/Ag core/shell nanorods. J. Phys. Chem. B, 2004, 108, 5882–5888.

[82] Zhu J, Huang L, Zhao J, et al. Shape dependent resonance light scattering properties of gold nanorods. Mater. Sci. Eng., 2005, 121: 199–203.

[83] Slaughter LS, Wu Y, Willingham BA, et al. Effects of symmetry breaking and conductive contact on the plasmon coupling in gold nanorod dimers. ACS Nano, 2010, 4: 4657–4666.

[84] Chen H, Ming T, Zhang S, et al. Effect of the dielectric properties of substrates on the scattering patterns of gold nanorods. ACS Nano, 2011, 5, 4865–4877.

[85] Ni W, Ambjörnsson T, Apell S P, et al. Observing plasmonic-molecular resonance coupling on single gold nanorods. Nano Lett., 2010, 10: 77–84.

[86] Huang X, El-Sayed I H, Qian W, et al. Cancer cells assemble and align gold nanorods conjugated to antibodies to produce highly enhanced, sharp, and polarized surface Raman spectra: A potential cancer diagnostic marker. Nano Lett., 2007, 7: 1591–1597.

[87] Zijlstra P, Chon JWM, Gu M. Five-dimensional optical recording mediated by surface plasmons in gold nanorod. Nature, 2009, 459: 410–413.

[88] Yavuz M S, Cheng Y, Chen J, et al. Gold nanocages covered by smart polymers for controlled release with near-infrared light. Nat. Mater., 2009, 8: 935–939.

[89] Gao L, Fei J, Zhao J, et al. Hypocrellin-loaded gold nanocages with high two-photon efficiency for photothermal/photodynamic cancer therapy in vitro. ACS Nano, 2012, 6: 8030–8040.

[90] Tian L, Gandra N, Singamaneni S. Monitoring controlled release of payload from gold nano-cages using surface enhanced Raman scattering. ACS Nano, 2013, 7: 4252–4260.

[91] Liu Y, Huang C Z. Screening sensitive nanosensors via the investigation of shape-dependent localized surface plasmon resonance of single Ag nanoparticles. Nanoscale, 2013, 5: 7458–7466

[92] Kamat P V, Flumiani M, Hartland G V. Picosecond dynamics of silver nanoclusters. photoejection of electrons and fragmentation. J. Phys. Chem. B, 1998, 102: 3123–3128.

[93] Ling J, Sang Y, Huang C Z. Visual colorimetric detection of berberine hydrochloride with silver nanoparticles. J. Pharm. Biomed. Anal., 2008, 47(4–5): 860–864.

[94] Wang H Y, Li Y F, Huang C Z. Detection of ferulic acid based on the plasmon resonance light scattering of silver nanoparticles. Talanta, 2007, 72(5): 1698–1703.

[95] Wu L P, Li Y F, Huang C Z, et al. Visual detection of Sudan dyes based on the plasmon resonance light scattering signals of silver nanoparticles. Anal. Chem., 2006, 78(15): 5570–5577.

[96] Yin Y, Li Z Y, Zhong Z, et al. Synthesis and characterization of stable aqueous dispersions of silver nanoparticles through the Tollens process. J. Mater. Chem., 2002, 12(3): 522–527.

[97] Huang J T, Zeng Q L, Yang X X, et al. A simple green route to prepare stable silver nanoparticles with pear juice and a new selective colorimetric method for detection of cysteine. Analyst, 2013, 138(18): 5296–5302.

[98] McGilvray K L, Fasciani C, Bueno-Alejo C J, et al. Photochemical strategies for the seed-mediated growth of gold and gold-silver nanoparticles. Langmuir, 2012, 28: 16148–16155.

[99] Ravi S S, Christena L R, SaiSubramanian N, et al. Green synthesized silver nanoparticles for selective colorimetric sensing of Hg^{2+} in aqueous solution at wide pH range. Analyst, 2013, 138: 4370–4377.

[100] Liu T, Su Y, Song H, et al. Microwave-assisted green synthesis of ultrasmall fluorescent water-soluble silver nanoclusters and its application in chiral recognition of amino acids. Analyst, 2013, 138: 6558–6564.

[101] Jana N R, Gearheart L, Murphy C J. Wet chemical synthesis of silver nanorods and nanowires of controllable aspect ratio. Chem. Commun., 2001, (7): 617–618.

[102] Li Z C, Zhou Q F, He X H, et al. Synthesis of silver nanocubes and nanorods through sodium-bromide-assisted polyol route. Micro Nano Lett., 2011, 6: 261–264.

[103] Liu Y, Huang C Z. Real-time dark-field scattering microscopic monitoring of the in situ growth of single Ag@Hg nanoalloys. ACS Nano, 2013, 7: 11026–11034.

[104] Jin R, Cao Y, Mirkin C A, et al. Photoinduced conversion of silver nanospheres to nanoprisms. Science, 2001, 294: 1901–1903.

[105] Sherry L J, Jin R, Mirkin C A, et al. Localized surface plasmon resonance spectroscopy of single silver triangular nanoprisms. Nano Lett., 2006, 6: 2060–2065.

[106] He R, Qian X F, Yin J, et al. Formation of silver nanoprisms and their optical properties. Chemical J. Chinese Univ., 2003, 24: 1341–1345.

[107] Métraux G S, Mirkin C A. Rapid thermal synthesis of silver nanoprisms with chemically tailorable thickness. Adv. Mater., 2005, 17: 412–415.

[108] Zhang L, Huang C Z, Li Y F, et al. Morphology control and structural characterization of Au crystals: From twinned tabular crystals and single-crystalline nanoplates to multitwinned decahedra. Cryst. Growth Des., 2009, 9: 3211–3217.

[109] Elghanian R, Storhoff J J, Mucic R C, et al. Selective colorimetric detection of polynucleotides based on the distance-dependent optical properties of gold nanoparticles. Science, 1997, 277: 1078–1081.

[110] Sato K, Hosokawa K, Maeda M. Rapid aggregation of gold nanoparticles induced by non-cross-linking DNA hybridization. J. Am. Chem. Soc., 2003, 125: 8102–8103.

[111] Liu Z D, Li Y F, Ling J, et al. A localized surface plasmon resonance light-scattering assay of mercury (II) on the basis of Hg^{2+}-DNA complex induced aggregation of gold nanoparticles. Environ. Sci. Technol., 2009, 43: 5022–5027.

[112] Qi W J, Wu D, Ling J, et al. Visual and light scattering spectrometric detections of melamine with polythymine-stabilized gold nanoparticles through specific triple hydrogen-bonding recognition. Chem. Commun., 2010, 46: 4893–4895.

[113] Li H, Rothberg L. Colorimetric detection of DNA sequences based on electrostatic interactions with unmodified gold nanoparticles. Proc. Nati. Acad. Sci. USA, 2004, 101: 14036–14039.

[114] Ai K, Liu Y, Lu L. Hydrogen-bonding recognition-induced color change of gold nanoparticles for visual detection of melamine in raw milk and infant formula. J. Am. Chem. Soc., 2009, 131(27): 9496–9497.

[115] Du J, Yin S, Jiang L, et al. A colorimetric logic gate based on free gold nanoparticles and the coordination strategy between melamine and mercury ions. Chem. Commun., 2013, 49(39): 4196–4198.

[116] He W, Li Y F, Huang C Z, et al. One-step label-free optical genosensing system for sequence-specific DNA related to the human immunodeficiency virus based on the measurements of light scattering signals of gold nanorods. Anal. Chem., 2008, 80: 8424–8430.

[117] Chen Z, Lei Y, Xu H, et al. Functionalized gold nanorods as an immunosensor probe for neuron specific enolase sensing via resonance light scattering. J. Mater. Chem. B, 2013, 1: 3031–3034.

[118] Zhang K, Shen X. Cancer antigen 125 detection using the plasmon resonance scattering properties of gold nanorods. Analyst, 2013, 138: 1828–1834.

Jian Wang, Jian Ling, Yue Liu, Peng Fei Gao

7 Nano light scattering spectrometry

As mentioned above, metallic and nonmetallic nanoparticles are great light scattering probes with their application prospects in analytical chemistry. Currently, many detection methods have been established based on their light scattering signal, especially localized plasmon resonant light scattering analysis method of noble metal nanoparticles have wide applications.

7.1 Basic principle

In the description of light scattering properties, absorption cross section and scattering cross section are involved. The both concepts describe probability density: the absorbance cross section is the absorption probability of interaction between a moving particle and a static one, and the scattering cross section is the scattering probability. The probability density is shown as a moving particle number in unit area that is vertical to the moving direction in unit time. For a spherical particle, when other parameters are constant, intensity is directly proportional to the scattering cross section. Thus, the scattering cross section can be expressed as follows [1]:

$$C_{\text{sca}} = \frac{8}{3} \pi r^2 x^4 \left| \frac{m^2 - 1}{m^2 + 2} \right|^2 \tag{7.1}$$

where C_{sca} is the scattering cross section, x is the particle parameter, which is equal to $2\pi r n_{\text{med}}/\lambda$; r is the spherical particle diameter; m is the ratio of particle refractive index (n_{sph}) with that of the environment(n_{med}). By introducing particle parameter, the equation (7.1) can be further expressed as follows:

$$C_{\text{sca}} = \frac{24 V^2 \pi^3 n_{med}^4}{\lambda^4} \cdot \left| \frac{m^2 - 1}{m^2 + 2} \right|^2 \tag{7.1a}$$

where V is the scattered particle volume.

It can be understood from equation (7.1) that particle size (r or V) greatly influences the scattering light intensity. In addition, the medium dielectric constant is also important. Various scattering analysis methods based on metal nanoparticle can be derived.

For example, construction of a nano-biosensor can be based on changes in the dielectric coefficient of nanoparticles induced by adsorption of biomolecules, and building a light scattering visible analysis and scattering imaging analysis can be

https://doi.org/10.1515/9783110573138-007

based on slight color changes in nanoparticles, or establishing an analytical method could be based on light scattering enhancement from nanoparticles aggregation. In the following chapters, we will discuss about how to use these properties of metallic nanoparticles to establish a light scattering method.

7.2 Nano light scattering spectrometry of noble metal nanoparticles

7.2.1 Biological labeling and staining by light scattering probes

As noble metal nanoparticles have special local surface plasmon resonance scattering properties, they can scatter different colors of light under white light illumination (see Figure 6.9). Similar to a fluorescence probe, well-prepared metal nanoparticle can be used as a light scattering probe for labeling and staining. The application of metal nanoparticles light scattering probes presents high sensitivity and simplicity, that has been widely used in immunoassay, chemical and biological sensing, and visual analysis. Related commercial products, such as urine, pregnancy, and HIV test papers are in use nowadays. The development and the use of these related kits are a source of income for the national economy and also is convenient for scientific studies, productive practice, and daily life.

Light scattering labeling and staining of metal nanoparticles have the following advantages:
(1) Compared with common fluorophores, such as fluorescent small organic molecules or fluorescent proteins, metallic nanoparticles have a high scattering quantum yield
(2) Fluorescent small organic molecules or fluorescent proteins have severe photobleaching, whereas plasmon resonance light scattering signal of noble metal nanoparticles do not have photobleaching
(3) Compared with fluorescent semiconductor quantum dots, noble metal nanoparticles show good biocompatibility, that have advantages in in-vivo analysis and biological tracer studies.

7.2.1.1 Light scattering labeling

Noble metal nanoparticles, such as gold nanoparticles can form Au–S bond with thiol, that allows labeling of nanoparticles with proteins, nucleic acids, or virus. If the target is moving, dynamic monitoring and real-time tracing can be achieved.

For example, Mirkin group used two different sizes of gold nanoparticles with different color for light scattering labeling. They selected 50 nm gold nanoparticles with green light scattering and 80 nm gold nanoparticles with yellow light scattering, and labeled target DNA complementary DNA sequences, respectively. After capturing of target DNA, the gold nanoparticles probes would attach on the solid phase. Thus, yellow and green color light scattered by the two kinds of gold nanoparticles can be seen clearly under white light illumination (as shown in Figure 7.1), and a simultaneous detection of two different target can be achieved [2].

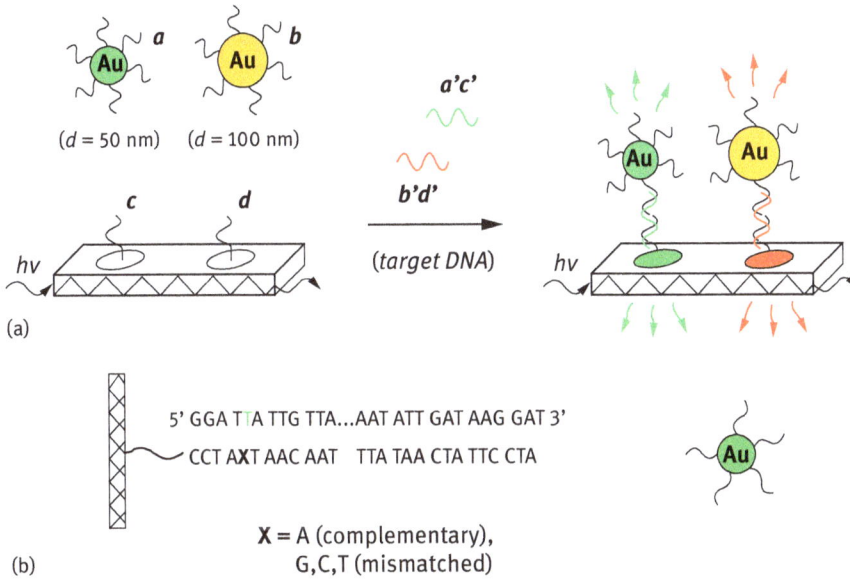

(a)

(b)

$X = A$ (complementary), G,C,T (mismatched)

Figure 7.1: Double-color light scattering analysis for DNA detection [2]. (A) Schematic diagram. (B) The used DNA sequence for detection.

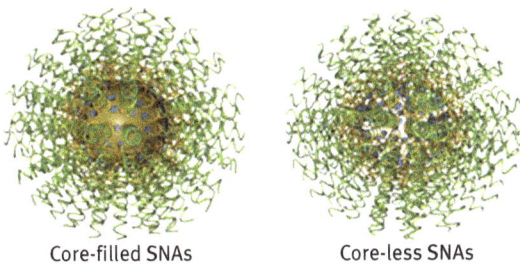

Core-filled SNAs   Core-less SNAs

Figure 7.2: Schematic diagram of spherical nucleic acid structure [3].

In the labeling process, there are many molecules that can bind to the surface of the nanoparticle. Chemical reaction conditions are needed for controlling its binding density. Based on these requirements, a labeling target and gold nanoparticles can be combined together to a new spectral probe, spherical nucleic acids (SNAs) (Figure 7.2) developed by Cutler is a good example [3]. Compared to ordinary fluorescent dyes, gold nanoparticle scattering probe has a higher sensitivity.

In addition to the labeling DNA, metal nanoparticles can be labeled with proteins and applied in immunoassay. For example, in the traditional sandwich immunoassay, antibody usually tagged with fluorescence, enzyme, and radioactive label. When a metal nanoparticle probe is used, its strong local plasmon resonance scattering property makes the assay has a similar or better sensitivity. Commercial immune scattering analysis products using gold nanoparticles has been widely applied for practice. If gold and silver nanoparticles are labeled on virus, real-time monitoring and analysis of infection trace are possible, which will be further described in Chapter 8.

7.2.1.2 Light scattering bio-staining

Noble metal nanoparticles has been widely used for bio-staining as its high sensitivity. For example, with strong plasmon resonance scattering properties of silver nanoparticles, its scattered light can be clearly seen with eyes irradiated by a common LED flashlight [4]. Detection of immune response can be realized through a simple method without using large-sized instruments.

Figure 7.3 is a classic example of light scattering staining. As silver nanoparticle only adsorbs on the slide binded with antibody, so immune response area can be directly observed with naked eyes on the slide. In addition, by observation and analysis of the silver nanoparticles adsorbed on the slide, it can be found that when the amount of antibody is less than the amount of the fixed antigen on the slide, then immune response occurs only in a partial area of the slide, whereas when antigen concentration is high, immune response occurs all over the slide. In other words, repeatability, uniformity, and stability of the solid-phase analysis need to be considered carefully.

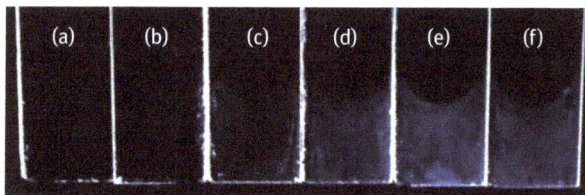

Figure 7.3: Light scattering staining [4]. Realization of visible detection of an immune response with LED flashlight on the slide.

7.2.2 Light scattering analysis based on particle growth

Based on light scattering principles, the increase in number and size of nanoparticles can enhance the light scattering signal. Analytical methods using for detection of Sudan red [5], glucose [6], small organic molecular drugs, and formaldehyde in the environment [7] can be built on these principles.

Sudan red is a chemical dye rather than a food additive. Its chemical component consists of an azo structure with certain carcinogenicity that is toxic to the human liver and kidneys. Most countries have banned it for food production. Research has found that in alkaline conditions, its naphthol structure undergoes isomerization, which can be further oxidized as a quinone structure with $AgNO_3$. In this process, $AgNO_3$ is reduced to silver atom and further forms to silver nanoparticles. The light scattering increase of the produced silver nanoparticles is related with the Sudan red concentration (Figure 7.4), and thus, a light scattering method for Sudan red can be established [5].

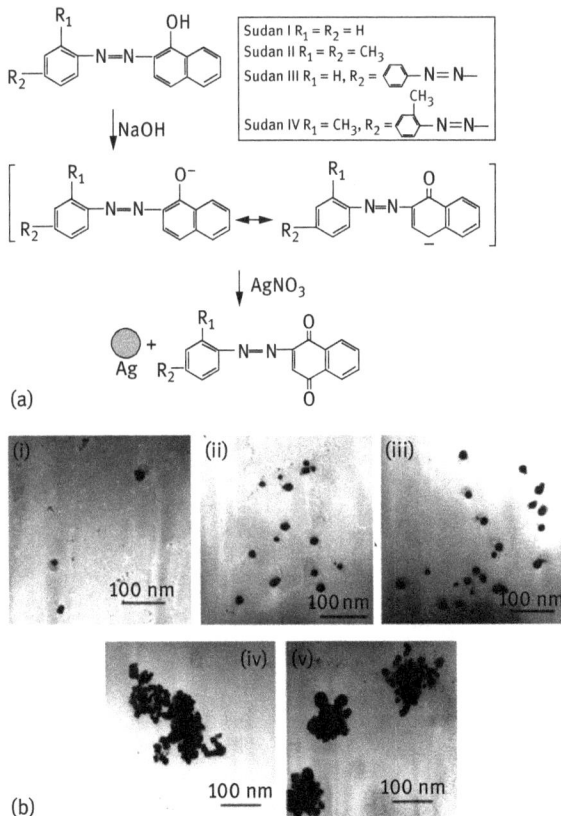

Figure 7.4: Sudan red detection in food. (a) Mechanism. (b) TEM image of silver nanoparticle.

Silver nitrate and ammonia molecules form silver ammonia solution. In the present of a reductant, silver nanoparticles could produce in the silver ammonia solution. The light scattering intensity of the solution would increase with the producing of silver nanoparticles, that can be used for establishing a simple, rapid, and sensitive analytical methods for reducing substance, such as formaldehyde [7].

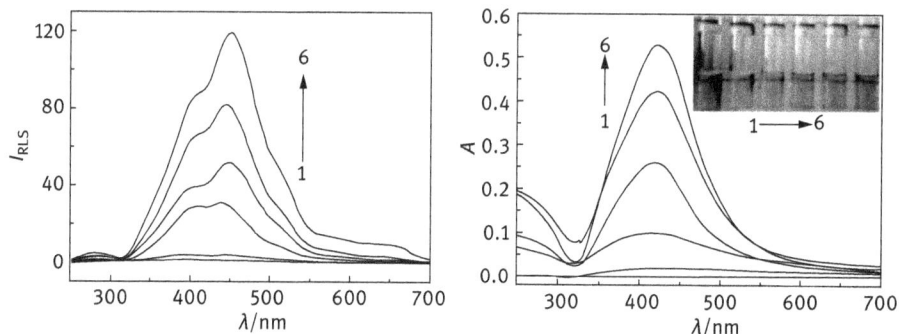

Figure 7.5: Light scattering and absorption spectra of silver nanoparticles produced in silver ammonia solution for the detection of formaldehyde [7]. 1–6 shown in arrow are formaldehyde concentrations (mol/L): 0, 1.0×10^{-6}, 5×10^{-6}, 1.0×10^{-5}, 1.5×10^{-5}, and 2.0×10^{-5}.

Diabetes is a metabolic disease caused by hyperglycemia of insulin secretion dysfunction or other dysfunction, or the involvement of both. Long-term hyperglycemia leads to chronic injury, dysfunction of eyes, kidneys, heart, blood vessels, and nerves. Hyperglycemia is a classic feature of diabetes. In hyperglycemia, blood sugar mainly refers to glucose. The detection of glucose can be realized by gold nanoparticles that produced by reducing of $HAuCl_4$ using glucose [6]. However, as the reducibility of glucose is weak, the pre-prepared small gold nanoparticles are added as crystal nucleus to catalyze reduction of large gold nanoparticles. The higher the glucose concentration is, the larger will be the particle size, and the stronger will be the plasmon resonance light scattering signal. Glucose analysis method can be built on this (Figure 7.6). Another

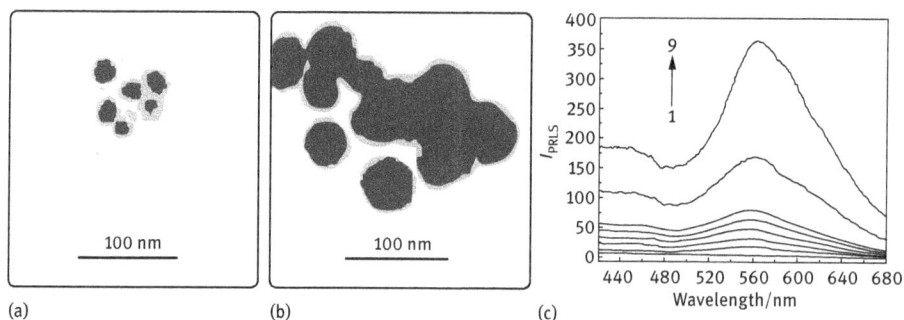

Figure 7.6: Light scattering detection of glucose [6]. (a, b): TEM image of different concentrations of glucose-reduced gold nanoparticles; (c)scattering signal curves, where the arrow 1–9 are glucose concentrations (μmol/L) 0,0,2.0,4.0,6.0,8.0,10.0,20.0,and 40.0, where 1 has $HAuCl_4$, 2 has no HAucl4; and 3–9 has 0.2 mmol/L $HAuCl_4$.

interesting phenomenon is that nanoparticle number is almost unchanged with its growth process.

It needs to be demonstrated that based on the oxidation–reduction principle, various nanomaterials can be prepared. Analysis and detection is a dynamic process in which the light scattering signal would change in the process of reaction.

Of course, there are many ways of nanoparticle production. As long as the target substance reacts specifically with the reagent to produce nanoparticles, an analysis method can be established through the enhanced light scattering signal from nanopartilces. This method is simple and sensitive, while its selectivity needs to be improved.

7.3 Single particle light scattering spectrum

Light scattering particles exist in nature in the form of dust in air, fog drop, micro-organisms in water, dye aggregate in solution, metal nanoparticles, and so on. By simultaneously scanning the excitation and emission monochromators of a common spectrofluorometer, we can obtain their scattering spectrum, or by using a dark-field microscopy, we can observe light scattering images of particles. Researchers have established a series of light scattering analysis methods with these particles.

As demonstrated in Chapter 5, the first resonance light scattering analysis methods is light scattering that comes from small organic molecular probes aggregates to analyze and detect nucleic acids and proteins (Section 5.2) [8, 9]. Afterward, light scattering change of nanoparticles aggregates induced by analytical target is used for analytical purpose (Section 5.4) [10, 11]. These methods were based on a overall system, referring to light scattering properties of the whole system, without reflecting the properties of single nanoparticles in the system. These methods also have disadvantages of high sample consumption, low detection sensitivity, low spectral resolution, and so on. These dis-advantages limit the applications of light scattering particles.

With the appearance of new optical detection instruments, especially optical ima-ging detection system, single particle scattering light can be easily detected. So single nanoparticles analysis (SNAs) has been built and applied for biochemical analysis. It has gradually become a hot research in the field of light scattering analysis. As SNAs has narrowed the detection unit to a single particle, or even a single nanoparticle level, it can effectively solve the disadvantages of the previously mentioned methods.

This chapter will introduce SNAs in recent years, including single particle count-ing, single particle sensing, single particle tracing imaging analysis, single particle reaction imaging monitoring, and single particle dark-field imaging studies. As plasmon nanoparticles have excellent local surface plasmon resonance light scatter-ing properties, the application of a single particle as a unit for biosensing has great prospect. So, this chapter mainly discusses the light scattering analysis and applica-tions of single plasmonic nanoparticles by the techniques of dark-field imaging and single particles light scattering spectroscopy.

7.3.1 Single particle counting analysis

Single particle counting is an analytical technology of single scattering particle statistics with a light scattering signal detection instrument. In 2003, authors of this book developed a light scattering imaging technology for counting of protein-induced $\alpha,\beta,\gamma,\delta$-4-sulfobenzoic porphyrin (TPPS) aggregation particle for protein detection [12]. This was a pioneering work in the light scattering imaging and image processing.

Without proteins, TPPS molecules do not aggregate, and the unbounded TPPS molecules that have quite weak molecular scattering signals; with proteins, TPPS molecules aggregate to form large particles with strong scattering light. For single particle counting, we built an light scattering imaging device. By using this device, the light scattering of TPPS aggregates excited by a laser from the side direction was observed through a fluorescence microscope and images were taken with a charge coupled device (CCD). In an imaging focal plane, scattering light from TPPS aggregates is easily observed by taking photos with a CCD. In these images, each light spot represents an aggregate of TPPS. With the increase in protein concentration, there are more light points in the images (Figure 7.7). As the point number has a linear relation with the protein concentration, single particle counting can be used to build a protein analysis method [12].

For reasonable counting, some parameters need to be set. As shown in Figure 7.7, the light points are mainly light scattering of aggregate particles on a focal plant, some are weak while others are strong. Scattering light of aggregate particles that are not present on a focal plant is weak, so counting of particles should have one standard. In our study, light intensity is used as a standard for data statistics.

It needs to be demonstrated that as there are many weak particles, which are not counted, because the light scattering has a high background (Figure 7.7). To improve sensitivity of this analysis, a dark-field microscopy system needs to be introduced to reduce the background. In recent years, dark-field optical microscopy greatly reduced the single particle counting detection limit because dark-field imaging has a very weak background, which is confirmed by current reports. Professor He Yan's group used a dark-field microscopy to count gold nanoparticles and their aggregates, and established a highly sensitive DNA detection method [13]. Monodisperse 13 nm gold nanoparticles have a weak scattering light, which can be rarely observed under a dark-field microscopy. When target DNA exists, sandwich-like DNA hybridization would lead to the aggregation the dispersed gold nanoparticles, thus the observed scattering light particles would change. By counting the amount of light scattering particles, they established a highly sensitive method for DNA detection with LOD of 0.1pmol/L.

For single particle counting, sensitivity of detector is an important factor for sensitivity of an analytical method. So, the development of a highly sensitive detector for particles counting can effectively lower the detection limit. Which is even lower to

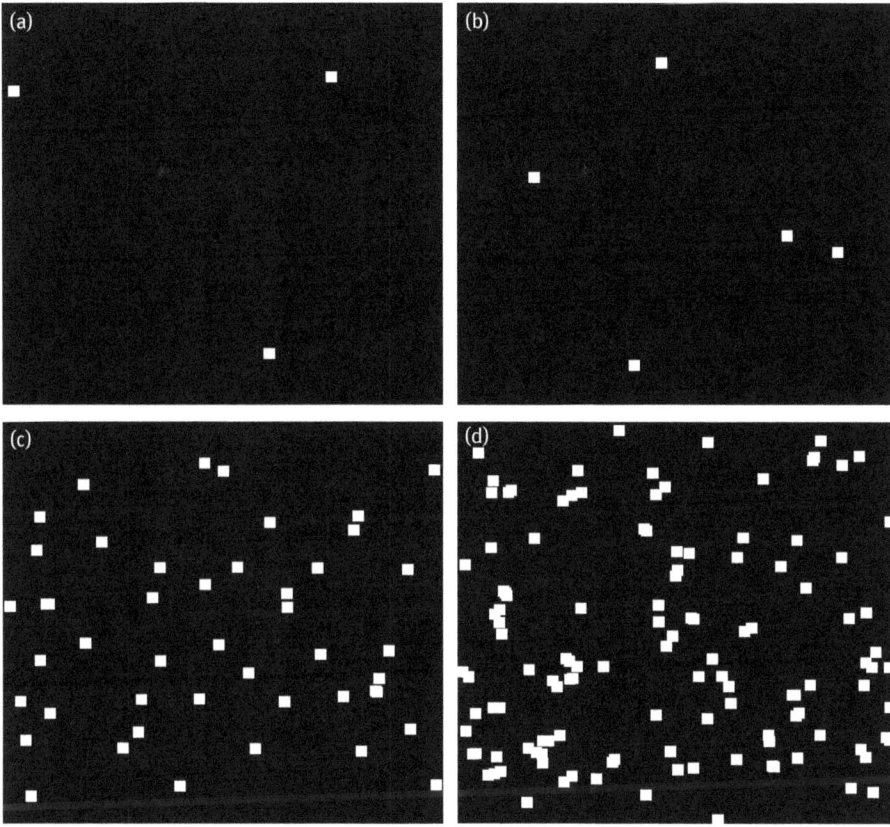

Figure 7.7: Particle counting of protein-induced porphyrin concentration. a–d, TPPS concentration (μmol/L) is 0.1, 0, 0.1, and 0.1; a–d, BSA concentration (ng/mL) 0, 2000.0, 60.0, and 180.0. All images are of 320×280 pixel size, which are about 0.02 cm². Images are taken from reference [12] with appropriate disposal.

single particle level. Yan X M research group of Xiamen University carried out a series of excellent researches in this filed. They built an ultrasensitive flow cytometry to detect the size of gold nanoparticles and their concentration through scattering light signal [14]. More innovative is that the method can detect nanoparticles at concentration of fmol/L level at sample volume of microliter level. So, the development of a highly sensitive detector is a future direction of single particle counting analysis. Combined with ultrasensitive light scattering particle counter, Li N research group from Peking University established a highly sensitive single particle counting method for detection of environmental and clinical bio-particles [15].

For single particle counting, interference light scattering produced by polluted particles (black products) in the medium or air should also be carefully considered [16, 17].

7.3.2 Single particle light scattering spectrometry

7.3.2.1 Features of single particle light scattering spectrometry

Metal nanoparticles have been utilized in the analytical detection of many biological molecules and environmental pollutants in clinical examination, drug and environmental analysis, and so on [2, 18, 19]. However, these methods were built mainly based on nanoparticle scattering light intensity changes induced by a target or a substance.

The single nanoparticles included in this chapter are all plasmonic nanoparticles, mainly include silver and gold nanoparticles. Plasmonic nanoparticles have excellent local plasmon resonant light absorption and scattering properties from the collective oscillation of free electrons on nanoparticle surface. As described in Section 6.3.1, noble metal nanoparticles have a local surface plasmon resonance (LSPR) property that are closely related to nanoparticle material, size, shape, surrounding medium, and plasmon coupling of nearby particles [20], so metal nanoparticles are widely used as LSPR sensor [21].

When combined with a dark-field microscopy imaging system (Figure 7.8), analytical sensing technology based on a single nanoparticle is the mainstream of current metal nanoparticle light scattering analysis, because this technology realized the characterization of LSPR property of a single nanoparticle. As described in Section 1.2, Tyndall found that the floating dust was clear in a dark background in his study of aerothermodynamics; firework display is prettier in a dark night. The same goes for a dark-field microscope. Single nanoparticle scattering signal has a high signal-to-noise ratio at low background, which greatly improves the sensitivity and spectral resolution of single nanoparticles spectrum. In addition, as single nanoparticle has a small physical size that produces a steady signal with high scattering efficiency, the advantage in bio-system applications stands out. For example, through real-time monitoring of a light scattering property change of a single noble metal nanoparticle in a bio-system, we can trace the overall process of nanoparticle cytophagy, understanding some key molecular events in a cell, providing new ways in the diagnosis of some rare diseases. Currently, single nanoparticle scattering analysis mainly focuses on the following aspects:
(1) Characterization and application of nanoparticle property changes
(2) Analysis of micro-environment changes by nanoparticle
(3) Analysis based on plasmon resonance coupling of nanoparticles.
(4) Plasmon resonance energy transfer study and application

7.3.2.2 Instrument system of a single particle light scattering spectrometry

As shown in Figure 7.8, a dark-field imaging spectral system is mainly composed of four parts: (1) dark-field microscopic observation, (2) true-color CCD imaging, (3)

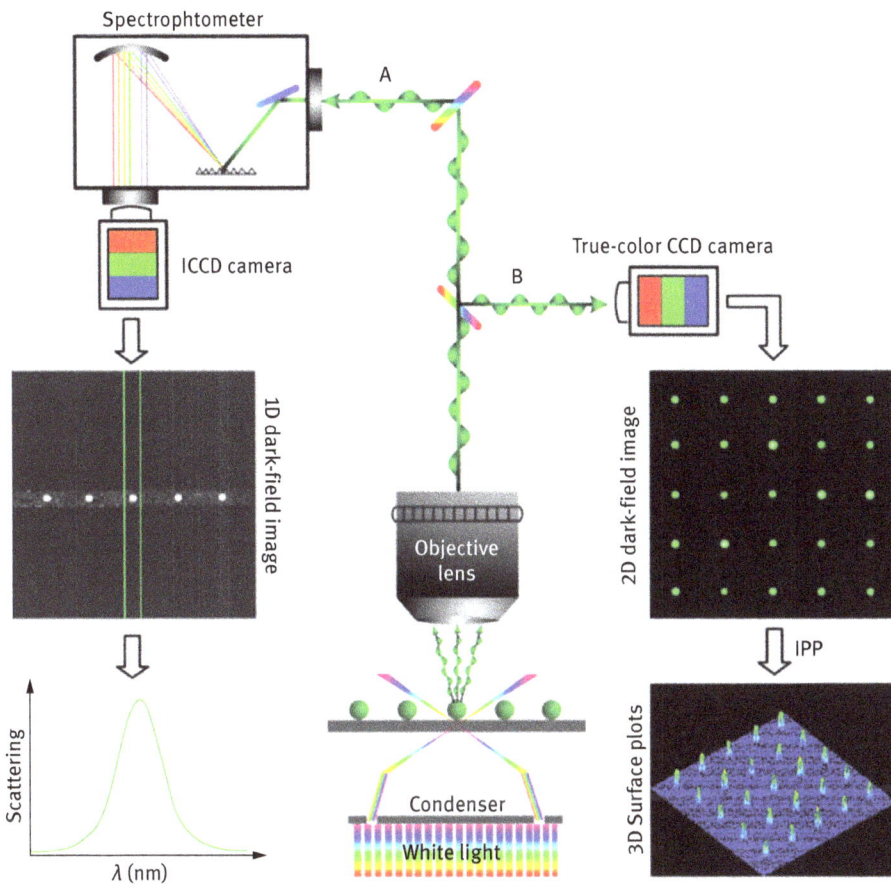

Figure 7.8: Instrument diagram of a single nanoparticle dark-field scattering imaging analysis.
(A) A single nanoparticle spectral scanning system, including micro-spectrometer and intensified charge-coupled device (ICCD) camera. After ICCD imaging, light selects a target particle and then performs spectral scanning.
(B) Photographs of dark-field scattering imaging system, with only one true-color CCD camera for a two-dimensional area of a metal nanoparticle scattering. Software such as Image-Pro Plus (IPP) and other IPP can be used for analyzing and obtaining a light intensity or a tricolor message from a single nanoparticle.

microspectrometer, and (4) micro-imaging spectral CCD analysis systems. Among these, dark-field microscopic observation system is most important, which is composed of an upright microscope (or inverted microscope, see Figure 3.12 and Figure 7.8) with high numerical aperture dark-field condenser and special high power objective. Its works on the mechanism of observing a target signal through an objective in concentrated light of a condenser (Figure 3.12). There are two imaging connectors of a microscope, which are connected to a microspectral system (Figure 7.8A) and a true-color CCD imaging system, respectively (Figure 7.8B). It needs to be noted that limited by resolution of the microcopy, the system obtains a light scattering signal of a target compound rather than the morphological features of scatterer itself.

To obtain scattering spectral features of target, a microspectrometer slit is needed to control its imaging zone for spectral resolution of microbiological tissues, organelles, or single scattering probe signals (Figure 7.8B). A split light signal is captured by another spectral intensified charge-coupled device (ICCD), where a linear spectrogram is obtained by a special software for micro-spectrum processing.

7.3.3 Real-time monitoring and analytical application of single nanoparticles

The LSPR property of noble metal nanoparticle is closely related to its composition, size, and shape (Section 6.3.1). At present, most precious metal nanomaterials are gold and silver nanoparticles. Their LSPR absorption and scattering are all located in the visible region. For example, 50 nm spherical AgNPs present a characteristic absorption peak near 410 nm, whereas 50 nm spherical AuNPs have a characteristic absorption peak near 530 nm; their LSPR optical properties of nano-alloy is quite different from that of single element nanoparticle. For another example, gold nanorod has two local plasma vibration absorption and scattering characteristics in transverse and longitudinal directions, but if silver nano shell is covered to form Au@Ag nanorods, the transverse and longitudinal LSPR absorption and scattering properties vary significantly [22]. So, element and environment greatly influence its LSPR optical properties.

In practical applications, top-level designs can be made according to the required specific optical properties, and then further synthesiz optical nanomaterials. A top-level design needs to consider three factors: The first one is nanoparticles with different elements, that is the material problems, which can be either a single metallic element or an alloy; Secondly, the LSPR properties of metal nanoparticles with the same element composition are significantly affected by the morphology and size of the nanoparticles [20]; and thirdly, for alloys, different proportion and shape are needed. Figure 7.9 shows that the resonance frequencies of plasmon resonance of nanoparticles with different sizes and shapes are discrepant, thus exhibiting very distrinct LSPR optical absorption and scattering properties. So, in theory, single nanoparticles can be used for detecting substances that cause the change of the composition, size, and shape of single nanoparticles.

In the preparation process of nanoparticles, the each growth stage of nanoparticles may be accompanied by the changes of element composition, shape and size. Therefore, the growth process of nanoparticles is also accompanied by the LSPR optical properties changes. In turn, we can observe the growth of nanoparticles in real time by monitoring the changes of LSPR properties during the growth of single nanoparticles.

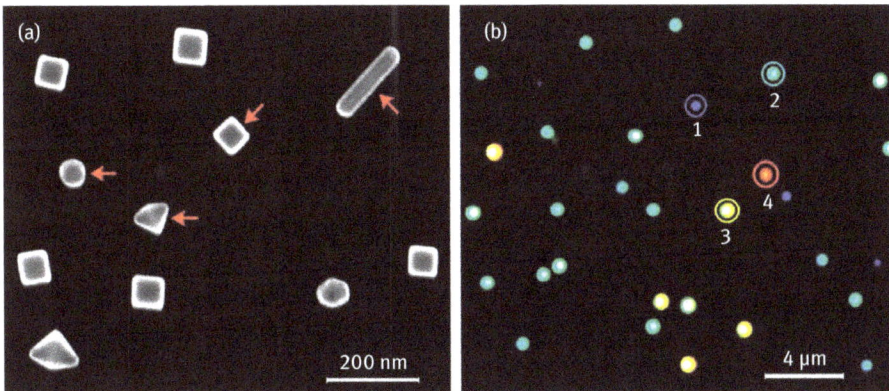

Figure 7.9: TEM of different shapes of silver nanoparticles (a) and corresponding light scattering images (b) [20] (refer to the colorful picture at the end-of-the-book) 1–4 in (b) are silver nanocubic (dark blue) in (a), silver nanorod (sky blue), silver nanosphere (yellow), and silver nano trigonal bipyramid (orange).

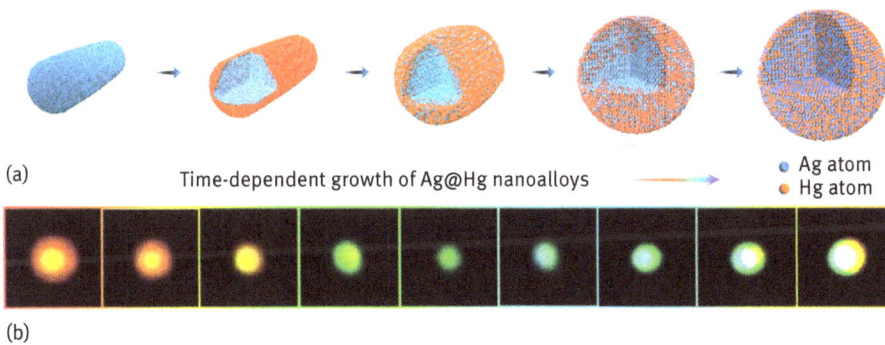

Figure 7.10: Time-dependent growth of Ag@Hg nanoalloys (a) Real-time monitoring of the in situ growth of a single Ag@Hg nanoalloy from a Ag nanorod by dark-field scattering microscopy and spectroscopy (b) [23].

For example, dark-field light scattering imaging was used to monitor the formation process of Ag@Hg nano-alloys and the growth of Ag@Hg nano-alloys by silver nanorods, silver triangular bipyramids, silver cubes and silver spheres (Figure 7.10). Thus, we determined that the formation mechanism of Ag@Hg nano alloy was directly through the algamation between AgNPs and mercury.

Ag–Hg nano-alloy formation goes through the following processes: (1) mercury ion rapidly adsorbs on the surface of AgNPs; (2) mercury ion diffusion and amalgamation reduce the aspect ratio and the sharpness of tips and edges and (3) mercury ion diffuses to form spherical Ag–Hg nano-alloy. The results of this study can be used to synthesize nano-materials with specific optical properties by monitoring the growth of nanoparticles, and to further apply in the fields of optical analysis and sensing. It should be pointed out that the monitoring process can not only rely on dark-field light scattering imaging, but also should be combined with scanning electron microscopy imaging, high-resolution transmission electron microscopy imaging, elemental analysis and other characterization methods to verify the formation of Ag@Hg nanoparticles.

For example, Long Y T research group studied the reduction of nicotinamide adenine dinucleotide (NAD$^+$, coenzyme I) to form NADH in vivo [25]. Because NAD$^+$/ NADH is an electronic carrier in the process of biochemical reactions in vivo, it is an extremely important reaction pathway in organisms and plays an important role in various enzymatic oxidation-reduction reactions. Balance in the cellular NAD$^+$ and NADH contents reflect the cell metabolism condition and health. With single AuNPs as sensor and monitoring of single AuNPs real-time light scattering spectrum of NADH/ Cu^{2+} oxidation–reduction reaction, Long Y T research group detected intracellular enzyme activity and studied anticancer drugs. The mechanism study shows that when both NADH and Cu^{2+} exist in the solution, Cu^0 is produced by oxidation-reduction reaction, and Cu^0 is deposited on the surface of gold nanoparticles to form Au@Cu core-shell structure nanoparticles, resulting in the red shift of the scattering spectrum of gold nanoparticles, and the degree of red shift is linearly related to the concentration of NADH. Red shift can be used to analyze reaction processes, further understanding the changes in cellular NADH.

Because the metabolism of cancer cells can be regulated by external anticancer drugs, Long Y T et al. [25] further used this kind of nanoprobe to detect the anticancer drug paclitaxel, which proved that paclitaxel can inhibit the metabolism of HeLa cells. These results indicate that precious metal nanoparticles as light scattering probes can be used in the screening of anticancer drugs, and can also be used in the monitoring of anticancer drug mechanism and pharmacological targeting process.

7.4 The microenvironment change sensing of nanoparticle

7.4.1 Light scattering sensing principle of single nanoparticle

All the above mentioned cases are the LSPR properties of nanoparticles are affected by their own material, morphology and size, but the influence of external environment on the LSPR of nanoparticles has not been involved. These are the following two situations:

(1) Previous studies have shown that the characteristic local plasmon resonance wavelength of a single AgNPs gradually shifts to a longer wavelength with the increase of the refractive index of the solvent medium around the nanoparticles, and there is a linear relationship between the displacement and the solvent refractive index [26].

(2) If biological or chemical molecules are adsorbed on the surface of noble metal nanoparticles, the microenvironment of the nanoparticles will change, leading to the change of LSPR properties.

Based on the above two cases, we can design biochemical analysis sensor unit based on a single nanoparticle light scattering. Because organic small molecule and biological macromolecule generally have larger refractive index than the environmental medium of nanoparticles, when the surface of nanoparticles is adsorbed with organic small molecule and biological macromolecule, the local microenvironment refractive index of nanoparticles will increase, the scattering light properties will be changed, and the scattering light red shift and scattering light intensity will increase. This is the basic theoretical basis for single nanoparticle analysis and sensing.

Scattering light spectral shift induced by biochemical molecular adsorption can be measured with the following equation:

$$\Delta\lambda_{max} = m(n_{absorbate} - n_{med})\left[1 - \exp\left(-\frac{2d}{l_d}\right)\right] \tag{7.2}$$

where m is the sensitivity factor, l_d is the attenuation length of the electromagnetic field, $n_{adsorbate}$ is the refractive index of the molecular adsorption layer, n_{med} is the medium refractive index, and d is the thickness of the adsorption molecule. Where m and l_d are two parameters decided by nanoparticle itself, such as material, size, shape, and so on.

For a given nanoparticle in a specific solvent, m, l_d, and n_{med} are constant, molecule-adsorbed-produced $\Delta\lambda_{max}$ is related to $n_{adsorbate}$ and d; whereas $n_{adsorbate}$ and d are decided by molecular size, so $\Delta\lambda_{max}$ is decided by the size of the molecule. Therefore, adsorption of biomacromolecule will generally cause greater spectral shift, and then establish the analytical method of biomacromolecule by wavelength shift.

7.4.2 Application of a single nanoparticle light scattering sensing

Single nanoparticle used for scattered light sensing can be performed in a medium of liquid-phase environment or gas-phase environment.

7.4.2.1 Liquid-phase analytical medium

By using single silver nanoparticles as probe, McFarland et al. studied the adsorption mechanism of thiol molecules on silver nanoparticles [26]. Results show that in different mediums, the characteristic local plasmon resonance scattering peak of the same silver nanoparticle has a red shift with an increase in the medium refractive

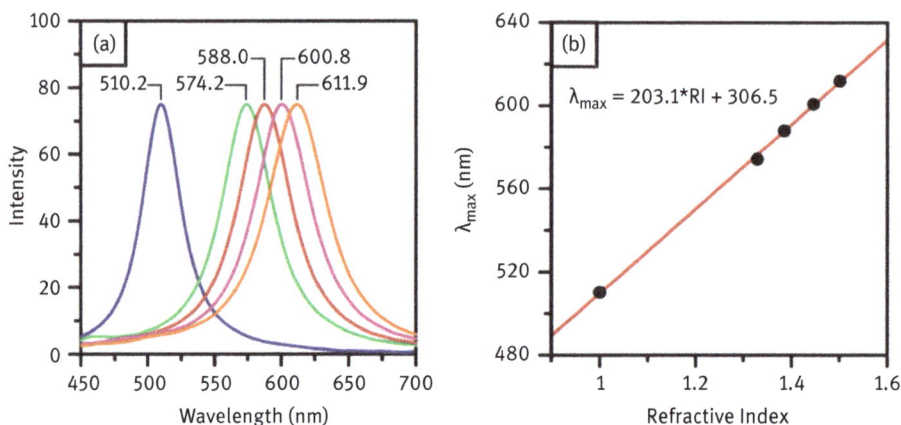

Figure 7.11: The relationship between scattering spectra of single nanoparticles and microenviron-ment. (a) Scattering spectra of single silver nanoparticles in different media; left to right: nitrogen, methanol, propanol, chloroform, and benzene (b) The relationship between scattering peak of silver nanoparticles and solvent refractive index. Photos are taken from reference [26].

index, and λ_{max} has a linear relationship with the refractive index (Figure 7.11). In order to further investigate the sensing ability of a single silver nanoparticle, the adsorption of 1-hexadecane mercaptan on its surface was studied. The results showed that the 1-hexadecane mercaptan monolayer formed on the surface of silver nanoparticles caused the red shift of 40 nm and the number of molecules adsorbed was about 6000.

Raschke et al. constructed a similar single nanoparticle sensing unit that was applied in bio-macromolecule analysis [27]. Biotinylated spherical gold nanoparticle is used as a sensing unit, specific adsorption of streptavidin molecule brings a red shift in the scattering spectra of gold nanoparticle; LOD of streptavidin is 1.0 μmol/L. Afterward, researchers used gold nanorod [28] and gold nanostar [29] as sensing units for the same analysis. The corresponding detection limits for gold nanorod and gold nanostar are 1.0 nmol/L and 0.1 nmol/L, respectively. Results show that the shape of the nanoparticle greatly influences the sensitivity of a single nanoparticle sensing. Similarly, a single nanoparticle sensor can be used as molecular scale plate for molecular length of DNA and bioenzyme activity test [30].

Hydrogen sulfide is a toxic gas that has a rotten egg smell, whereas human hydrogen sulfide content is related to many diseases, such as Alzheimer's disease, hyperglycemia, and arterial ischemia, and so on. So, hydrogen sulfide detection is quite meaningful. He Yan research group constructed a single nanoparticle system for imaging analysis of intracellular sulfides [31]. They synthesized silver shell gold core nanorod as a sensing unit, in which silver shell reacts with sulfide, serving as a sensing substance, whereas gold core serves as a signal reporting unit. Silver works with sulfide to form silver sulfide, and gold nanorod refractive index changes from 0.17 to 2.2, bringing a red shift to the light scattering spectra. The red shift wavelength of gold nanorods is linearly related to the exponential concentration of sulfides. Based on this, a method for determination of sulfide content can be established. Furthermore, the single nanoparticle analysis system was applied to real-time analysis of sulfide in cells. By monitoring the scattering spectra of silver-coated gold nanorods, the changes of intracellular sulfide concentration can be observed in real time.

He Yan research group further introduced gold nanoparticles with strong light scattering and a narrow spectrum to realize the hydrogen detection and perform lithography of gold nanoparticles and palladium nanoparticles in the substrate, thus maintaining a small distance. At this time, palladium nanoparticles play a role in affecting the gold nanoparticle environment. When palladium nanoparticles interact with hydrogen to form palladium hydride, the dielectric environment of gold nanoparticles changes, resulting in the scattering spectral shift of gold nanoparticles. Based on this, the analysis and detection of hydrogen can be realized. Afterward, they studied gold–palladium nanoparticle adsorption process of hydrogen to find that shape, crystal plane and thickness of palladium shell influence the hydrogen adsorption [32]. This study can guide the synthesis of a highly sensitive hydrogen sensing unit.

It should be pointed out that gold nanorods as light scattering spectroscopic probes and further applied to biomedical analysis have very important prospects. Due to the plasmon resonance absorption and scattering characteristics of gold nanords in both longitudinal and transverse directions, the polarization probe with excellent light scattering properties can be constructed by the polarization characteristics of light scattering.

7.4.2.2 Gas-phase analysis medium

The single nanoparticle sensor platform can be used not only for the analysis and detection of biochemical molecules in homogeneous solution, but also for the determination of gas content. When the hydrogen content in the air exceeds 4%, it can cause an explosion, so detecting hydrogen is a matter of public concern and great significance. Alivisatos research group designed a hydrogen sensing platform with single nanoparticles [33]. As Pd forms hydrogenated palladium with hydrogen, in theory, nanoparticles can be used as a sensing unit for hydrogen analysis. However, the light scattering of Pd nanoparticles is not in the visible light region and the scattering spectrum is very wide, so it is not suitable to act as a single nanoparticle sensing platform.

7.4.3 Coding of three primary colors in a single nanoparticle

In the above works, it is necessary to use a single nanoparticle scattering spectral scanning system to obtain the light scattering spectra characteristics of single nanoparticle. However, scanning single nanoparticle scattering spectrum is a very cumber-some and time-consuming process. Therefore, we used Image Pro-Plus (IPP) software to introduce a three-primary colors system to encode the color of a single nanoparticle by measuring the chromaticity value of its scattered light, red, green and blue (RGB).

Figure 7.12 shows three primary colors and the corresponding chromaticity value expression [34]. Red and green mix to produce yellow, red and blue mix to produce purple, and green and blue mix to produce sky blue, when red, green and blue mix to become white. In turn, each color can be expressed in data group (R, G, B); pure red can be expressed as (255, 0, 0); pure green can be expressed as (0, 255, 0); and pure blue can be expressed as (0, 0, 255). Each color component can be calculated through equation (7.3a) to equation (7.3c).

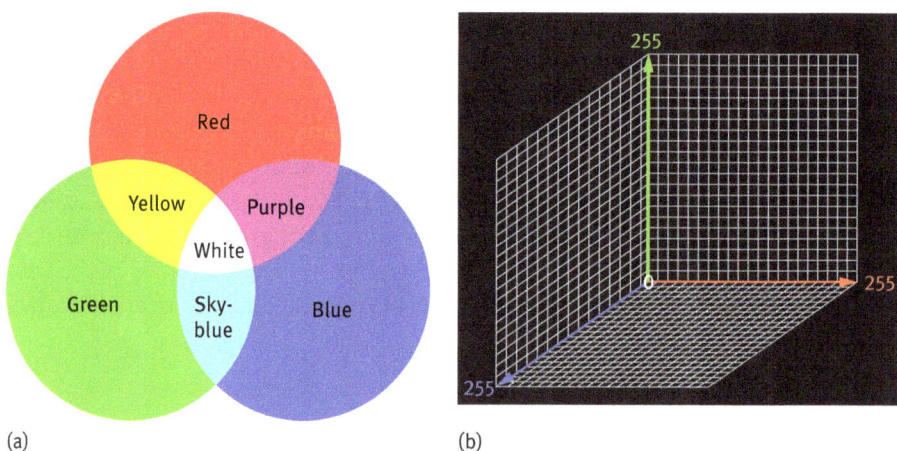

Figure 7.12: The expressions of three primary colors (a) and chromaticity values (b).

$$P_R = V_R/(V_R + V_G + V_B) \tag{7.3a}$$

$$P_G = V_G/(V_R + V_G + V_B) \tag{7.3b}$$

$$P_B = V_B/(V_R + V_G + V_B) \tag{7.3c}$$

where P_R, P_G, and P_B are proportions of red, green, and blue in a complex color, whereas V_R, V_G, and V_B are red, green, and blue data value (Figure 7.13).

Figure 7.14 is a thiol molecule adsorbed on gold nanoparticle surface of dark-field light scattering imaging taken by a true-color CCD camera (a, b). Through IPP software, each scattering facula of gold nanoparticle has its RGB information (c, d). The experimental results showed that the light scattering RGB value of the same nanoparticles changed before and after adsorption, which was used to detect thiol molecules. It

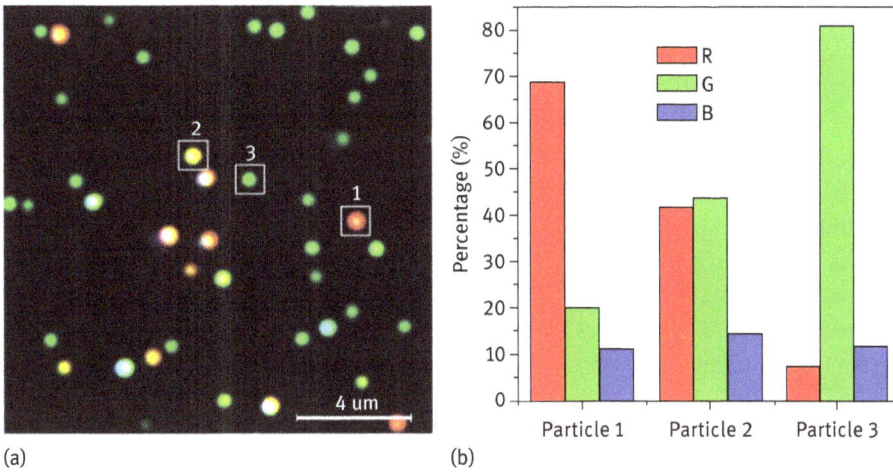

Figure 7.13: Silver nanoparticles light scattering and RGB value expression [34]. Boxes 1–3 are silver nanoparticles with red, green, and yellow scattering light.

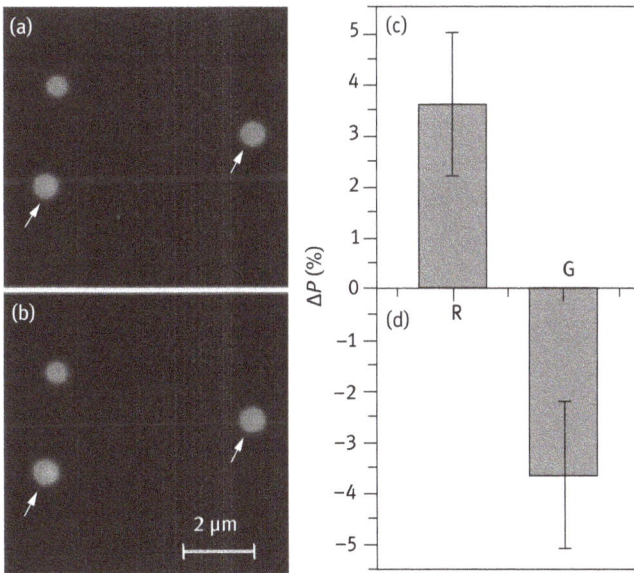

Figure 7.14: Numeralization of light scattering imaging after the interaction of gold nanospheres with thiol compounds. (a) and (b) are images of gold nanoparticles before and after reaction; (c) and (d) are red (R) and green (G) color changes of 12 gold nanoparticles before and after reaction [35].

needs to be demonstrated that as thiol molecules cause a shift in the nanoparticle light scattering wavelength and also on its light scattering intensity, thus the analysis method based on a single nanoparticle light intensity can be built [35].

In addition, the hue, saturation, and brightness composed of HIS color coding method is also successfully used for the quantitative analysis of a single-particle scattering imaging. H value (hue) in HIS color system can be simply used for imaging in the field of chemistry and biochemistry. The feasibility of this method can be verified by the accurate analysis of the change of light scattering properties caused by the change of ambient medium around gold nanoparticles and the change of refractive index caused by DNA hybridization [36].

As mentioned above, the single nanoparticle light scattering analysis method is simple to operate, and only needs camera photography and conventional imaging processing software, and does not need large instruments and equipment, so it can be completed in ordinary laboratories, and can be popularized.

7.5 Plasmon resonance coupling and molecular ruler

7.5.1 Plasmon resonance coupling and its influencing factors

When two or more noble metal nanoparticles are close to each other, their LSPR interact and plasmon coupling occurs. If the LSPR frequencies of nanoparticles are equal, resonance coupling between electromagnetic fields will occur, so that the plasma coupling between nanoparticles will be enhanced, and the electromagnetic field will be greatly enhanced.

The plasma resonance coupling between nanoparticles is affected by the distance, shape and size, spatial arrangement, polarization direction of incident light and other factors. Zhang et al. discussed the relationship of plasmon resonance coupling features and the distance of two oval AuNPs theoretically and experimentally [37]. They found that coupling brings a red shift to nanoparticle LSPR absorption and scattering peaks; the larger the distance, the smaller the red shift is, as shown in the equation relation of Figure 7.15. When the distance between the two nanoparticles is greater than 2.5 times of the short axis length of the oval nanoparticles, the plasma coupling between the nanoparticles disappears.

The local plasma resonance coupling between nanoparticles is not only affected by the distance between them, but also by the size of the particles, and the limiting distance of the interaction between the nanoparticles of different sizes is also different. Research shows that the larger particle leads to the greater coupling distance.

Noble metal nanoparticle with anisotropy, such as gold nanorod, are coupled by plasmon, which is also closely related to the spatial distribution of nanoparticles. For example, the plasmon coupling of two different assembling modes of gold nanorods exhibit completely different properties [38]. In addition, incident light polarization direction also greatly influences the coupling [39].

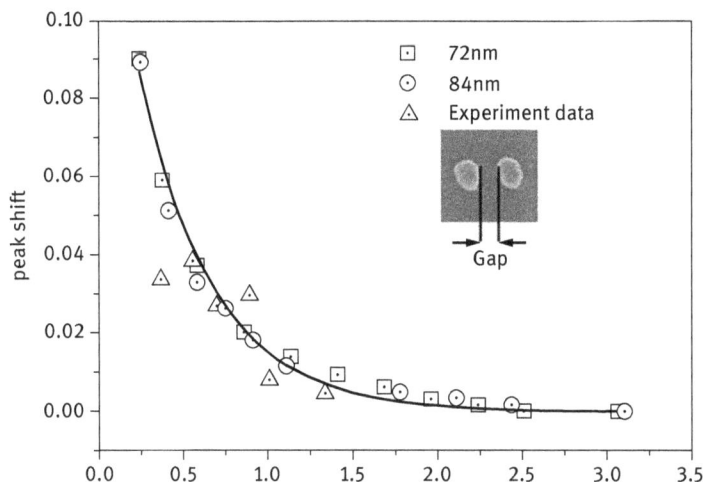

Figure 7.15: Relationship between scattering spectral shift and nanoparticle spacing. Comparison of calculated estimation result (square and circle) with the experimental result (triangle). Photo is taken from reference [37].

The coupling of nanoparticles can lead to two characteristic changes in the LSPR spectra of nanoparticles: one is the shift of the characteristic absorption and scattering peaks, and the other is the enhancement of the intensity at these peaks. These changes can be distinguished with an ordinary dark-field optical microscope. Professor He Yan research group established a highly sensitive DNA analysis method based on light scattering color change of a single nanoparticle [40].

Sandwich DNA hybridization makes monodisperse noble metal nanoparticles adjacent and forms a dimer or a polymer, so that there is plasmon resonance coupling in nanoparticles. Using a dark-field light scattering microscope for observing nanoparticle light scattering color changes before and after hybridization is developed for target DNA analysis. On the contrary, coupled nanoparticle depolymerization also brings clear changes in the nanoparticle light scattering, establishing a protein scattering visible analysis method [41].

In addition, the fluorescence and Raman scattering signals of fluorescent molecules or Raman molecules in plasmon resonantly coupled electromagnetic fields will be greatly enhanced, so that it exhibits good application prospects in surface-enhanced Raman scattering analysis field.

7.5.2 Plasmon molecule ruler

The traditional Foster resonance energy transfer (FRET) has been used to study the molecular interaction and conformational change at nanoscale, but it is limited to the FRET occurring within 10 nm. Theoretically, the spectral changes can be detected when the nanoparticle spacing is less than 2.5 times the particle size. However, the size of gold and silver nanoparticles, which can be imaged by scattering light in dark-field imaging, is usually over 20 nm, so the 2.5 times of the size of gold and silver nanoparticles is usually larger than 50 nm or even more than 100 nm. So, based on the coupling distance of a nanoparticle and sensitivity of its external surrounding environment, researchers designed a plasmon molecular ruler to discuss quantitative analysis through coupling and distance relation, or even a biological macromolecular conformation.

Plasmon resonance coupling of a long-distance molecular ruler breaks the distance limit of traditional FRET molecular ruler. For example, Alivisatos research group designed a ruler of gold or silver nanoparticles for the dynamic process of DNA hybridization from biotin–streptomycin affinity interaction [42], and separated nanoparticles with DNA sequence of 33 bases. Compared with a single nanoparticle, gold or silver nanoparticles present a longer scattering wavelength and stronger intensity, that is, the spectrum has a quite obvious red shift. They also found that, at the same distance, silver nanoparticle has a red shift of 102 nm, whereas gold nanoparticle has red shift of only 23 nm. It indicates that silver nanoparticles are more suitable for the construction of plasma molecular scale. Because the double-stranded DNA is more rigid than single-stranded DNA, the distance between the nanoparticles is pulled apart from each other and the coupling effect is weakened after hybridization, showing a blue shift in the spectrum. By monitoring the shift of the spectrum, DNA hybridization can be real-time analyzed.

To analyze a single molecule imaging of caspase, Alivisatos research group also designed one cell-satellite plasma molecular ruler [43]. As shown in Figure 7.16 (a), through specific functions of biotin–streptomycin affinity, assemble streptavidin–gold core and biotinylated gold satellite nanoparticles to form a core-satellite plasmon structure. In this structure, a caspase-specific cutting polypeptide is used to adjust the distance.

In this structure, there are approximately 3–6 gold satellite nanoparticles surrounding the gold-core nanoparticle (Figure 7.16 [b]). As gold satellite and gold core nanoparticles have a strong local plasmon resonance coupling, an orange scattering light that is 44 times higher than a single gold nanoparticle intensity is shown in a core-satellite nanoparticle system (Figure 7.16 [c]), together with 75 nm red shift in a spectrum (Figure 7.16 [d]). Its occurrence can be confirmed by an in-vitro experiment. For example, when there is caspase, the gold satellite nanoparticles are far away from the gold core nanoparticles due to the cut-off of the peptide sequence, the plasmon resonance coupling effect weakens or disappears, the scattering spectrum of the nanoparticles blue shifts, and the scattering light intensity decreases. Because caspase can cut off the polypeptide sequence, the plasma molecular scale can be further used to study the activity of caspase in living cells. As shown in Figure 7.16 (e –g), with the prolonged incubation time of TNF-α and CHX, which can induce apoptotic protease production, the orange-red scattering light spots in the cells gradually

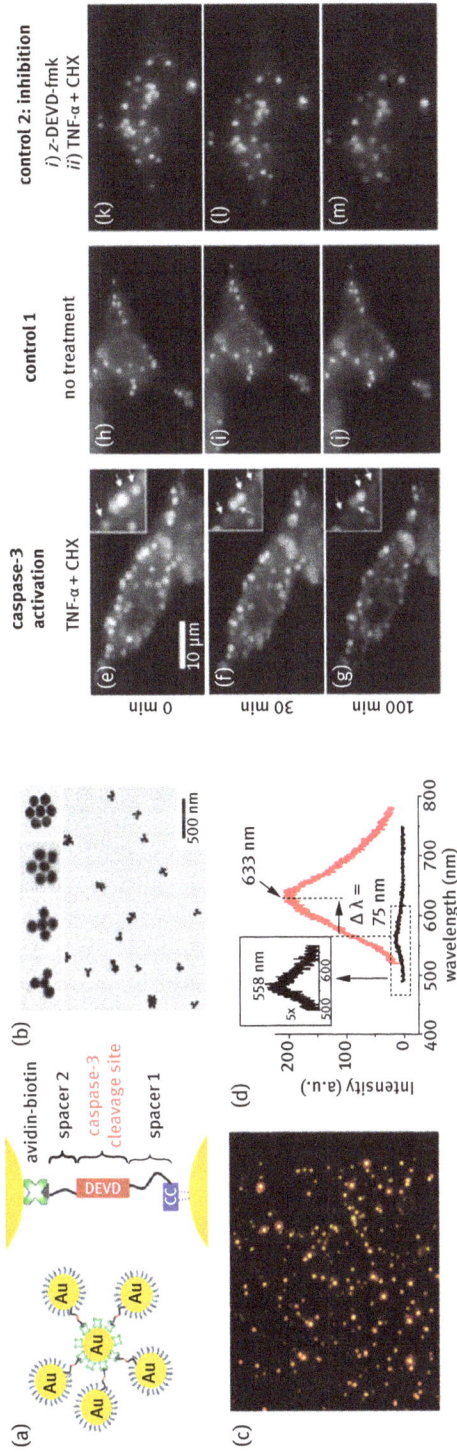

Figure 7.16: Core-satellite nano plasmon molecular ruler (refer to the colored picture at the end-of-the-book) Representative (a–d) Cell caspase activity imaging analysis (e–m) (a) Core-satellite schematic diagram, (b) TEM imaging of core-satellite nanoprobe, (c) dark-field scattering image of core-satellite nanoparticles (orange), (d) single core-satellite nanoparticle (red curve) and gold nanoparticle (black curve) scattering spectra, (e–g) TNF-α and CHX-induced apoptotic cell, (h–j) cell with no disposal, (k–m) dark-field light scattering imagings of cells disposed with caspase inhibitors, THF-α and CHX. Photos are taken from reference [43].

become light red or green, indicating that the apoptotic protease in the cells has cut off the polypeptide sequence and the core- satellite nanoparticles have been far away from the surface of gold nanoparticles.

7.6 Plasma resonance energy transfer and its applications

In traditional FRET system, there are multiple energy donors, such as small organic molecules, quantum dots, and polymers. While most of them are fluorescent organic molecules with weak anti-photobleaching ability, causing disadvantages of using FRET in the biochemical analysis application. It affects the accuracy of quantitative analysis and reliability of imaging, especially for long-time analysis. So energy donors with photobleaching resistance is of theoretical research significance and practical application prospect for constructing new resonance energy transfer system in a biochemical field.

Quantum dots solved the above problems to a certain degree; however, they are toxic and hard to modify. Noble metal nanoparticles have excellent local surface plasmon resonance light absorption and scattering characteristics, attracting wider attention in constructing energy transfer, gradually becoming a mainstream in the analysis and detection field. While metal nanoparticles only work as an energy acceptor, they are unable to properly solve energy donor problems. It has been found that precious metal nanoparticles scatter light is at a specific wavelength under the light excitation, so they can also be used as energy donors in theory. Moreover, the scattering and photobleaching of metal nanoparticles are superior to other luminescent materials. So, for using metal nanoparticle as energy donor to build energy transfer system can compensate for the disadvantages of other energy transfer technology as well as realizing long-time image analysis. In addition, energy transfer and application with metal nanoparticle as energy donor is a novel field of research study with the necessity of study and development.

7.6.1 Plasmon resonance energy transfer

In 2007, Lee research group put forward one new energy transfer form to build noble metal nanoparticles based plasmon resonance energy transfer (PRET) [44], where noble metal nanoparticles can serve as energy donor, and energy receptors can be small organic molecules, biological macromolecules, or even other nanoparticles with certain light absorption capacity.

Similar to FRET, as a new form of energy transfer [44], PRET naturally requires a series of resonance transfer conditions. At least three points are included.

(1) The scattering spectra of metal nanoparticles overlap with their surface molecular absorption spectra. The greater the overlapping, the higher the energy transfer efficiency.

(2) A certain suitable distance range is needed for PRET

(3) PRET occurrence is related to the morphology and orientation of an acceptor and a donor. The reason is that the donor is nanoparticles. Unlike FRET, nanoparticles are larger than small organic molecules or even some biological macromolecules.

The above spectral overlapping is a necessary condition for the resonance energy transfer, which can be illustrated in Figure 7.17. Lee research group's result shows that when cytochrome c is adsorbed on the surface of a gold nanoparticle, its resonance plasma spectrum starts sinking, which is consistent with 520–580 nm wave band of cytochrome c, satisfying spectral overlapping condition of energy transfer. To prove that sinking of gold nanoparticle scattering is caused by cytochrome c, they compared peptide adsorption on gold nanoparticle of the wave band and cytochrome c adsorption on polystyrene nanoparticles, no PRET was observed. It means that spectral overlapping and free electron are essential conditions for PRET.

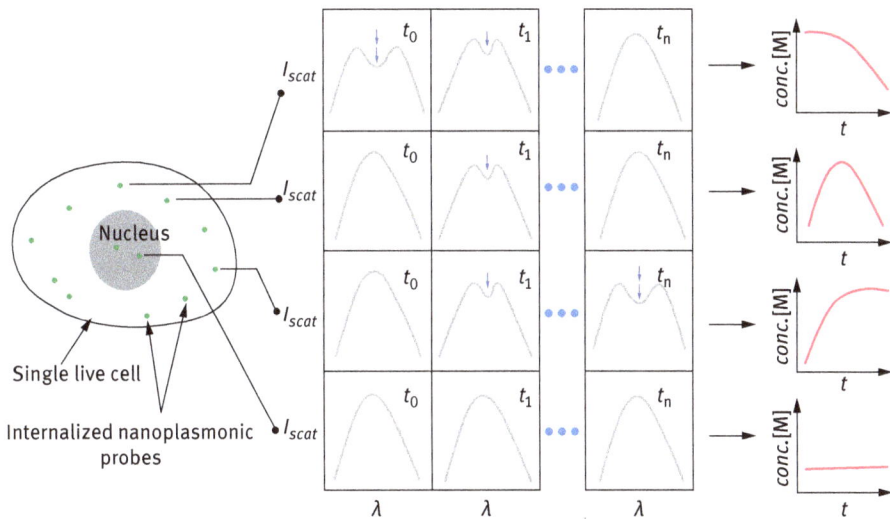

Figure 7.17: Real time analysis of cytochrome c molecules in living cells based on PRET. Cell apoptosis was induced by ethanol. Mitochondria released cytochrome c molecules into cytoplasm. The concentration of cytochrome c molecules in different regions of cell nucleus and cytoplasm could be observed by real-time monitoring the scattering spectra of gold nanoparticles in different parts of cell. Photos are taken from reference [44] with a suitable disposal.

It needs to be demonstrated that, by far, PRET mechanism is still unclear. However, the most reasonable explanation is the interaction of metal nanoparticle plasma resonant dipole electromagnetic field and surface molecular dipole electromagnetic field.

7.6.2 Plasma resonance energy transfer application

Cytochrome c serves as the energy transfer mediated material and shows important functions in the transfer of energy production, matter and energy metabolism, cell apoptosis, and other processes. In cell apoptosis, permeability of the mitochondrial outer membrane increases and releases inner cytochrome c into the cytoplasm. Therefore, it is of great significance to study the real-time changes of cytochrome c in vivo and its transport processes.

Based on PRET principle, Lee group used a real-time imaging to analyze cytochrome c in alcohol-induced cell apoptosis [45]. Consistent with the results observed in (Figure 7.17), when 3-mercaptopropionic acid modified gold nanoparticles were incubated with ethanol-treated HepG2 cells for 6 hours, a depression at 550 nm was observed in the scattering spectra of gold nanoparticles in the cytoplasm. Scattering spectral depression is positively correlated with the concentration of cytochrome c molecule because mitochondria release cytochrome c molecule into cells. To confirm the release of cytochrome c, the authors compared the scattering spectra of gold nanoparticles in the nucleus and determined that the cytochrome c content in the nucleus was almost unchanged. Therefore, by monitoring the scattering spectrum of gold nanoparticles in real time, we can detect the concentration of cytochrome c in cells and understand the apoptosis state of cells.

Generally speaking, the analysis method based on PRET involves spectral resolution and is based on dark field imaging, so it usually has high spectral resolution, high sensitivity and high selectivity. For example, the PRET system involves the coordination reaction of the ligand molecules in many cases, which determines that the method usually has good selectivity. Lee research group introduced ligand with ethylenediamine group, which can coordinate with Cu^{2+}. The molecules have no absorption in the AuNPs scattering spectral region; After the coordination with Cu^{2+}, broad absorption occurs in the AuNPs scattering spectral region, and thus PRET happens. Quantitative analysis detection of Cu^{2+} based on this method has a high sensitivity and high selectivity, which can be used for Cu^{2+} imaging analysis in cell [46] and the detection of 2,4,6-trinitrotoluene [47].

PRET is one nonradiative energy transfer form of noble metal nanoparticles and particle surface molecules, whereas there is another energy transfer form between an acceptor and a donor, that is, nano-metal-surface energy transfer (NSET). When there are two or more energy transfer processes, most of them are cascade energy transfer in one direction, which are mainly limited by wavelength match and Stokes shift.

We chose gold nanoparticles and organic dye molecules tetramethylrhodamine (TAMRA) as an energy acceptor and donor (Figure 7.18). By fluorescence imaging, single particle scattering imaging and spectroscopy combined with quasi-static approximation and finite time domain difference simulation, we confirmed their double-feedback energy transfer system [48]. This work is a good exploration of double-feedback energy transfer system.

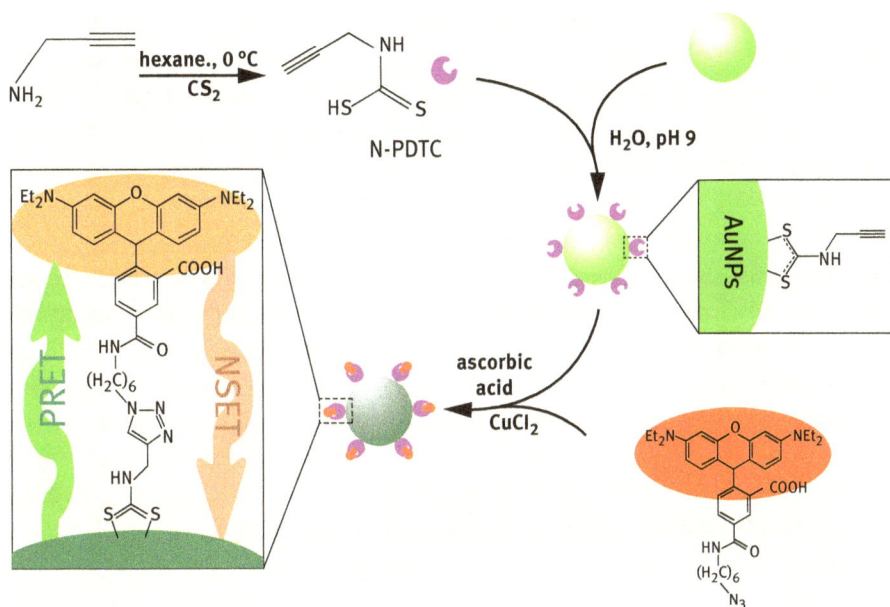

Figure 7.18: PRET and NSET double directional energy transfer between Au nanoparticles and TAMRA [48].

7.7 Application of single nanoparticle light scattering in imaging analysis

Noble metal nanoparticles as light scattering imaging probes have the following advantages:

(1) Noble metal nanoparticles have a localized plasmon resonance and strong scattering light. For example, 60 nm AuNPs have a molar extinction coefficient of $6.33 \times 10^{-11}(mol/L)^{-1} \cdot cm^{-1}$, scattering efficiency of 31.3%, and scattering capacity of $1.98 \times 10^{-11}(mol/L)^{-1} \cdot cm^{-1}$, which is about five order magnitudes to that of fluorescence quantum yield of fluorescein molecules.

(2) The color of the light scattering is adjustable with accommodation modes of changing shape, size, surface modification, and medium. In Section 6.3, we

described the optical properties of nanoparticle LSPR and its close relation to the elemental composition, shape, and size in details. For example, 40 nm AgNPs scatter blue light, 50 nm AuNPs scatter green light, and 100 nm silver triangular nanoparticles display red light scattering, which are consistent with the results of theoretical simulation. Therefore, through the top-level design, and according to the need to synthesize different forms of nanoparticles with diverse scattering colors can be obtained.

(3) Scattering light intensity of noble metal nanoparticles is stable and difficult to bleach, which are suitable for long-time imaging analysis and tracing.

(4) Noble metal nanoparticles have high biocompatibility and are easy to modify, which are incomparable with quantum dots. Quantum dots have been widely used as real-time imaging tracer optical probes, but they are toxic and difficult to modify.

Noble metal nanoparticles have become very promising optical imaging probes because of the above characteristics. For example, by tracking the light scattering properties of a single nanoparticle in real time, people can obtain the location information of nanoparticles in living cells, helping to clarify the path of nanoparticles into cells, and further for bioimaging analysis and disease diagnosis.

7.7.1 Single nanoparticle imaging for monitoring molecular reactions

In-situ imaging monitoring of nano-scale reaction process using single nanoparticles as sensing units can obtain more detailed local reaction information combined with spectral measurements. Due to the strong light scattering signal of noble metal nanoparticles, single nanoparticles can be used as an independent reaction monitoring platform, which makes the reaction monitoring with high spatial resolution. The change of the nanoparticles material, the surrounding medium or the nanoparticles coupling degree induced by the reaction process can cause the scattering signal change of the single particles, which can be used as the observation signal for the real-time in-situ monitoring of the reaction process.

The oxidation or sulfidation of the surface of noble metal nanoparticles can increase the refractive index of the surface part of the particles, resulting in the red shift of the LSPR wavelength and the change of the corresponding color of the scattered light, accompanied by the change of the intensity of the light scattering signal. The light-triggered breaking of silver-dithiocarbamate bonds on the surface of silver nanoparticles by chemically adsorbed molecules containing dithiocarbamate structure was studied. It was shown that the different degradation modes of dithiocarbamate molecules at different pH caused the cause the variation of scattering

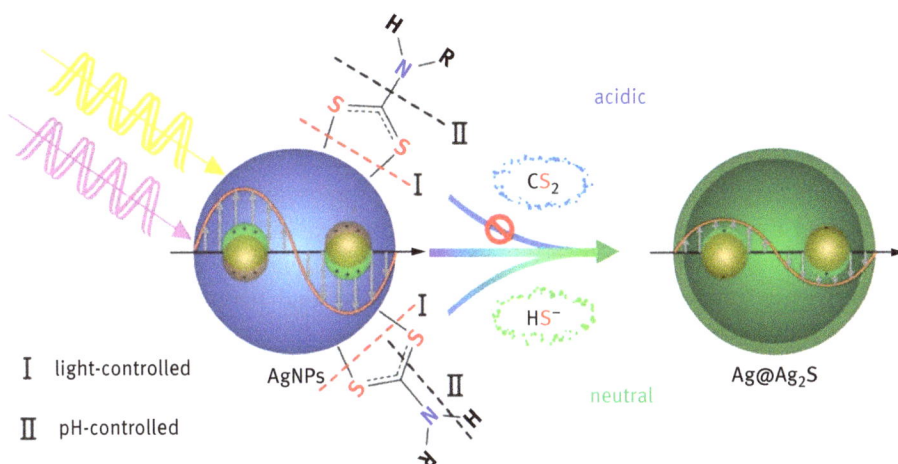

Figure 7.19: Silver nanoparticle as a probe to monitor the light triggered reaction of chemical bonds of silver and aminodithioformic acid. Aminodithioformic acid degradation products under different pH conditions decide differences in the particle signal. Photo is taken from reference [49].

signals of nanoparticles (Figure 7.19). Based on this, we realized imaging monitoring of chemical bond breakage and triggering in Ag–aminodithioformic acid at a single nanoparticle level. We also found and confirmed an important silver ion oxidization process. In terms of degree of obvious color changes, spherical nanoparticle is more suitable for monitoring compared to rod-shaped silver nanoparticles because it has clear color changes and wavelength red shift [49].

Based on the similar principles, we further used silver nanoparticle as a single particle imaging probe to study the degradation reaction of fungicide dimethyl aminodithioformic acid (sodium formamide, also known as Dibam). It was found that in alkaline condition, Dibam molecule was more easily degraded, which showed that the scattering signal of nanoparticles had a significant intensity reduction and spectral red shift. Meanwhile, if copper ions are introduced to form a steady coordination structure with Dibam negative ions, the degradation reaction is difficult to occur [50].

Changes in the surface of nanoparticles, such as corrosion reactions, can also cause changes in the shape and surface structure of nanoparticles, resulting in significant differences in light scattering properties, so as to achieve the dynamic monitoring process of the fusion reaction of nanoparticles with different materials. For example, with the aid of a silver nanocube with uniform shape and particle size as a single nanoparticle scattering imaging probe, we studied a strong oxidant sodium persulfate corrosion reaction of silver nanoparticle and put forward the oxidation corrosion mechanism [51]. Co-positioning of dark-field imaging and SEM confirmed that after corrosion, silver nanocubic transfers to spherical silver nanoparticles, which attributes to nanocubic sharp angle having highest electric field intensity and energy, and strongest reactivity. At the same time, this study confirmed that the uniform shape and size of nanoparticles as a probe can effectively reduce the demand of co-location imaging characterization of dark field and electron microscopic in the reaction monitoring.

Iodine corrosion for gold nanoparticle is another popular study. Authors performed an in-situ imaging monitoring of this reaction with a single nanoparticle scattering imaging technology. The corrosion process of gold nanoprobe by KI/I_2 was studied with spherical gold nanoparticles, rod-like gold nanoparticles and gold triangle nanosheets as probes. The spectral properties show a significant blue shift and the scattering intensity decreases gradually. The reaction performs on a mechanism of cutting corners and layer-by-layer peeling. Eventually, the probe tends to be transformed into an isotropic spherical structure [52].

In addition to the above-mentioned monitoring of reactions that cause significant changes in the composition and shape of nanoparticles, in-situ monitoring of reactions can also be effectively achieved by changing the coupling state of nanoparticles. Professor Long Y T group made a successful attempt. They performed azidation modification on a 60 nm gold nanoparticle fixed on a glass slide and acetylene modification on a 14 nm gold nanoparticle in solution, a cuprous ion-catalyzed click reaction in a dark-field microscope was observed. It is found that once a click chemical reaction occurs, the coupling of two gold nanoparticles happens, which results in a significant red shift in the spectra, thus realizing in situ monitoring of organic chemical reaction by single particle scattering imaging [53].

Due to the ubiquity of scattering, non-plasma micro-nanoparticles can also be used as a single particle sactterring probe for monitoring of a reaction process. Huang C Z research group used polyphenylene diamine leaf-shaped microparticles as an imaging sensing unit for performing in-situ real-time monitoring of corrosion reactions with cobalt ions. By means of infrared spectroscopy, dynamic light scattering, and scanning electron microscopy, they studied the reaction dynamic process in detail and put forward the mechanism of ion that gradually permeates in to the inner particle, causing deformation and gradual disintegration of particle [54]. This study broadens the scope of the single particle dark field scattering imaging sensor probe and provides a new research idea for different types of reaction monitoring.

Operational errors are difficult to eliminate in the process of dark field imaging, which brings serious problems to the accuracy of imaging measurement, especially when slides need to be taken and put back to the objective table for dark-field imaging quantitative analysis. Whether in quantitative analysis or in the dynamic monitoring process of reaction, there are large differences in imaging results in the same batch of slide samples, so it has always troubled researchers about what method to be adopted for reducing result errors and improving their accuracy. For this, gold nanoparticles with good stability and optical properties can be used as the internal standard of imaging observation, and as a standard to calibrate the scattering intensity of imaging probe, so as to improve the accuracy of imaging analysis [55]. The method has been successfully used to calibrate and monitor the change of scattering intensity and quantitative analysis of glucose in the oxidation process of silver nanoparticles at room temperature (Figure 7.20), and quantitative analysis of scattering intensity of gold nanoparticles in plasmon resonance energy transfer process, confirming the validity of this method.

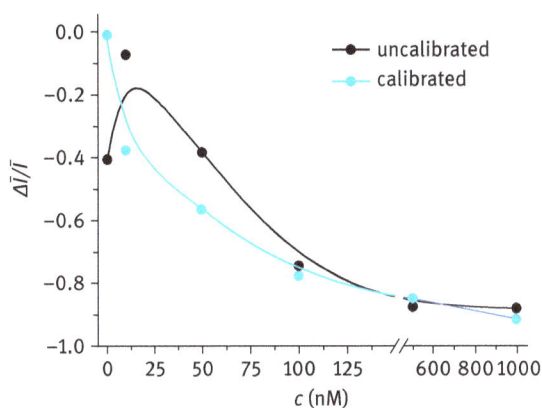

Figure 7.20: Glucose detection with silver nanoparticle as a probe.

7.7.2 Single nanoparticle image tracing analysis

One of the prerequisites of single particle tracing is that it can target and bind specifically, so it is necessary to modify the single particle specifically. Nam et al. developed a acceptor–donor interaction analysis method through real-time tracing of single gold nanoparticle [56]. Authors constructed a lipid bilayer simulating cell membrane on the slide surface, then connected a gold nanoparticle to the bilayer. As lipid bilayer is flowing, the movement of single nanoparticle can be observed in a dark-field microscope. In the presence of cholera toxin, the fluidity of the bilayer decreases and the diffusion coefficient of gold nanoparticles decreases due to its specific binding to ganglioside GM1 in the phospholipid bilayer. This property constitutes the theoretical basis for the analysis of receptor ligand interactions using single nanoparticle. Further study has found that univalent gold nanoparticles have a higher diffusion coefficient compared with that of multivalent gold nanoparticles. The detection limit of cholera toxin was 10-100 pmol/L, sensitivity 100 times higher than fluorescence analysis.

For real-time tracing of single gold nanoparticle spectrum and spatial position, Asahi research group developed a confocal scattering spectral imaging system to study nanoparticles and mouse fibroblast interaction [57]. In cell, different mediums have different refractive indexes, then nanoparticles show different scattering light properties in different places.

The results showed that the scattering spectra of gold nanoparticles changed significantly when they entered the cell, and the scattering properties of gold nanoparticles changed when they entered or left a organelle. This phenomenon lays a theoretical foundation for the study of intracellular transport of single nanoparticles. Combined with a confocal scattering spectral imaging system, Asahi et al. observed the movement of 80 nm nanoparticles in fibroblasts. The scattering peak shifts recorded in real time reflected the real-time microenvironment of nanoparticles and the interaction mechanism between nanoparticles and cells. They observed that the scattering spectrum of nanoparticles shifted to 12 nm at 28 min and returned to its original state after 5 min. This phenomenon can be explained by the fact that when the nanoparticles penetrate into the cell membrane or enter the connotation, the surface of the nanoparticles is coated by a phospholipid bilayer, and the refractive index of the surrounding media increases.

Cell membrane surface molecules play an important role in living activities, such as endocytosis, signal delivery and cell–cell communication, and real-time monitoring of surface molecule diffusion and translocation provide vital information for the series of activities. Xu's research group monitored the diffusion and permeation of silver nanoparticles on the surface of bacterial (Pesudomonas aeruginosa) cell membranes under a dark-field microscope using the superstrong plasmon light scattering characteristics of silver nanoparticles. They found that silver nanoparticles with diameters smaller than 80 nm can pass through the membrane and accumulate in the cytoplasm, suggesting that the hole in the membrane might be over 80 nm. Meanwhile, the study found that chloramphenicol increased membrane permeability, causing nanoparticles to pass through the cell. In addition, this method was applied for studying bacterial resistance.

In recent years, Xu group prepared silver nanoparticles with various sizes, and their optical properties and single particle spectra were characterized in detail. In the reseach of zebrafish embryos, the movement of silver nanoparticles on the surface and inside of the embryos was observed directly by dark-field microimaging, and the movement tracking of a single silver nanoparticle on the surface of the embryonic egg membrane and the movement process through the egg membrane were recorded, thus showing that the silver nanoparticles were single particles, which are used as single particle nano-optical probe for living animal imaging.

In addition, the light scattering imaging of plasma nanoparticles has also made gratifying progress in the imaging monitoring of viral invasion of cells. The authors used biotin–streptomycin affinity modification method to perform respiratory syncytial virus (RSV) labeling on gold nanoparticles (Figure 7.21) that was used for real-time monitoring process of instructing virus invading cell [58]. With overall

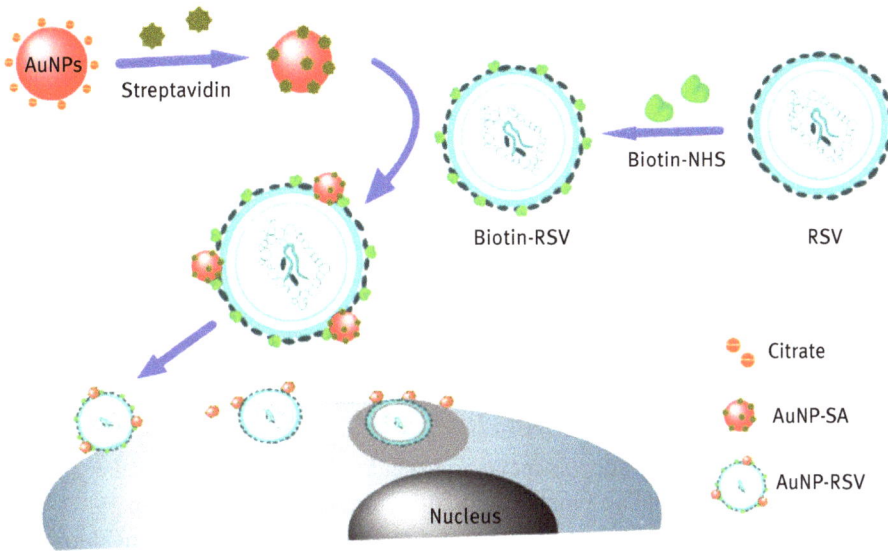

Figure 7.21: Gold nanoparticle-labelled instruction of respiratory syncytial virus (RSV) invading cell [58].

consideration of virus virulence, light scattering signal intensity and RSV virus modification, gold nanoparticle imaging and tracing effect of 13 nm was obviously better than gold nanoparticles of 6.5, 30, and 50 nm. Real-time dynamic analysis of a single nanoparticle light scattering imaging in virus invading cell is important for better understanding of this mechanism.

7.8 Light concentration effect and application of plasmon nanoparticles

The oblique illumination mode is used to separate the incident light source from the imaging area, and the advantages of low background and high signal-to-noise ratio are obtained. However, the oblique illumination mode has a higher requirement on the numerical aperture, and has a certain impact on the resolution.

To overcome this problem, many researchers have made different explorations in the illumination mode of scattering imaging. Novotny et al. performed a clever and effective work. They used two incident lights with specific angle and frequency of ω_1 and ω_2 as light sources, tested mixed-wave frequency $\omega_{4wm}=2\omega_1-\omega_2$ as an evanescent field to obtain a high-resolution sample imaging (Figure 7.22) [59]. This method is a nonlinear dark-field imaging illumination mode, which can study the linear sample and nonlinear optical properties. This method is widely used for transparent and

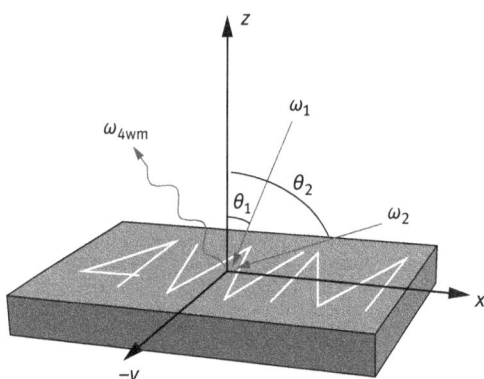

Figure 7.22: Schematic diagram of an optical path of nonlinear dark-field microscope [59].

nontransparent samples, such as metal and semiconductors, which are also applied for bio-imaging, material and instrument fault analysis fields, and so on.

Plasmon nanoparticles have light concentration effects, that is, the incident light can be converged to a small area near the nanoparticles, forming a strong photoelectric magnetic field. This is attributed to the fact that the plasma nanomaterials are much larger than the cross sections of their physical cross sections. In this way, the light scatterring around the nanoparticles will be stronger than the incident light, showing an enhancement effect of light, which is contrary to the widely studied phenomenon of "invisible cloak"

This property of plasmon plays an important role in the dark-field imaging process, which is characterized by strong light scattering signal and visibility of plasmon nanoparticles. In widely developed optical studies, such as metal-enhanced fluorescence, plasma surface-enhanced Raman scattering, and so on, plasma nanomaterial light concentration effects can be indirectly perceived by optical signals that are more significantly enhanced than the fluorescence or Raman signals of dye molecules themselves.

Combined with the current dark-field oblique illumination mode, the authors introduced a monochromatic illumination source with vertical incidenc, and constructed one composite illumination imaging system (Figure 7.23). The introduced intensity and wavelength of the light source can be effectively controlled by a group of center density attenuation film and optical filter, thus effectively improving the visibility of imaging. Studies have found that when an introduced monochromatic energy matches with LSPR energy of a plasmon nanoparticle, the complex field has best improving effect for plasmon nanoparticle scattering image. This phenomenon widely exists in plasmon particles at different scattering wavelength. In addition, a suitable monochromatic intensity is key to effectively improve the visibility. Excessive monochromatic light intensity will bring high scattering imaging background, which is easy to annihilate the scattering signal of single plasma particle. Excessive weak monochromatic light intensity is difficult to play an enhanced role.

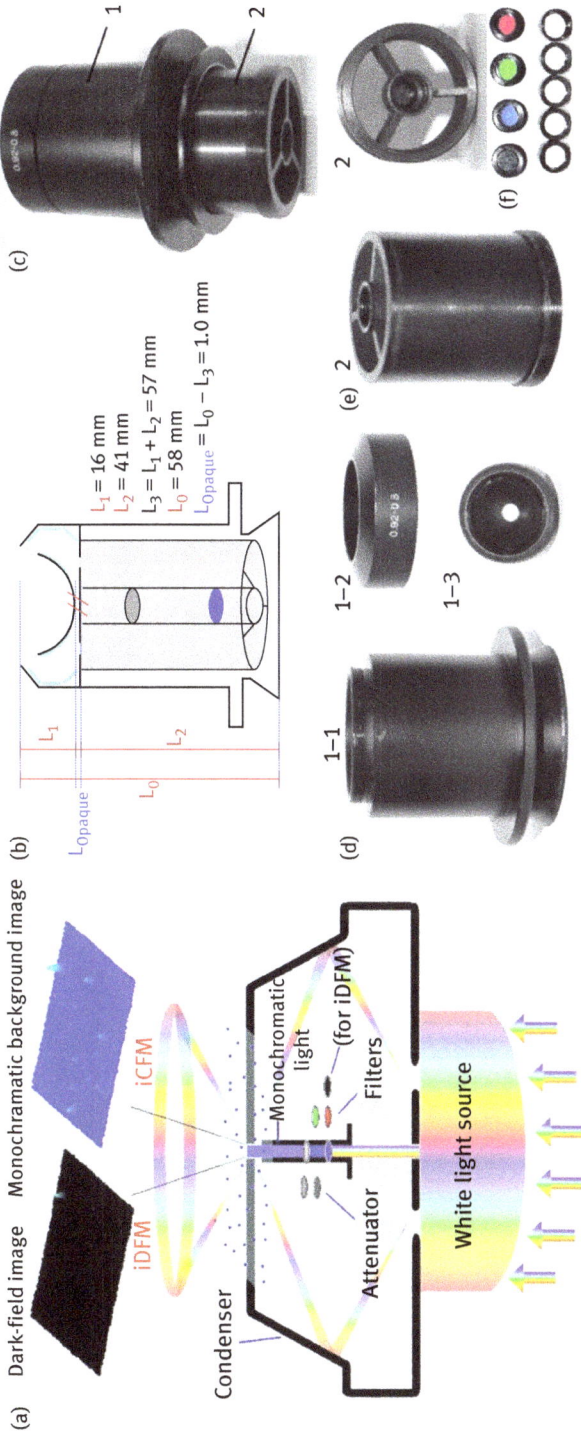

Figure 7.23: Schematic diagram and physical diagram of composite field illumination microscopic imaging system. Optical schematic diagram of composite field illumination microscopic imaging system (a) Schematic diagram of structural profile of dark field condenser (b) Figures (c–f) are the object of the condenser after the modification of the composite field illumination circuit where 1 is original lens, 1–1 is the original lens cone, 1–2 is a lense cone round cap, 1–3 is the curved optical glass element. 2 is the metallic frame structure, which is used for introducing a filter in the construction of a new light pathway (on f) and neutral density polarizer. (below f). Photos are from taken reference [60] with number adjustment.

The reason why these two methods work is that the "light concentration" effect of plasma nanoparticles is closely related to the LSPR and scattering efficiency of plasma nanoparticles. Previous experimental studies and finite-time-domain difference (FDTD) simulations have confirmed that both LSPR and scattering efficiency of plasma nanoparticles have significant wavelength dependence.

Compared with cell and other bio-samples, plasmon nanoparticles are more sensitive to wavelength, which determines that this method is also suitable for imaging of biological samples with complex background. The authors used a constructed complex field to perform imaging analysis on human Hep-2 [60]. It was found that the Ag nanoparticles modified by BSA and untreated Ag nanoparticles exhibited significantly different aggregation and dispersion during incubation with Hep-2 cells. The silver nanoparticles modified by BSA are dispersed, so they still emit blue scattering light, while the unmodified silver nanoparticles are aggregated, thus emitting red scattering light. The enhanced visibility of both scattered light silver nanoparticles is demonstrated by the composite field illumination mode (Figure 7.24). Especially for particles with weaker self scattering

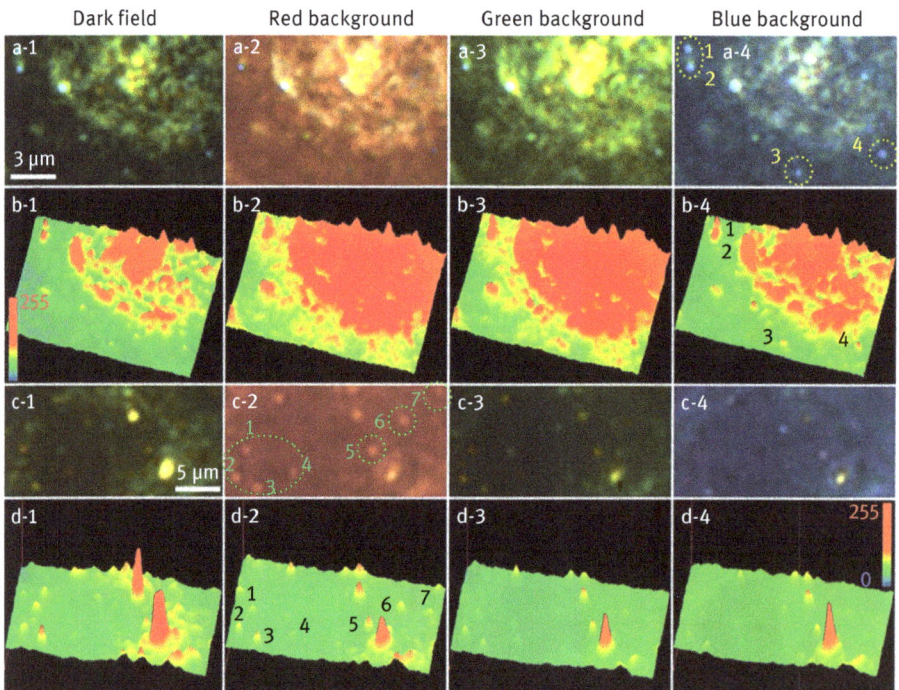

Figure 7.24: Composite field illumination for scattering imaging of plasma nanoparticles in cell samples [60]. BSA-modified silver nanoparticles (a,b) and unmodified silver nanoparticles (c,d) in the dark-field imaging and complex field images of monochromatic incident light, where b, d is the corresponding three-dimension mode.

signal, it shows a more significant improvement. In addition, authors have used a high pass filtered imaging process to effectively filter low-frequency monochrome background by composit field illumination. At the same time, the original intensity of the scattering signal of plasma particles remains unchanged.

In general, dark-field microscope imaging technology still has a large improvement space through reasonable adjustment of imaging of light path, physical and chemical methods to adjust the particle scattering point spread function, and so on. It is hopeful to further improve the visibility and resolution of imaging. In addition, combined with advantages and features of various probes, it will be effective to improve scattering image effect and studying new imaging modes.

7.9 Development trend of light scattering analysis method for single nanoparticle

7.9.1 PRET theory study

Since Lee put forward the theory that plasma resonance energy transfer is one new form of energy transfer, related applications have attracted much attention. While the application still has many disadvantages, its theoretical basis is not well developed for a single possible reason that it is difficult to find a suitable acceptor. In addition, the study of donor–acceptor distance and acceptor molecular orientation effect for energy transfer efficiency is not enough.

From current reports, PRET has great potential in terms of analysis, detection, and living cell imaging, which will be a vital study in the future, mainly involving two aspects in theory:

7.9.1.1 Acceptor–donor spectral overlapping

Current study finds that the precondition of PRET is a large-degree overlap of acceptor absorption spectrum and donor scattering spectrum. The exact degree and overlap method is not supported by enough experimental data. The current FRET theory can be used for reference, but they are two identically different energy transfer forms, and similarity is not reliable. To design an effective PRET system, overlap degree influence is meaningful for PRET theory study; of course, it involves physical interaction mechanism of acceptor and donor, whether it is donor plasma electromagnetic field or a donor electromagnetic coupling is still worth discussing.

In addition, the system is a specific energy transfer process of nano scattering energy to small organic molecules or other acceptors with absorbance ability, which is different from nanosurface energy transfer (NSET). The latter usually uses a fluorescent molecule as an energy donor and metal nanoparticle surface as an energy acceptor. Although double-mode and double-direction PRET and NSET energy transfer processes have been proven [48], their feedback and interaction influence recognition is still worth studying further.

7.9.1.2 Donor–acceptor distance

Compared with FRET, dipole–dipole interaction in Föster fluorescence resonance energy transfer efficiency is inversely proportional to the sixth-power of distance, whereas for dipole–metal surface energy transfer, efficiency is inversely proportional to the fourth-power of distance. By far, there is no report about distance and PRET efficiency.

7.9.2 PRET application prospect

The analysis and application of PRET have been reported frequently and attracted the attention of analysts. However, it is not the mainstream at present. The reason may be related to the difficulty of pairing donor and recipient of PRET. As a PRET acceptor, the following features need to be considered:
(1) Obvious changes of absorption spectrum induced by the samples.
(2) The absorption spectrum should overlap with the scattering spectrum of the donor.
(3) The interaction between receptor molecules and the tested substances must be specific.

Because of the above three problems, how to establish a universal PRET analysis and detection system has become a difficult problem. As a new concept, PRET research needs a long process, which requires researchers to explore and improve all aspects of the content thta will be mature. It is believed that PRET will have a place in the field of in vitro analysis and detection, and biological imaging analysis, and so on.

It should be noted that the application of precious metal nanoparticles in the field of long-term real-time tracer imaging analysis is far from ideal due to the limitation of optical imaging technology. The development of plasmon nanoparticles with excellent luminescent properties and smaller particle size will greatly promote the development of imaging applications. In addition, the higher imaging resolution is also the research direction pursued by single particle real-time imaging analysis.

References

[1] Pasternack R F, Bustamante C, Collings P J, et al. Porphyrin assemblies on DNA as studied by a resonance light-scattering technique. J. Am. Chem. Soc., 1993, 115: 5393–5399.

[2] Taton T A, Lu G, Mirkin C A. Two-color labeling of oligonucleotide arrays via size-selective scattering of nanoparticle probes. J. Am. Chem. Soc., 2001, 123: 5164–5165.

[3] Cutler J I, Auyeung E, Mirkin C A. Spherical nucleic acids. J. Am. Chem. Soc., 2012, 134(3): 1376–1391.

[4] Ling J, Li Y F, Huang C Z. A label-free visual immunoassay on solid support with silver nanoparticles as plasmon resonance scattering indicator. Anal. Biochem., 2008, 383: 168–173.

[5] Wu L P, Li Y F, Huang C Z, et al. Visual detection of Sudan dyes based on the plasmon resonance light scattering signals of silver nanoparticles. Anal. Chem., 2006, 78(15): 5570–5577.

[6] Shen X W, Huang C Z, Li Y F. Localized surface plasmon resonance sensing detection of glucose in the serum samples of diabetes sufferers based on the redox reaction of chlorauric acid. Talanta, 2007, 72: 1432–1437.

[7] Long Y F, Zhou L P, Han M. Determination of trace formaldehyde based on formation of silver nanoparticles. Chinese J. Appl. Chem., 2009, 26(11): 1349–1352.

[8] Pasternack R F, Collings P J. Resonance light scattering: A new technique for studying chromophore aggregation. Science, 1995, 269(5226): 935–939.

[9] Huang C Z, Li K A, Tong S Y. Determination of nucleic acids by a resonance light-scattering technique with α,β,γ,δ-tetrakis [4-(trimethylammoniumyl)phenyl] porphine. Anal. Chem., 1996, 68(13): 2259–2263.

[10] He W, Li Y F, Huang C Z, et al. One-step label-free optical genosensing system for sequence-specific DNA related to the human immunodeficiency virus based on the measurements of light scattering signals of gold nanorods. Anal. Chem., 2008, 80(22): 8424–8430.

[11] Qi W J, Wu D, Ling J, et al. Visual and light scattering spectrometric detections of melamine with polythymine-stabilized gold nanoparticles through specific triple hydrogen-bonding recognition. Chem. Commun., 2010, 46(27): 4893–4895.

[12] Huang C Z, Liu Y, Wang Y H, et al. Resonance light scattering imaging detection of proteins with α, β, γ, δ-tetrakis(p-sulfophenyl)porphyrin. Anal. Biochem., 2003, 321(2): 236–243.

[13] Yuan Z, Cheng J, Cheng X, et al. Highly sensitive DNA hybridization detection with single nanoparticle flash-lamp darkfield microscopy. Analyst, 2012, 137(13): 2930–2932.

[14] Zhu S, Yang L, Long Y, et al. Size differentiation and absolute quantification of gold nanoparticles via single particle detection with a laboratory-built high-sensitivity flow cytometer. J. Am. Chem. Soc., 2010, 132(35): 12176–12178.

[15] Xu X, Chen Y, Wei H, et al. Counting bacteria using functionalized gold nanoparticles as the light-scattering reporter. Anal. Chem., 2012, 84(22): 9721–9728.

[16] Yao J, Larson D R, Vishwasrao H D, et al. Blinking and nonradiant dark fraction of water-soluble quantum dots in aqueous solution. Proc. Natl. Acad. Sci. USA, 2005 102(40): 14284–14289.

[17] Bu X, Chen H, Gai H, et al. Scattering imaging of single quantum dots with dark-field microscopy. Anal. Chem. 2009, 81(17): 7507–7509.

[18] Storhoff J J, Lucas A D, Garimella V. Homogeneous detection of unamplified genomic DNA sequences based on colorimetric scatter of gold nanoparticle probes. Nat. Biotechnol., 2004, 22(7): 883–887.

[19] Liu Z D, Li Y F, Ling J, et al. A localized surface plasmon resonance light-scattering assay of mercury (II) on the basis of Hg^{2+}-DNA complex induced aggregation of gold nanoparticles. Environ. Sci. Technol., 2009, 43(13): 5022–5027.

[20] Liu Y, Huang C Z. Screening sensitive nanosensors via the investigation of shape-dependent localized surface plasmon resonance of single Ag nanoparticles. Nanoscale, 2013, 5(16): 7458–7466.

[21] Anker J N, Hall W P, Lyandres O, et al. Biosensing with plasmonic nanosensors. Nat. Mater., 2008, 7(6): 442–453.

[22] Becker J, Zins I, Jakab A, et al. Plasmonic focusing reduces ensemble linewidth of silver-coated gold nanorods. Nano Lett., 2008, 8(6): 1719–1723.

[23] Liu Y, Huang C Z. Real-time dark-field scattering microscopic monitoring of the in situ growth of single Ag@Hg nanoalloys. ACS Nano, 2013, 7(12): 11026–11034.

[24] Novo C, Funston A M, Mulvaney P. Direct observation of chemical reactions on single gold nanocrystals using surface plasmon spectroscopy. Nat. Nanotechnol., 2008, 3(10): 598–602.

[25] Zhang L, Li Y, Li D W, et al. Single gold nanoparticles as real-time optical probes for the detection of NADH-dependent intracellular metabolic enzymatic pathways. Angew. Chem. Int. Ed., 2011, 123(30): 6921–6924.

[26] McFarland A D, van Duyne R P. Single silver nanoparticles as real-time optical sensors with zeptomole sensitivity. Nano Lett., 2003, 3(8): 1057–1062.

[27] Raschke G, Kowarik S, Franzl T, et al. Biomolecular recognition based on single gold nanoparticle light scattering. Nano Lett., 2003, 3(7): 935–938.

[28] Nusz G J, Marinakos S M, Curry A C, et al. Label-free plasmonic detection of biomolecular binding by a single gold nanorod. Anal. Chem., 2008, 80(4): 984–989.

[29] Dondapati S K, Sau T K, Hrelescu C, et al. Label-free biosensing based on single gold nanostars as plasmonic transducers. ACS Nano, 2010, 4(11): 6318–6322.

[30] Liu G L, Yin Y, Kunchakarra S, et al. A nanoplasmonic molecular ruler for measuring nuclease activity and DNA footprinting. Nat. Nanotechnol., 2006, 1(1): 47–52.

[31] Xiong B, Zhou R, Hao J, et al. Highly sensitive sulphide mapping in live cells by kinetic spectral analysis of single Au-Ag core-shell nanoparticles. Nat. Commun., 2013, 4: 1708.

[32] Tang M L, Liu N, Dionne J A, et al. Observations of shape-dependent hydrogen uptake trajectories from single nanocrystals. J. Am. Chem. Soc., 2011, 133(34): 13220–13223.

[33] Liu N, Tang M L, Hentschel M, et al. Nanoantenna-enhanced gas sensing in a single tailored nanofocus.Nat. Mater., 2011, 10(8): 631–636.

[34] Liu Y, Ling J, Huang C Z. Individually color-coded plasmonic nanoparticles for RGB analysis. Chem. Commun., 2011, 47(28): 8121–8123.

[35] Liu Y, Huang C Z. Digitized single scattering nanoparticles for probing molecular binding.Chem. Commun., 2013, 49(74): 8262–8264.

[36] Zhou J, Lei G, Zheng L L, et al. HSI colour-coded analysis of scattered light of single plasmonic nanoparticles. Nanoscale, 2016, 8(22): 11467–11471.

[37] Su K H, Wei Q H, Zhang X, et al. Interparticle coupling effects on plasmon resonances of nanogold particles. Nano Lett., 2003, 3(8): 1087–1090.

[38] Funston A M, Novo C, Davis T J, et al. Plasmon coupling of gold nanorods at short distances and in different geometries. Nano Lett., 2009, 9(4): 1651–1658.

[39] Yang S C, Kobori H, He C L, et al. Plasmon hybridization in individual gold nanocrystal dimers: Direct observation of bright and dark modes. Nano Lett., 2010, 10(2): 632–637.

[40] Xiao L, Wei L, He Y, et al. Single molecule biosensing using color coded plasmon resonant metal nanoparticles. Anal. Chem., 2010, 82(14): 6308–6314.

[41] Waldeisen J R, Wang T, Ross B M, et al. Disassembly of a core-satellite nanoassembled substrate for colorimetric biomolecular detection. ACS Nano, 2011, 5(7): 5383–5389.

[42] Sönnichsen C, Reinhard B M, Liphardt J, et al. A molecular ruler based on plasmon coupling of single gold and silver nanoparticles. Nat. Biotechnol., 2005, 23(6): 741–745.

[43] Jun Y W, Sheikholeslami S, Hostetter D R, et al. Continuous imaging of plasmon rulers in live cells reveals early-stage caspase-3 activation at the single-molecule level. Proc. Natl. Acad. Sci. U S A, 2009,106(42): 17735–17740.

[44] Liu G L, Long Y T, Choi Y, et al. Quantized plasmon quenching dips nanospectroscopy via plasmon resonance energy transfer. Nat. Methods, 2007, 4(12): 1015–1017.

[45] Choi Y, Kang T, Lee L P. Plasmon resonance energy transfer(PRET)-based molecular imaging of cytochrome c in living cells. Nano Lett., 2009, 9(1): 85–90.

[46] Choi Y, Park Y, Kang T, et al. Selective and sensitive detection of metal ions by plasmonic resonance energy transfer-based nanospectroscopy. Nat. Nanotechnol., 2009, 4(11): 742–746.

[47] Qu W G, Deng B, Zhong S L, et al. Plasmonic resonance energy transfer-based nanospectro-scopy for sensitive and selective detection of 2,4,6-trinitrotoluene (TNT). Chem. Commun., 2011, 47(4): 1237–1239.

[48] Gao M X, Zou H Y, Gao P F, et al. Insight into a reversible energy transfer system. Nanoscale, 2016, 8(36): 16236–16242.

[49] Gao P F, Yuan B F, Gao M X, et al. Visual identification of light-driven breakage of the silver-dithiocarbamate bond by single plasmonic nanoprobes. Sci. Rep., 2015, 5: 15427.

[50] Lei G, Gao P F, Liu H, et al. Real-time scattered light dark-field microscopic imaging of the dynamic degradation process of sodium dimethyldithiocarbamate. Nanoscale, 2015, 7(48): 20709–20716.

[51] Wang Y, Zou H Y, Huang C Z. Real-time monitoring of oxidative etching on single Ag nanocubes via light-scattering dark-field microscopy imaging. Nanoscale, 2015, 7(37): 15209–15213.

[52] Sun S S, Gao M X, Lei G, et al. Visually monitoring the etching process of gold nanoparticles by KI/I2 at single-nanoparticle level using scattered-light dark-field microscopic imaging. Nano Res., 2016, 9(4): 1125–1134.

[53] Shi L, Jing C, Ma W, et al. Plasmon resonance scattering spectroscopy at the single- nanopar-ticle level: Real-time monitoring of a click reaction. Angew. Chem. Int. Ed., 2013, 52(23): 6011–6014.

[54] Yang H, Liu Y, Gao P F, et al. A dark-field light scattering platform for real-time monitoring of the erosion of microparticles by Co^{2+}. Analyst, 2014, 139(11): 2783–2787.

[55] Ma J, Liu Y, Gao P F, et al. Precision improvement in dark-field microscopy imaging by using gold nanoparticles as an internal reference: A combined theoretical and experimental study. Nanoscale, 2016,8(16): 8729–8736.

[56] Yang Y H, Nam J M. Single nanoparticle tracking-based detection of membrane receptor-ligand interactions. Anal. Chem., 2009, 81(7): 2564–2568.

[57] Louit G, Asahi T, Tanaka G, et al. Spectral and 3-dimensional tracking of single gold nanopar-ticles in living cells studied by Rayleigh light scattering microscopy. J. Phys. Chem. C, 2009, 113 (27): 11766–11772.

[58] Wan X Y, Zheng L L, Gao P F, et al. Real-time light scattering tracking of gold nanoparticles-bioconjugated respiratory syncytial virus infecting HEp-2 cells. Sci. Rep., 2014, 4: 4529.

[59] Harutyunyan H, Palomba S, Renger J, et al. Nonlinear dark-field microscopy. Nano Lett., 2010, 10(12): 5076–5079.

[60] Gao P F, Gao M X, Zou H Y, et al. Plasmon-induced light concentration enhanced imaging visibility as observed by a composite-field microscopy imaging system. Chem. Sci., 2016, 7(8): 5477–5483.

[61] K. L. Kelly, E. Coronado, L. L. Zhao and G. C. Schatz, The optical properties of metal nanoparticles: The influence of size, shape, and dielectric environment. J. Phys. Chem. B, 2003, 107, 668–677.

Li Qiang Chen, Cheng Zhi Huang

8 Light scattering microscopic bioimaging

We exhibited the light scattering signal from a single noble nanoparticle that can be used for monitoring the dynamics of intracellular virus invading and tracing analysis in Section 7.7. The light intensity difference led by absorption and scattering can be used for analysing the cell morphology, intracellular dynamics, and imaging of biological tissue structures.

As light scattering signal from a cell is strong and quantitative, optical microscopic imaging technology based on light scattering signal has been successfully applied in the imaging of living organisms and detection of fine structure of the cell. What is more important is that through a combination of modern scattering spectrum and microscopic imaging system, the obtained cell and organism signals can provide an important biological information for clinical histopathological detection, live imaging, new drug development, basic biological study, and other applications. Based on these, this chapter will provide an overview on the latest developments and achievements of label-free light scattering cell imaging and nanoparticle as contrast for light scattering imaging, as well as a discussion on the furthure and emerging directions in the field.

8.1 Microscopic bioimaging techniques

8.1.1 Optical micoscopic imaging in biology

In the history of the development of biology, the foundation and development of an optical microscopy was essential for the study of fine structure and function of an organism. In 1665–1677, British scientists Robert Hooke (1635–1703) and Antonie van Leeuwenhoek (1632–1723) were the first to use a self-made microscope to obtain outstanding achievements in observing the microstructures of animal cell, plant cell and other orginsms. In the second half of the nineteenth century, Robert Koch (1843–1910) and Louis Pasteur (1822–1895) and other scientists made great achievements in the field of microorganism discovery, and their structural study by the means of optical microscopic imaging techniques.

With the development of microscopic and biostaining techniques, optical microscopy is widely used in the fields of biology, medicine, and chemistry, becoming an important tool for promoting the rapid development of science in first half of the twentieth century. Duo to the diffration limit of light and the diffration limit of light, optical microscope has a resolution limit and the imaging magnification cannot be amplified infinitely, which make traditional optical microscopy cannot image within higher resolution to satisfy the requirements of biological application. So, it is difficult

https://doi.org/10.1515/9783110573138-008

for people to have in- situ, living, and real-time understanding of interaction of substances in the life activities process on the scale of single cell or single molecular level.

Benefited from the laser technique of 1960s and later photoelectric technology, computer video and image processing technology, as well as the appearance of new lens, condenser, and ultrasensitive detector, light microscopy entered a new developmental phase. The invention and development of laser scanning confocal microtechnique, near-field scanning optical microscopy, and the total internal reflection fluorescence microscopy are greatly influenced for the advancement of cell biology, molecular biology, biomedicine, and other fields.

8.1.2 Light scattering bioimaging

Light scattering bioimaging techniques are a series of microimaging techniques that are developed on the basis of light scattering signal from a living body. Currently, widely used techniques mainly include reflected confocal reflectance microscopy (RCRM), dark-field microscopy (DFM), optical coherence tomography (OCT), optical scattering imaging based on Fourier filtering (OCIF), quantitative phase microscopy (QPM), and so on.

In recent years, computed tomography based on computer data disposal has been used for light scattering imaging of biological tissue and its spectral analysis, especially with good prospect in three-dimensional reconstruction of tissues and computational biology. These methods mainly include (1) finite-difference time-domain (FDTD) simulation for biosampling of light scattering properties; (2) use of interferometric synthetic aperture microscopy (ISAM) to improve the three-dimensional imaging quality. The principles, applications, and advantages of the optical scattering imaging technology will not be discussed in this chapter. If wish, please read the related overview [1].

In bioimaging, the analysis quality depends on four parameters: (1) imaging velocity, (2) sensitivity, (3) resolution, and (4) specificity. Since light scattering has a larger section, with constant intensity and photon detection sensitivity, light scattering signals have a higher output than other optical signals, so light scattering imaging might have a higher signal-to-noise ratio. Although resolution is limited by the diffraction limit, advanced high-level technology can be used to exponentially improve the imaging resolution. For example, reflection confocal microscopy could reach a spatial resolution of 0.5 μm, whereas the OCT only has a resolution of a micrometer level [2]. In addition, as elastic scattering signal of light is sensitive to the physical properties of tissues, the alternation of measurement, dimensions, optical path length, and refractive index fluctuation of light scattering signal can be used to further improve the imaging quality. This shows that without any modification on a biological sample, optical scattering imaging technique can aquire highly senstive optical signal and high-resolution image in the living body structures.

It is worth attention that the optical scattering imaging is not specific; it works on all biostructures and components with scattering properties that are different from fluorescence imaging, which specifically works only on tissues or organelles with modification. So, one important developmental direction in the field is to develop probes with excellent properties for specifically labeled imaging of the target tissue. On the other hand, even though light scattering has no specificity, the obtained light scattering signal from tissue or cell (such as reflection confocal microscopy or optic coherence tomography) can still be used for visible analysis of the biological structure, with advantages of safety, fast speed, and no need of dye markers. More important is that these imaging technologies can be directly used for real-time monitoring of the slight changes in the cells or tissues, providing effective methods for noninvasive study of tissue or cell interaction in living organisms or unknown structural changes.

8.2 Label-free light scattering cell imaging techniques

Cells, the basic unit of all living bodies, can only be observed through a microscope because of its small size. Cells have strong light scattering signals, especially structures of nucleus and intracellular vesicles, which can be directly observed through a light scattering microscope. However, it is still challenging to directly analyze the fine structure of an unlabeled organelle because the light scattering signal in a cell is composed of multicomponent, multidimensional elastic scattering signals, and highly effective signal collection and analysis methods need to be developed to separate the organelle signals of interest from complex background signals.

8.2.1 Confocal light absorption and scattering spectroscopic microscopy

Itzkan et al. developed one confocal light absorption and scattering spectroscopic microscopy (CLASS), which was able to detect an organelle distribution in an individual living cell noninvasively [3]. This technology combines good features of the confocal scanning microscope and scattering spectrum, which is able to identify the cellular particle size, refractive index, and its shape without any exogenous labeling. As shown in Figure 8.1, time-lapse imaging of a single living cell with DHA disposal is successfully realized with this technology; the analysis is performed with size, shape, and the spatial distribution of nucleus and organelle as lysosomes.

8.2.2 Optical coherence tomography

Optical coherence tomography (OCT) is a imaging method based on an after-detection light scattering signal. As OCT can perform deep-level tissue detection, it is

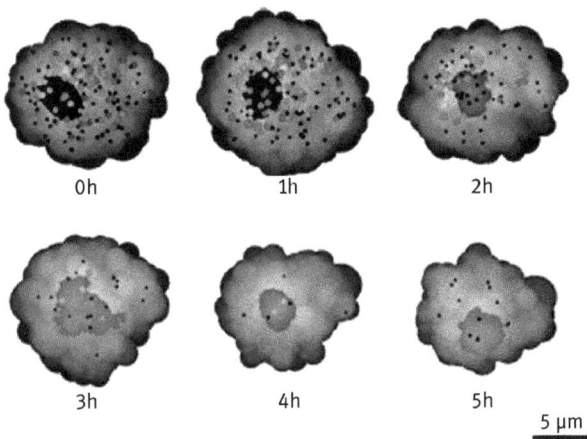

Figure 8.1: Time sequence CLASS simulation changes in a single-cell structure.

widely used in clinical medical imaging, especially in the diagnosis of disease of ophthalmology, cardiology, dermatology, and oncology. Currently, OCT is successfully applied in cell neuropsychobiological studies, especially for slight changes in the structure of a neuron styloid in the transmission of nerve signals. Lazebnik used this technology to analyze slight changes in the transmission of nerve signals in a neuron, showing that the rearrangement of water and ion is the reason for nerve fiber expansion, and changes in the OCT imaging signal [4]. Graf used a rapid optical interference imaging to confirm that nerve impulse signal transduction in *Aplysia californica* neuron is the same as changes in the cell light scattering signal [5].

8.2.3 Optical scattering imaging based on Fourier filtering

Optical scattering imaging based on Fourier filtering (OCIF) has a great performance in cell imaging analysis, especially changes in particle size in cell apoptosis. Pasternack performed an imaging analysis with shape structures of two organelles with homologous gene and different apoptotic signals, confirming their differences in the mitochondrial length and aspect ratio [6]. Although OCIF is based on a light scattering signal, as OCIF not only depends on fine imaging mode and morphological measurement of the obtained figure, it is also suitable for the parametric measurement of unstained living cells in nonspherical organelles; especially, it has wide application prospects in high throughput single-cell analysis with wide imaging range.

8.2.4 Quantitative phase microscopy

Quantitative phase microscopy (QPM) is especially alluring for biological imaging studies because this technology provides a high-sensitive and high-resolution quantized phase imaging for biological samples, whereas this information can provide an important cell biology reference for cellular dynamic characteristics, subcellular structures, and organelle distribution in cells [7].

In 1996, Oldenbourg used QPM to successfully realize the measurement of molecular stunting in tissues and canalized development [8]. In addition, QPM has an excellent imaging and analytical result for highly ordered structures, such as actin, microtubules, and collagen (Figure 8.2) [9, 10].

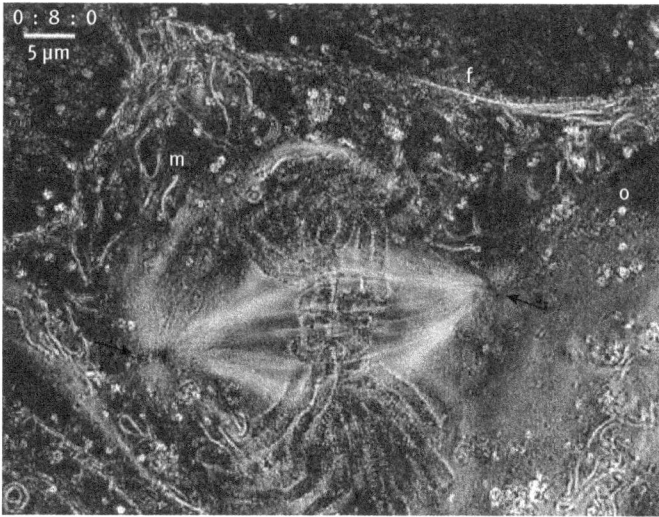

Figure 8.2: OPM imaging of mitosis in salamander lung epithelial cells. This image shows that the micro-tube formed by white spindle fibers distribute between two spindle bodies, white arrow points toward a spindle body; m: mitochondria; f: spindle fiber; o: round organelle.

It can be seen that label-free scattering techniques can be directly used for imaging of cell structure and organelle distribution and for the analysis of aspect ratio of the organelle, particle size, distribution, or organelle's physiological functions.

8.3 Nanoparticle as contrast for light scattering cell imaging

With strong plasma resonance scattering light, noble metal nanoparticles of gold and silver became excellent probes for developing non-photobleaching fluorescent probes that are widely used for biological labeling and imaging [11]. Noble metal nanoparticle usually binds to the cell surface through nonspecific adsorption. To prepare nano optical probes with specific recognition functions, nanoparticles must be combined with biomolecular recognition groups through surface functional modification.

Widely used nano-probe biorecognition ligands include antigen–antibody, polypeptide, oligonucleotide, nucleic acid aptamer, and various small organic molecules. Since multiple nanoparticles can serve as imaging probes, this section will emphasize on the applications of gold and silver nanoparticles with the widest application in cell labeling and imaging analysis.

8.3.1 Silver nanoparticle as contrast for cell imaging

Thanks to super strong plasma resonant light scattering properties, silver nanoparticle with 10 nm size has a scattered light that can be clearly observed in a dark-field microscope. Meanwhile, silver nanoparticles with different diameters exhibit the size-dependent color changes under the microscope. For example, silver nanoparticles with diameter under 20 nm are usually blue in color under the microscope, Meanwhile, silver nanoparticles with diameter near 60 nm are green, and when size of the silver nanoparticle increases to 90 nm and above, it is red. These features make silver nanoparticles develop into multi-colored probes that can be used for cell labeling and light scattering imaging analysis.

8.3.1.1 Intercellular molecular imaging

Schultz found that silver nanoparticles scatter a strong colored light without any photobleaching under a dark-field microscope, which makes them pose great potential to be developed as a light scattering probe. Through monitoring of single nanoparticle light scattering image and spectral analysis, in-situ hybridization and single-molecule analysis are performed in the chromosome of Drosophila [12]. Since 2004, Xu's group performed systematic studies in terms of cell labeling and imaging using silver nanoparticles as optical contrast. On the basis of characterization of the optical property and single particle spectrum of sivler nanoparticle probe, they observed that drugs influence on combined antibacterial activity and membrane permeability of *Pseudomonas aeruginosa* [13, 14]. Results showed that silver nanoparticles of different sizes show pleochromatism in a dark-field microscope. A small-sized nanoparticle enters the cytoplasm through the cell wall, whereas a large-sized nanoparticle only adsorbs and aggregates on the surface of the cell membrane. Xu's study also found that surface particles of the membrane have brighter light spots compared with those in cells. The study results demonstrate that silver nanoparticles can be used as a non-photobleaching probe for nanoscale real-time dynamic analysis in a living cell. There are other reports showing that silver nanoparticles with diameter of 10 nm and concentration of

1.3 pmol/L have no obvious toxicity for *P. aeruginosa*, can be applied for the dynamic study of nanoparticle penetration for cell membrane [15].

In 2007, Xu's group coupled the antibodies of cell surface proteins with silver nanoparticle, and successfully performed a specific labeling and imaging of intercellular membrane proteins through specific recognition bewteen antibody and protein [16]. With dark-field scattering image, they investigated the membrane protein distribution and dynamic characteristics in detail on the surface of a single living cell. With the toxicity investigation of mouse embryonic fibroblasts with 11.3 nm ± 2.3 nm silver nanoparticles, they showed that the cell toxicity is dependent on their concentration. After a 12 hour interaction, dark-field imaging shows that silver nanoparticle enters into the cytoplasm and the nucleus, causing toxicological response as cell growth inhibitory, blocking of cell differentiation and multicellular nucleus effect [17].

On the basis of detection of plasma light scattering properties and single-particle spectrum of silver nanoparticle, the authors have successfully applied the properties of light scattering from silver nanoparticle for drug detection and protein immunoassay. In addition, we have developed aptamer-labelled silver nanoparticles as a novel optical probe for simultaneous intracellular protein imaging and single nanoparticle spectrometry, wherein AgNPs were combined with proteins by electrostatic incorportaiton. It was found that streptavidin-conjugated and aptamer-functionalized AgNPs show satisfactory biocompatibility and stability in cell culture medium, and single nanoparticle spectrometry demonstrates that the spectrum is closey related to the intercellular miroenvironment, suggesting a potential application for sensitive detection of the changes in cellular microenvironment [18].

Wang, from Boston University, combining dark-field light scattering imaging with multispectral analysis technique, analyzed silver nanoparticle-marked spatial distribution of epidermal growth factor receptor on the cell surface of A431 through the acquisition and inergral analysis of multiple wavelength channel of scattering images (red channel: 510/494 nm, green channel: 473/445 nm, and blue channel: 427 nm) [19]. Since concentration of the surface immunolabeling silver nanoparticle is directly relevant to the receptor concentration, the method is of significant for analysis of concentration changes of the surface acceptor and spatial contribution, as well as provides a new platform for real-time analysis of surface acceptor dynamic process with simple imaging technology.

8.3.1.2 Organisms and tissues labeling imaging

Besides imaging at cell level, silver nanoparticles also have outstanding performance as probe in marking imaging in living animals. For example, based on the preparation and detailed representation of optical properties of different size of nanoparticles, Xu's group used a dark-field imaging technique to directly observe the moving of silver nanoparticles both on the surface and in intervals of zebrafish embryos, and recording movements of locus and oolemma cross-moving process of an individual silver nanoparticle (Figure 8.3) [20, 21]. Through histology and dark-field imaging, the distribution of silver nanoparticles in organs of *Danio rerio* is directly observed. This not only shows excellent property of silver nanoparticles as optical contrast for imaging of living animals, but also provides a new insight into nanomaterial bioabsorption, distribution, and metabolism.

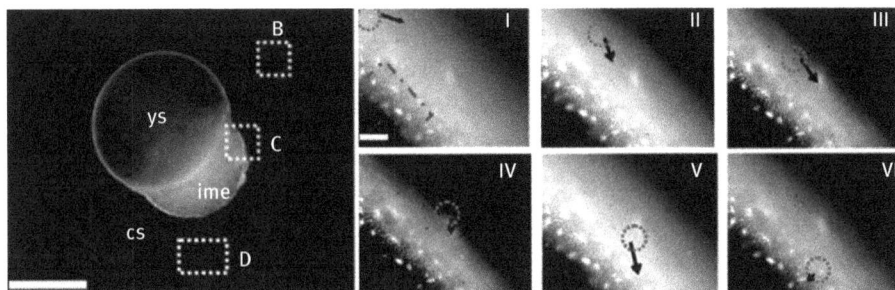

Figure 8.3: Dark-field light scattering image of silver nanoparticles on the surface of chorion of zebrafish embryo [20, 21]. Left side is a dark-field image of zebrafish embryo, ys: yolk sac; ime: internal embryo cells; cs: chorionic spaces; Right side of the figure shows a serial photograph of a single silver nanoparticle moving on the chorion of embryo, black arrow is the direction of the moving single particle with a scale bar of 400 μm.

8.3.2 Gold nanoparticle as optical contrast for intercellular protein imaging and cancer cell recognition

Cancer is a malignant disease, taking away thousands of lives. As cancer has a strong elusive property, obvious clinical pathological symptoms are not shown until terminal stages of cancer, causing difficulties in the early stage diagnosis and treatment, so successful early stage diagnosis is essential for cancer treatment. Modern nanotechnique development has provided useful tool and a new platform for early stage diagnosis of cancer, especially recognition and rapid detection of cancer cells.

Gold nanoparticles have special optical properties, including ultra-strong scattering signal, non-photobleaching property, and plasma resonance coupling. These good optical properties and excellent biocompatibility make gold nanoparticles serve as an optical contrast, with special advantages in labeling imaging and tracing in the target protein of the living cell.

In 2005, El-Sayed from University of California, Los Angeles modified an antibody of an epidermal growth factor on the surface of gold nanoparticles and used them as probes for labeling and recognition of oral epithelium cancer cells [22]. It is interesting that normal cells (human keratinocyte HaCaT) and cancer cells (human oral epithelium cancer cells, HSC/HOC) both have weak scattering signals and are dark green in color under the microscope. The authors thought that these signals are probably from an intracellular organelle or cell membrane. With unmodified gold nanoparticles, the cells bring some of gold nanoparticles inside them through nonspecific adsorption and endocytosis and enhancing light scattering signal of cells, whereas there is no obvious difference between the normal cells and cancer cells. When modified gold nanoparticles are added, there are obvious changes in the signal of different types of cells. Cancer cell surface is adsorbed by a large number of gold nanoparticles with super strong light scattering signal, and these particles are aggregated on the surface of cells, forming the outline of an obvious shape of the cancer cell. In contrast, non-cancer cell surface particles are randomly distributed with difficulty in distinguishing the cell outline (Figure 8.4[a]). The results show that gold nanoparticles are excellent scattering probes, which can be used for recognition and detection of cancer cell.

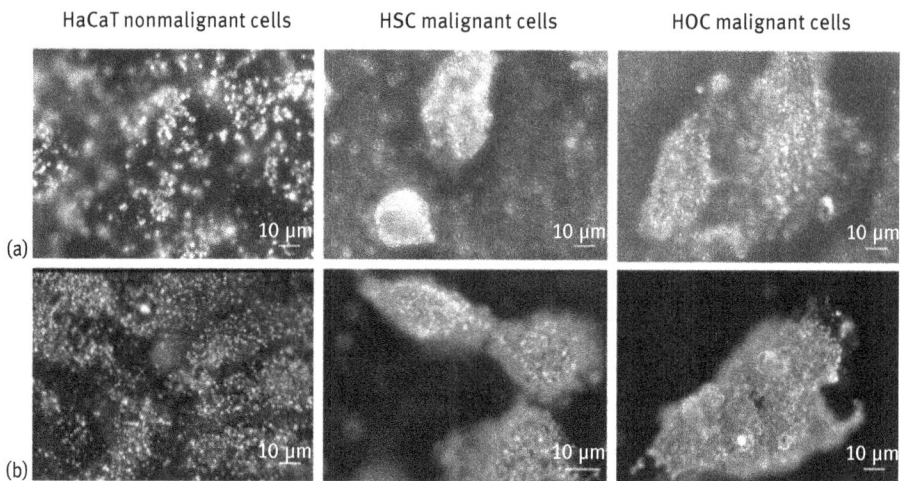

Figure 8.4: Light scattering images of cancer cells associated with gold nanoparticles and gold nanorods [23]. (a) Gold nanoparticles; and (b) gold nanorods.

Based on similar principles, Huang et al. modified epidermal factor antibody on the surface of gold nanorod to study the labeling and recognition of cancer cells [23]. As gold nanorods exhibit red scattering light after aggregation, cancer cells labeling by gold nanorods are dark red in color under a dark-field microscope (Figure 8.4[b]). Meanwhile, as nanorods have strong light absorption features, the nanorods based-probes can be successfully applied in the photothermal therapy of cancer cells. In addition, combining dark-field scattering imaging with Raman spectrum, Huang's group used this probe to study cancer cell-sensitive detection [24]. The results showed that gold nanoparticle-based cancer cell recognition and labeling can be potentially applied in early stage detection of cancer and novel technique development in the diagnosis of cancer.

Kumar et al. modified polyethylene glycol and target protein specific antibody on the surface of a gold nanoparticle to construct a multifunctional nano-biological sensor [25]. With features of target recognition, cytoplasmic transfer function, good biocompatibility, and other qualities, the sensor can enter the cytoplasm through the cell membrane and specifically recognize the actin. Dark-field scattering imaging clearly shows the combinations of gold nano-probe with actin and cell synapse, with obvious yellow aggregates. It can also be observed that the labeled actin delivery in the synapse in living cells through real-time dynamic imaging (Figure 8.5). In addition, Yu et al. modified cell-specific monoclonal antibody on gold nanorod in breast cancer and studied specific recognition and labeling of three cytokinal subtypes on the surface of cancer cells simultaneously with dark-field scattering imaging [26].

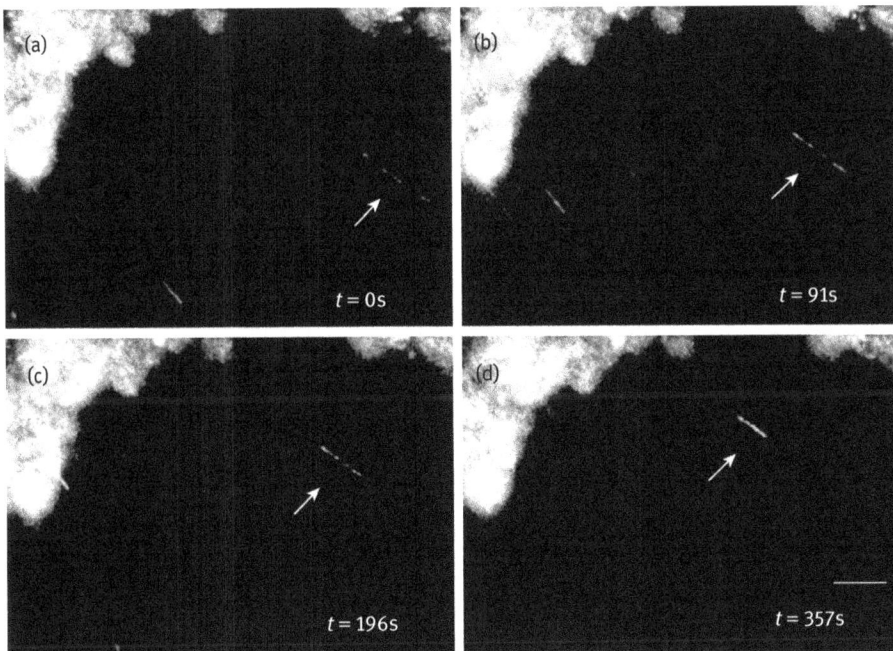

Figure 8.5: Time-lapse images of gold nanoparticle-labeled actin moving on the filapodia of live cells. While the light zone points out the cell body, the arrows indicate actin filaments labeled by gold nanoparticle. Scale bar is 20 μm.

In addition to antibody, other molecular-specific ligands can also be used for imaging of target recognition gene for cell labeling. For example, oligonucleotide-coupled gold nanoparticles are successfully used in cancer cell labeling and scattering imaging [27]. Polypeptide gold nanorods are used in cell nucleus labeling and scattering imaging [28]. Aptamer-modified gold nanoparticles can realize prostate-specific antigen labeling and cancer cell scattering imaging [29]. As many recognition mechanisms can be used for surface target molecule and protein recognition, so it is thought that the modification of various recognition genes on the surface of gold nanoparticles and their use as an optical imaging probe for analysis will have wide application prospects in fields of cell surface molecule labeling, protein dynamics tracing, cancer cell recognition, labeling, and so on.

8.3.3 Other nanomaterals-based cell imaging with light scattering signals

In addition to wide applications of gold, silver nanoparticles as probe, some other nano-materials have been developed into optical scattering probes for cancer cell-labeled analysis and cell imaging of molecular assembly. For example, folic acid-modified platinum nanoparticles have a strong recognition for cancer cells, which can enter cells showing bright white spots [30]. Antibiotic-mediated gold nanoparticle's net complex can also be used as an optical probe for dark-field scattering imaging of a cell and Raman scattering spectrometry [31].

Our study also shows that a complex of carbon nanotube/gold nanorods and carbon nanotubes/silver nanoparticles have a strong light scattering property, which can be used as excellent probes for imaging analysis [32]. Similarly, cuprous oxide nanoparticles also have stronger scattering properties, modified with characteristic antibody of the surface protein, which can be used for cancer cell-specific recognition labeling and scattering imaging [33]. Polyphenylenediamine crystal is a leafy micrometer particle with specific recognition for cobalt ions; we develop it into a reagent that can specifically detect cobalt ions in solution and biotissues. When polyphenylenediamine particles combine with cobalt ions in fish tissues, ions corrode the particles, bringing large-sized polyphenylenediamine breakage, forming small particles dispersing in tissues, and bringing an orange color to their scattering imaging. So polyphenylenediamine crystals also serve as potential scattering probe for biotissue imaging analysis [34].

8.3.4 Real-time monitoring of dynamics of intracellular molecules

As shown in Section 7.5, when two metal nanoparticles are placed adjacently, the couple of their local plasma resonance will happen, bringing obvious changes to the local surface plasma resonance spectrum and in the properties of light scattering of nanoparticles. For example, nanoparticles' dimer or polymer resonance spectrum having a red shift and imaging color will also change; this change in the spectrum and color can be used for ultrasensitive detection of molecules or changes in real-time monitoring of intracellular substance concentration.

For example, based on the fact that a single 40 nm silver nanoparticle has a resonance scattering peak at 420 nm, when two silver nanoparticles get nearer and form a dimer, the resonance scattering peak of two nanoparticles has a red shift near

500 nm, whereas scattering light color changes from blue to green. Recently, Sokolov group used gold nanoparticles with the specificity to label an epidermal growth factor receptor on the surface of the living cell. When receptors aggregate together during their internalization process or moving on the surface of cells, the distance between nanoparticles is shortened, then color change from blue to yellow and then brown can be observed in a dark-field microscope imaging, which is consistent with the receptor vesicle endocytosis period and gold nanoparticle aggregation degree [35]. This study fully shows the application of metallic nanoparticle plasma resonance couling and a dark-field optical imaging in the real-time monitoring of biomolecular activity and their concentration changes in the living cell.

Based on the principles of plasmon resonance energy transfer (PRET), Lee's group from Korea developed a novel method for noninvasive imaging of intracellular biomolecules in living cell biomolecule imaging [36]. When a gold nanoparticle get colse to an acceptor with a specific distance during its internaliztion into cytoplasm, energy is delivered as plasma resonance from gold nanoparticles as donor to the receptor of cytochrome c (Cyt c) because they have adjecent resonance freqency. The PRET between gold nanoparticle and Cyt c shows as the quenching dip in the over-lapping of gold nanoparticle's Rayleigh scattering spectrum and Cyt c absorption spectrum. The shape change in the spectrum is closely related to the distance between donor and receptor, as well as their concentrations in the cell. Choosing a resonance frequency-matched gold nanoparticle and biomolecule Cyt c as energy transfer donor and receptor, the dynamic process of visual observation in alcohol-induced apoptosis process in HepG2 internal Cyt c is successfully established.

8.4 Furthure and emerging directions for optical scattering imaging

8.4.1 Design and preparation of new contrast agents for optical scattering imaging

Spectral features of a natural light provide a tool to effectively detect living body structure and function at the cellular and molecular level. Although fluorescent and bioluminescent probes have been the mainstream of molecular optical imaging, the development and application of lots of optical probes with other features, such as scattering, absorption, and plasma resonance effect, will also improve the overall development of an optical imaging technique.

As described above, optical scattering probes usually have a high efficiency; it can achieve optical sensing through direct detection of scattering signal changes or indirect detection of weakening degree of an incident light. For example, in OCT imaging analysis, probes changes the refractive index of the specific area of organelle or tissue, alterating the intensity of afterward scattering light of the organelle labeling by nano-probe, labeling, providing necessary information for imaging of the biological structures change. Currently, bubble micro-structures, such as buble microsphere, and protein micro-particles covered with nanoparticles, which exposed through an intravenous injection or cell culture, have been used as OCT imaging probes and are applied in the analysis of biological structure and organelle formation. Especially for small-sized nanoparticles, as it can easily enter the cancer cells through blood circulation, many researchers have prepared the small-sized nano-shell or nanoparticles as a probe with capability of specific recognition of molecular acceptor in cancer cells. When these probes are delivered into a cancer cell, the light scattering signal of the target cell is greatly enhanced, which are convenient for the detection of cancer cells or tissues [37].

Based on the Rayleigh scattering theory, section area of the particle is closely related to the scattering intensity. So, it is challenging to prepare a nano-probe with a small sectional area and enough scattering intensity for detection in a cell or in a living organism. Metallic nanostructures, including nanospheres, nanorods, nanoshells, and nanocages with advantages of super-strong scattering and unique plasma resonance features, small sizes, and permeation through a biobarrier, are widely used as optical probes for light scattering imaging [38]. So, the development trends of the furtuer optical probes design and preparation are to develop the optical scattering probe with specific recognition ability and good biocompatibility, and utilizaiton of photothermal properties of nanoparticles for cancer cell recognition, identification, and diagnosis.

Currently, in vivo imaging for cancer recognition and cancerous organism identification using a nano-scattering probe is still in the laboratory phase. The imaging ability of nano-probe still needs to be improved for application in clinical testing and cancer treatment. Meanwhile, nano-probe biocompatibility and in-vitro pharmacokinetic properties still need further investigation.

8.4.2 Monitoring of biological process in cell biology

In high-resolution microscopic imaging, any slight change in the structure of the living organism brings relative changes in its scattering signal. So, the normal cellular biological and pathological processes can be monitored through imaging anlysis of the dynamic change of light scattering singals.

For example, cell apoptosis is an essential biological process. People can rapidly and invasively monitor the changes of intracellular structures and their related pathological process through detction of the changes in the light scattering signal of cell during cell apoptosis. Studies have shown that in the initial hour of cell apoptosis, changes in the scattering signal are basically consistent with the average size of a particle. While this change is controlled by one anti-apoptosis protein activity in mitochondria, so the acitivity of anti-apoptosis protein can be indirectly controlled through the monitoring of changes in the scattering signal of initial stage of cell apoptosis stage scattering signal [39]. There are other studies indicate that cytochrome c release during cell apoptosis is basically consistent with intercellular light scattering signals change, further confirming that changes in the scattering signal caused by mitochondrial changes is the vital basis for measuring cell apoptosis [40].

Another important biological application of optical scattering imaging is to monitor a single neuron or tissue nerve activity through changes in the scattering signal. In a nerve impulse signal conduction process, water and ion rearrangement causes nerve fiber expansion, bringing scattering signal changes in an OCT imaging [4]. Rapid optical interference imaging study has confirmed that in nerve cell division process of *Aplysia californica*, light scattering signal changes are basically consistent with nerve impulse signal conduction [5]. Stepnoski hypothesis that the signal changes are from cell membrane dipole rearrangement caused by the changes in neuron's local refractive index [41]. Yao also believed that the signal change of a polarization function is likely to provide a highly sensitive signal source for monitoring of nerve cell activity [42].

Combining dark-field imaging with Raman spectrum, Kang et al. successfully monitored real-time changes of amino acid in chromosome cycle (Figure 8.6) [43]. It can also be used to construct a laser confocal light scattering imaging system to improve the resolution (Figure 8.7) [44].

8.4.3 Enhanced dark-field light scattering imaging

Usually, photographs can be expressed in hue, brightness, and saturation. To improve the imaging quality, there are three ways to improve the image resolution and the contrast ratio. Among these, microphysical optics design (Section 7.8) is the most fundamental but limited by optical diffraction limits. Excellent results can be obtained with the combination of nonlinear optical properties, or disposed with computer software (as single nanoparticle three-color coding in Section 7.4.3); chemical reaction can be used to improve the quality with one case history of using fluorescent molecules to skillfully avoid optical diffraction limit, winning the 2014 Nobel Prize in chemistry for their micro-optical imaging results.

As shown in Figure 8.8, carbon nanotube/gold nanorod complex, carbon nanotube/silver nanoparticle complex might greatly reduce the background scattering signal, whereas they can be better used as scattering probes in cell imaging analysis [32].

By using aggregates of the chromogenic substrate diaminobenzidine (DAB), which is most sensitive to the enzyme horseradish peroxidase, and through overexpression of breast cancer cell and traditional immunohistochemistry, a new enzyme catalysis-enhanced dark-field imaging technology can be constructed [45].

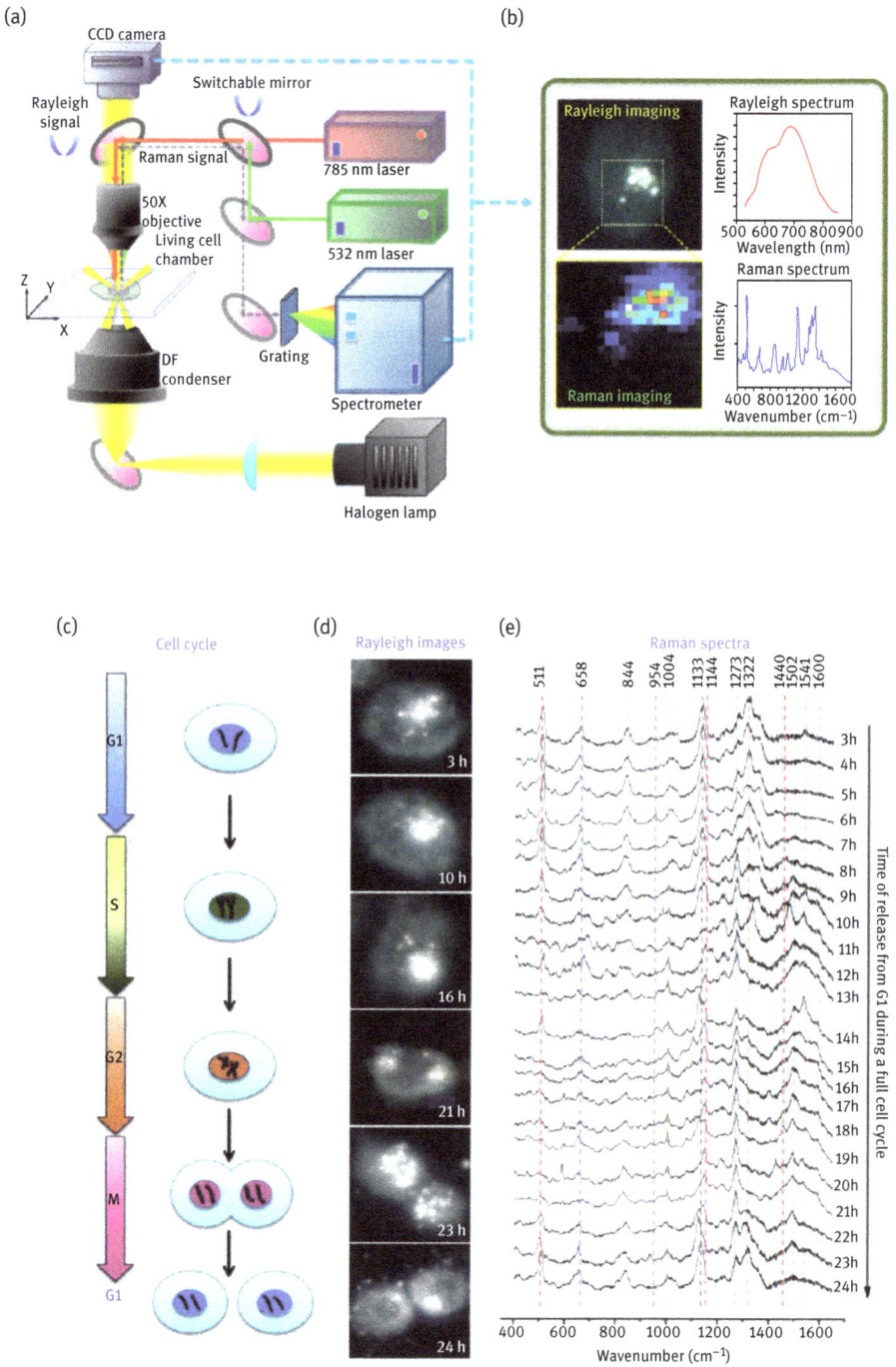

Figure 8.6: Combined dark-field scattering imaging and Raman spectra for monitoring of amino acid real-time changing process in a cell division process [43].

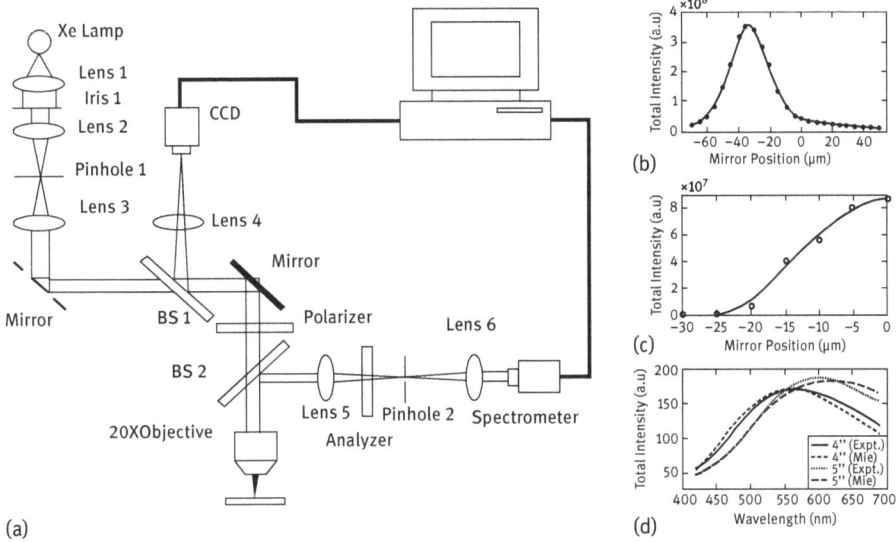

Figure 8.7: Confocal light scattering spectroscopic imaging system [44].

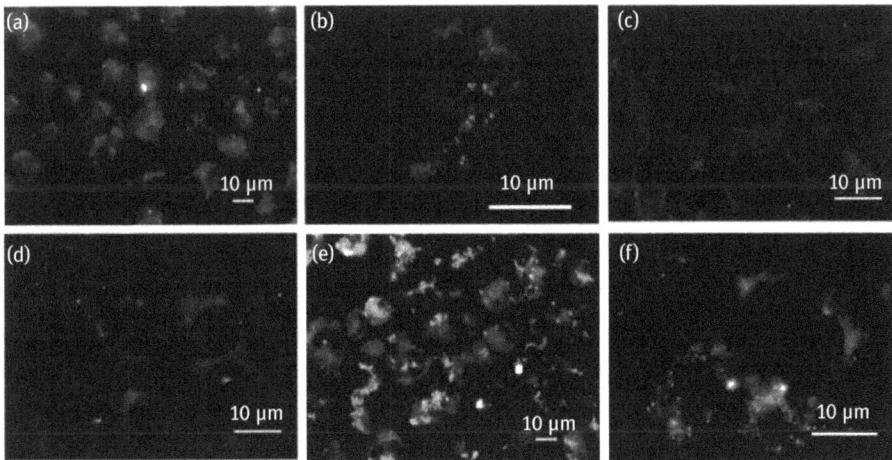

Figure 8.8: Silver nanosphere–CNTs complex used as a dark-field scattering imaging probe in lung cancer cells [32]. (a, b): PAD covered AgNS/CNT hybrid body under different magnification times; (c,d): comparison of CNTs and AgNSs; (e,f): citric acid covered AgNS/CNT hybrid and its interaction with cancer cell .

8.4.4 Two-photon light scattering imaging

Ray et al. [46], constructed a two-photon Rayleigh scattering (TPRS) analytical method based on the incresing of electric dipole from aggregation of gold nanorods. Light scattering intensity of TPRS can be expressed with equation (8.1) [46, 47]:

$$I_{\text{TPRS}} = G(N_{\text{w}}\beta_{w}^2 + N_{\text{nano}}\beta_{\text{nano}}^2)I_{\omega}^2 e^{-N_{\text{nano}}\varepsilon_{2\omega}L} \tag{8.1}$$

where G is the geometrical factor; N_{w} and N_{nano} are the number of water molecules and gold nanorod number in a unit volume, respectively; β_{w} and β_{nano} are quadratic hyperpolarizabilities of single molecules and single nanoparticles; $\varepsilon_{2\omega}$ is the nanorod molar absorption coefficient at 2ω; I_{ω} is the matrix intensity; and L is the optical path length.

The results show that when *Escherichia coli* antibody-connected gold nanorod is mixed with *E. coli* O157:H7 at different concentrations, its TPRS intensity increased by 40 times, providing a good platform for quantitative analysis of *E.coli* [46]. Further study has shown that TPRS technology is used in the detection of the protein tau, a biomarker of Alzheimer's disease by using monoclonal anti-tau antibody-covered gold nanoparticle. This method achieved a detection level of 1 pg/mL, almost two-order magnitudes lower than cerebrospinal fluid analysis [48]. If DNA molecules are used to make gold nano-antenna within the range of 25 nm, it can build a long-range two-photon scattering spectroscopy ruler, in which twice the distance larger than FRET optical ruler, further realizing high selectivity of prostate cancer cell selection [49].

In general, light scattering signal of a cell changes in microscopic imaging can not only be used for the investigation of cell apoptosis and nerve cell signal transduction, but also can be used for monitoring inner structural changes of multiple important biological processes, such as cell signaling transduction, metabolism, and cellular response.

References

[1]   Boustany N N, Boppart S A, Backman V. Microscopic imaging and spectroscopy with scattered light. Annu. Rev. of Biomed. Eng., 2010, 12: 285–314.
[2]   Fujimoto J G. Optical coherence tomography for ultrahigh resolution in vivo imaging. Nat. Biotech., 2003, 21: 1361–1367.
[3]   Itzkan I, Qiu L, Fang H, et al. Confocal light absorption and scattering spectroscopic microscopy monitors organelles in live cells with no exogenous labels. Proc. Natl. Acad. Sci. USA, 2007, 104:17255–17260.
[4]   Lazebnik M, Marks D L, Potgieter K, et al. Functional optical coherence tomography for detecting neural activity through scattering changes. Opt. Lett., 2003, 28: 1218–1220.
[5]   Graf B W, Ralston T S, Ko H J, et al. Detecting intrinsic scattering changes correlated to neuron action potentials using optical coherence imaging. Opt. Express., 2009, 17: 13447–13457.
[6]   Pasternack R M, Qian Z, Zheng J Y, et al. Measurement of subcellular texture by optical Gabor-like filtering with a digital micromirror device. Opt. Lett., 2008, 33: 2209–2211.
[7]   Popescu G, Park Y, Dasari R R, et al. Coherence properties of red blood cell membrane motions. Phys. Rev. E, 2007, 76: 031902.
[8]   Oldenbourg R. A new view on polarization microscopy. Nature, 1996, 381:811–812.

[9]   Katoh K, Hammar K, Smith P J S, et al. Birefringence imaging directly reveals architectural dynamics of filamentous actin in living growth cones. Mol. Biol. Cell, 1999, 10: 197–210.

[10]   Oldenbourg R, Salmon E D, Tran P T. Birefringence of single and bundled microtubules. Biophys. J., 1998, 74: 645–654.

[11]   Murphy C J, Gole A M, Stone J W, et al. Gold nanoparticles in biology: Beyond toxicity to cellular imaging. Acc. Chem. Res., 2008, 41: 1721–1730.

[12]   Schultz S, Smith D R, Mock J J, et al. Single-target molecule detection with nonbleaching multicolor optical immunolabels. Proc. Natl. Acad. Sci. USA, 2000, 97: 996–1001.

[13]   Kyriacou S V, Brownlow W J, Xu X H N. Using nanoparticle optics assay for direct observation of the function of antimicrobial agents in single live bacterial cells. Biochemistry, 2004, 43: 140–147.

[14]   Xu X H N, Brownlow W J, Kyriacou S V, et al. Real-time probing of membrane transport in living microbial cells using single nanoparticle optics and living cell imaging. Biochemistry, 2004, 43: 10400–10413.

[15]   Nallathamby P D, Lee K J, Desai T, et al. Study of the multidrug membrane transporter of single living Pseudomonas aeruginosa cells using size-dependent plasmonic nanoparticle optical probes. Biochemistry, 2010, 49: 5942–5953.

[16]   Huang T, Nallathamby P D, Gillet D, et al. Design and synthesis of single-nanoparticle optical biosensors for imaging and characterization of single receptor molecules on single living cells. Anal. Chem., 2007, 79: 7708–7718.

[17]   Nallathamby P D, Xu X H N. Study of cytotoxic and therapeutic effects of stable and purified silver nanoparticles on tumor cells. Nanoscale, 2010, 2: 942–952.

[18]   Chen L Q, Xiao S J, Peng L, et al. Aptamer-based silver nanoparticles used for intracellular protein imaging and single nanoparticle spectral analysis. J. Phys. Chem. B, 2010, 114: 3655–3659.

[19]   Wang H, Rong G, Yan B, et al. Optical sizing of immunolabel clusters through multispectral plasmon coupling microscopy. Nano Lett., 2011, 11: 498–504.

[20]   Lee K J, Nallathamby P D, Browning L M, et al. In vivo imaging of transport and biocompatibility of single silver nanoparticles in early development of zebrafish embryos. ACS Nano, 2007, 1: 133–143.

[21]   Nallathamby P D, Lee K J, Xu X H N. Design of stable and uniform single nanoparticle photonics for in vivo dynamics imaging of nanoenvironments of zebrafish embryonic fluids. Acs Nano, 2008, 2: 1371–1380.

[22]   El-Sayed I H, Huang X, El-Sayed M A. Surface plasmon resonance scattering and absorption of anti-EGFR antibody conjugated gold nanoparticles in cancer diagnostics: Applications in oral cancer. Nano Lett., 2005, 5: 829–834.

[23]   Huang X, Ei-Sayed I H, Qian W, et al. Cancer cell imaging and photothermal therapy in the near-infrared region by using gold nanorods. J. Am. Chem. Soc., 2006, 128: 2115–2120.

[24]   Huang X, El-Sayed I H, Qian W, et al. Cancer cells assemble and align gold nanorods conjugated to antibodies to produce highly enhanced, sharp, and polarized surface Raman spectra: A potential cancer diagnostic marker. Nano Lett., 2007, 7: 1591–1597.

[25]   Kumar S, Harrison N, Richards-Kortum R, et al. Plasmonic nanosensors for imaging intracellular biomarkers in live cells. Nano Lett., 2007, 7: 1338–1343.

[26]   Yu C, Nakshatri H, Irudayaraj J. Identity profiling of cell surface markers by multiplex gold nanorod probes. Nano Lett., 2007, 7: 2300–2306.

[27]   Nitin N, Javier D J, Richards-Kortum R. Oligonucleotide-coated metallic nanoparticles as a flexible platform for molecular imaging agents. Bioconjugate Chem., 2007, 18: 2090–2096.

[28]   Oyelere A K, Chen P C, Huang X, et al. Peptide-conjugated gold nanorods for nuclear targeting. Bioconjugate Chem., 2007, 18: 1490–1497.

[29] Javier D J, Nitin N, Levy M, et al. Aptamer-targeted gold nanoparticles as molecular-specific contrast agents for reflectance imaging. Bioconjugate Chem., 2008, 19: 1309–1312.

[30] Teow Y, Valiyaveettil S. Active targeting of cancer cells using folic acid-conjugated platinum nanoparticles. Nanoscale, 2010, 2: 2607–2613.

[31] Souza G R, Christianson D R, Staquicini F I, et al. Networks of gold nanoparticles and bacteriophage as biological sensors and cell-targeting agents. Proc. Natl. Acad. Sci. USA, 2006, 103: 1215–1220.

[32] Zhang L, Zhen S J, Sang Y, et al. Controllable preparation of metal nanoparticle/carbon nanotube hybrids as efficient dark field light scattering agents for cell imaging. Chem. Commun., 2010, 46:4303–4305.

[33] Qi W J, Huang C Z, Chen L Q. Cuprous oxide nanospheres as probes for light scattering imaging analysis of live cells and for conformation identification of proteins. Talanta, 2010, 80: 1400–1405.

[34] Zhen S J, Guo F L, Chen L Q, et al. Visual detection of cobalt(II) ion in vitro and tissue with a new type of leaf-like molecular microcrystal. Chem. Commun., 2011, 47: 2562–2564.

[35] Aaron J, Travis K, Harrison N, et al. Dynamic imaging of molecular assemblies in live cells based on nanoparticle plasmon resonance coupling. Nano Lett., 2009, 9: 3612–3618.

[36] Choi Y, Kang T, Lee L P. Plasmon resonance energy transfer (PRET)-based molecular imaging of cytochrome c in living cells. Nano Lett., 2009, 9: 85–90.

[37] Sokolov K, Follen M, Aaron J, et al. Real-time vital optical imaging of precancer using anti-epidermal growth factor receptor antibodies conjugated to gold nanoparticles. Cancer Res., 2003, 63: 1999–2004.

[38] Jain P K, Huang X, El-Sayed I H, et al. Noble metals on the nanoscale: Optical and photothermal properties and some applications in imaging, sensing, biology, and medicine. Acc. Chem. Res., 2008, 41:1578–1586.

[39] Boustany N N, Tsai Y C, Pfister B, et al. BCL-xL-dependent light scattering by apoptotic cells. Biophys. J., 2004, 87: 4163–4171.

[40] Wilson J D, Giesselman B R, Mitra S, et al. Lysosome-damage-induced scattering changes coincide with release of cytochrome c. Opt. Lett., 2007, 32: 2517–2519.

[41] Stepnoski R A, LaPorta A, Raccuia-Behling F, et al. Noninvasive detection of changes in membrane potential in cultured neurons by light scattering. Proc. Natl. Acad. Sci. USA, 1991, 88: 9382–9386.

[42] Yao X C, Foust A, Rector D M, et al. Cross-polarized reflected light measurement of fast optical responses associated with neural activation. Biophys. J., 2005, 88: 4170–4177.

[43] Kang B, Austin L A, El-Sayed M A. Real-time molecular imaging throughout the entire cell cycle by targeted plasmonic-enhanced Rayleigh/Raman spectroscopy. Nano Lett., 2012, 12: 5369–5375.

[44] Huang P, Hunter M, Georgakoudi I. Confocal light scattering spectroscopic imaging system for in situ tissue characterization. Appl. Opt., 2009, 48: 2595–2599.

[45] Fan L, Tian Y, Yin R, et al. Enzyme catalysis enhanced dark-field imaging as a novel immuno-histochemical method. Nanoscale, 2016, 8: 8553–8558.

[46] Singh A K, Senapati D, Wang S, et al. Gold nanorod based selective identification of Escherichia coli bacteria using two-photon Rayleigh scattering spectroscopy. ACS Nano, 2009, 3: 1906–1912.

[47] Ray P C. Label-free diagnostics of single base-mismatch DNA hybridization on gold nanoparticles using hyper-Rayleigh scattering technique. Angew. Chem., Int. Ed, 2006, 45: 1151–1154.

[48] Neely A, Perry C, Varisli B, et al. Ultrasensitive and highly selective detection of Alzheimer's disease biomarker using two-photon Rayleigh scattering properties of gold nanoparticle. ACS Nano, 2009, 3:2834–2840.

[49] Sinha S S, Paul D K, Kanchanapally R, et al. Long-range two-photon scattering spectroscopy ruler for screening prostate cancer cells. Chem. Sci., 2015, 6: 2411–2418.

Jian Wang
9 Light scattering spectrometry of inorganic ions

It can be concluded from Chapter 4 that light scattering spectrometry has a high sensitivity, which has been widely used in inorganic chemistry [1], organic chemistry [2], biology, and bioscience related fields [3, 4]. Inorganic ions, as important compositions of life, including metal ions and negative ions, are of great significance for the environment, human health, and so on. So, this chapter will focus on the applications of light scattering spectrometry in the field of common inorganic ions, including some life-related metal ions and negative ions, whereas scattering analysis of related heavy metal ions will be discussed in Chapter 14, "Aquatic environmental light scattering spectrometry".

9.1 Basic principles of light scattering spectrometry of inorganic ions

Both metallic and nonmetallic ions present small volumes, thus their light scattering spectrometry is built mainly on the basis of equation (9.1):

$$I = I_0 k c_p \tag{9.1}$$

Where, k is a constant, and its influencing factors had been discussed in Chapter 4, including differences in the particle size, medium, and refractive index, namely:

$$k = \frac{16\pi^4 r^6 n_{\text{med}}^4}{d^2 \lambda_0^4} \cdot \left| \frac{m^2 - 1}{m^2 + 2} \right|^2 \cdot \sin^2 \Phi \tag{9.1a}$$

Under fixed instrumental conditions (detection angle Φ and distance d), to obtain a higher sensitivity the particle size, medium, and refractive index should be considered:

$$m = \frac{n}{n_{\text{med}}} = \frac{n_{\text{rel}} + i n_{\text{im}}}{n_{\text{med}}} \tag{9.2}$$

So the following three aspects need to be considered to build a highly sensitive analysis method of inorganic ions:

https://doi.org/10.1515/9783110573138-009

(1) The chosen probe reagent forms large-sized scattering particles with inorganic ions
(2) The formed particle possesses great absorbance ability at specific wavelengths together with the obvious changes in refractive index
(3) The analysis method requires high selectivity, which needs to consider the specific interaction of probe reagents with the testing ions.

9.2 The detection of metal ions

Inorganic metal ions play important roles both in the individual life and the ecological environment. For example, free metal ions play direct roles in living activities, whereas some metal ions directly participate in protein coordination, becoming a living activity center. Statistics indicate that approximately 30% of the enzyme's metal ions serve as a living activity center; on the one hand, metalloenzyme serves as a catalyst in the living body; on the other hand, metal ions can also play the role of fixing the protein structure.

Currently, light scattering spectrometry of metal ions is mainly based on the formation of larger scale ionic associations or precipitation particles that are insoluble in a solvent medium in the testing system. The former involves the mechanism of the formation of cation complex between metal ions and coordination reagents then ion-pair complexes are formed with an acidic dye. For example, Ag^+ forms a cation complex with 1,10-phenanthrolinemonohydrate, then it forms ion-pair complexes with acid dyes, such as bromophenol blue (BPB), bromcresol green (BCG), and bromcresol purple (BCP) enhancing scattering to measure Ag [5]. Cationic surfactants also play important role in detection of inorganic ions, which can form ion-pair complexes instead of dyes. Inorganic ions, such as $[PdI_4]^{2-}$ forms complex with cetyltrimethylammonium bromide (CTMAB) to detect palladium in the form of $[PdI_4]^{2-}$–CTMAB [6], whereas $[AgI_2]^-$–CTMAB system can be used to detect Ag [7].

In recent years, new analysis methods have been developed with the synthesis of new reagents, and the selectivity has been greatly enhanced. Then scattering analysis of the common metal ions will be discussed in details.

9.2.1 Silver ions

Silver is a common industrial raw material. In photographic industry, silver ion is used as a color developing agent, but when excess silver or silver salts enter into the human body will cause deposition and lesion on the skin, eye, and mucosa. So, it needs to strictly detect and control the concentration of silver ions in living organisms.

In addition to the common formation of association to detect silver ions with scattering methods (Table 9.1), there are DNA-nanoparticle visualization methods [8].

Table 9.1: Determination of silver ion with light scattering methd.

System	Linear range	Detection limit	Application	Reference
Ag^+–NPs–DNA	0.13–1.12 µmol/L	62 nmol/L	Water sample	[8]
Ag–phen–acidic triphenylmethane	9.6×10^{-3}–0.60 µg/mL	0.72 ng/mL	Water sample	[5]
Ag–phen–alizarin red	4.2×10^{-3}–0.8 µg/mL	1.25 ng/mL	Environmental water sample	[9]
Ag–phen–eosin	8.0×10^{-3}–0.6 µg/mL	2.40 ng/mL	Environmental water sample	[10]
Ag(I)–EGTA chelating system	5.4×10^{-3}–2.7 µg/mL	0.17 ng/mL	Waste film	[11]
Ag–UV	0.1–4 µg/mL	0.05µg/mL	Waste water	[12]
Ag–sodium sulfide –CTAB	0.0–6.0 µg/mL	0.041 µg/mL	Environmental water sample	[13]
Ag–tetrachlorotetrabromo fluorescein	5.40×10^{-3}–1.30 µg/mL (Tetrachlorotetrabromo fluorescein)	0.0813 ng/mL (Tetrachlorotetrabromo fluorescein)	Waste film	[14]
Ag–tetrachlorotetrabromo fluorescein	5.40×10^{-3}–0.864 µg/mL (Tetrachlorotetrabromo fluorescein)	0.40 ng/mL(Tetrachlorotetrabromo fluorescein)		
Ag(I)–SDBS–Alizarin red	0.022–2.160 µg/mL	0.139 ng/mL	Waste film	[15]
Ag^+–eosin B (EB)or eosin Y (EY)	0.010 8–1.30 µg/mL	74.2 ng/L(EB)86.5 ng/L(EY)	Waste film	[16]
Ag(I)–iodide–CTAB	0–0.6 µg/mL	0.005 µg/mL	Waste water	[7]
Ag^+–Cl^-	0.015–1.2 µg/mL	0.015 µg/mL	Silver plating waste water	[17]

A C-bases rich single-strain of DNA is able to adsorb on the surface of gold nanoparticles to improve their stability while particles are dispersed in salt. Similar to Hg^{2+} and T forms of T–Hg^{2+}–T mismatch structure, C-base specifically forms C–Ag^+–C complex with silver ions, turning a single-strain DNA into a rigid double-stranded DNA, which no longer adsorbs on the surface of gold nanoparticle. Then, gold nanoparticles aggregate with salts, enhancing the scattering signal and achieving quantitative determination of silver ions (Figure 9.1).

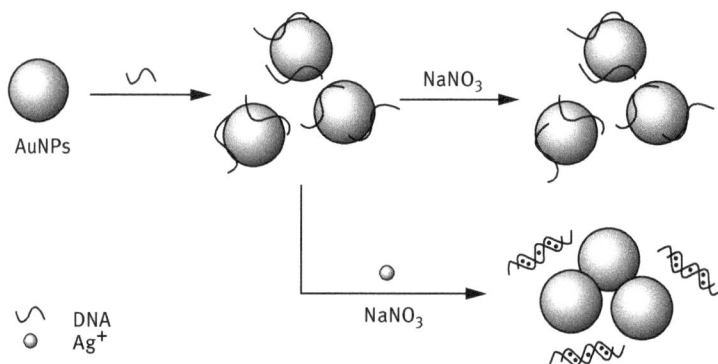

Figure 9.1: Selective detection of silver ions with gold nanoparticles as light scattering probe [8].

9.2.2 Zinc ions

Zinc is said to be the "flower of life" with essential physiological and biochemical functions in the human body. As a key component of enzymes, transcription factor, and excitatory nerve ending vesicles, zinc presents in serum with a concentration of approximately 12 µmol/L [18]. Zinc deficiency causes stunted growth, poor intelligence, low immunity and lack of appetite in kids. For people with sufficient amount of zinc, extra zinc supplement may lead to excessive amounts, influencing metabolism of Cu and Fe ions, causing Cu deficiency syndrome. So, it is necessary to accurately detect zinc in food or drinking water. Scattering method of detecting Zn^{2+} is mainly based on the enhancement of scattering signal by the formation of association [19–25].

In a HAc–NaAc buffer solution and polyvinyl alcohol, Zn^{2+} forms a complex anion $[Zn(SCN)_4]^{2-}$ with thiocyanate, then it forms an ionic association with neutral red cation. The association has a stronger resonance Rayleigh scattering signal [23]. Sheng Li et al. [24] built a resonance Rayleigh scattering analysis detection of Zn^{2+} with linear range of 0–0.5 µg/mL, and a detection limit of 1.9×10^{-2} ng/mL. This method was used to detect Zn^{2+} in water sample.

Li Y M et al. used a resonance light scattering for detection of zinc content in food [25]. In NH_3–NH_4Cl buffer solution at pH 8.5, Zn^{2+} reacts with phenanthrene and Jaune Brilliant that greatly enhances resonance light scattering at 359.4 nm to build the light scattering method of detecting Zn^{2+}. The enhanced resonance light scattering intensity presents a good linear relationship with the mass concentration of Zn^{2+} in the 0–10 µg/L range with a detection limit of 0.11 µg/L. This method can be used to detect zinc content in banana, cucumber, and laver, which are consistent with that of atomic absorption spectroscopy.

Zhai Haoying et al. used resonance light scattering to detect trace zinc content in zinc supplements and nutrient salts [26]. In NaAc–HAc medium at pH 5.0 and tween 80 solution, $[Zn(SCN)_4]^{2-}$ forms ion association particles with crystal purple, enhancing the resonance scattering signal. In 0.020–0.80 µg/mL range, the enhanced scattering intensity and Zn^{2+} concentration present a good linearity with a detection limit of 8.35×10^{-3} µg/L. This method can be used to detect zinc content in zinc supplements and nutrient salts with a recovery of 94.8%–98.3%.

9.2.3 Copper ions

Copper ion is a trace element that is closely related to health with catalysis as the main biochemical function in organism. Many copper metalloenzymes work as oxidase to participate in the oxidation–reduction process in human body, especially, they could reduce oxygen molecule into water. Many copper metalloenzymes have been confirmed with vital physiological functions in human body. For example, copper ion is constitute of Cu-containing enzyme and Cu-binding proteins, maintains hematopoietic function, promotes desmoplasia, maintains central nervous system health promotes the formation of melanin, maintains hair structure, protects cells from superoxide anion damage, affects on lipid and sugar metabolism, and so on. It is important to control the cellular copper ion concentration, although the total ion concentration is quite high within the range of 10^{-5}–10^{-4} mol/L, the dissociative copper ion concentration is usually less than 10^{-18} mol/L. If the instantaneous overload of copper ion is too high, such as the uptake is over 64 mg per day, which will cause copper poisoning, if the uptake of copper is 1000 times higher than the normal level, it will be a fatal risk.

There are many detection methods for copper ions, such as fluorescence and atomic absorption. Herein, resonance light scattering method will be mainly introduced. Table 9.2 lists the resonance light scattering methods, which are mainly based on the enhanced light scattering caused by the formation of associations.

Classic resonance light scattering spectroscopy for copper detection is shown in Figure 9.2(a), which is a cysteine adjustment method [34]. This method used a single-stranded oligonucleotide rich in C-bases to synthesize silver nanoclusters, which displayed an excellent fluorescence spectrum. When cysteine was introduced, the intensity

Table 9.2: Analysis of copper ion light scattering detection.

System	Linear range	Detection limit	Application	Reference
Cu^{2+}–ascorbic acid copper reagent–AuNPs	0.02–2.54 µg/mL	0.001 µg/mL	Waste water	[26]
Cu^{2+}–tungstate–butyl rhodamine B	0.02–0.50 µg/25mL	0.93 ng/mL	White wine	[27]
Cu^{2+}–tungstate–rhodamine 6G	0.05–1.0 ng/mL	0.026 ng/mL	Flour and tea	[28]
Cu^{2+}–ammonium molybdate–rhodamine B	0.002–0.40 µg/mL	1.2 ng/mL	Tap water and river water	[29, 30]
Cu^{2+}–sulfide–CTAB	0.25–3.50µg/mL	0.017 µg/mL	Environmental water sample	[31]
Cu^{2+}–KI–CTAB	0.0–0.6 µg/mL	0.00097 µg/mL	Fenghuang tea	[32]
Flowing injection scattering detection of Cu	0.0–400 µg/mL	3.266 µg/mL	Water sample	[33]

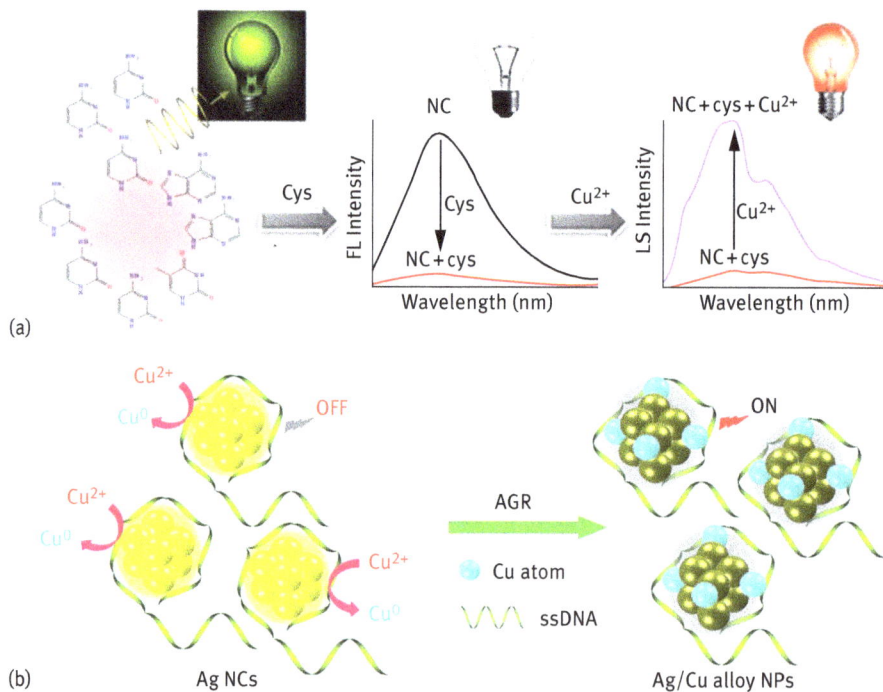

Figure 9.2: Silver nanoclusters are used for copper ion detection. (a) Cysteine regulation method [34] (b) Anti-Galvani reduction method (ARG) [35].

of silver nanocluster emission was obviously reduced. While copper ions were added, weak signals were significantly enhanced, which could be developed to detect copper.

Figure 9.2 (b) shows another strategy to detect copper ion using silver nanoclusters as optical probe [35]. Single-stranded oligonucleotide rich in C-bases synthesized silver nanoclusters can directly reduce Cu^{2+} into Cu^0, which spontaneously formed copper nanoparticles. Large diameter silver/copper alloy nanoparticle is formed to enhance the scattering signal of nanoclusters, realizing the quantitative determination of copper ions. This method has high sensitivity, good selectivity, low cost, and is easy to repeat.

Professor Jiang's group developed a DNAzyme cracking-nanogold resonance Rayleigh scattering spectral method for the determination of trace Cu^{2+}. This study finds that in a buffer solution of pH 7.0 and 0.19 mol/L NaCl, substrate stranded DNA (SS) and enzyme stranded DNA (ES) hydridized into a double-stranded DNA (dsDNA) at 80°C. The substrate chain of dsDNA could be cracked by Cu^{2+}, and the released single-stranded DNA (ssDNA) were adsorbed on the nanogold surface to produce a stable nanogold-ssDNA conjugate. The unprotected nanogold was aggregated to form nano-gold aggregation that exhibited a resonance Rayleigh scattering peak at 627 nm. When the Cu^{2+} was added, the nanogold-ssDNA increased, and the nanogold aggregates reduced that caused the RRS intensity decreasing at 627 nm, which presented a linear relation with Cu^{2+} in 15–250 nmol/L. Based on this, the resonance Rayleigh scattering spectrometry of Cu^{2+} was developed. This method presented a high selectivity because of the specific cracking of dsDNA by Cu^{2+}.

Another common method is to induce the decrease of scattering signal of gold nanoparticles. For example, when cupric is reduced to cuprous ion by ascorbic acid, it forms a yellow complex with neocuproine, which has a strong absorbance for the resonance light scattering nanogold at 580 nm, thus reducing light scattering intensity. The reduction value ΔI has a linear relation with Cu^{2+} [37].

9.2.4 Potassium ion

Potassium ion is a vital living metal ion, and 98% of it is found in cells, which majority of the cation of intracellular fluid. Usually, potassium content in the daily diet is enough to satisfy body requirements to avoid deficiency. Myocardial and neuromuscular irritabilities can be avoided with a constant potassium concentration. If the serum potassium is higher than 5.5 mmol/L, hyperkalemia is caused. Too high serum potassium inhibits myocardium, stopping heartbeat in the relaxation period. Too low serum potassium brings cardiac excitability, stopping heartbeat in systole period. When serum potassium is lower than 3.5 mmol/L, it will cause hypokalemia. When suffer from potassium deficiency, body becomes lethargic, listless, and restless; serious potassium deficiency brings arrhythmia, rhabdomyolysis, renal dysfunction, or even death. Thus, it is of meaningful physiological significance to detect potassium ions.

Jiang Z L and his group developed resonance light scattering methods for potassium ion detection with aptamers as recognition probe [38, 39]. As shown in Figure 9.3,

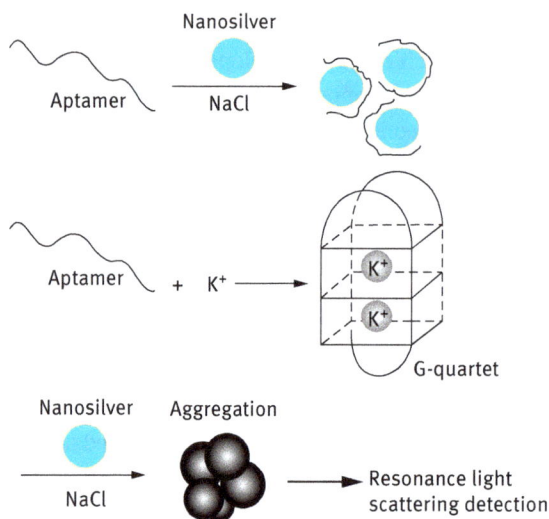

Figure 9.3: A highly sensitive and selective resonance scattering spectral assay for potassium ion in serum based on aptamer and nanosilver aggregation reaction [38].

aptamer is adsorbed on the surface of silver nanoparticles (AgNPs), forming a steady aptamer–AgNPs complex to avoid aggregation induced by aggregated by NaCl. When in the presence of K^+, it forms a steady G-quadruplex structure complex with aptamer, which is no longer adsorbsed on the surface of silver nanoparticles. When high salt is present, silver nanoparticles form large-scale aggregates, enhancing the scattering intensity, which can be used to detect K^+ [38]. Based on a similar principle, they used gold nanoparticles as a probe to test K^+ [39] in serum.

In addition, sodium tetraphenylborate reacts with K^+ interaction to produce insoluble salts with a surfactant to keep the suspicion steady, which could be developed for K^+ detection based on the enhancement of light scattering [40, 41].

9.2.5 Iron ions

Fe is an essential trace mineral for humans, and is found in the metallically active center of hemoglobin, myoglobin, and various enzymes. Thus, too-high or too-low iron content in the body causes abnormality. For example, when iron is excessive, acute and chronic iron poisoning is caused so that iron cannot be eliminated from the body on time while accumulating in the liver, pancreas, heart, and skin, leading to hemoglobin deposition, abnormal liver function, myocardial injury and diabetes, tumor, osteoporosis, and so on. If human body lacks enough iron, synthesis of hemoglobin and myoglobin will be influenced, reducing the enzyme activity of cytochrome c, ribonucleotide reductase, succinic dehydrogenase, and so on, thus affecting normal process of biological phenomena of biooxidation, tissue respiration, neurotransmitter decomposition and synthesis

[42]. So, iron is an essential trace mineral in humans for metabolism, which plays important functions in cell growth, proliferation, biooxidation, and biological transformation, thus iron an important marker to evaluate the health state.

Wang H M et al. synthesized 4-methyl piperazine-1,8-naphthalimide-modified β-cyclodextrin, which was able to well identify Fe^{3+} and Fe^{2+}[43]. When an appropriate amounts of ammonium hydroxide was added into the compound aqueous solution, the fluorescence of the above compound completely quenched. If Fe^{3+} and Fe^{2+} were added, both the fluorescence signal at 520 nm and resonance Rayleigh scattering peak at 415 nm were enhanced. The difference of the intensity ratio between fluorescence and resonance Rayleigh scattering induced by Fe^{3+} and Fe^{2+} can be used to identify them. Meanwhile, this method presented a high sensitivity, the linear range of detection of Fe^{3+} was 1.5×10^{-5}–2.2×10^{-5} mol/L with a limit of detection (LOD) of 1.1×10^{-5} mol/L; Linear range of Fe^{2+} was 5.0×10^{-6}–4.2×10^{-5} mol/L with a LOD of 3.5×10^{-6} mol/L.

Electrostatic reaction and coordination are able to form large associations or an adsorbable complex, which can be developed to detect iron with resonance light scattering method. Fan W X et al. [44] used Fe^{3+} to enhance the resonance light scattering of gold orange G to detect Fe^{3+}, which was applied in trace level of Fe^{3+} in water. Zhai H Y [45] used Fe(III) and carboxymethyl chitosan (CMCS) coordination to form the adsorption complex with strong signals to detect Fe^{3+}. The method is used for Fe(III) content in cloth bag ash and slag samples with minor relative standard deviation (RSD) and recovery of 96% and 108%. Li Y M et al. [46] used Fe(II) to enhance the resonance light scattering of phenanthroline and brilliant yellow in ammonia–ammonium chloride buffer solution to establish the method of detecting iron content in edible oil. This method was sensitive, accurate, simple, rapid with high precision, effective, and environmental friendly, which satisfied the requirements of detection of iron content in edible oil.

In addition, there are indirect method of detecting Fe(III) based on the oxidizability of Fe(III) [47]. In an acidic medium, Fe(III) oxidizes I^- to form I_2 which further reacts with excessive I^- to form I_3^-. Then, I_3^- electrostatically attracts acridine red (AR) to form AR–I_3^- ion association, which further spontaneously aggregates to form $(AR-I_3^-)_n$ associate particle, greatly enhancing resonance light scattering intensity. Results show that the intensity enhancement has a good linear relation in 0.08–0.56 mg/L with a detection range of 5.76×10^{-7} g/L. This method displays a good selectivity and can be used for detection of trace level Fe(III) in water sample, which is consistent with the results detected by atomic absorption method.

9.2.6 Cobalt ions

Cobalt is an essential trace mineral in humans, which is a metallically active center of life, largely existing in hydrogenase, methyl-coenzyme M reducing ferment, carbon monoxide dehydrogenase, superoxide dismutase, and urease. The amount of cobalt is about 1.1–1.5 mg in human body. Cobalt in food is absorbed in the small intestine into the plasma and combines with three trans-cobalt proteins to the liver and all over the body, and finally is discharged by urine. When human body suffers from the deficiency of intrinsic factor and cobalt transport protein, as well as the insufficient uptake of cobalt, or absorption interference by digestive system disease cause deficiency of cobalt and vitamin B_{12}. The amount of serum cobalt reduces when suffers from pernicious anemia, acute leukemia. However, the amount of serum cobalt increases when suffers from chronic granulocytic leukemia. In addition, cobalt coordination complexes have a stronger antibacterial activity for *Pseudomonas aeruginosa*, *Staphylococcus aureus*, *Staphylococcus*, and *E. coli*.

The authors developed cobalt detection methods with dark-field scattering [48, 49] and Raman scattering [50], respectively. Investigation have found that under mild condition, poly (*p*-phenylenediamine), AgNO₃ and, polyvinylpyrrolidone (PVP) and can synthesize leaf-like microcrystals, whereas Co^{2+} can specifically corrode the surface of leaf-like microcrystals. It has been successfully applied for visual detection of external and tissue Co^{2+} under a dark-field microscope (Figure 9.4[a]) [48]. Figure 9.4(b) shows the principle of Raman scattering for cobalt detection. The as-prepared leaf-like microcrystals can greatly enhance the Raman scattering signal; but after specific corrosion,

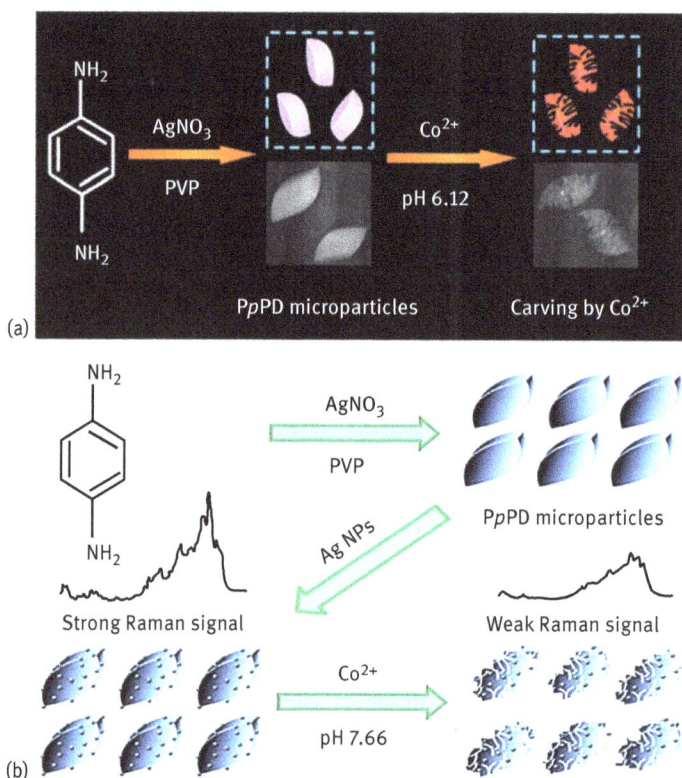

Figure 9.4: Light scattering analysis of cobalt ion. (a) Dark-field scattering [48]. (b) Raman scattering method [50].

Raman signal is reduced, which could be developed to detect cobalt base on this principle [50].

Currently, resonance scattering signal method for cobalt detection is mainly based on scattering signal enhancement of ion associates. For example, Co (II) and alizarin red, and cetyl pyridinium bromide form large-scale hydrophobic ternary ion association through electrostatic force and hydrophobic force, and thus leading to the enhancement of light scattering signal [51]. In a acidic condition and polyvinyl alcohol (PVA) solution, complex anion $[Co(SCN)_4]^{2-}$ and neutral red form ion associates to enhance the light scattering signal [52]; in Tris-HCl buffer solution at pH 9.0, Co(II) and 1-(2-pyridine-azo)-2-naphthol form binary cationic chelates with positive charges, which further reacts with an anionic surfactant of sodium dodecyl benzene sulfonate to form ion associates to enhance the light scattering [53].

9.2.7 Nickel ion

Nickel is an essential living element, which particites in hormone action, biomacromolecule structure stability, and metabolism process. Usually, the amount of nickel ion in adult human body is approximately 10 mg, and the daily requirement is about 0.3 mg. Nickel deficiency causes diabetes, anemia, cirrhosis, uremia, kidney failure, and liver lipid metabolism and abnormal phospholipid metabolism. Nickel is also common sensitized metal. If the daily uptake of soluble nickel over 250 mg, it will lead to poisoning with special symptoms of dermatitis, respiratory disorders, and respiratory cancer [54].

The resonance ligt scattering method of detecting nickel is mainly based on ion associate or polymer cause the enhancement of light scattering signal. For example, in a Tris-HCl buffer at pH 8.0, sodium dodecyl benzene sulfonate sensibilizes nickel and 1-(2-pyridineazo)-2-naphthol to form ion associate and enhance the light scattering, which could be developed for nickel detection in water sample [55, 56].

In addition, the detection of nickel could be achieved based on the enhanced light scattering of insoluble dimethylglyoxime-nickel. Zhong M H used flowing injection – light scattering couple technique with dimethylglyoxime as optical probe to test trace levels of Ni, overcoming the unstability of precipitation. This method presents a high sensitivity, good selectivity, and rapid analysis speed, which is useful for determination of nickel content in wastewater [57].

9.2.8 Aluminum ion

Aluminum is a chemical element that is frequently contacted in daily life, which are widely used in food additives, coagulants, drugs, various containers, and cookers [58]. Aluminum-related biological effect is not very clear, whereas there is no final conclusion about whether aluminum is an essential trace mineral. It is known that the amount of aluminum in the brain of patients with senile dementia or psychotic disorder is about

10–30 times higher than that in normal people brain. There are several reports on aluminum accumulation, concluding that cerebral neurodegeneration, memory loss, intelligence, and personality disorders will occur. Too high aluminum content in food causes early aging, or even senile dementia.

Recently, some Al^{3+} resonance scattering methods have been developed. Long [59] found that in a weak acid medium, ethyl violet (EV) and DNA would result in great enhancement of resonance Rayleigh scattering intensity Al^{3+} has a competitive reaction with DNA, which no longer binds with EV, thus reducing the scattering intensity, and then to detect aluminum in biosynthetic samples. Based on the similar principle, Liu F H et al. [60] found that in a sodium acetate–acetic acid medium, DNA and histone complex produce scattering attenuation, which could be developed for aluminum detection with high sensitivity, simplicity and rapidness.

In addition, Fan W X et al. [61, 62] used an Al(III) resonance spectral enhancement effect on eosin Y and golden orange G to establish quantitative determination of Al^{3+}, which is used for the determination for trace amounts of Al^{3+} in tea.

9.3 Anion measurements

Metal ions are important for human health, simultaneously, anions are significantly for human, and a certain amount of anions are needed to maintain normal volume, distribution, and steady electrolyte concentration for metabolism.

Small organic molecules of cationic dye can be used to detect anion. Among them, basic dyes the most widespread. Common basic dyes include crystal violet (CV), ethyl violet (EV), brilliant green (BG), malachite green (MG), iodide green (IG), Vitoria blue 4R (VB4R), rhodamine B (RhB), rhodamine 6G (RhG), rhodamine S (RhS), ethyl rhodamine B (ERhB), butyl rhodamine B (b-RhB), and so on. These dyes form associates with metal complex anions, increasing the RLS for anion detection. For example, molybdophosphate 2RS system can be used to detect phosphorus [63].

9.3.1 Nitrite

Own to the strong oxidizing property, nitrite oxidizes hemoglobin into methemoglobin in the blood, losing its oxygen carrying capacity, which induce the oxygen deficiency of tissues. Nitrite takes vasodilator action on the surrounding blood vessels, which is also able to convert secondary amine substances in food or stomach into nitrosamine with strong carcinogenesis. So, the detection of nitrite salt is closely related to human health [64].

There are many references related to the detection of nitrite [64–74]. Table 9.3 compares sensitivities of nitrite determination with resonance light scattering methods. Most of them are based on the formation of ion associate to enhance the light scattering signal. Among these, the iodide–basic dye systems are commonly used to detect nitrite [75]. For example, Cr(VI) [76] can oxidize I^- into I_3^-, and then I_3 further forms an ion associate with basic dye to increase the light scattering intensity. In addition, NO_2 can perform catalytic oxidation of I^- into I_3^- by dissolved oxygen. Then, I_3^- forms an associate with basic dye to enhance the scattering intensity (Figure 9.5). Based on this, NO_2^- [77] detection can be achieved.

Table 9.3: Determination of nitrite with light scattering method.

System	Linear range	Detection limit	Application	Reference
O_2–I^-—crystal purple	2.20–340 µg/L	0.68 µg/L	Enviromental water sample	[65]
EDTA–AgNPs	1.8×10^{-4}–3.0 µg/mL	1.8×10^{-4} µg/mL	Gastric juice	[66]
I_3^--rhodamine 6G	6.4×10^{-4}–3.0 µg/mL	1.8×10^{-4} µg/mL	Gastric juice	[67]
Neutral red-8-oxychinolin	0–480×10^{-6} g/L	8.9×10^{-6} g/L	Vegetable	[68]
Sulfonamide N, N-dimethylaniline.	0–1.2×10^{-3} g/L	8.0×10^{-5} g/L	Vegetable	[69]
Nuclear fast red	2.0×10^{-8}–2.8×10^{-7} g/mL	9.34×10^{-9} g/mL	Enviromental water sample	[70]
Rhodamine B	10.0–180 µg/L	3.88 µg/L	Enviromental water sample	[71]
Potassium bromate–Janus green–SDBS	0.83–41.67 ng/mL	0.70 ng/mL	Enviromental water sample	[72]
Naphthalene ethylenediamine	0–800 ng/mL	7.3 ng/mL	Vegetable	[73]
Aminobenzene sulfonic acid	0.01–0.70 µg/L	7.2 ng/mL	Tap water	[74]

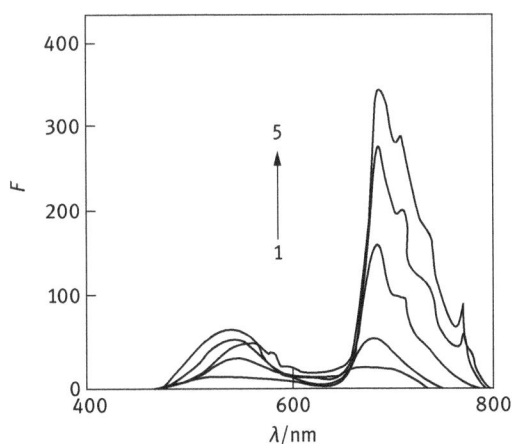

Figure 9.5: Resonance light scattering for NO_2^- detection [77]. 1. Crystal purple–NO_2^-; 2. Crystal purple–I^-; 3–5. Crystal purple–NO_2^-–H^+–I^- (The concentrations of NO_2^- are 40 µg/L, 80 µg/L, and 200 µg/L, respectively.).

Furthermore, in the scattering analysis of nitrite, the particularity of nitrite, including diazotization [64, 69], catalysis [77], and nitrification [69], could be taken into consideration to detect nitrite.

9.3.2 Chloride anion and chlorate group

Chlorine takes important role during relaesing oxygen process in the photosynthesis system II, whereas excessive chlorine is harmful for crops. Different crops have diversity in resistance to chlorine. Tobacco, peaches, avocados, and some legumes are the most chloride-sensitive crops. Chlorine exist in many forms, such as Cl^-, ClO^-, ClO_2^-, ClO_3^-, ClO_4^-, and so on.

The general principle for chlorine test is base on the the formation of AgCl colloid resulted from silver nitrate and chloride ion, which enhances the light scattering signal [79–82]. Li S W used this principle to detect chloride in beer chloridions [79], in mepiquat soluble concentrate [80], and in acid copper plating bath solution [81, 82] with high sensitivity and precision, easy operation, and being able to satisfy the requirements of actual production.

Zhang B M et al. measured [83, 84] ClO^- with a resonance light scattering method. The principle is that ClO^- reacts with excessive I^- to form I_3^-, which forms associated microparticles with RhS, RhB, b-RhB, and Rh6G, reducing the system fluorescence and enhancing light scattering signal, which has been successfully used for ClO^- determination in decolourant, bleaching agent, and bleaching powder.

Kang C Y et al. [85] used resonance scattering spectrum of cationic surfactant to detect chlorite ion. In sodium acetate–hydrochloric acid buffer solution, chlorite oxidizes I^- to I_2, and then excessive I^- and I_2 forms I_3^-. Some cationic surfactant (CS), such as tetradecyl dibenzyl ammonium chloride, bromotetradecylpyridine, cetyl trimethyl ammonium bromide, and tetrabutylammonium iodide can form associate particles with I_3^- and enhance resonance light scattering at 467 nm, which could be develop for ClO^- detection in water sample.

The authors have tested perchlorate with probe cetyl trimethyl ammonium bromide (CTAB) [86]. In acidic conditions, perchlorate forms an ion associate with CTAB through electrostatic interaction, thus enhancing the system scattering intensity. The common anions in water sample, such as Cl^-, Br^-, ClO_3^-, NO_3^-, and PO_4^{3-} display weak scattering intensity, whereas in the presence of perchlorate, the system scattering intensity changes due to the synergistic effect.

9.3.3 Phosphate

Phosphorous is important for life activities, which exist in many forms, such as organic and inorganic phosphorus. Many life-related substances, including the key substance of adenosine triphosphate (ATP) for energy transfer, genetic material of nucleic acids, biofilm phospholipids, and many enzymes, contain phosphorus. In addition, phosphorus is the important component for bones and teeth, which participate in daily activities in the form of phosphate group.

The resonance light scattering methods of detecting phosphorus are mainly based on the interaction between phosphate and molybdate to form heteropolyacid. Jiang's group developed a resonance light scattering method in H_2SO_4 medium, where sodium phosphate and sodium molybdate and antimony potassium tartrate react to form a yellow phosphorous antimony molybdenum heteropolyacid. A stronger resonance scattering peak is produced at 710 nm when in the presence of gold nanoparticles, while ascorbic acid reduces heteropolyacid to heteropoly blue, which decreases the resonance light scattering at 710 nm [87]. Phosphorus concentration is in good linear relationship with resonance scattering intensity. This method can be applied in detecting phosphorous in water sample, which shows consistent results with that of spectrophotometry.

Oshima et al. determined trace amounts of phosphate by detecting the light-scattering intensity with a spectrofluorimeter. Used fluorospectrophotometer to detect the scattering intensity for trace level phosphate [88]. The ion associate of molybdophosphate with a chloro derivative of Malachite Green (Cl-MG) or Rhodamine B (RhB) formed aggregates in an acidic solution. The intensities were obtained at 460 nm for Cl-MG and 580 nm for RhB. The diameter of the aggregate was determined to be mainly 650 nm by using a laser light-scattering detector. The number of ion associates in the aggregate was roughly estimated to be 4.3×10^7. The method was applied to flow injection, where the calibration graph was rectilinear from 2×10^{-7} to 1×10^{-6} mol/L and the detection limit was 6×10^{-8} mol/L. This method shows high selectivity.

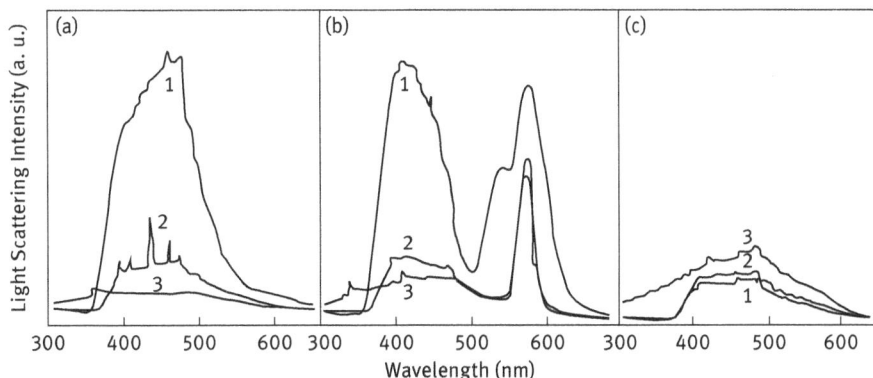

Figure 9.6: Resonance light scattering signal for phosphoric acid detection. (a) Cl–MG; (b) RB; and (c)no dye. (1) 4×10^{-7}mol/L phosphate; (2) reagent blank; and (3) molybdic acid free salt.

9.3.4 Sulfate group

Sulfate is a common anion, potassium sulfate is a common potash fertilizer. And barium sulfate is utilized in barium meal, which can be used for X-ray detection of digestive system. It is also a used in protecting coating, plaster is used for fixing in medicine, which is also used in home decors and fast setting mixed concretes.

Resonance light scattering method for sulfate detection is mainly performed by interacting Ba^{2+} with SO_4^{2-} to form $BaSO_4$ [89, 90]. Jiang Zhiliang group found that in HCl–Triton X 100 medium, $BaCl_2$ solution has weak scattering. When in the presence of SO_4^{2-}, the formed $(BaSO_4)_n$ particle has a stronger scattering peak at 470 nm. Synchronous scattering spectrometry is built for the detection of sulfate anion in cement [89]. In this method, SO_4^{2-} and Ba^{2+} form $BaSO_4$ molecule through electrostatic attraction. In addition, $BaSO_4$ has stronger hydrophobic forces and intermolecular forces, aggregating to form $(BaSO_4)_n$ particles and solid–liquid interface, which are the primary reason for the enhancement of synchronous light scattering.

Zhao et al. proposed a sulfate salt detection method based on Rayleigh scattering [90]. After the formation of barium sulfate particle, there is strong scattering in water solution (Figure 9.7). In the range of 8–400 µg/mL, the light scattering intensity has a linear relation with the sulfate concentration with a detection limit of 0.3 µg/mL. To determine its validity, water, drinks, and vegetable samples were analyzed, the results of which were consistent with that of the standard turbidimetry.

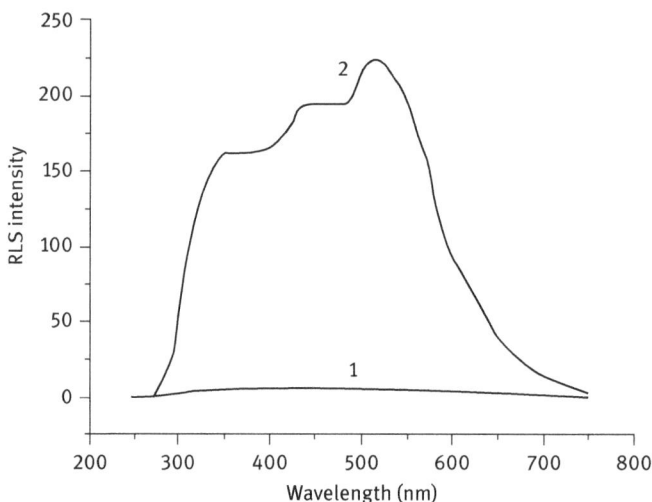

Figure 9.7: Resonance light scattering method for sulfate salt detection. (1) Reagent blank; (2) $BaSO_4$ suspension.

9.3.5 Halogen anion

In addition to the above-mentioned anions, sensitive detection of F^-, Br^-, and I^- can be preformed by the resonance light scattering technique [91–93]. With the development of nanotechnology, the detection strategy will be improved. Moreover, the sensitivity and selectivity of this method will be further improved, detection targets will be expanded, and it will be able to gradually satisfy the requirements of daily production and life.

9.3.6 Thiocyanate group

Thiocyanate group is an important anion and its aqueous solution is neutral. It forms a dark red color sanguine thiocyanate iron with ferric salts, and forms dark-blue rhodanide cobalt with cobalt salt, white silver rhodanide precipitate with silver salt, or black cuprous rhodanide precipitate with copper salts. Thiocyanate group reacts with acids to produce a toxic gas that damages the thyroid gland with maximum allowable concentration of **50 mg/m^3** [94].

Zhong M H used a flowing injection-resonance light scattering method to detect thiocyanate group content. In acetic acid buffer and SCN^- mixture, ascorbic acid reduces Cu (II) into Cu (I), which further forms CuSCN precipitate with SCN^-. This precipitation reaction is coupled with a flowing injection to build a resonance light scattering method of detecting of SCN^-. This method is simple with wide linear range and small amount of reagent, which can be used for detecting SCN^- content in vegetables [95].

References

[1] Liu Z D, Li Y F, Ling J, et al. A localized surface plasmon resonance light-scattering assay of mercury (II) on the basis of Hg^{2+}-DNA complex induced aggregation of gold nanoparticles. Environ. Sci. Technol., 2009, 43: 5022–5027.

[2] Wu L P, Li Y F, Huang C Z, et al. Visual detection of Sudan dyes based on the plasmon resonance light scattering signals of silver nanoparticles. Anal. Chem., 2006, 78: 5570–5577.

[3] He W, Huang C Z, Li Y F, et al. One-step label-free optical genosensing system for sequence-specific DNA related to the human immunodeficiency virus based on the measurements of light scattering signals of gold nanorods. Anal. Chem., 2008, 80: 8424–8430.

[4] Du B A, Li Z P, Liu C H. One-step homogeneous detection of DNA hybridization with gold nanoparticle probes by using a linear light-scattering technique. Angew. Chem. Int. Ed., 2006, 45: 8022–8025.

[5] Han Z H, Lv C Y, Yang S Y, et al. Determination of silver with Ag(I)-phenanthroline-acidic triphenylmethane dye systems by resonance Rayleigh scattering method J. Instr. Anal., 2006, 25(1): 69–72.

[6] Zhong M H. Determination of palladium in electroplating waste water with iodide-Cetyltrimethyl ammonium bromide system by resonance light scattering methods. Chem. Anal. Meterage, 2006, 15(1): 18–20.

[7] Zhong M H. Resonance light scattering spectrometric method for the determination of silver in waste water with Ag(I)-iodide-cetyltrimethyl ammonium bromide system Metal. Anal., 2006, 26(3): 50–52.

[8] Wu C, Xiong C, Wang L, et al. Sensitive and selective localized surface plasmon resonance light-scattering sensor for Ag$^+$ with unmodified gold nanoparticles. Analyst, 2010, 135(10): 2682–2687.

[9] Han Z H, Gao S Q, Xiao X L. Determination of silver with Ag(I)-Phenanthroline-Alizarin system by resonance rayleigh scattering method. Chin. J. Health Lab. Technol., 2005, 15(12): 1424–1426.

[10] Determination of silver with Ag(I)-Phenanthroline-Eosin system by resonance Rayleigh scattering method. Stud. Trace Elements Health., 2006, 23(3): 36–37.

[11] Zhai H Y, Yu PR J, Zhu Y P. Resonance scattering spectrometric determination of trace silver in Ag(I)-EGTA chelate system J. Anal. Sci., 2009, 25, (3): 363–365.

[12] Jiang Z L, Yang M M, Mo Q. Photochemical resonance scattering spectrophotometric determination of microamount of silver. Prec. Metal 2000, 21(3): 34–37.

[13] Zhong M H, Huang J S, Lin Y R. Resonance light scattering method for the determination of silver with the sulfide-cetyltrimethyl ammonium bromide system. J. Hanshan Norm. Univ. (Nat. Sci. Ed.), 2007, 28(6): 60–63.

[14] Zhai H Y, Han Z Z. Resonance scattering method for determination of trace silver in the silver-tetrachlorotetrabromofluororescein system. Chem. Reag. 2010, 32(8): 715–717.

[15] Zhai H Y, Ruan S Q, Zhu Y P. Determination of silver (I) in Alizarin Red-Ag (I) system by the resonance scattering spectra. Chemistry, 2009, 72(4): 374–376.

[16] Zhai H Y, Zhang X Y, Ruan S Q, et al. Eosin resonance scattering method for the determination of trace silver. Metall. Anal., 2009, 29(12): 17–20.

[17] Zhong M H. Resonance light scattering spectrometric method for the determination of silver in waste water with Ag(I)-iodide-cetyltrimethyl ammonium bromide system. Metall. Anal., 2006, 26(3): 50–52.

[18] Zhong M H, Huang J S. Determination of trance silver ion by resonance light-scattering. Instr. Anal. Mon., 2005, 21(3): 36–37.

[19] Hanaoka K, Kikuchi K, Urano Y, Nagano T. Selective sensing of zinc ions with a novel magnetic resonance imaging contrast agent. J. Chem. Soc. Perkin Trans., 2001(9): 1840–1843.

[20] Yan H T, Li N. Determination of Zn^{2+} in human serum under conditions of scattering backgrounds by laser thermal lens spectrometry. Chinese J. Anal. Lab., 2006, 25(7): 46–49.

[21] Li S D. RLS Determination of Zn in environmental water by resonance Rayleigh light-scattering method. Chinese J. Spectrosc. Lab., 2007, 24(4): 671–674.

[22] Li Y Z, Li S D, Huang M Y, et al. Determination of zinc(II) in environmental water with Zn(SCN)$_4$$^{2-}$–HSA ystem by resonance Rayleigh light-scattering method. J. Instr. Anal., 2008, 27(2): 206–208.

[23] Gao J H, Lin P, Chen B, et al. Antidouble scattering spectra of zinc(II)-thiocyanate-Rhodamine dye system and their analytical applications. Metall. Anal., 2000, 20(4): 1–3.

[24] Sheng L, Mi Y, Tao C H. Resonance Rayleigh scattering method for the determination of trace amount of zinc(II) with potassium thiocyanate and neutral red dye system. Chinese J. Anal. Lab., 2014(6): 684–686.

[25] Li Y M, Li R Y, Shi P F. Resonance light scattering method for determination of zinc content in food. J. Food Sci., 2011, 32(24): 204–207.

[26] Zhai H Y, Zhang Y X, Wang B. Resonance scattering determination of trace zinc with Zn^{2+}-thiocyanate-crystalline violet system. Chinese J. Anal. Lab., 2011, 30(3): 94–97.

[27] Zhao X, Li Z G, Ding Z T, et al. Determination of trace copper ion in liquor by resonance light-scattering. Liq-Mak. Sci. Technol., 2007, (10): 88–89, 92.

[28] Shao Y T, Mao Z C, Wang P L, et al. Determination of copper in flour and tea by resonance light-scattering technique with Cu^{2+}-tungstate-Rhodamine 6G system. J. Food Sci., 2015, 36(4): 110–113.

[29] Yang J, Liao L F, Yang J L, et al. Determination of trace copper by photochemical catalytic kinetics resonance light scattering method. Chinese J. Anal. Lab., 2007, 26(3): 110–113.

[30] Yang J. Determination of trace copper (II), cobalt (II) and nickel (II) in environment. Hengyang: Master's thesis of Nanhua University, 2007.

[31] Zhong M H. Resonance light scattering method for the determination of copper with the sulfide-cetyltrimethyl ammonium bromide system. Chinese J. Health Lab. Technol., 2005, 15: 1286–1287.

[32] Zeng X Y, Huang W Y, Wu X Q, et al. Determination of trace copper ion by light scattering. Jiangxi Chem. Ind., 2005, 4: 117–119, 126.

[33] Zhong M H, Yan Z K, Hao G X. A study on flow injection light scattering determination of copper (II). J. Hanshan Teachers Coll., 2010, 31(3): 66–69.

[34] Liu G L, Feng D Q, Chen T F, et al. DNA-templated formation of silver nanoclusters as a novel light-scattering sensor for label-free copper ions detection. J. Mater. Chem., 2012, 22(39): 20885–20888.

[35] Feng D Q, Liu G L, Wang W. A novel biosensor for copper (II) ions based on turn-on resonance light scattering of ssDNA templated silver nanoclusters. J. Mater. Chem. B, 2015, 3(10): 2083–2088.

[36] Wang S M, Wu M, Liang A H, et al. DNAzyme cracking-nanogold resonance Rayleigh scattering spectral method for the determination of trace Cu^{2+}. Spectrosc. Spectr. Anal., 2013, 33(1): 147–150.

[37] Wen G Q, Liang A H, Tan M N, et al. Analysis of Cu in wastewater by nanogold resonance scattering spectrometry. Environ. Sci. Technol., 2009, 32(6): 101–102.

[38] Cai W, Fan Y, Jiang Z, et al. A highly sensitive and selective resonance scattering spectral assay for potassium ion based on aptamer and nanosilver aggregation reactions. Talanta, 2010, 81 (4–5): 1810–1815.

[39] Liu Q Y, Fan Y Y, Li T S, et al. Resonance scattering spectral detection of trace K^+ by aptamer-modified nanogold probe. Spectrosc. Anal., 2010, 30(11): 3115–3118.

[40] Wang X H, Li S W, Su X D, et al. Determination of potassium (K) by resonance light scattering turbidimetry (RLST). Chem. Res. Appl., 2004, 16(6): 837–838.

[41] Xiao J, Chen J, Ren F, et al. Highly sensitive determination of trace potassium ion in serum using the resonance light scattering technique with sodium tetraphenylboron. Microchim. Acta, 2007, 159(3): 287–292.

[42] Pandya A, Sutariya P G, Lodha A, et al. A novel calix[4]arene thiol functionalized silver nanoprobe for selective recognition of ferric ion with nanomolar sensitivity via DLS selectivity in human biological fluid. Nanoscale. 2013; 5(6):2364–2371.

[43] Wang H M, Zhao C D, Liu X L, et al. A novel recognition method for Fe^{3+} and Fe^{2+} based on the resonance Rayleigh scattering and the fluorescence enhancement. Chem. Bull., 2014, 77(6): 572–572.

[44] Fan W X, Li S Q, Gao X M, et al. Determination of trace iron(III) in water by iron(III)-orange G resonance light scattering spectrometry. Chinese Inorg. Anal. Chem., 2013, 3(2): 19–21.

[45] Zhai H Y, Zou H, Chen Y, et al. Determination of iron (III) by resonance scattering method with carboxymethyl chitosan. Metall. Anal., 2015, 35(9): 58–61.

[46] Li Y M, Li R Y, Shi P F. Resonance light scattering method for determination of iron content in edible oil. J. Chinese Cereals Oils Assoc., 2013, 28(4): 103–107.

[47] Song Z R, Zou R, He J H, et al. Resonance light scattering spectral method for determination of trace ferrum(III) with acridine red. Chinese J. Appl. Chem., 2012, 29(5): 597–603.

[48] Zhen S J, Guo F L, Chen L Q, et al. Visual detection of cobalt (II) ion in vitro and tissue with a new type of leaf-like molecular microcrystal. Chem. Commun., 2011, 47(9): 2562–2564.

[49] Yang H, Liu Y, Gao P F, et al. A dark-field light scattering platform for real-time monitoring of the erosion of microparticles by Co^{2+}. Analyst, 2014, 139(3): 2783–2787.

[50] Yang L, Zhen S J, Liu Z D, et al. Raman scattering detection of cobalt (II) ions based on their specific etching effect on leaf-like poly(p-phenylenediamine) microcrystals. Anal. Methods, 2014, 6(14): 5054–5058.

[51] Zou R, He J H, Li G Q. Determination of cobalt (II) by resonance Rayleigh scattering and resonance non-linear scattering with alizarin red S as probe. Metall. Anal., 2015, 35(8): 22–28.

[52] Qiu Y H, Liu Q L, Liao Q, et al. A resonance Rayleigh scattering method for determination of cobalt with Co(II)-SCN–NR-PVA system. J. Anal. Sci., 2010, 26(3): 332–334

[53] Liu Y M, Yang J, Gao Z P. Determination of Co(II) in Co(II)-PAN-SDBS system by resonance Rayleigh scattering. Chinese J. Appl. Chem., 2007, 24(12): 1467–1469.

[54] Sun P Z. Bioinformatics analysis of differentially expressed genes based on Ni^{2+} cytotoxicity gene expression microarray. Nanjing: Master's thesis of Southeast University, 2008.

[55] Qian C. Resonance scattering spectrometric determination of trace amounts of nickel by its reaction with PAN sensitized with SDBS, Phys. Test. Chem. Anal. B, 2009, (4): 431–433.

[56] Shan Z G. Determination of Trace Nickel by micellar sensitization resonance light scattering. Chinese J. Anal. Lab., 2008, 27(S1): 261–263.

[57] Zhong M H. Flow injection light scattering determination of micro nickel (II). Chinese J. Spectrosc. Lab., 2009, 26(6): 1473–1476.

[58] Sheng M C. Review of the research progress on the effect of aluminum on human health. Anhui J. Preven. Med., 2006, (1): 46–48.

[59] Long X, Li D, Wang N, et al. A novel and sensitive method for recognition and indirect determination of Al III in biological fluid based on the quenching of resonance Rayleigh scattering intensities of "Al^{III}-EV-DNA" complexing system. Spectrochim. Acta A, 2008, 69(1): 142–147.

[60] Liu F H, Du X Y, Zheng J, et al. The determination of aluminum based on light-scattering decreased in sodium acetate-acetate buffer solution, DNA and histone system. Chinese J. Health Lab. Technol., 2005, 15(6): 646–647.

[61] Fan W X, Xu H J, Li S Q,et al. Determination of traces of aluminum by resonance Rayleigh scattering,second order scattering and frequency doubling scattering spectrometry of eosin Y. Phys. Chem. Inspect. (chemical classification), 2014, 50(4): 409–412.

[62] Fan W X, Li S Q, Gao X M, et al. RLS method for trace level of Al(III) detection. J. Anal. Instr., 2012, (05): 48–50.

[63] Jiang Z L, Li F, Liang H. Resonance scattering spectroscopic study of the sSystem of PMo12 heteropoly acid and rhodamin S, Acta Chim. Sin., 2000, 58(8): 1059–1062.

[64] Ianoul A, Coleman T, Asher S A. UV resonance Raman spectroscopic detection of nitrate and nitrite in wastewater treatment processes. Anal. Chem., 2002, 74(6): 1458–1461.

[65] Li G R, Wang Y S, He D X. Determination of trace nitrite by catalytic resonance light scattering method. Chinese J. Anal. Chem., 2005, 33(9): 1304–1306.

[66] Wang C C, Luconi M O, Masi A N, et al. Derivatized silver nanoparticles as sensor for ultra-trace nitrate determination based on light scattering phenomenon. Talanta, 2009, 77(3): 1238–1243.

[67] Jiang Z L, Sun S J, Kang C Y, et al. A new and sensitive resonance-scattering method for determination of trace nitrite in water with rhodamine 6G. Anal. Bioanal. Chem., 2005, 381(4): 896–900.

[68] Wu X H, Dong C Z. Resonance light scattering of neutral red-8-hydroxyquinoline-nitrite system and its analytical application. Chinese J. Anal. Lab., 2007, 26(10): 37–40.

[69] Wu X H, Dong C Z. Application and study on resonance scattering spectra of diazotization-coupling reaction. Chinese J. Pub. Health, 2007, 23(4): 483–484.

[70] Dong C Z. Determination of trace nitrite by resonance light scattering method with nuclear fast red. Chinese J. Health Lab. Technol., 2007, 17(11): 1989–1990

[71] He D X, Li G R, Yang H X, et al. Determination of trace nitrite in water samples by resonance light-scattering with Rhodamine B. Chinese J. Health Lab. Technol., 2004, 14(6): 689–690.

[72] Tan G H, Li G R, Han L J, et al. Determination of trace nitrite by catalytic resonance light scattering method with surfactant as sensitizer. Chinese J. Health Lab. Technol., 2006, 16(7): 790–791.

[73] Wu X H, Dong C Z. Resonance light scattering of naphthyl ethylenediamine – nitrite system and its analytical application. Chinese J. Health Lab. Technol., 2006, 16(12): 1420–1421.

[74] Gao X, Li S W, Ma Y J, et al. Research on determination of nitrite by resonance light scattering spectra. J. Guizhou Chem. Ind., 2012, 37(1): 20–21.

[75] Feng Y H, Yang B Y, Qiao G J, et al. Application of resonance light scattering technique for the inorganic analysis. Metal. Anal., 2008, 28(3): 20–29.

[76] Shen J S, Li X R. Determination of Cr(VI)and Cr(III) in environmental water by resonance scattering with the ion association complex of rhodamine B and I_3^-. Chinese J. Anal. Chem., 2001, 29(8): 944–946.

[77] Li G R, Wang Y S, He D X. Determination of trace nitrite by catalytic resonance light scattering method. Chinese J. Anal. Chem., 2005, 33(9): 1304–1306.

[78] Xiao N, Yu C X. Rapid-response and highly sensitive noncross-linking colorimetric nitrite sensor using 4-aminothiophenol modified gold nanorods. Anal. Chem., 2010, 82(9): 3659–3663.

[79] Xie A L, Li S W, Zhang S, et al. Determination of chloride ions in beer by resonance light scattering turbidimetry method. Beer Sci. Technol., 2014(7): 20–23.

[80] Zhang T, Chen D D, Ren L P, et al. Determining chlorine in mepiquat soluble concentrate by a resonance light scattering method. Agrochemicals, 2006, 45(2): 111–112.

[81] Su X D, Li S W, Wang X H, et al. Determination of chloride ions in acid copper plating bath by spectroanalysis of resonance light scattering. Electroplat. Finish., 2003, 22(6): 45–47.

[82] Su X D, Li S W, Wang X H, et al. Determination of chloride ion by Resonance light Scattering Method. Chem. Res. Appl., 2004, 16(4): 535–536.

[83] Zhang B M, Jiang Z L. A method for the determination of trace hypochlorite based on the fluorescence quenching effect of the rhodamine S particles. J. Guangxi Nor. Univ. (Nat. Sci. Ed.), 2005, 23(4): 69–72.

[84] Liang A H, Zhang B M. Determination of trace hypochlorite with rhodamine 6G by fluorescence quenching method. J. Guilin Univ. Technol., 2008, 28(2): 212–215.

[85] Kang C Y, Liao Z H, Xi D L, et al. Resonance scattering spectrometric determination of trace chlorite using cationic surfactants. J. Anal. Sci., 2008, 24(2): 177–180.

[86] Xu J L, Li Y F. Interaction of perchlorate and cetyltrimethylammonium bromide investigated by light scattering technique. Chinese J. Appl. Chem., 2010, 27(8): 935–938.

[87] Fan Y Y, Liang A H, Jiang Z L. Resonance light scattering quenching determination of phosphorus with nanogold. Metal. Anal., 2009, 29(5): 3–65.

[88] Li Z P, Hao L T, Cheng Y Q. Determination of micro amounts of phosphorous with resonance light-scattering technique by using the system of phosphomolybdate heteropoly acid-quinine. J. Hebei Univ. (Natural. Science. Edition.), 2008, 28(02): 158–161.

[88] Oshima M, Goto N, Susanto J P, et al. Determination of phosphate as aggregates of ion associates by light-scattering detection and application to flow injection. Analyst, 1996, 121(8):1085–1088.

[89] Deng J G, Tang G S, Jiang Z L. Barium sulfate particle synchronous scattering spectral determination of sulfate ion in cement sample. Metall. Anal., 2005, 25(2): 13–16.

[90] Ruan W W. Determination of sulfate radical in water samples by resonance light scattering turbidimetry. Channel Sci., 2009, (6): 175–176.

[91] Lv C Y, Huang M Y, Li Y Z, et al. Determination of fluoride ion by resonance Rayleigh light scattering with human serum albumin. J. Instr. Anal., 2004, 23(5): 56–59.

[92] Jiang Z L, Ye L L, Liang A H. A resonance Rayleigh scattering method for the determination of bromine ion: China, CN103604792A. 2014

[93] Li Y M, Li R Y, Shi P F. Determination of iodine in food by resonance light scattering. J. Food Sci., 2012, (08): 151–154.

[94] Liu Q Y. Monitoring of thiocyanate in fresh milk from different regions. Heilongjiang Anim Husband. Vet. Med, 2015, (21): 282–283.

[95] Zhong M H. Study on the determination of thiocyanate content through coupling technique of flow injection analysis with resonance light scattering. J. Anhui Agricult. Sci., 2008, 36(19): 7983–7984.

Zhong De Liu, Jian Wang

10 Light scattering spectrometry of organic small molecules and drugs

Light scattering applications in pharmaceutical analysis and drug quality control are extensive and classic, which normally are used for solution turbidity, that is, to get clarity of the solution to determine insoluble impurities in drugs. This is an significant indicator for the injection crude drug purity, which is an important parameter to reflect the quality and technique level of the drug. The solution clarity is usually checked by hydrolysis of urotropine into formaldehyde under acidic condition, usually which condensates with hydrazine to produce insoluble white benzalhydrazine for standard turbidity solution. At room temperature, the test solution and the same amount of standard solution are mixed in matched turbid tubes, which will help understand whether clarity conforms to the regulation.

As the resonance light scattering (RLS) method [1–3] has high sensitivity and does not require any complex instrument, it has developed rapidly along with the advancement of preparation, properties, and applications of nanoparticles. With the interference of an external magnetic field, free electrons of the metal nanoparticles can form the surface plasma and can act as a strong scatterer, resulting in the scattering signals for the analysis and applications of nanoparticles and can also be used for discussing scattering features of a single nanoparticle, nanoparticle aggregate, and assembly. Aiming at addressing some practical problems in food and drug safety industries, abundant, rapid, simple, and sensitive RLS methods, have been built. In this chapter, we discuss about clinical drugs as the analytic targets to give an overview about pharmaceutical analysis principle, probe, strategies, and so on in terms of organic small molecules scattering probe and nanoparticle scattering probe.

10.1 Light scattering principles for drug analysis

Drugs are with special commodity property that is closely related to human health. The pharmaceutical analysis runs through a series of research and development, production, management, and applications. Before 1970s, the volumetric analysis was in the leading position in pharmaceutical analysis technique. Afterward, with the development and advancement of spectroscopy, chromatography and electrochemistry, pharmaceutical analysis also takes into account automation and intelligence to a large degree with surprising development in terms of sensitivity and high throughput. Trace level drug identification and pharmaceutical detection, in vivo test and metabolism of pharmaceuticals, overall analysis and control of drug complex system have all reached a new level.

https://doi.org/10.1515/9783110573138-010

As light scattering technique has several merits such as simple operation and high sensitivity, it has been widely applied in analysis. Normally, pharmaceutical molecule displays a weak light scattering signal, which can also bind other materials, such as analytical reagents or probes through electrostatic force and hydrophobic force to form larger scattering particles or self-aggregate under certain conditions to enhance system's scattering signal, thus it can be used for pharmaceutical analysis.

In pharmaceutical analysis with light scattering technique, four kinds of conditions can be divided based on the enhancement of light scattering signal and the features of pharmaceutical molecules.

(1) Under specific acidic conditions, pharmaceutical molecules have corresponding charges, the charge type and density are related to their molecular structures. Under certain conditions, the pharmaceutical molecules with charges could bind organic dye molecules [4–8], surfactant [9], inorganic acid anion [10, 11], and some nanoparticles [12–14] with opposite charge through electrostatic force and hydrophobic force, producing larger particles or inducing nanoparticle aggregation, thus enhancing light scattering signal, which can be applied in pharmaceutical analysis.

(2) Pharmaceutical molecules are usually composed of N and O in the forms of carbonyl, hydroxy, and C=N double bond, which could coordinate with metal ions with strong binding ability, thus producing scattering particles to enhance light scattering signal. On the other hand, the complex of pharmaceutical molecules and metal ions are charged, which could bind dyes with opposite charge to produce scattering particles to enhance light scattering signal. For example, La (III) and tanshinone II A sodium sulfonate to form complex to enhance resonance Rayleigh scattering [15]. And other ternary systems, such as Cd(II)-methotrexate chelate and methylene blue [16], furosemide-Pd(II) chelate and some basic triphenylmethane dyes [17], Cu(II)-fluoroquinolone antibiotics and erythrosine [18], enhance light scattering, which are applicable to pharmaceutical analysis.

(3) There are some drugs with the reduction property, such as aldehydes (including streptomycin, pyridoxal, and glucose), phenols (including levodopa, adrenaline, morphine, physostigmine, tocopherol, and sodium salicylate, and so on.), alcohol and enols (including inositol, mannitol, cortisone, ephedrine, proxyphylline, and vitamin C), hydrazines and amines (isoniazid, hydrazine pyridazine hydrochloride, procaine hydrochloride, sodium p-amino salicylate and sulfonamides), thiols and sulfides (including mercaptopurine, captopril, dimercaptopol, cysteine, propylthiouracil, and lipoic acid), low-valence metal organic drug (antimony potassium tartrate and sodium antimony gallate, etc.), carbon–carbon double bond and conjugate double bond systems (including vitamin A, amphotericin B and other drugs with conjugate double bond, zinc undecylenate and other unsaturated fatty acids, pinene, terpene, etc.), and heterocyclic pyrazolone derivatives such as analgin, benzothiazide derivatives such as isopropyl hydrochloride and chlorpromazine hydrochloride. Under suitable conditions,

silver nitrate and chloroauric acid with oxidability can be reduced by above-mentioned reducing drugs to produce corresponding metal nanoparticles, significantly enhancing the scattering signal. The intensity of plasma resonance scattering signal of nanoparticles could help in quantitative determination for drugs [19, 20].

(4) Some structures hydrophilic groups such as $-COO^-$, $-OSO_3^-$, and $-OH$, which are able to protonate to greatly reduce their hydrophily. Under suitable conditions, it aggregates through Van der Waals force, hydrogen bond, and hydrophobic effect, thus forming larger aggregates, to significantly enhance light scattering signal.

For example, three cholates such as sodium taurocholate, sodium cholate, and sodium deoxycholate could self-aggregate to form larger agggregates in certain concentration of HCl, H_2SO_4, or HNO_3 solution to produce strong light scattering signal [21]. It is interesting that different cholates have similar spectral signatures in the same kind of medium, showing a good relation with the acid medium anion or acidity. For example, in an HNO_3 medium, the scattering wavelength is at 349 nm, and in a H_2SO_4 medium it is at 359 nm, while it is at 369 nm in an HCl medium.

Under pH 3.0, dextrin self-aggregates with a characteristic scattering peak at 393 nm [22].

10.2 Scattering probes for pharmaceutical analysis

As pharmaceutical molecules display a weak scattering signal, they interact with other substances to produce new scattering particles with larger volume or larger refractive index changes, to enhance scattering signal to build pharmaceutical analysis method. Therefore, based on the structures and property features of drugs, it is quite vital to choose suitable scattering probes. The involved scattering probes for pharmaceutical analysis mainly include organic molecule and nanoparticle. In the following parts, we will introduce applications of the above-mentioned two kinds of probes for pharmaceutical analysis.

10.2.1 Organic molecule scattering probe

10.2.1.1 Small molecule with positive charges for pharmaceutical analysis

In a suitable medium, many drug molecules present positive charges to bind acidic dye and anionic surfactant through electrostatic interaction, producing large scattering particle and enhancing light scattering signal, which are applicable to pharmaceutical analysis.

These drugs are mainly included in the following categories:
(1) Aminoglycoside antibiotics: such as kanamycin sulfate, netilmicin sulfate, tobramycin sulfate, micronomicin sulfate, gentamicin sulfate, neomycin sulfate, amikacin, and so on.
(2) Alkaloid drugs: such as berberine hydrochloride, atropine sulfate, quinine sulfate, racine anisodamine, and other drugs with complex azacyclo structures.
(3) Drugs with nitrogen atoms: such as doxepine hydrochloride, amitriptyline hydrochloride, diphenhydramine hydrochloride, chlorphenamine maleate, tetracaine hydrochloride, procaine hydrochloride, lidocaine hydrochloride, pioglitazone hydrochloride, promethazine hydrochloride, chlorpromazine hydrochloride, ciprofloxacin, and so on.

10.2.1.2 Small molecule with negative charges for pharmaceutical analysis

Table 10.1 lists the applications of organic molecule scattering probe in pharmaceutical analysis.

Under certain acidity, drugs with negative charges bind with basic dys, cationic surfactants through electrostatic interaction, to produce large scattering particle and enhance scattering signal, which are applicable to pharmaceutical analysis.

These pharmaceutical molecules with negative charges mainly contain sulfonate, carboxyl group and so on, such as heparin sodium, sodium hyaluronate, sodium alginate, derma sulfate, sodium carboxymethyl cellulose, and sodium dextran sulfate. Figure 10.1 shows a resonance light scattering spectrum of a surfactant and heparin. Heparin could enhance light scattering signal of five kinds of surfactants under similar experimental conditions and with similar peak shapes but with a difference in detection sensitivity [46].

10.2.2 Nanoparticle light scattering probe

As nanoparticles especially for noble metal nanoparticles have unique optical and electrical characteristics (refer to Chapter 6), their free electrons form surface plasma with interference of external magnetic field, which can act as strong scatterers. Especially during local surface plasmon resonance (LSPR) coupling aggregate and assembly process [58–60] that changes nanoparticle plasma resonance scattering wavelength and scattering intensity [58], has become widespread analysis and strategy in pharmaceutical analysis. In other words, by detecting the light scattering signal of the nanoparticles after the reaction of drug molecules and reagents, or detecting the changes of light scattering before and after the aggregation and assembly of the nanoparticles, the light scattering method for pharmaceutical analysis can be established. It also provides a simple, fast and universal method for preparing new nanoparticles and constructing nanoparticle assemblies.

Table 10.1: Applications of organic molecule scattering probe in pharmaceutical analysis.

Probe	Drug	pH	λ_{max} (nm)	Linear range (µg/mL)	Detection limit (ng/mL)	Reference
Erythrosine	Diphenhydramine	4.5	580	0.0067–2.0	2.0	[23]
Erythrosine	Ciprofloxacin	4.4	333	0.057–5.6	17	[24]
Erythrosine	Norfloxacin	4.4	374	0.060–4.4	18	[24]
Erythrosine	Levofloxacin	4.4	333	0.047–2.4	14	[24]
Erythrosine	Lomefloxacin	4.4	333	0.061–3.2	27	[24]
Erythrosine	Chlorphenamine	4.4	578	0.0080–3.6	3.6	[25]
Erythrosine- Pd(II)	Lincomycin	5.2	287	0.050–2.7	1.5	[26]
Aniline blue-UO^{2+}	Doxycycline	5.8	343	0.10–6.7	28.5	[27]
Aniline blue -UO^{2+}	Chlortetracycline	5.8	343	0.10–6.2	38.6	[27]
Aniline blue -UO^{2+}	Oxytetracycline	5.8	343	0.10–7.5	39.9	[27]
Aniline blue -UO^{2+}	Tetracycline	5.8	343	0.10–7.2	44.9	[27]
Titan yellow	Amikacin	5.5	482	0.4–2.4	860	[28]
Diphenylnaphthyl methane	Sodium hyaluronic	3.6	280	0–13.0	13.7	[29]
Dichlorofluorescein	Famotidine	3.2	470	0.0087–1.25	2.6	[30]
Dibromofluorescein	Famotidine	3.8	540	0.0034–1.0	1.0	[30]
Dibromofluorescein-Pd(II)	Ceftazidime	3.5	291	0.042–0.12	12.7	[31]
Dibromofluorescein-Pd(II)	Ceftriaxone	3.5	291	0.045–0.12	13.6	[31]
Dibromofluorescein -Pd(II)	Cefoperazone	3.5	291	0.053–0.18	16	[31]
Dibromofluorescein-Pd(II)	Cefotaxime	3.5	291	0.113–0.24	33.9	[31]
Diiodofluorescein	Famotidine	4.4	553	0.0068–1.5	2.0	[30]
Congo red	Epirubicin	2.5	370	0.050–12.0	54	[6]
Congo red	Daunorubicin	2.5	370	0.050–12.0	58	[6]
Congo red	Mitoxantrone	2.5	370	0.040–7.5	33	[6]

(continued)

Table 10.1 (Continued)

Probe	Drug	pH	λ_{max} (nm)	Linear range (µg/mL)	Detection limit (ng/mL)	Reference
Congo red	Amikacin	5.5	563	0.040–6.0	4.0	[8]
Congo red	Doxorubicin	2.6	380	1.0–12	50	[7]
Congo red	Quinoline sulfate	4.5	562	0.20–8.4	12.0	[32]
Congo red -Co(II)	Ciprofloxacin	5.4	560	0.026–2.64	7.68	[33]
Congo red -Co(II)	Norfloxacin	5.4	560	0.045–3.2	13	[33]
Congo red -Co(II)	Ofloxacin	5.4	560	0.037–4.0	11.24	[33]
Malachite green -Hg(II)	Cefotaxime	5.3	340	0.016–3.5	4.7	[34]
Toluidine blue	Heparin	5.4	329	0–0.8	26	[35]
Methyl-red	Octreotide	3.8	310	0–0.54	3.6	[36]
Methyl-blue	Chondroitin sulfate	5.5	314	0.24–3.0	0.024	[37]
Methyl purple-Pd(II)	Furosemide	6.1	324	0.016–2.5	4.9	[17]
Methyl green-Pd(II)	Furosemide	5.7	340	0.013–2.0	4.0	[17]
Crystal violet-Pd(II)	Furosemide	6.1	324	0.054–4.0	3.4	[17]
Crystal violet–UO$_2^{2+}$	Folic acid	4.7	328	0.014–3.5	4.7	[38]
Nile blue sulfate	Carboxymethyl cellulose sodium	6.0	298	0.050–3.0	1.8	[39]
Hexadecylpyridinium chloride	Chondroitin sulfate	4.4	340	0.020–8.0	13	[40]
Trypan red	Kanamycin sulfate	3.2	400	0.013–6.0	12.9	[41]
Trypan red	Neomycin sulfate	3.2	400	0.014–6.0	14.1	[41]
Trypan red	Gentamicin sulfate	3.2	400	0.018–6.0	17.6	[41]
Trypan red	Tobramycin sulfate	3.2	400	0.016–6.0	16.2	[41]
Alizarin violet	Ofloxacin	6.9	405	0.050–3.0	21	[4]
Trypan blue	Kanamycin	4.0	580	0.01–6.0	6.1	[42]
Trypan blue	Gentamicin	4.0	580	0.020–6.0	6.8	[42]
Trypan blue	Ambermicin	4.0	580	0.020–6.0	7.3	[42]
Trypan blue	Neomycin	4.0	580	0.020–6.0	8.1	[42]

SDBS	Dequalinium chloride	0.5	392	0.096–2.88	2.98	[43]
SDBS	Vitamin B1	3.29	375	0.0012–0.85	1.2	[44]
Sodium dodecyl sulfate-eosin Y	Amikacin	5.5	525	4.0–20	10.8	[45]
Tetradecyl dimethyl benzonium chloride	Heparin	6.09	470	0–8.0	11	[46]
Hexadecyl dimethyl benzonium chloride	Heparin	6.09	470	0–16	12	[46]
Cetyl pyridine - eosin B	Acyclovir	2.8	364	0.40–24	25	[47]
Eosin B	Promethazine hydrochloride	3.85	363	0.075–4.5	18	[48]
Eosin Y-Pd(II)	Ciprofloxacin	4.5	368	0–2.4	9.4	[49]
Eosin Y-Pd(II)	Norfloxacin	4.5	368	0–2.4	12.8	[49]
Eosin Y-Pd(II)	Ofloxacin	4.5	368	0–2.2	12.2	[49]
Victoria blue B	Sodium alginate	4.2	557	0.075–3.0	22	[50]
Victoria blue 4R	Sodium alginate	5.0	526	0.085–4.0	25.6	[50]
Bromocresol green	Meropenem	1.9	572	0.01–14	14.6	[51]
Methylene blue-Pd(II)	Methotrexate	5.0	342	0.008–2.0	2.0	[16]
Acridine orange	Carboxymethyl cellulose sodium	5.0	315	0–3.0	41.7	[52]
Acridine yellow	Carboxymethyl cellulose sodium	5.0	285	0–1.5	6.0	[52]
Acridine yellow	Adenosine disodium triphosphate	9.2	325	0.80–20.0	86	[53]
Ethyl violet	1- hydroxyvaline	8.0	396	4.0–982	120	[54]
Ethyl violet	Sodium hyaluronic	4.93	326	0.40–48.0	96	[55]
Night blue	Sodium aescinate	5.0	416	0.025–20	7.5	[56]
Night blue	Sodium hyaluronic	3.6	408	0–1.5	13.7	[57]

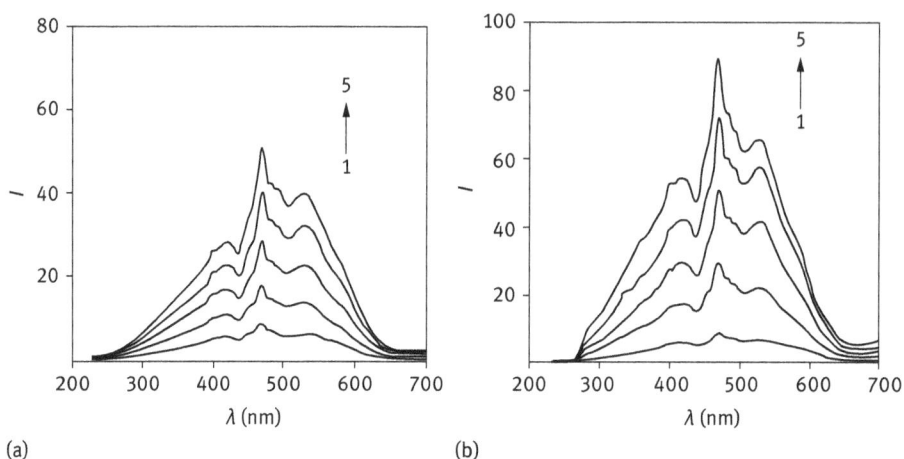

Figure 10.1: Resonance Rayleigh scattering spectra of heparin with some cationic surfactants [46]. (a) Heparin–cetyldimethyl benzylammonium chloride (CDBAC) system. CDBAC concentration: 10 µg/mL, pH 6.09; heparin concentration: 1, 0 µg/mL; 2, 0.2 µg/mL; 3, 0.4 µg/mL; 4, 0.6 µg/mL; 5, 0.8 µg/mL. (b) Tetradecyl dimethyl benzyl ammonium chloride (TDBAC) heparin system. TDBAC concentration: 30 µg/mL; pH 6.09; heparin concentration: 1, 0 µg/mL; 2, 0.4 µg/mL; 3, 0.8 µg/mL; 4, 1.2 µg/mL; 5, 1.6 µg/mL.

Nanoparticles used as the scattering probe for pharmaceutical analysis include metal nanomaterials, such as Au, Ag, Cu nanoparticles, inorganic semiconductor nanoparticles mainly consist of IIB–VIA group elements and the most studied CdX (X = S, Se, Te). By measuring the intensity of the light scattering signal of the newly generated nanoparticles or the change of the intensity of the light scattering signal before and after the aggregation of nanoparticles, the aim of quantitative analysis of the drugs is achieved. Table 10.2 summarizes some applications of nanoparticle light scattering probes in pharmaceutical analysis.

On the basis of nanoparticle aggregates, drugs with thiol group such as captopril (Figure 10.2) [66] and cysteine [68] can be detected by light scattering method, which involve two mechanisms, including of the formation of Au-S covalent bond between gold and thiol group and the electrostatic interaction or hydrogen bonding among pharmaceutical molecules, resulting in the reduce of distance between nanoparticles to form aggregates, leading to the red-shift of absorption and the enhancement of light scattering intensity.

There is also another common pharmaceutical analysis method by producing nanoparticles through the redox reaction between reducing drugs and metal ions, such as glucose [20] and ferulic acid [19] (Figure 10.3). The method is based on determinating of the local plasmon resonance signals of metal nanoparticles formed by reduction of metal ions by reductive drugs.

Wang et al. [73] established an 8-hydroxyl-2-deoxy guanosine (8-OHdG) detection method based on specific recognition between aptamer and target (Figure 10.4).

Table 10.2: Applications of nanoparticle light scattering probes in pharmaceutical analysis.

Probe reagent	Drugs	pH	λ_{max} (nm)	Linear range (µg/mL)	Detection limit (ng/mL)	References
CdS quantum dots	Streptomycin sulfate	6.0	310	0.0147–0.2	4.4	[61]
CdS quantum dots	Streptomycin sulfate	6.0	310	0.0057–0.64	1.7	[62]
CdSe quantum dots	Chitosan	6.0	370	0.042–3.0	1.2	[62]
Gold nanoparticles	Cysteine	5.0	556	0.01–0.25	2.0	[63]
Gold nanoparticles	Heparin	7.0	552	0.01–0.8	2.0	[64]
Gold nanoparticles	Methimazole	5.2	555	0–1.5*	0.002*	[65]
Gold nanoparticles	Captopril	2.09	553	0.10–1.7	32	[66]
Gold nanoparticles	Kanamycin sulfate	6.7	560	0.02–0.8*	0.002*	[14]
Gold nanoparticles	Tetracycline	2.36	570	4.0–26*	0.006*	[67]
Gold nanoparticles	Glucose	11.2	from 553 to 563	2.0–250*	0.21*	[20]
Gold nanoparticles	Cysteine	4.2	566	0.01–0.40	2.9	[68]
Gold nanoparticles	Penicillin	5.4	560	0.0075–1.7	0.78	[69]
Gold nanoparticles	Dodecyl gallate	4.0	540	0.02–0.3	0.007	[70]
Gold nanorods	Heparin	5.33	565	0.02–0.70	8.0	[71]
Silver nanoparticles	Berberine hydrochloride	3.42	500	0.05–0.4*	0.05*	[72]
Silver nanoparticles	Ferulic acid	NaOH	470	0.2–2.0*	0.0152	[19]

*µmol/L.

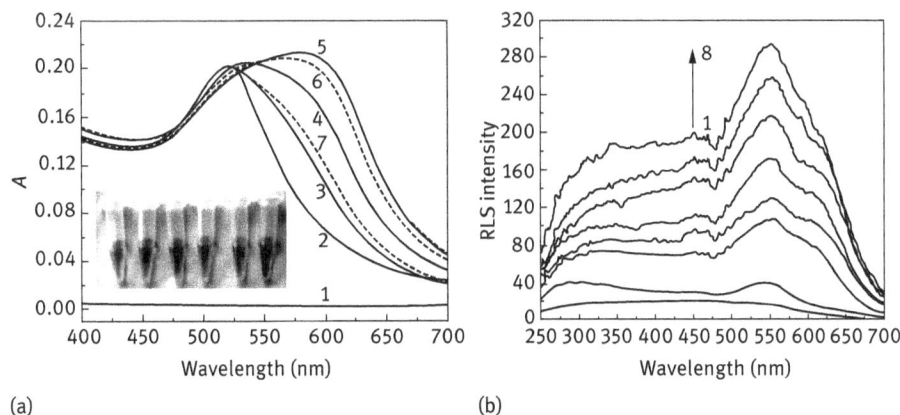

(a)   (b)

Figure 10.2: Determination of captopril based on the aggregation of gold nanoparticles [68].
(a) Absorbance spectra: 1. Captopril; 2. gold nanoparticles; 3–7. Captopril+gold nanoparticles, gold nanoparticles concentration: 1.2×10^{-4} mol/L; Captopril concentration ($\times 10^{-6}$ mol/L): 1–7. 40.0, 0, 0.5, 1.0, 2.0, 4.0, and 6.0 (pH 2.09). 6 and 7 are sample 5 sonicated for 15 minutes and adjusted to pH 3.78. (b) Scattering spectra: 1. Captopril; 2. gold nanoparticles; 3–8. Captopril+gold nanoparticles, gold nanoparticles concentration: 1.2×10^{-4} mol/L; Captopril concentration: ($\times 10^{-6}$ mol/L): 1–8, 40.0, 0, 0.5, 1.0, 2.0, 4.0, 6.0, and 8.0 (pH 2.09).

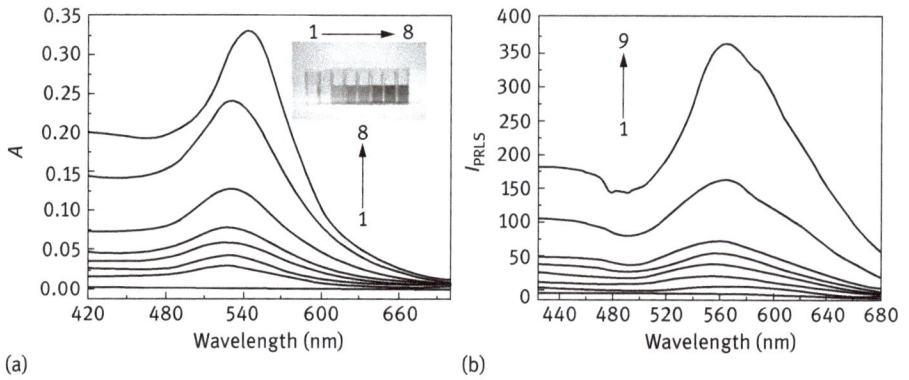

(a)　　　　　　　　　　　　　　(b)

Figure 10.3: Glucose detection based the growth of gold nanoparticles [20]. (a) Absorbance spectra: 1. HAuCl$_4$; 2–8. HAuCl$_4$+ glucose, glucose concentration (×10^{-6} mol/L): 1–8, 0.0, 2.0, 4.0, 6.0, 8.0, 10.0, 20.0, 40.0 (pH 11.20). HAuCl$_4$ concentration: 0.2 mmol/L. (b) Scattering spectra: 1. HAuCl$_4$; 3–9, HAuCl$_4$+glucose, HAuCl$_4$ concentration: 0.2 mmol/L, glucose concentration (×10^{-6} mol/L): 1–9, 0.0, 0.0, 2.0, 4.0, 6.0, 8.0, 10.0, 20.0, and 40.0.

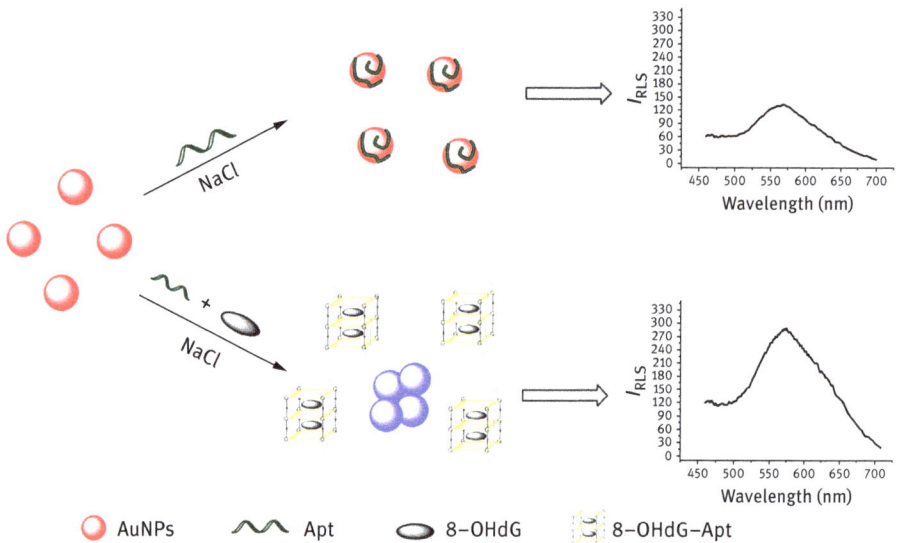

Figure 10.4: Detection of 8-OHdG based on aptamer recognition induced aggregation of gold nanoparticles [73].

A single-stranded DNA adsorbs onto the surface of gold nanoparticles to improve their stability. When in the presence of NaCl, gold nanoparticles do not aggregate thus present weak scattering signal. However, in the presence of target (8-OHdG), the aptamer changes from single strand to G-quartet structure. At this time, aptamer no longer adsorbs on the gold nanoparticle surface, leading to a lower stability. When NaCl is present, gold nanoparticles form aggregates with stronger scattering signal, which could be developed for 8-OHdG detection.

In summary, the light scattering technique has been widely used in pharmaceutical analysis. Own to its high sensitivity, low detection limit, and simple and rapid operation. It provides a new way for the content analysis and the detection of trace drugs in biological fluids. At the same time, with the development of research, some problems of light scattering technology have been exposed, such as low selectivity and poor reproducibility. Therefore, it is particularly important to overcome the above problems by developing new strategies and methods for light scattering analysis.

10.3 Development strategy of light scattering spectral method for pharmaceutical analysis

The traditional light scattering analysis usually studies on the light scattering characteristics of the ordinary solution system, using organic molecules and nanoparticles as light scattering probes, mainly by the electrostatic attraction, hydrophobicity and redox reaction between the drugs and probes to produce scattering particles. The scattering signals of scattering particles are excited and emitted with monochromator by synchronous scanning an ordinary fluorescence spectrophotometer. However, due to the interference of various substances in the solution, the light scattering spectral method of pharmaceutical analysis generally suffers from low selectivity and poor reproducibility. Therefore, a variety of improved light scattering analysis techniques have been established and have good prospects for pharmaceutical analysis.

10.3.1 Change the analysis and detection instruments

As discussed in Chapter 3, when the light spreads from an optically denser medium to an optically thinner medium with an incident angle larger than that of the total internal reflection, there is 100–200 nm thick evanescent wave with longitudinal propagation and exponential damping on the side of the optically thinner medium. Thus, when there is a total internal reflection, only the substance on the interface interacts with the light to produce the scattering signal while other substances are kept in the solution. Thus, the selective analysis of substances that only react to the interface can be realized. On the other hand, the interface inrichs the tested substance, making it more concentrated than in solution, so it can also improve the detection sensitivity.

For example, Feng et al. [74] found that on water/carbon tetrachloride liquid–liquid interface, aureomycin–europium–trioctylphosphine complex can be formed, increasing total internal reflection (TIR)-RLS signal at 340 nm. The enhanced signal intensity is in proportional to aureomycin concentration in the range of 0.98×10^{-7}–20.0×10^{-7} mol/L, with a detection limit of 9.8 nmol/L, which is used for aureomycin detection in urine, blood, and milk, with a recovery of 95.4%–106.4%. Later, TIR-RLS-based detection for drugs such as berberine [75] and penicillin [76] were reported, further confirming its advantages in improving selectivity and sensitivity.

10.3.2 Improvement in the injection modes

As an on-line disposal and quantitative flow analysis technique for the sample in nonequilibrium conditions, flow injection analysis (FIA) present the features of high automation and good injection reproducibility. As discussed in Chapter 4 (Section 4.6), combining FIA technique with light scattering technique can avoid error due to manmade injection and operation; meanwhile, by combining high-precision of FIA with high-sensitivity of light scattering, the analysis speed can be improved, which also realize automation of scattering analysis.

Currently, the FIA technique coupled with light scattering technique has been used for the analysis of heparin [77], chlorpromazine hydrochloride, and promethazine hydrochloride [78], as well as the local anesthetics such as tetracaine, procaine, and lidocaine [79]. The results show that FIA improved reproducibility, obviously decreased the relative standard deviation (RSD), and simplified operation, proving its advantage in improving analysis precision and accuracy as well as an increase in the analysis speed. Meanwhile, this combination is helpful in applications and expansion of light scattering technique in automatic analysis and microfluidic chip.

10.3.3 Improvement of signal processing

In light scattering analysis with an ordinary fluorescence spectrophotometer, it normally tests the signal intensity of the spectrum at a single wavelength. However, the fluctuation of the signal intensity, the fluctuation of the voltage on the photomultiplier tube, as well as the small changes in the material concentration and experimental conditions, will lead to the fluctuation of the spectral signal at the single wavelength, and then affect the precision of the experimental data and the accuracy of the method.

To solve these problems, dual wavelength ratiometric method is developed. By selecting two suitable scattering wavelengths, or selecting a scattering wavelength and a fluorescence emission wavelength, the ratio of the spectral signals at the selected two wavelengths is used as the basis for quantitative analysis. At this time, even if the external causes the absolute value of the spectral signal to fluctuate, the ratio of the spectral signal is usually the same because the wavelengths at the two wavelengths are basically constant. Thus, the accuracy of dual wavelength ratiometric method is not affected by the signal intensity changes at the emission wavelength, or limited by the the concentration changes of chosen probe. Thus, RSD changes of detection results are obviously decreased.

For example, Long et al. [80] used organic small molecules probes and dual wavelength ratiometric method to detect heparin, which showed better productivity and a wider linear range than single wavelength light scattering analysis. While, Liao [81] used CdS quantum dot as optical probe, and aminoglycoside antibiotics such as tobramycin, kanamycin and gentamicin induced CdS quantum dots aggregation, causing quantum dot fluorescence quenching and scattering signal enhancement. The scattering fluorescence ratiometric method is used for the detection of aminoglycoside antibiotics. Compared with single wavelength light scattering analysis, the results further prove the advantages of the dual dual wavelength ratiometric method.

References

[1]   Pasternack R F, Bustamante C, Collings P J, et al. Porphyrin assemblies on DNA as studied by resonance light-scattering techniques. J. Am. Chem. Soc., 1993, 115: 5393–5399.

[2]   Huang C Z, Li K A, Tong S Y. Determination of nucleic acids by a resonance light-scattering technique with α, β, γ, δ -tetrakis [4-(trimethylammoniumyl) phenyl] porphine. Anal. Chem., 1996, 68: 2259–2263.

[3]   Huang C Z, Li K A, Tong S Y. Determination of nanograms of nucleic acids by their enhancement effect on the resonance light scattering of the cobalt(II)/4-[(5-chloro-2-pyridyl)azo]-1,3-diaminobenzene complex. Anal. Chem., 1997, 69: 514–520.

[4]   Chen Z G, Peng Y R, Chen J H, et al. Determination of antibacterial ofloxacin in human serum samples by a resonance light scattering technique with alizarin violet 3B. Anal. Sci., 2009, 25(7): 891–896.

[5]   Qin Z H, Zhao X H, Niu W F. Resonance Rayleigh scattering spectra of the interaction between alizarin green and tolterodine tartrate. Spectrosc. Lett., 2009, 42(2): 100–107.

[6]   Wang F, Liu Z F, Liu S P, et al. Study on the interaction between three anthracycline anticancer drugs and Congo red by absorption, fluorescence and resonance Rayleigh scattering spectra. Acta Chim. Sin., 2005, 63(21): 1991–1998.

[7]   Hu C M, Wang C, Zhu L, et al. Resonance rayleigh scattering method for the determination of free doxorubicin content in doxorubicin liposome. Chinese J. Pharm. Anal., 2008, 25(5): 718–721.

[8]   Liu Z F, Liu S P, Hu X L. Resonance Rayleigh scattering and resonance nonlinear scattering spectra of Congo red Amikacin system and their analytical applications. Sci. China Ser. B., 2006, 36(4): 317–325.

[9]   Liu S P, Wang F, Liu Z F, et al. Resonance Rayleigh scattering spectra for studying the interaction of anthracycline antineoplastic antibiotics with some anionic surfactants and their analytical applications. Anal. Chim. Acta, 2007, 601(1): 101–107.

[10] Wang F, Liu Z F, Liu S P, et al. Determination of mitoxantrone via ammonium molybdate using high sensitive resonance Rayleigh scattering method. Chem. J. Chinese U., 2006, 27(8): 1459–1461.

[11] Zou J M, Jiang H L, Wang L S, et al. Resonance Rayleigh scattering spectra of berberine-tetraphenylboron association nanoparticle and its analytical application. Acta Pharm. Sin., 2003, 35(7): 530–533.

[12] Wang Q, Liu Z F, Liu S P. Resonance Rayleigh scattering spectra method for the determination of anthracycline anticancer drugs with CdS nanoparticles as probe. Chem. J. Chinese U., 2007, 28(5): 837–842.

[13] Li T S, Liu Z F, Liu S P, et al. Fluorescence and Resonance Rayleigh scattering characteristics of CdTe nanocrystalline solution and interaction between CdTe nanocrystals and aminoglycoside antibiotics. Sci. China Ser. B., 2008, 38(9): 798–807.

[14] Wang X Y, Zou M J, Xu X, et al. Determination of human urinary kanamycin in one step using urea-enhanced surface plasmon resonance light scattering of gold nanoparticles. Anal. Bioanal. Chem., 2009, 395(7): 2397–2403.

[15] Peng J D, Liu Z F, Liu S P, et al. Resonance Rayleigh scattering spectra of lanthanum (III) and sodium tanshinone A sulfonate chelate and Its Analytical Application. Sci. China Ser. B., 2006, 36(3): 227–233.

[16] Xi C X, Liu Z F, Liu S P, et al. Resonance Rayleigh scattering spectra of the interaction of Pd(II)-methotrexate chelate with methylene blue and their analytical application. Chem. J. Chinese U., 2008, 29(3): 510–514.

[17] Li C X, Liu Z F, Liu S P, et al. Resonance Rayleigh scattering and resonance nonlinear scattering spectra of furosemide -Pd (II) chelates and some basic triphenyl methane dyes and their analytical applications. Sci. China Ser. B., 2010, 40(12): 1862–1873.

[18] Wang J, Liu Z F, Liu S P, et al. Absorption, fluorescence and resonance Rayleigh scattering spectra of copper (II) - fluoroquinolone antibiotic chelates and erythrin red system and their analytical applications. Sci. China Ser. B., 2007, 37(5): 453–462.

[19] Wang H Y, Li Y F, Huang C Z. Detection of ferulic acid based on the plasmon resonance light scattering of silver nanoparticles. Talanta, 2007, 72(5): 1698–1703.

[20] Shen X W, Huang C Z, Li Y F. Localized surface plasmon resonance sensing detection of glucose in the serum samples of diabetes sufferers based on the redox reaction of chlorauric acid. Talanta, 2007, 72(4): 1432–1437.

[21] Nan H J, Liu Z F, Liu S P. Resonance Rayleigh scattering spectra of self-aggregation for bile salts in acidic media and their analytical application. Acta Chim. Sin., 2006, 64(12): 1253–1259.

[22] Chen Z G, Zhu L, Song T H, et al. Determination of dextrin based on its self-aggregation by resonance light scattering technique. Anal. Chim. Acta, 2009, 635(2): 202–206.

[23] Tang X L, Liu Z F, Liu S P, et al. Absorption, fluorescence and Resonance Rayleigh scattering spectra of the interaction between diphenhydramine and erythrin and their analytical applications. Sci. China Ser. B., 2007, 37(1): 68–75.

[24] Wang J, Liu Z, Liu J, et al. Study on the interaction between fluoroquinolones and erythrosine by absorption, fluorescence and resonance Rayleigh scattering spectra and their application. Spectrochim. Acta A, 2008, 69: 956–963.

[25] Yang J D, Zhou S. Fluorescence quenching and resonance scattering enhancement of Chlorphenamine Maleate and scarlet red. Chinese Sci. Bull., 2008, 53(16): 1892–1896.

[26] Yi A, Liu Z, Liu S, et al. Study on the interaction between palladium (II)-lincomycin chelate and erythosine by absorption, fluorescence and resonance Rayleigh scattering spectra and its analytical applications. Luminescence, 2009, 24: 23–29.

[27] Wei X Q, Liu Z F, Liu S P. Resonance Rayleigh scattering method for the determination of tetracycline antibiotics with uranyl acetate and water blue. Anal. Biochem., 2005, 346: 330–332.

[28] Hao Q Y, Wang J Z. Determination of amikin by resonance light scattering method. J. Chem. Res., 2008, 19(3): 87–90.

[29] Chen L H, Jian Y, Luo H Q, et al. Determination of sodium hyaluronate with some basic bisphenylnaphthylmethane dyes by resonance Rayleigh scattering method. Chinese Chem. Lett., 2007, 18: 1099–1102.

[30] Chen P L, Liu Z F, Liu S P, et al. Resonance Rayleigh scattering spectra of palladium (II) and three element ionic association nanoparticles of famotidine and halogenated fluorescein dyes and their analytical applications. Sci. China Ser. B., 2011, 41(6): 1059–1067.

[31] Fu S, Liu Z, Liu S, et al. Study on the resonance Rayleigh scattering spectra of the interactions of palladium (II)-cephalosporins chelates with 4,5-dibromofluorescein and their analytical applications. Anal. Chim. Acta, 2007, 599: 271–278.

[32] Zeng Y, Cai L, Wang H, et al. Resonance light scattering study on the interaction between quinidine sulfate and congo red and its analytical application. Luminescence, 2010, 25: 30–35.

[33] Wang J, Liu Z F, Liu S P, et al. Study on fluoroquinolone antibiotics-cobalt(II)-Congo red systems by resonance Rayleigh scattering spectra and their analytical applications. Acta Chim. Sin., 2008, 66(11): 1337–1343.

[34] Fu S H, Liu Z F, Liu S P, et al. Resonance Rayleigh scattering spectra of the interactions of $HgCl_2$ cefotaxime anionic chelate with basic triphenylmethane dye and their analytical applications. Chem. J. Chinese U., 2008, 29(11): 2164–2170.

[35] Luo H Q, Liu S P, Liu Z F, et al. Resonance Rayleigh scattering spectra for studying the interaction of heparin with some basic phenothiazine dyes and their analytical applications. Anal. Chim. Acta, 2001, 449: 261–270.

[36] Fan L, Li D, Sun Y, et al. Determination of octreotide by resonance Rayleigh scattering method with methyl red. Chinese J. Appl. Chem., 2010, 27(5): 585–589.

[37] Zhang L, Li N, Zhao F, et al. Spectroscopic study on the interaction between methylene blue and chondroitin 4-sulfate and its analytical application. Anal. Sci., 2004, 20: 445–450.

[38] Xi C, Liu Z, Kong L, et al. Effects of interaction of folic acid with uranium (VI) and basic triphenylmethane dyes on resonance Rayleigh scattering spectra and their analytical applications. Anal. Chim. Acta, 2008, 613: 83–90.

[39] Kong L, Liu Z F, Liu S P. Resonance light scattering method for the determination of sodium carboxymethyl cellulose with nile blue sulphate. Chinese J. Appl. Chem., 2006, 23(7): 708–712.

[40] Jiang Z L, Zou M J, Liang A H, et al. Resonance scattering spectral study of cationic surfactant-chondroitin sulfate association particle systems and its analytical application. Acta Chim. Sin., 2006, 64(2): 111–116.

[41] Hu X L, Liu S P., Luo H Q. Resonance Rayleigh scattering spectra of interaction of aminoglycoside antibiotics with trypan red and their analytical applications. Acta Chim. Sin., 2003, 61(8): 1287–1293.

[42] Liu S P, Hu X L, Li N B. Resonance Rayleigh scattering method for the determination of aminoglycoside antibiotics with Trypan Blue. Anal. Lett., 2003, 36(13): 2805–2821.

[43] Chen Z, Peng Y, Chen J, et al. Determination of antibacterial quaternary ammonium compound in lozenges and human serum by resonance light scattering technique. J. Pharm. Biomed. Anal., 2008, 48(3): 946–950.

[44] Lu W, Huang C Z, Li Y F. Novel assay of thiamine based on its enhancement of total internal reflected resonance light scattering signals of sodium dodecylbenzene sulfonate at the water/ tetrachloromethane interface. Anal. Chim. Acta, 2003, 475(12): 151–161.

[45] Hao Q Y, Wang J Z, Zhang Y F. Determination of amikin with eosin Y by resonance light scattering method using SDS as sensitizer. J. Anal. Sci., 2009, 25(1): 111–113.

[46] Liu S P, Luo H Q, Xu H, et al. Resonance Rayleigh scattering study of interaction of heparin with some cationic surfactants and their analytical application. Spectrochim. Acta A, 2005, 61: 861–867.

[47] Feng S L, Wu C K, Zeng C. Resonance light scattering spectrometric method for the determination of aciclovir in tablets. Chinese J. Antibiot., 2009, 34(6): 355–358.

[48] Feng S L, Pan Z H, Fan J. Resonance light scattering technique for determination of trace promethazine hydrochloride in the presence of eosin B. Chinese J. Appl. Chem., 2008, 25(12): 1444–1447.

[49] Yang Z P, Liu Z F, Hu X L, et al. Resonance Rayleigh scattering spectra of some fluoroquinolones antibiotics-Pd(II)-eosin Y systems and their analytical application. Chinese J. Appl. Chem., 2007, 24(3): 261–267.

[50] Yang J D, Wang X H, Liu Z F, et al. Determination of sodium alginate with basic bisphenyl-naphthylmethane dyes by resonance Rayleigh scattering. Chinese J. Appl. Chem., 2008, 25(4): 385–390.

[51] Liu Q L, Qiu Y H, Li S Q. Resonance Rayleigh scattering spectrum of bromocresol green-meropenem system and its analytical application. J. Anal. Sci., 2010, 26(5): 559–562.

[52] Liu S P, Chen S, Liu Z F, et al. Resonance Rayleigh scattering spectra of interaction of sodium carboxymethylcellulose with cationic acridine dyes and their analytical applications. Anal. Chim. Acta, 2005, 535: 169–175.

[53] Feng S, Shi H. Spectroscopic study on the interaction of acridine yellow with adenosine disodium triphosphate and its analytical application. Spectrochim. Acta A, 2007, 68: 244–249.

[54] Ou-Yang Y F, Wang Y S, Mi X, et al. Resonance light scattering of 1-hydroxypyrene-ethyl violet-anionic surfactant system and its analytical application. Anal. Sci., 2007, 23(5): 533–537.

[55] Luo H Q, Li N B, Liu S P. Resonance Rayleigh scattering study of interaction of hyaluronic acid with ethyl violet dye and its analytical application. Biosens. Bioelectron., 2006, 21: 1186–1194.

[56] Wang X H, Yang J D, Liu S P, et al. Resonance Rayleigh scattering method for determination of sodium aescinate by means of night blue. Chinese J. Appl. Chem., 2008, 25(6): 656–660.

[57] Chen L H, Liu S P, Luo H Q, et al. Determination of sodium hyaluronate with night blue by resonance Rayleigh scattering method. J. Anal. Sci., 2005, 21(3): 304–306.

[58] Sonnichsen C, Reinhard B M, Liphardt J, et al. A molecular ruler based on plasmon coupling of single gold and silver nanoparticles. Nat. Biotechnol., 2005, 23: 741–745.

[59] Funston A M, Novo C, Davis T J, et al. Plasmon coupling of gold nanorods at short distances and in different geometries. Nano. Lett., 2009, 9: 1651–1658.

[60] Pralnod P, Thomas K G. Plasmon coupling in dimers of Au nanorods. Adv. Mater., 2008, 20: 4300–4305.

[61] Liu Z, Liu S, Wang L, et al. Resonance Rayleigh scattering and resonance non-linear scattering method for the determination of aminoglycoside antibiotics with water solubility CdS quantum dots as probe. Spectrochim. Acta A, 2009, 74: 36–41.

[62] Peng J, Liu S, Wang L, et al. Study on the interaction between CdSe quantum dots and chitosan by scattering spectra. J. Coll. Interf. Sci., 2009, 338: 578–583.

[63] Li Z P, Duan X R, Liu C H, et al. Selective determination of cysteine by resonance light scattering technique based on self-assembly of gold nanoparticles. Anal. Biochem., 2006, 351(1): 18–25.

[64] Xiang M, Xu X, Li D, et al. Selective enhancement of resonance light-scattering of gold nanoparticles by glycogen. Talanta, 2008, 76: 1207–1211.

[65] Liu X L, Yuan H, Pang D W, et al. Resonance light scattering spectroscopy study of interaction between gold colloid and thiamazole and its analytical application. Spectrochim. Acta A, 2004, 60(1–2): 385–389.

[66] Liu Z D, Huang C, Li Y F, et al. Enhanced plasmon resonance light scattering signals of colloidal gold resulted from its interactions with organic small molecules using captopril as an example. Anal. Chim. Acta, 2006, 577(2): 244–249.

[67] Hu B, Huang C Z, Zhang L. Plasma resonance light scattering analysis of tetracycline hydrochloride based on multi walled carbon nanotubes. Sci. China Ser. B., 2008, 38(7): 578–583.

[68] Wang J, Li Y F, Huang C Z, et al. Rapid and selective detection of cysteine based on its induced aggregates of cetyltrimethylammonium bromide capped gold nanoparticles. Anal. Chim. Acta, 2008,626: 37–43.

[69] Jiang Z, Li Y, Liang A, et al. A sensitive and selective immuno-nanogold resonance-scattering spectral method for the determination of trace penicillin G. Luminescence, 2008, 23: 157–162.

[70] Andreu N A, Fernandez R J M, Gomez H A. Determination of antioxidant additives in foodstuffs by direct measurement of gold nanoparticle formation using resonance light scattering detection. Anal. Chim. Acta, 2011, 695(1–2): 11–17.

[71] He W, Li Y F, Tan K J, et al. Characterization of interaction between gold nanorods and heparin as well as the plasma resonance light scattering analysis of heparin. Chinese Sci. Bull., 2007, 52(24): 2840–2845.

[72] Ling J, Sang Y, Huang C Z. Visual colorimetric detection of berberine hydrochloride with silver nanoparticles. J. Pharm. Biomed. Anal., 2008, 47: 860–864.

[73] Wang J C, Wang Y S, Xue J H, et al. An ultrasensitive label-free assay of 8-hydroxy-2'-deoxyguanosine based on the conformational switching of aptamer. Biosens. Bioelectron., 2014, 58: 22–26.

[74] Feng P, Shu W Q, Huang C Z, et al. Total internal reflected resonance light scattering determination of chlortetracycline in body fluid with complex cation of chlortetracycline-europium-trioctyl phosphine oxide at the water/tetrachloromethane interface. Anal. Chem. 2001, 73: 4307–4310.

[75] Pang X B, Huang C Z. A selective and sensitive assay of berberine using total internal reflected resonance light scattering technique with fuorescein at the water/1, 2-dichloroethane interface. J. Pharm. Biomed. Anal. 2004, 35: 185–191.

[76] Huang C Z, Feng P, Li Y F, et al. Adsorption of penicillin-berberine ion associates at a water/tetrachloromethane interface and determination of penicillin based on total internal reflected resonance light scattering measurements. Anal. Chim. Acta., 2005, 538: 337–343.

[77] Dai X X, Li Y F, Huang C Z. Analysis of heparin by combining technique of flow injection analysis with resonance light scattering. Chinese J. Anal. Chem., 2005, 33(11): 1535–1538.

[78] Chen P L, Liu Z F, Liu S P, et al. Determination of chlorpromazine hydrochloride and pPromethazine hydrochloride by resonance Rayleigh scattering method coupled with flow injection technique. Chinese J. Anal. Chem., 2010, 38(7): 1007–1010.

[79] Hu X L, Xu D P, Liu Z F, et al. Determination of some local anesthetics by resonance Rayleigh scattering method coupled with flow injection analysis. Chem. J. Chinese U., 2008, 29(10): 1963–1968.

[80] Long Y J, Li Y F, Huang C Z. A wide dynamic range detection of biopolymer medicines with resonance light scattering and absorption ratiometry. Anal. Chim. Acta., 2005, 552: 175–181.

[81] Liao Q G, Li Y F, Huang C Z. A light scattering and fluorescence emission coupled ratiometry using the interaction of functional CdS quantum dots with aminoglycoside antibiotics as a model system. Talanta, 2007, 71(2): 567–572.

Li Zhang, Jian Wang
11 Light scattering spectrometry of nucleic acids

Nucleic acid is the key biomacromolecule in living beings and has a function of storage and delivery of genetic information. With considerable research of gene structure and function, it has been determined that many genetic diseases are related to nucleic acid molecule base sequence mutation and position, and degree of differences leads to different diseases. Thus, it is quite useful for the diagnosis and treatment of a disease on molecular level for blood, fluid, and tissue-specific sequence nucleic acid test.

The detection of DNA-specific sequence usually adopts a hybrid method. A DNA probe needs necessary chemical modifications such as element modification, fluorescence modification, increasing cost to a certain degree, and bringing tedious preprocessing for practical medical applications. Hence, an effective, sensitive, label-free hybrid analysis method is a growing demand in the field of DNA analysis. With the development of technique, there are many DNA hybridization analysis methods. Scattering analysis for DNA quantification has some unique features, such as high sensitivity, simple instrument, and no need to label. Combining the scattering technique with modern nanotechnology further promotes effective DNA analysis process. A resonance scattering technique in nucleic acid analysis and application will be further demonstrated.

11.1 Principle and method of nucleic acid light scattering spectrum

Nucleic acid is a biomacromolecule with a basic unit of nucleotide, where a nucleotide has a base, a pentose, and a phosphoric acid group. A nucleotide is a special secondary or high-level structured biomacromolecule through a linkage of phosphate diester bonds.

Nucleic acid has its characteristic absorption peak at 260 nm and can be detected with UV–Vis absorption. In addition, using other properties of nucleic acid, fluorescence [1] and electrochemistry methods [2] can be developed.

11.1.1 Structural features for nucleic acid used for light scattering analysis

Similar to the analysis of protein with light scattering method, nucleic acid analysis is based on its special structure and chemical properties, as well as on probe's structure and its charges.

https://doi.org/10.1515/9783110573138-011

First, as nucleic acid has a phosphate group, it can hydrolyze in neutral conditions to bring negative charges, then detection of DNA can be realized based on electrostatic interaction.

For example, hexadecyl trimethyl ammonium bromide (CTAB) and many cationic surfactants bind with nucleic acid through electrostatic interaction [3], forming nucleic acid-centered large particle aggregates or complexes. Compared with individual size of nucleic acid and probe, these aggregates have enhanced scattering signal, and the enhancement has a linear relation with the concentration of nucleic acid concentration. The nucleic acid scattering method can be built on this. However, these methods, on the basis of electrostatic interaction, have poor selectivity. For nucleic acid with negative charges, these methods cannot be selective. Thus, most scattering methods of nucleic acid electrostatic interaction cannot distinguish common nucleic acids such as fish sperm DNA (fsDNA), calf thymus DNA (ctDNA), yeastRNA, and so on. Later, researchers gradually changed their strategy to develop some new methods for nucleic acid test with specific sequence.

Secondly, nucleic acid has a special structure. DNA has two main chains, twining in right-hand direction with one core axis and parallel, forming double helix configuration in the opposite direction. DNA outward is the backbone connected by deoxyribose and phosphoric acid. Most DNA exists in double helix configuration with major groove and minor groove; once heated or disposed by alkaline, they get into single strand state. With the structural difference, a strategy can be designed for the identification of single/double strand nucleic acid. For example, some dyes such as acridine orange, ethidium bromide, and methylene blue [4] could inset double-stranded DNA groove structure, assembling on double-stranded DNA. As these dyes do not assemble on single-stranded DNA, there is no scattering signal difference, thus DNA of different shapes can be distinguished. Consider another example: spermine insets double-stranded DNA major groove as well as binds in-strand and across-strand with DNA base and phosphate groups, forming dense loop helical structure, bringing DNA condensation, thereby considerably enhancing DNA resonance light scattering [5]. Combined with nanotechnology, the nucleic acid object is no longer limited to ordinary nucleic acid such as fsDNA, ctDNA, yeastRNA, and so on, while special sequences in biology and clinical medicine, such as HIV, are focused. The single-stranded nucleic acid adsorbs on gold nanoparticle and carbon tubes, while double-stranded nucleic acid cannot adsorb; there are differences with regard to scattering signal intensity, and so nucleic acid can be analyzed.

11.1.2 Features of nucleic acid light scattering method

Apparently, nucleic acid light scattering analysis method is the main binding way of a probe and nucleic acid. As described in 5.5.2, small molecules and nucleic acid combines to phosphate backbone, base and pentose through irreversible covalent binding, reversible noncovalent bonding and shear bonding, while combines to electron-rich probes as carbon nanotubes, graphene through Van der Waals force, π-π stacking, and so on. Thus, different scattering enhancements are produced with various probes with distrinct mechanisms.

In the detection method, other analysis methods can be combined with light scattering analysis method to improve the detection stability and operability. For example, combining the magnetic separation technique with light scattering technique for multicomponent analysis improves the detection stability and reproducibility.

In general, the nucleic acid light scattering analysis method has considerable advantages compared with other methods, resulting in development in the field.

11.2 Nucleic acid light scattering probe

11.2.1 Cation light scattering probe

As nucleic acid has negative charges, organic small molecules with positive charges will have an electrostatic interaction with nucleic acid. There are a lot of related researches, including organic cation dye, cation metal complex, and so on.

11.2.1.1 Organic cation dye

With cationic porphyrin as a nucleic acid scattering spectral probe, Pasternack (1993) studied the aggregation of porphyrins on nucleic acid [6]. Pasternack found that porphyrin and its metal derivatives would have ordered stacking in nucleic acid molecules, forming long range chiral aggregates with obvious circular dichroism in porphyrin Soret band. The aggregate induces enhanced scattering signal, in direct proportion to the molecular concentration, pioneer in the applicaiton of resonance light scattering (RLS) in biological macromolecules aggregation. Since then, the RLS method for the interaction of organic dyes and nucleic acid has been widely studied.

The first report about nucleic acid analysis based on RLS method uses α, β, γ, δ-4(p-trimethylamine) porphyrin (TAPP) as the RLS probe [7]. TAPP interacts with ctDNA and fsDNA, and a rapid and sensitive acid quantitative analysis method is built based on the enhanced RLS signals caused by the aggregation of porphyrin on nucleic acid. In certain range of acidity and ion strength, there are two interaction modes between TAPP and nucleic acid. When their molar ratio is lower than 4:1, porphyrin has obvious hypochromic effect with reduced fluorescence intensity and enhanced RLS signal; when the ratio is higher than 4:1, the mechanisms for porphyrin-porphyrin and porphyrin-nucleic acid interactions change forming a new fluorescent complex.

Chen et al. used a similar principle based on the enhanced RLS signals caused by the interaction between ferriporphyrin (FeTPPCl) and nucleic acid, and developed various methods for highly sensitive quantitative analysis of different nucleic acid molecules. By comparing the scattering signals of porphyrin-nucleic acid complex and porphyrin-denatured nucleic acid complex, it was proven that porphyrin mainly reacts with duplex DNA through imbedding, inducing porphyrin for ordered aggregate [8].

If organic small molecules contain suitable conjugation groups, they can be embedded into a groove or bases. Owing to this, different structural dyes have different interaction mechanisms with nucleic acid. For example, Janus green B (JGB) binds the DNA backbone phosphate group by an electrostatic interaction, forming H-type long-distance assembly and enhancing the scattering signal. When the molar ratio of JGB and nucleic acid changes, they have different binding modes and JGB is in the bases with decreased scattering signals when the concentration of nucleic acid increase. By controlling JGB concentration, a DNA quantitative analysis platform can be built based on the linear relationship of signal and molecular concentration.

Ethidium bromide (EB) is a widely used cation dye that embeds into a helix structure and changes the system's scattering signal. Cheng et al. used a competitive interaction of ionic liquid and EB with nucleic acid to build a simple nucleic acid analysis method. As shown in Figure 11.1, EB embeds into a nucleic helix structure, significantly enhancing EB fluorescence and scattering signal. When nucleic acid is extracted with [Bmim]PF$_6$, the cation group [Bmim]$^+$ embeds into the DNA helix structure with P–O bond interaction, decreasing the base gap while changing DNA configuration and inhibiting EB and DNA embedding. It is hard for

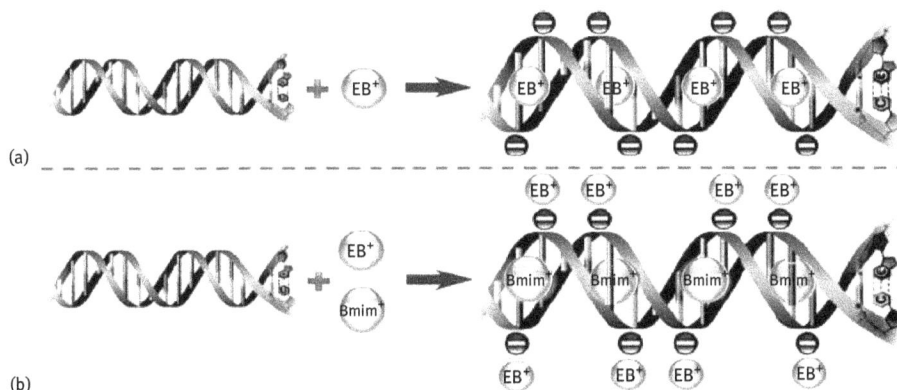

Figure 11.1: EB and nucleic acid interaction differences in an aqueous solution (a) and ionic liquid (b).

EB to insert into double strand DNA in the presence of ionic liquid, and EB adsorbs on DNA surface with monomer form through electrostatic and hydrogen bond. At this time, EB has strong absorbency and the inner-filter effect reduces system's RLS signal. With this phenomenon, we can extract human blood DNA in ionic liquid and direct quantitative analysis of DNA can be further realized [10].

Similar cation fluorescent dyes interacting with nucleic acid with insert, groove, or electrostatic effect and induce enhanced resonance scattering. Numerous RLS methods for nucleic acid analysis are built based on Alcan blue 8GX [11], reddish black B [12], pentamethoxy red [13], crystal purple [14], Victoria blue [15], thionine [16], phenosafranin [17], neutral red [18, 19], and Nile blue sulfate [20].

With the different interaction of single- or double-stranded DNA with dye, Long built a new type of DNA analysis method by the enhanced scattering signal caused by aggregation induced enhancement [21]. The authors used scattering spectra and transmission electron microscope to observe the interaction of single- and double-stranded DNA with quinone–imine type dyes, such as acridine yellow, neutral red, acridine orange, bright cresol blue, thionine, azure A, azure B, and methylene blue. They found that this type of dyes binds double-stranded DNA through grooving function and the binding affinity is closely related to dye concentration and GC bases content of DNA. Further study found that acridine yellow interaction with double-stranded DNA has the most significant signal enhancement. With enhanced signal, a label-free method can be built for DNA hybridization (Figure 11.2). The method has good selectivity and is able to effectively distinguish single-base, double-base, and triple-base mismatch, which shows a great promise for the detection of gene mutation.

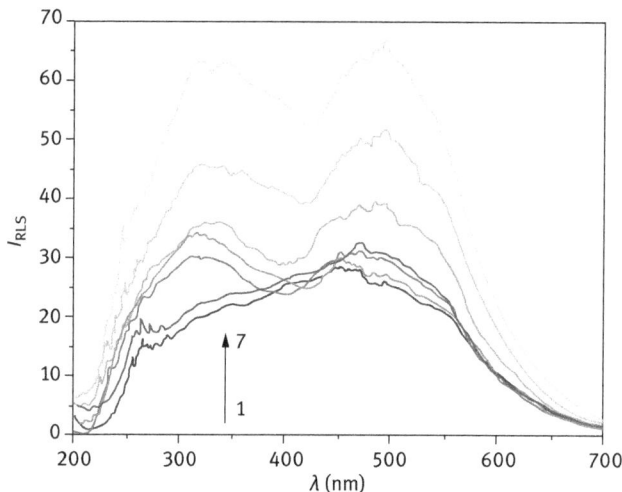

Figure 11.2: The light scattering spectra of single- and double-stranded DNA and acridine yellow interaction. 1. DNA single-stranded probe; 2. DNA single-stranded probe–target DNA; 3. acridine yellow; 4. acridine yellow-DNA single-stranded probe; 5–7. DNA single-stranded probe–target DNA-acridine yellow [22].

Drugs and macromolecule interaction could be used for illustration the drug target mechanism in vitro on the one hand, and on the other hand, it can be used to build a macromolecule analysis method. For example, cationic drugs have special interaction with DNA and caused an enhanced RLS signal, which can be used for nucleic acid analysis. Researches by Chen also indicated that nucleic acid significantly enhances the scattering of norfloxacin. SEM further confirms that the particles of norfloxacin increases in the presence of nucleic acid, and their interaction mechanism is related to the molecular structure of norfloxacin. Norfloxacin is vertical to DNA double-helix axis, binding with the grooving interaction, that is to say, they use DNA molecule as the template, where norfloxacin aggregates and the signal is enhanced [22]. In a slightly acidic condition, kanamycin has positive charges, while nucleic acid molecules have negative charges, and so kanamycin stacks on the nucleic acid surface by electrostatic interaction, producing an enhanced scattering signal [23]. Drugs perform long-distance assembly on the nucleic acid surface by groove reaction and electrostatic effect; meanwhile, molecules such as adriamycin amycin embed in DNA bases to produce a strong RLS signal [24].

In addition to the aforementioned drugs, some pesticides, such as imidacloprid, and bio-small molecules, such as histidine, can also be used for quantitative determination of DNA [25, 26].

11.2.1.2 Metal complex probe

After complex formation of metal ions with some ligands, the coordination ability of metal ions is not saturated; they could coordinate with nitrogen on nucleic acid bases. A metal complex may have positive charges to occur an electrostatic interaction with a phosphate group with negative charges.

For example, cobalt ion and cobalt reagent (5-Cl-PADAB) complex use nucleic acid as the template to perform stacking on its surface, bringing obvious changes to Co(II)/5-Cl-PADAB complex absorption spectrum and significant enhancement for the system's RLS signal (Figure 11.3). Its degree of enhancement is directly proportional to the concentration of nucleic acid [27]. However, their interaction mechanisms involve metal ion and nucleic acid are unclear [28, 29]. He group studied the interactions of four Ru complexes and ctDNA, and find that DNA could enhance system scattering to different degrees [30, 31].

11.2.1.3 Multicomplex light scattering probe

On the basis of previous studies, Guo introduced bromopyrogallol red (BPR) and TAPP into nucleic acid quantitative analysis. TAPP can perform ordered stacking on nucleic acid, thereby enhancing the scattering signal to a certain degree; BPR leads to the

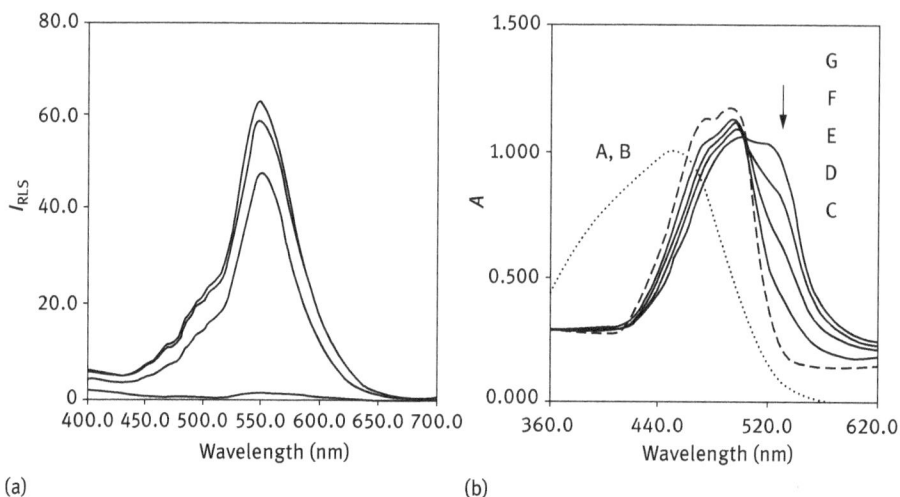

Figure 11.3: The RLS spectra and absorption spectra of nucleic acid and Co(II)/5-Cl-PADAB interaction [27]. (a) The scattering spectra, from top to bottom: fsDNA, ctDNA, yDNA; and (b) the absorption spectra: 5-Cl-PADAB(A), DNA + 5-Cl-PADAB(B), Co(II)+ 5-Cl-PADAB(C), Co(II)+ 5-Cl-PADAB +DNA (D–G).

formation of BRP–TAPP aggregate. The aggregate can easily perform a long-distance assembly on the nucleic acid surface, which is beneficial for DNA superhelical structure, and so the scattering signal is significantly enhanced when compared with the binary system; and the quantitative analysis sensitivity of nucleic acid is improved [32].

With surfactant, organic small molecules have synergistic effect as scattering probe. For that, we first need to discuss about the cationic surfactant and nucleic acid interaction. By dynamic light scattering, Cárdenas et al. found that the interaction between DNA and CTAB changes the DNA configuration from semiflexible crimp condition to a compacted sphere, thus significantly reduce its particles [33]. Marchetti et al. used similar methods to study the interaction mechnisms of CTAB, dodecyldimethylamine oxide (DDAO), and ctDNA, and they found that these surfactants can easily interact with nucleic acid for a soluble complex with configuration changing from free curl to spherical DNA. When the molar ratio of a surfactant and phosphate group of DNA is 1:1, the formation trend of a complex is maximum, and when the ratio is lower than that, free curl and spherical DNA coexist in the system [34]. Further study found that the compacted-sphere configuration DNA has better heat stability when compared with free-curly DNA; its thermal denaturation temperature is remarkably increased when compared with that of the free-curl configuration [35]. Through electrostatic interaction of negative charge DNA and cationic surfactant CTAB, the formed ionic associate has a strong RLS signal enhancement [36].

In the presence of surfactants, organic small molecular dyes as light scattering probes for nucleic acids also have synergistic effects. For example, Liu et al. used the synergistic effect between CTAB and morin dye for sensitive detection of nucleic acid. As morin water solubility is not good, it can first be extracted into CTAB micelle, thereby greatly enhancing microenvironment morin concentration and leading to the aggregation of morin in CTAB micelle with the result of enhancing the scattering signal. The introduction of nucleic acid changes the charge property of the system, where nucleic acid with negative charges binds CTAB with electrostatic interaction, further promoting morin self-assembly, and greatly enhancing its scattering signal (Figure 11.4), thus establishing a sensitive quantitative analysis method [37]. With similar principles, a brilliant green combination with CTAB can also be used for sensitive detection of ctDNA and fsDNA (Figure 11.4(B)) [38].

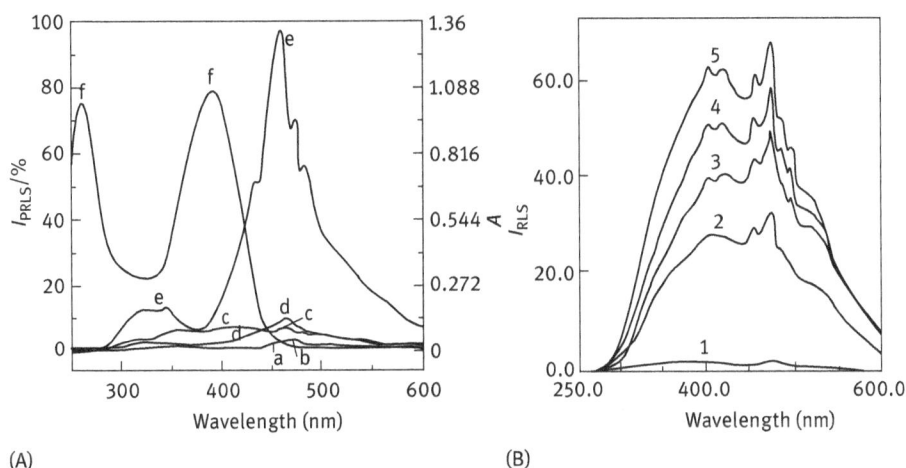

Figure 11.4: The RLS spectra of morin, brilliant green, and DNA. (A) (a) morin, (b) Morin–ctDNA, (c) CTAB–ctDNA, (d) morin–CTAB, (e) morin–CTAB–ctDNA; (a)–(e) scattering spectra, and (f) absorbance spectrum of morin [37]. (B) RLS spectra of BG, CTAB and ctDNA, [38] where 1–5 ctDNA concentration (mg/L) 0, 0.6,0.8, 1.0, 1.2; other conditions: BG: 2.0×10^{-5} mol/L, CTAB: 2.0×10^{-5} mol/L, pH: 11.58, ion strength: 0.02 M.

11.2.2 Nanoparticles light scattering probe

With continuous development in the field of nanotechnology, quantitative analysis of nucleic acid with nanoparticles has gradually become easier. A highly sensitive and rapid detection method can be built on the basis of the scattering signal changes of nanoparticles with and without target. So, Cheng used Fe nanoparticles with positive charges and nucleic acid with negative charges to develop a simple but effective DNA analysis method since the electrostatic interaction between nanoparticles and nucleic acid leads the aggregation of nanoparticles with the result of enhancing the scattering signals in the presence of nucleic acid (Figure 11.5) [39].

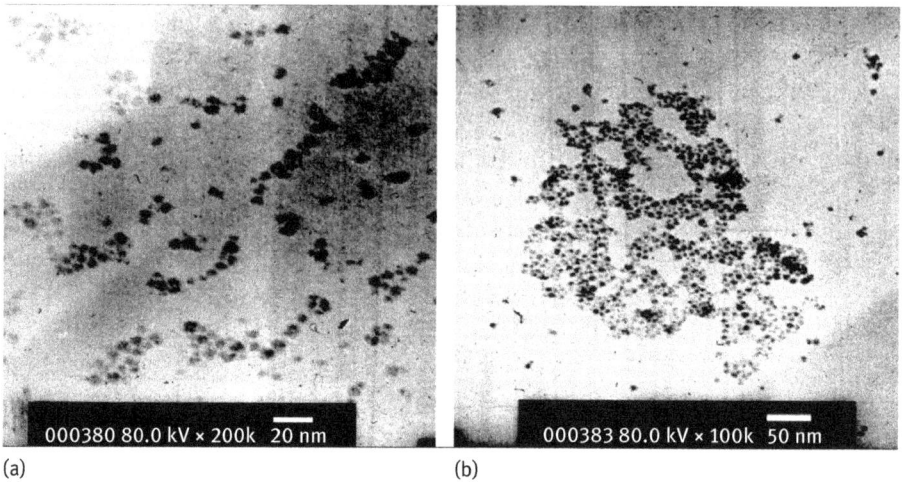

(a)   (b)

Figure 11.5: SEM imaging of DNA-induced iron nanoparticle aggregates. (a) Dispersed iron nanoparticles and (b) DNA-induced iron nanoparticle aggregate [40].

On the basis of a similar mechanism, Zou et al. successfully synthesized polyhedral oligomeric silsesquioxane (POSS) oligomer nanoparticle, that is to say, the electrostatic interaction of nanoparticles and nucleic acid leads to POSS aggregation, enhancing system's scattering signal. A quantitative detection platform for nucleic acid can be built on that [40].

11.3 Nucleic acid analysis based on the plasma resonnace light scattering of nobel metal nanoparticles

Hybridization of nucleic acid is a steady homologous or heterogeneous double-stranded molecule formation process through noncovalent bonding of complementary nucleotide sequences (DNA and DNA, DNA and RNA, and RNA and RNA), based on the Watson–Crick base pairing principle. The basic process is denaturation and renaturation. Hybridization of nucleic acid has biological significance and important applications for clinical molecular diagnosis of diseases.

As noble metal nanoparticle has plasma resonance absorbance and scattering; its light scattering has advantages for nucleic acid hybridization analysis, especially in visual analysis. DNA molecular design has far exceeded scopes of technique, technology, and science. Instead, the design of DNA has arrived at the stage of arbitrary and ingenuity. Hence, there is a new word can be associated with the DNA study, and it is

DNA art (or state of DNA art). In these studies, a stuffed or sandwich structure is the main idea, while metal nanoparticles usually serve as the signal reporter.

11.3.1 Gold nanoparticle

Currently, most used gold nanoparticles include gold nanosphere and gold nanorod. On the one hand, they can be easily synthesized with stable properties, and on the other hand, they have special properties. For gold nanosphere and nanorod, their two properties are mainly used: lights scattering signal enhancement before and after hybridization, plasma resonance absorption and scattering. Gold nanosphere shows more of the former property, while gold nanorod applications are more of the latter property, especially its plasma resonance properties. In this chapter, we will mainly discuss their light scattering signal changes before and after hybridization, while the properties of related plasma resonance scattering spectrum will be further discussed in the analysis of single nanoparticle.

When compared with the fluorescence method, the light scattering method has higher sensitivity and is better for small sample detection. For example, Bao et al. designed a group of experiments to perform systematic comparisons of the two methods. As shown in Figure 11.6, target gene is fixed on array surface, while probe DNA modified with biotin and Cy3 fluorescent dye at the same time, incubate array and biotin modified gold nanoparticles, then detect with fluorescent scanner in the white light, and detect the scattering signal through CCD imaging system. The results show that when the probe concentration is rather low, compared with that in the fluorescent

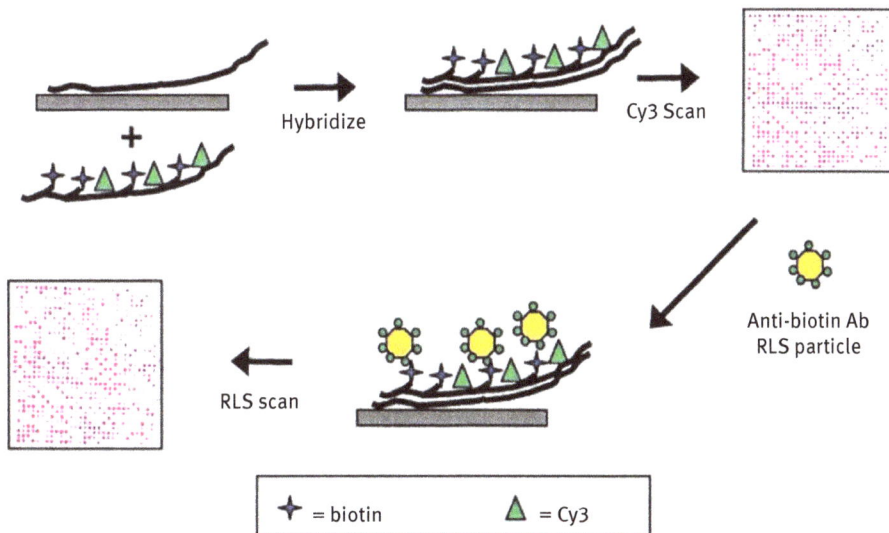

Figure 11.6: A diagram of the scattering method for DNA hybridization [41].

method, the light scattering method can detect more target genes, and so it has higher sensitivity, which is more beneficial when compared with the detection of low-concentration target gene [41].

DNA-modified gold nanoparticle serves as a light scattering probe for gold nanoparticle–DNA–gold nanoparticle sandwich structure. Based on the relation between the scattering intensity and the target concentration, a sensitive DNA analysis method can be built. As shown in Figure 11.7, two DNA probes corresponding to the target are individually modified on a gold nanoparticle surface. When the target gene exists, the helix structure is formed, leading to gold nanoparticle aggregate and formation of sandwich structure, thus the gold nanoparticle plasma resonance scattering signal is greatly enhanced. Qualitative and quantitative analysis of the target gene can be realized based on this. The method does not involve any tedious washing procedure, which can detect a target gene as low as 0.1 pmol/L. When compared with the gold nanoparticles colorimetric method, the sensitivity has improved at least by four times [42]. Du et al. further improved the method, fully exploiting the advantages of its selectivity for single-base mismatch and three-bases mismatch recognition, with important meaning to the study of single nucleotide polymorphism [43].

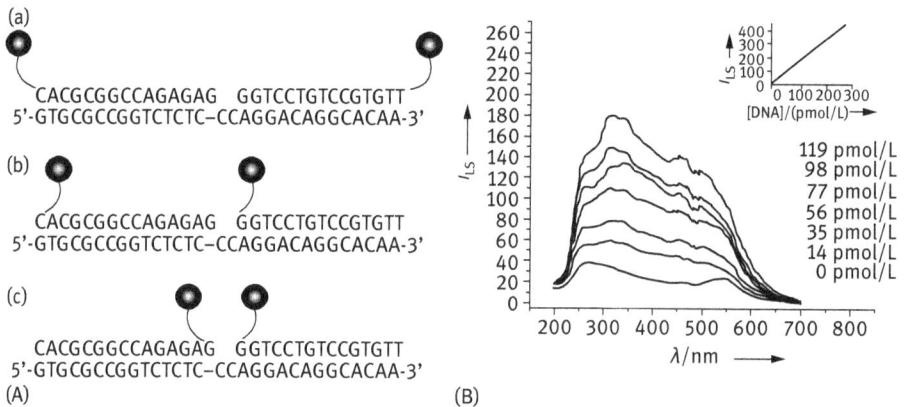

Figure 11.7: Surface plasma resonance scattering method for the detection of DNA hybridization [43]. (A) hybridization strategy and (B) scattering spectra.

Dai et al. adopted similar designs to build DNA analysis method with dynamic light scattering. As shown in Figure 11.8, with complementary DNA sequences, gold nanoparticles form aggregates, and their surface plasma resonance spectrum and particle diameter change considerably. By dynamic light scattering monitoring of particle change and target gene concentration relation, a simple and sensitive DNA analysis method is established. The method could effectively recognize single base mismatch [44].

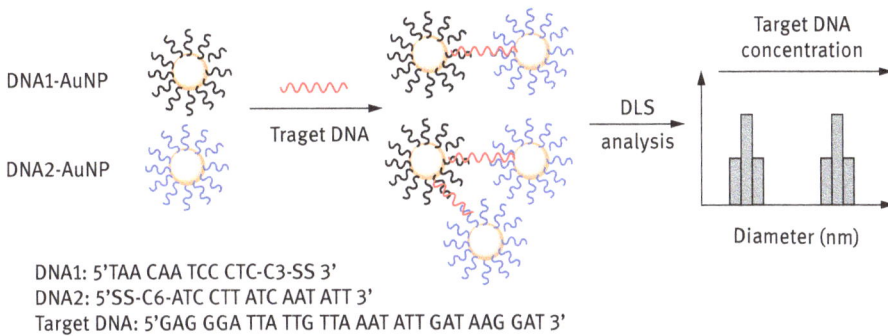

DNA1: 5'TAA CAA TCC CTC-C3-SS 3'
DNA2: 5'SS-C6-ATC CTT ATC AAT ATT 3'
Target DNA: 5'GAG GGA TTA TTG TTA AAT ATT GAT AAG GAT 3'

Figure 11.8: A diagram of dynamic light scattering for DNA hybridization [44].

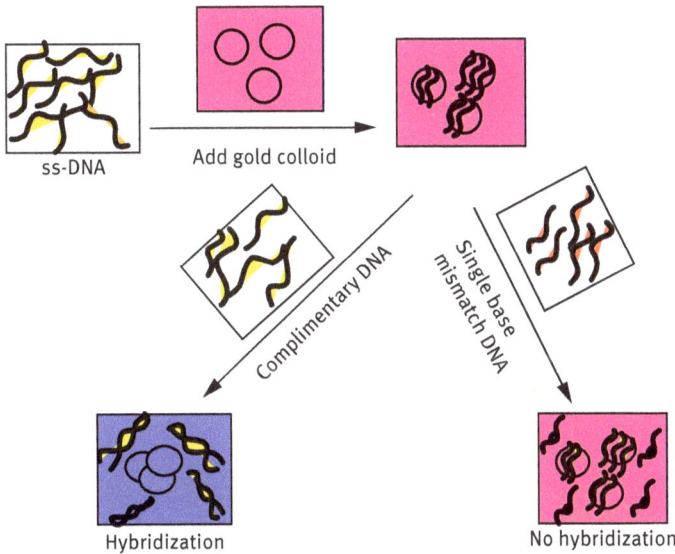

Figure 11.9: A diagram for DNA detection based on single- and double-stranded DNA and gold nanoparticles interaction induced different light scattering signal [45].

On the basis of interaction differences of single- and double-stranded DNA and noble metal nanoparticles, Ray built an ultrasensitive DNA analysis method. The mechanism is similar to that of DNA colorimetric analysis method by Mirkin' group; that is single-stranded DNA adsorbs to gold colloid surface through Van der Waals electrostatic force, enhancing repulsive force of nanoparticles. The electrostatic force is from the dipole interaction of base and gold nanoparticles, and the intensity depends on single-stranded DNA configuration and arrangement. With the target DNA, hybridization forms the DNA helix structure while the base is covered in DNA backbone, thereby reducing the dipole interaction. The repulsive force of the phosphate skeleton and gold nanoparticle surface citrate makes double-stranded DNA difficult to adsorb; with NaCl charge shield-ing effect, gold colloid aggregates with nonlinearly enhanced scattering signal, and DNA hybridization analysis method is built on this [45].

On the basis of a similar mechanism, Ray further used gold nanorod as the probe to establish a label-free recognition method for HIV. He first mixed a CTAB-covered gold nanorod and poly(sodium-p-styrenesulfonate) (PSS) solution. Through electrostatic interaction, PSS with negative charges was adsorbed on the surface of the gold nanorod. As single- and double-stranded DNA have different adsorption behavior on gold nanorod surface, with its different nonlinear scattering intensity before and after aggregation, an effective and sensitive method is built for qualitative and quanti-tative analysis of HIV gene (Figure 11.10). The study also made detailed theoretical discussion about gold nanorod enhancement, illustrating the theoretical basis of applications with the scattering method [46].

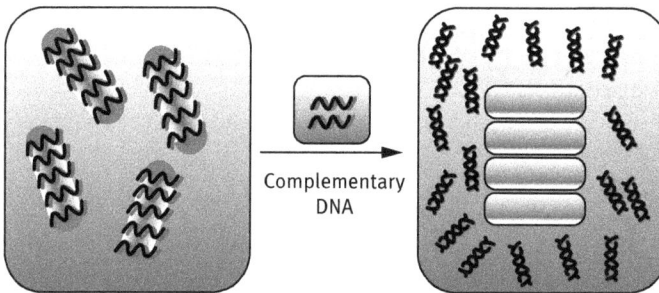

Figure 11.10: A diagram for DNA hybridization based on single- and double-stranded DNA and gold nanorods with adsorption differences of negative charges [46].

A CTAB-covered gold nanorod with positive charges can be used for DNA light scattering analysis by another mechanism, as discussed in Chapter 6 (Section 6.4; Figure 6.24). Base complementary pairing forms a double helix structure, with double-stranded DNA as the bridge, gold nanorods aggregate, surface plasma reso-nance scatteringsignal is greatly enhanced, and qualitative and quantitative testing of HIV gene section can be built on this [47].

11.3.2 Silver nanoparticle scattering probe

In fact, the application of noble metal nanoparticle probe could use enhanced scattering signals before and after aggregation for DNA analysis, and could use its plasma resonance scattering properties directly as a signal report unit.

For example, as the silver nanoparticles have a larger scattering section com-pared with other nanomaterials, it can be used on a gold nanoparticle surface to further enlarge scattering signal through silver staining; the silver coated gold nanoparticle serves as a good scattering probe and is widely used in sensitive detection of biomolecules, such as DNA and protein.

Li et al. constructed a solid phase light scattering analysis method as shown in Figure 11.11. They first modified the fourth-generation polyamidoamine (PAMAM) dendrimer on an aldehyde slide by the Schiff base reaction and then reduced with $NaBH_4$ to improve the immobilization stability of PAMAM; further, with glutaraldehyde (GA) activation, a three-dimensional glass substact is formed. An amino-modified probe DNA can be connected to the polymer surface by covalent bonding. When the target DNA exists, a part of double-stranded structure is formed. Another probe-modified gold nanoparticle is combined to three-dimensional polymer surface by hybridization, then silver staining is used to further enlarge the scattering signal, with Au@Ag complex nanoparticle as the light scattering probe, and a simple flat-bed scanner can be used to realize sensitive detection of target DNA [48].

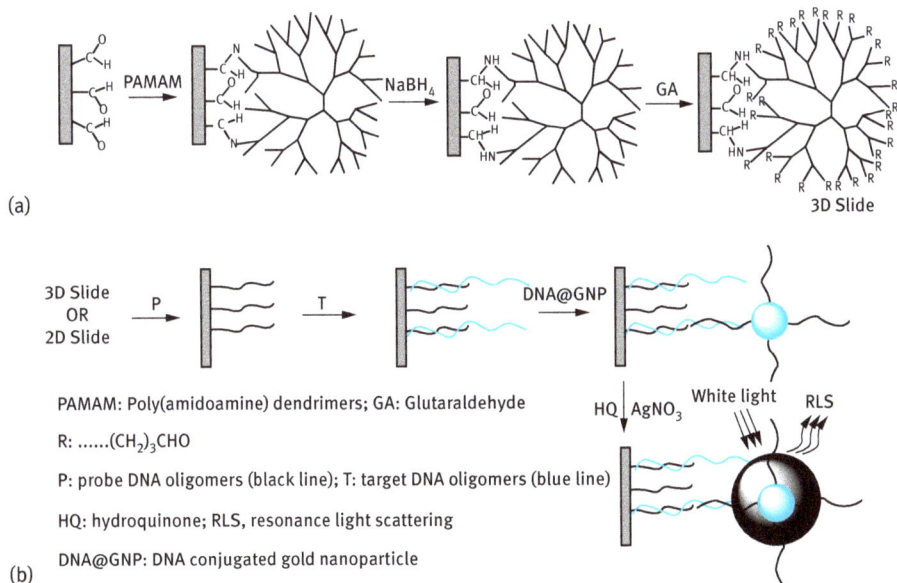

(a)

3D Slide

(b)

PAMAM: Poly(amidoamine) dendrimers; GA: Glutaraldehyde

R: ......(CH$_2$)$_3$CHO

P: probe DNA oligomers (black line); T: target DNA oligomers (blue line)

HQ: hydroquinone; RLS, resonance light scattering

DNA@GNP: DNA conjugated gold nanoparticle

Figure 11.11: PAMAM dendronized polymer-modified slide and light scattering method for the detection of DNA [48].

Xu et al. combined high sensitivity of the scattering method and separation properties of magnetic nanoparticles for high-sensitivity, high-selectivity analysis of DNA. As shown in Figure 11.12, a section of probe DNA is modified on the magnetic nanoparticle surface and another probe sequence is fixed on the gold nanoparticle surface. When the target DNA exists, the magnetic nanoparticle–DNA–gold

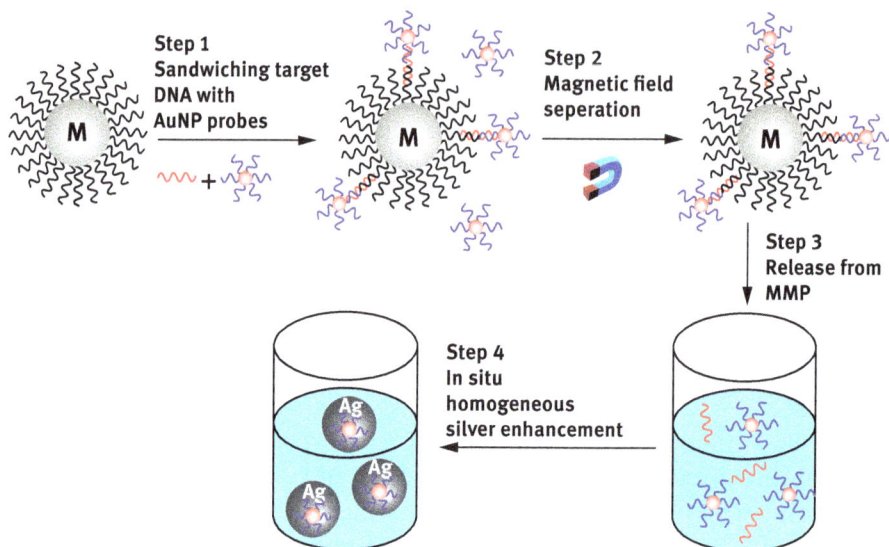

Figure 11.12: A diagram of DNA hybridization detection based on silver staining enlarged scattering signal [50].

nanoparticle sandwich structure is formed based on the principle of complementary base pairing. The magnetic separation is used to remove unbound gold nanoparticles, and heating denaturation is used to release gold nanoparticles from the sandwich structure. Silver staining is used to enlarge the metal nanoparticle scattering signal. On the basis of linear relationship of scattering intensity and target DNA, an effective DNA analysis sensing platform can be built [49].

11.3.3 Other nanoparticle light scattering probes

Not only noble nanoparticles can serve as a probe for the analysis of nucleic acid, but other nanoparticles, such as silicon nanoparticles and carbon nanotubes, also have been used for the analysis of DNA.

For example, Piao et al. used the light scattering signal caused by silicon nanoparticle aggregation for the development of human papillomavirus gene chip, as shown in Figure 11.13. First, the probe DNA is coupled to the chip and another probe DNA is modified on the silicon nanoparticle surface, and a chip–DNA silicon nanoparticle sandwich structure is formed with the existence of target DNA. With silicon nanoparticle's strong light scattering signal, visible detection of target with a microscopic imaging system or flat-bed scanner can be achieved. Connecting another silicon nanoparticle probe by a linker DNA can further enlarge scattering signal for the sensitive detection of the virus gene. Further study finds that silicon nanoparticles of different sizes greatly influence the sensitivity: the larger the

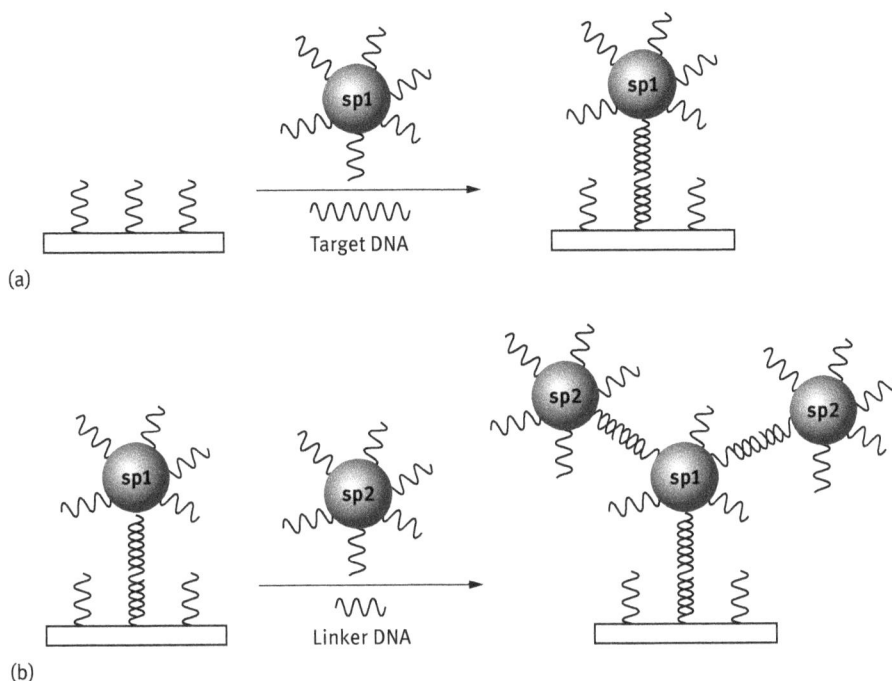

Figure 11.13: A diagram of a DNA sensor chip construction with silicon nanoparticle as the probe [50].

particle, the higher is the sensitivity. A 286 nm silicon nanoparticle can realize a visible detection of target DNA as low as 200 pmol/L.

With interaction differences of carbon nanotube and single- and double-strand DNA, and using carbon nanotube as probe, a label-free DNA analysis can be achieved. A carbon nanotube is composed of a five-membered ring and a six-membered ring with a number of delocalized π electrons on the surface. Single-stranded DNA has a free-curve configuration with all bases exposed, which adsorbs to the carbon nanotube surface by π electrons' conjugate action; double-stranded DNA is steady double-helix structure with bases covered in a skeleton, unable to adsorb to the carbon nanotube. Hence, on the basis of the interaction of carbon nanotube and different DNA configurations, many sensitive DNA fluorescent methods were built [51–53]. These methods have good selectivity and high sensitivity; however, the DNA probes need tedious fluorescent modification, complicating the experimental process, while the light scattering method compensates the aforementioned disadvantages to a certain degree. With carbon nanotube as the scattering probe, a highly sensitive, selective, label-free DNA analysis can be achieved.

For example, single stranded DNA wraps around carbon nanotube surface with higher stability, while dsDNA could not wrap along the sidewall of carbon nanotubes

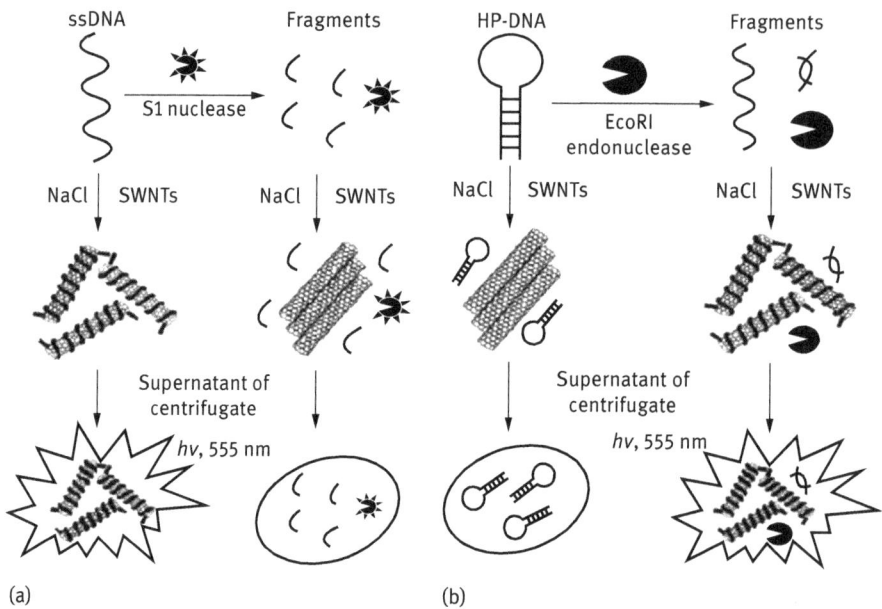

Figure 11.14: The diagram of DNA enzyme detection with a carbon nanotube adsorption difference with single- and double-stranded DNA [55].

and no π–π stacking interaction exists between dsDNA and carbon nanotubes. Thus, a simple, effective, label-free DNA analysis method can be built based on the different light scattering of supernatants before and after hybridization [54].

Similar principles can be used for the analysis of enzymes. As shown in Figure 11.14, a single-stranded DNA protects a carbon nanotube to steadily exist with certain NaCl concentration. While S1 nuclease cuts DNA, DNA pieces cannot wind on nanotube surface, NaCl charge shielding effect induces carbon nanotubes to aggregate, related enzyme analysis can be realized based on the changes of scattering of supernatants after centrifugation. We further design DNA into a stem-loop structure with a specific sequence. The structure focuses on double strain with weaker adsorption on the carbon nanotube surface. When the restriction enzyme EcoRI exists, corresponding spots of DNA strand are cut and the resulting single-stranded DNA adsorbs on carbon nanotube which leads to a different degree of aggregation of the carbon nanotube. Through monitoring of the scattering intensity of supernatants, enzyme analysis comes true [55]. Common sensing platforms can be developed for DNA analysis and enzyme detection, laying a good foundation for bioanalysis and further applications.

(Zhang L, Wang J)

References

[1] Wang Q, Wang W, Lei J, et al. Fluorescence quenching of carbon nitride nanosheet through its interaction with DNA for versatile fluorescence sensing. Anal. Chem., 2013, 85: 12182–81218.

[2] Drummond T G, Hill M G, Barton J K. Electrochemical DNA sensors. Nat. Biotech., 2003 21(10): 1192–1199

[3] Liu S P, Hu X L, Luo H Q, et al. Resonance Rayleigh scattering spectral properties and application of cation surfactants and nucleic acid reaction. Sci. China Ser. B, 2002, 32(1): 18–26.

[4] Huang X P, Li Y Q, Huang X, et al. Combining a loop-stem aptamer sequence with methylene blue: A simple assay for thrombin detection by resonance light scattering technique. RSC Adv., 2015, 5: 30268–30274.

[5] Wu Q H, Li Z P, Wang C, et al. Interaction of polyamines with deoxyribonucleic acid and selective determination of spermine by resonance light scattering technique. Chinese J. Anal. Chem., 2006, 34: 1299–1302.

[6] Pasternack R F, Bustamante C, Collings P J, et al. Porphyrin assemblies on DNA as studied by a resonance light-scattering technique. J. Am. Chem. Soc., 1993, 115: 5393–5399.

[7] Huang C Z, Li K A, Tong S Y. Determination of nucleic acids by a resonance light-scattering technique with $\alpha,\beta,\gamma,\delta$-tetrakis [4- (trimethylammoniumyl) phenyl]porphine. Anal. Chem., 1996, 68: 2259–2263.

[8] Chen Z, Liu J, Zhu L, et al. Development of a sensitive and rapid nucleic acid assay with tetraphenyl porphyrinatoiron chloride by a resonance light scattering technique. Luminescence, 2007, 22: 493–500.

[9] Huang C Z, Li Y F, Huang X H, et al. Interactions of Janus Green B with double stranded DNA and the determination of DNA based on the measurement of enhanced resonance light scattering. Analyst, 2000, 125(7): 1267–1272.

[10] Cheng D H, Chen X W, Wang J H, et al. An abnormal resonance light scattering arising from ionic-liquid/DNA/ethidium interactions. Chem.-Eur. J., 2007, 13: 4833–4839.

[11] Li Y F, Huang C Z, Huang X H, et al. Enhanced resonance light scattering of Alcian blue 8GX as an assay of DNA. Anal. Lett., 2001, 34: 1117–1132.

[12] Li Y F, Huang C Z, Li M. Study of the interaction of Azur B with DNA and the determination of DNA based on resonance light scattering measurements. Anal. Chim. Acta, 2002, 452: 285–294.

[13] Wei Q, Zhang H, Du B, et al. Sensitive determination of DNA by resonance light scattering with Pentamethoxyl Red. Microchim. Acta, 2005, 151: 59–65.

[14] Zhang W, Xu H, Wu S, et al. Determination of nucleic acids with Crystal Violet by a resonance light-scattering technique. Analyst, 2001, 126: 513–517.

[15] Su J S, Chen X M, Long Y F, et al. Resonance light scattering study on the interaction of Victory blue-B with deoxyribonucleic acid and application to DNA assay. Spectrosc. Spectr. Anal., 2004, 24: 37–40.

[16] Li Y F, Huang C Z, Tan K J. Interaction of thionine with DNA and the method for the determination of deoxyribonucleic acid based on intensity ratio of light-scattering and fluorescence emissions. Chinese J. Anal. Chem., 2008, 36: 529–532.

[17] Jielili D M L, Huang C Z. Resonance light-scattering study on the interactions of phenosafranine with deoxyribonucleic acid and the light-scattering determination of trace deoxyribonucleic acid. Chinese J. Anal. Chem., 1999, 27: 1204–1207.

[18] Huang C Z, Li Y F, Huang X H, et al. Spectra of the supramolecular interaction of neutral red with double-stranded DNA in a neutral medium. Acta Phys.-Chim. Sin., 2001, 17: 222–228.

[19] Huang C Z, Li Y F, Feng P, et al. The aggregation of neutral red encouraged by deoxyribonucleic acids and its depending factors. Chinese J. Anal. Chem., 2000, 28, 682–686.

[20] Huang C Z, Li Y F, Li N B, et al. Long range assembly of nile blue sulphate on the molecular surfaces of nucleic acids and the determination of nucleic acids by triple wavelength resonance light-scattering technique. Chinese J. Anal. Chem., 1999, 27: 1241–1247.

[21] Long Y F, Huang C Z, Li Y F. Hybridization detection of DNA by measuring organic small molecule amplified resonance light scattering signals. J. Phys. Chem. B, 2007, 111: 4535–4538.

[22] Chen Z, Zhang T, Chen X, et al. Interactions of norfloxacin with DNA and determination of DNA at nanogram levels based on the measurement of enhanced resonance light scattering. Microchim. Acta, 2007, 157: 107–112.

[23] Long X, Miao Q, Bi S, et al. Resonance Rayleigh scattering method for the recognition and determination of double-stranded DNA using amikacin. Talanta, 2004, 64: 366–372.

[24] Feng S, Li Z P, Zhang S H, et al. Study on interaction of adriamycin and DNA with a resonance light scattering technique and its DNA analytical application. Spectrosc. Spectr. Anal., 2006, 26: 707–710.

[25] Jia G, Wang P, Qiu J, et al. Determination of DNA with imidacloprid by a resonance light scattering technique at nanogram levels and its application. Anal. Lett., 2004, 37: 1339–1354.

[26] Du X, Sasaki S, Nakamura H, et al. The determination of DNA based on light-scattering of a complex formed with histone. Talanta, 2001, 55: 93–98.

[27] Huang C Z, Li K A, Tong S Y. Determination of nanograms of nucleic acids by their enhancement effect on the resonance light scattering of the cobalt(II)/4-[(5-chloro-2-pyridyl)azo]-1,3-diaminobenzene complex. Anal. Chem., 1997, 69: 514–520.

[28] Yang C X, Li Y F, Feng P, et al. Study on the interaction of aluminum ion with deoxyribonucleic acid by resonance light scattering measurement. Chinese J. Anal. Chem., 2002, 30: 473–477.

[29] Li Z P, Li Y K, Wang Y C. Study of the interaction of hexa-amine cobalt (III) ion with DNA by a resonance light scattering technique and its analytical application. Luminescence, 2005, 20: 282–286.

[30] Chen F, Huang J, Ai X, et al. Determination of DNA by Rayleigh light scattering enhancement of molecular "light switches". Analyst, 2003, 128: 1462–1466.

[31] Huang J, Chen F, He Z. A resonance light scattering method for determination of DNA using Ru(bpy)$_2$PIP(V)$^{2+}$. Microchim. Acta, 2007, 157: 181–187.

[32] Guo Z X, Li L, Shen H X, et al. Bromopyrogallol red enhanced resonance light-scattering spectroscopic determination of DNA with 5,10,15,20-tetrakis [4-(trimethylammoniumyl)phenyl] porphine. Anal. Chim. Acta, 1999, 379: 45–51.

[33] Cárdenas M, Schillen K, Nylander T, et al. DNA compaction by cationic surfactant in solution and at polystyrene particle solution interfaces: A dynamic light scattering study. Phys. Chem. Chem. Phys., 2004, 6: 1603–1607.

[34] Marchetti S, Onori G, Cametti C. DNA condensation induced by cationic surfactant: A viscosimetry and dynamic light scattering study. J. Phys. Chem. B, 2005, 109: 3676–3680.

[35] Marchetti S, Onori G, Cametti C. Calorimetric and dynamic light-scattering investigation of cationic surfactant-DNA complexes. J. Phys. Chem. B, 2006, 110: 24761–24765.

[36] Li Y F, Shu W Q, Feng P, et al. Determination of DNA with cetyltrimethylammonium bromide by the measurement of resonance light scattering, Anal. Sci., 2001, 17: 693–696.

[37] Liu R, Yang J, Wu X, et al. Interaction of morin-cetyltrimethylammonium bromide with nucleic acids and determination of nucleic acids at nanograms per milliliter levels based on the enhancement of preresonance light scattering. Analyst, 2001, 126: 1367–1371.

[38] Huang X H, Shu W Q, Li Y F, et al. The sensitizing effect of cetyltrimethylammonium bromide on the resonance light-scattering enhancement between the interaction of deoxyribonucleic acids and brilliant green. Chinese J. Anal. Chem., 2001, 29, 271–275.

[39] Cheng Y, Li Z Su Y, et al. Ferric nanoparticle-based resonance light scattering determination of DNA at nanogram levels. Talanta, 2007, 71: 1757–1761.

[40] Zou Q C, Yan Q J, Song G W, et al. Detection of DNA using cationic polyhedral oligomeric silsesquioxane nanoparticles as the probe by resonance light scattering technique. Biosens. Bioelectron., 2007, 22: 1461–1465.

[41] Bao P, Frutos A G, Greef C, et al. High-sensitivity detection of DNA hybridization on microarrays using resonance light scattering. Anal. Chem., 2002, 74: 1792–1797.

[42] Elghanian R, Storhoff J J, Mucic R. C, et al. Selective colorimetric detection of polynucleotides based on the distance-dependent optical properties of gold nanoparticles. Science, 1997, 277: 1078–1081.

[43] Du B A, Li Z P, Liu C H. One-step homogeneous detection of DNA hybridization with gold nanoparticle probes by using a linear light-scattering technique. Angew. Chem. Int. Ed., 2006, 45: 8022–8025.

[44] Dai Q, Liu X, Coutts J, et al. A one-step highly sensitive method for DNA detection using dynamic light scattering. J. Am. Chem. Soc., 2008, 130: 8138–8139.

[45] Ray P C. Diagnostics of single base-mismatch DNA hybridization on gold nanoparticles by using the hyper-Rayleigh scattering technique. Angew. Chem. Int. Ed., 2006, 45: 1151–1154.

[46] Darbha G K, Rai U S, Singh A K, et al. Gold-nanorod-based sensing of sequence specific HIV-1 virus DNA by using hyper-Rayleigh scattering spectroscop. Chem. Eur. J., 2008, 14: 3896–3903.

[47] He W, Huang C Z, Li Y F, et al. One-step label-free optical genosensing system for sequence-specific DNA related to the human immunodeficiency virus based on the measurements of light scattering signals of gold nanorods. Anal. Chem., 2008, 80: 8424–8430.

[48] Li X, Gao J, Liu D, et al. Developing oligonucleotide microarray-based resonance light scattering assay for DNA detection on the PAMAM dendrimer modified surface. Anal. Methods, 2010, 2: 1008–1012.

[49] Xu X, Georganopoulou D G, Hill H D, et al. Homogeneous detection of nucleic acids based upon the light scattering properties of silver-coated nanoparticle probes. Anal. Chem., 2007, 79: 6650–6654.

[50] Piao J Y, Park E H, Choi K, et al. Direct visual detection of DNA based on the light scattering of silica nanoparticles on a human papillomavirus DNA chip. Talanta, 2009, 80: 967–973.

[51] He S, Song B, Li D, et al. A graphene nanoprobe for rapid, sensitive, and multicolor fluorescent DNA analysis. Adv. Funct. Mater., 2010, 20: 453–459.

[52] Lu C H, Li J, Liu J J, et al. Increasing the sensitivity and single-base mismatch selectivity of the molecular beacon using graphene oxide as the "nanoquencher". Chem. Eur. J., 2010, 16: 4889–4894.

[53] Lu C H, Yang H H, Zhu C L, et al. A graphene platform for sensing biomolecules. Angew. Chem. Int. Ed., 2009, 48: 4785–4787.

[54] Zhang L, Huang C Z, Li Y F, et al. Label-free detection of sequence-specific DNA with multiwalled carbon nanotubes and their light scattering signals. J. Phys. Chem. B, 2008, 112: 7120–7122.

[55] Zhao C, Qu K, Song Y, et al. A universal, label-free, and sensitive optical enzyme-sensing system for nuclease and methyltransferase activity based on light scattering of carbon nano-tubes. Adv. Funct. Mater., 2011, 21: 583–590.

Jian Ling, Yun Fei Long

12 Light scattering spectrometry of proteins

Compared with light scattering analysis of inorganic ions, the analysis of bioma-cromolecules, including nucleic acids, proteins, and carbohydrates has its unique features. Firstly, inorganic ions are small, whereas biomacromolecules are large. Secondly, inorganic ions are highly polarized in water that they have the coordination or hydrogen bonding interaction with water or hydroxyl; while, biomacromo-lecules has not only hydrogen bonding but also hydrophobic interaction with their surroundings. So, RLS method for biomacromolecules analysis has its own features compared to that of inorganic ions. Although nucleic acids and proteins are both important biomacromolecules in RLS analysis, the structure and property of them are completely different. Thus, the strategy for the establishment of RLS method for protein analysis is different from that of nucleic acids.

12.1 Principles of protein analysis by light scattering

Protein is a key component in a living body, participating in body construction and adjustment of metabolism, which are closely related to all living activities. Protein is also directly expressed in various signs and conditions of life; the content of different proteins directly reflects health conditions of the living body. In a clinical examination, protein analysis and detection is important evidence of early stage diagnosis of diseases. So, protein detection has always been one of the frontier issues that are related to developments in life sciences.

Proteins are important biomacromolecules with a specific three-dimensional structure, consisting of one or more long chains of amino acid residues connected by peptide linkage. The amino acids residues in proteins have characteristic spectral properties. For example, tyrosine, phenylalanine, and tryptophan of proteins have characteristic absorption peaks in the UV region; tyrosine and tryptophan have stronger fluorescence emission with ultraviolet excitation. As most proteins have these amino acids, proteins can be detected by the UV-visible absorption spectroscopy and fluorescence spectroscopy. In addition, based on other properties of proteins, electrochemistry, chemiluminescence, and other methods can be used for protein analysis.

12.1.1 Structural features of protein for light scattering analysis

To establish the analysis of a highly sensitive protein light scattering, two aspects of proteins and probe should be considered. In terms of protein, it has a special structure

https://doi.org/10.1515/9783110573138-012

and a large molecular weight, diversified chemical properties, and special amphiphilic properties; it can be simply analyzed as follows:

– First, protein is a biomacromolecule made up of units of amino acids with high molecular weight. The surface of proteins made up of amino acids has many dissociated groups, such as phenolic hydroxyl, amino, carboxyl, imidazolyl, and so on that brings an amphiprotic property to the protein. Thus, the same as amino acids, protein carries different electrical charge at different pH environment. There is a specific pH, which the molecule carries no net electrical charge or is electrically neutral in the statistical mean, called the isoelectric point (pI). Different proteins have different amino acid components, and thus the isoelectric points are different. So, it is easy to change the charge carries of proteins to positive, negative, or no net charge by adjusting the surrounding pH.

– In addition, the protein surface has many hydrophilic groups, forming one layer of hydration shell on the protein surface. The shell makes protein molecules unable to get closer; they are dispersed to make the protein solution steady.

– Finally, a protein is formed by the peptide linkage connection in amino acid units, so it has many dissociated groups and amphipathic characteristics, forming a double electrical layer with its surrounding opposite ions at certain acidity. The hydration shell and double electrical layer of a protein surface makes the protein solution a hydrophilic colloidal solution.

12.1.2 Characteristics of protein analysis by light scattering

As proteins have high molecular weights and adjustable electrical properties, many hydrophobic regions and molecular recognition region, together with functional groups of high chemical activities, such as thiol the principle of light scattering method for protein analysis is based on the electrostatic attraction and molecular recognition.

12.1.2.1 Electrostatic attraction

Based on the electrical charge properties of proteins, light scattering probes, such as dyes, small molecules, and nanoparticles at suitable pH, can interact with proteins by electrostatic attraction or other chemical bonding. The interaction produces protein-centered large aggregates or complexes that have great light scattering signal enhancement. If the signal enhancement has a linear relation with a concentration, a light scattering method for protein can be established.

Based on the principle of protein light scattering method, the binary system, which directly based on the electrostatic attraction of the probe and target protein to enhance the signal of the system, is commonly used. This method is simple and rapid. Light scattering probes carried electrical charge opposite to target protein at a certain pH environment can be chosen as probes for protein analysis. In addition, the protein light scattering method is built on the basis of interaction of the protein–probe ternary system. For example, in a neutral environment, bovine serum albumin (BSA) and human serum albumin (HSA) have negative charges. When an anionic dye probe is used, the cationic surfactant needs to be added as a bridge to form a protein–-anionic dye-cationic surfactant ternary complex, enhancing the system scattering signal [1].

These protein light scattering methods based on electrostatic attraction have a problem of poor selectivity. For proteins with similar electrical charge properties, these methods cannot distinguish them. Most protein light scattering methods cannot distinguish common proteins, such as BSA, HSA, and immunoglobulin. However, these are some methods that are able to distinguish different proteins through selective choice of special probes or unique light scattering methods.

For example, with special interaction of rare earth metal ions and HSA and the coordination interaction of sodium dodecylbenzene sulphonate (SDBS), a protein–thorium complex can be extracted to the liquid–liquid interface of water–carbon tetrachloride. With a liquid-liquid interface total internal reflection light scattering device, the light scattering signal of the complex from the liquid-liquid interface can be selectively detected, and thus, HSA can be highly selective detected by light scattering method [2]. For other proteins, such as immune globulin in water solution, it cannot be detected as it cannot be extracted in the liquid–liquid interface. This method does not need separate pretreatment for other proteins in a sample; for this, HSA quantitative detection can be performed directly.

12.1.2.2 Molecular recognition

Molecular recognition has a high specificity, where immune response is one typical example. Research field expands with the development of light scattering technology, from normal proteins such as HSA and BSA to lysozyme [3–5], thrombin [6, 7], prion [8, 9], insulin [10], and other important proteins. In terms of working mechanism, aptamer is used as a recognition unit that greatly improves the selectivity of protein analysis.

For detection methods, other analytical methods can be combined with light scattering analysis methods to improve protein detection sensitivity, selectivity, and easy operation. For example, combining flowing injection with light scattering technology, protein detection can be automatically realized with improved stability and repeatability [4]. In general, compared with other optical analysis methods, protein light scattering analysis method is rapid, simple, and sensitive, thus gaining a good detection in the development of protein light scattering.

12.2 Light scattering probes for protein analysis

With the complexity of protein structure, there are many types of probes that can be used for protein analysis. In general, there are small organic molecular dyes, nano-materials, surfactants, and ionic associations. Till now, there are more applications of small molecular dyes in protein analysis, whereas nanoparticles probes have wide prospects in protein analysis.

12.2.1 Small organic molecular dye probes

In Section 5.5.3, we have discussed that in acidic conditions, proteins possess positive charges, which are able to form large complexes with organic dye molecules with negative charge. The small organic molecular dyes are actually probes that are initially used for protein detection. The light scattering signal of dye molecules is weak, however, when a dye molecule aggregates with proteins through electrostatic and hydrophobic interaction or accumulates on the protein surface, a dramatic RLS signal enhancement can be observed.

Light scattering study of dye molecule aggregates started in 1960s, however, the study of interaction between dyes and proteins through covalent or electrostatic effect that form large molecular aggregates and the applications for proteins determination became popular at the beginning of 20th century. That is the time that light scattering technique was introduced in biochemical analysis.

Small organic molecular probes with strong dissociative functional groups, such as sulfonic acid groups and carboxyl groups have negative charges in acidic medium, whereas most proteins have positive charges. An organic dye–protein complex can be formed with electrostatic forces. So, the interaction of these organic dyes and proteins is based on electrostatic interaction. With an optimum pH, dye molecules form dye molecule aggregates with the "skeleton" of the protein through electrostatic attraction, producing a strong RLS enhancement signal.

At certain concentration of dye, the protein concentration determines the aggregate quantity, which decides the degree of signal enhancement. Based on the detected enhanced signal with an ordinary fluorescence spectrophotometer, the protein quantitative analysis method can be built. Compared with ordinary UV-visible absorption and fluorescence spectrometry, this method has improved its sensitivity from 10 to 1,000 times. It needs to be demonstrated that protein charges that depend on its own amino acid sequence, there is always a suitable pH range for electrostatic interaction of negatively charged small organic molecules and proteins.

For example, the use of triaromatic methane-like dye chlorophenol red when combines with a protein at pH 4.0, the light scattering signal is greatly enhanced and the protein detection method is built on this. Linear range of BSA and HSA is 0–1.0 mg/L and 0–0.75 mg/L, respectively, and the detection limit is 0.020 mg/L

[11]. When Fast green FCF is used for BSA and HSA, pepsin and chymotrypsin detection, the detection limit is 4.54, 0.6, 22.8, and 4.32 µg/L [12]; at pH4.0, Solochrome cyanine R (SCR) is used for 400 nm wavelength scattering signal enhancement for BSA detection, linear range is 0–5.0 mg/L with a detection limit of 4.44×10^{-2} mg/L [13]. In a buffer solution of pH 3.54, light analysis method based on chromazurine S (CAS)-protein light scattering enhancement has sensitivity 50 times higher than that of Coomassie brilliant blue [14]. If tetraiodophenol sulfonphthalein (TIPSP) is used to detect HSA, the linear range is 0.34–12.24 mg/L [15]. In a weak acidic condition, the detection ranges for BSA in proteins and Chromazol KS (CALKS), Acid Chrome Dark Blue (ACDB), Chrome Blue SE (CBSE), Acid Chrome Blue K (ACBK), Chlorophosphonazo I (CPAI), Arsennazo I (AAI) and Chromotrope 2R (CT2R), and other monoazo dye signals are 0.010–0.030mg/L; the sensitivity range in sequence is CALKS>AAI>CPAI>ACBK> CT2R>ACDB> CBSE [16].

Li studied that in an acidic medium with pH 0.6–2.0, the enhanced light scattering signal for proteins and azo dyes, such as orange red G, methyl orange, methyl red, and orange yellow G for the detection of BSA, HSA, and alpha-chymotrypsin (Chy), detection limits are 2.6×10^{-3}, 3.4×10^{-3}, and 7.1×10^{-3} mg/L, respectively, where orange red G has the highest sensitivity [17]. Use of 3-(4-sulfophenylazo)-4, 5-dihydroxy-2, 7-naphthalene disulfonic acid and protein-interacted signal enhancement for BSA detection, the detection linear range is 0.125–14.9 mg/L [18]. If Eriochrome Black T (EBT) [20] is used, the detection limit can be as low as 0.039 mg/L. Li also studied that for the interaction of m-acetylchlorophosphonazo (CPA-mA) and protein, HSA detection linear range is 0.5–35.0 mg/L with a detection limit of 0.104 mg/L [21]; if m-carboxychlorophosphonazo (CPAmK) is used [22], the HSA detection linear range is 0.25–40.0 mg/L.

In a study of the interaction of Fast Red VR (FRV)–protein, the authors [23,24] have found that at 287 nm wavelength, the detection limit of BSA, HSA, and IgG is 0.025 mg/L. However, if Ponceau G (PG) [25,26] combines with protein, the detection limit is also 0.025 mg/L. As shown in Figure 12.1, nonionic surfactant alkylphenol polyoxyethylene (OP) at pH 2.35, RY-protein light scattering enhancement for BSA and HEM detection is used, the limit can be as low as 1.04×10^{-2} and 1.14×10^{-2} mg/L, respectively [27]. Similarly, in BR buffer solution of pH 4.10, beryllium interacts with protein only when after weak scattering signal there will be strong light scattering at 370 nm wavelength; the light scattering analysis of this method is simple, rapid, and can be used for protein detection in HSA, with consistent results compared to the classic Coomassie brilliant blue staining method [28].

Other small organic molecular probes such as titanium cyanine, including tetrasulfonic aluminum titanium (AlS_4Pc) [29] and tetrasulfonic manganese titanium (MnTSPc) [30] have negative charges in certain conditions, creating an enhanced scattering signal through electrostatic interaction with proteins. Hydroxy anthraquinone dyes, such as alizarin red S (ARS) [31,32] and natural pigment quercetin (QT), are also applied in protein light scattering analysis. Especially at pH 3.6, interaction of ARS

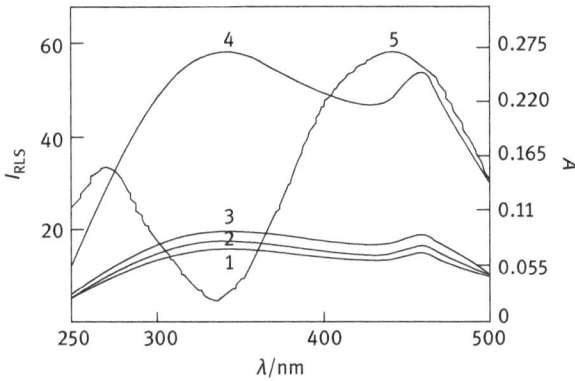

Figure 12.1: Interaction of an organic dye and protein with a nonionic surfactant. Scattering spectra (1–4): 1, RY; 2, RY–OP; 3, RY–BSA; 4, RY–OP–BSA. Absorption spectrum, 5, RY. Conditions: RY: 2.2×10^{-5} mol/L; OP, 3.0×10^{-5} mol/L; BSA: 2 µg/mL; pH 2.35.

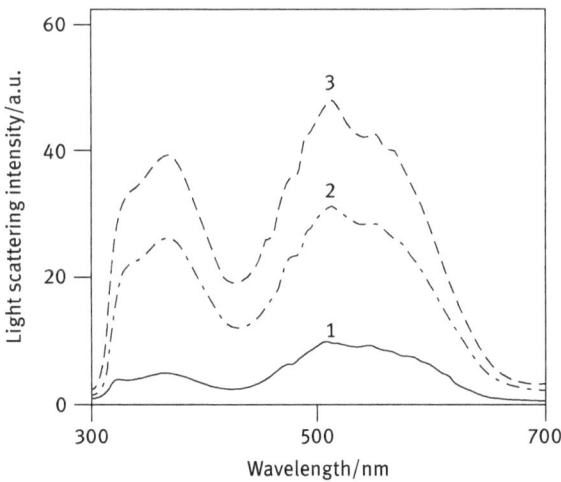

Figure 12.2: ARS and protein interaction scattering spectra. 1. ARS (2.0×10^{-4} mol/L) 2. ARS (2.0×10^{-4} mol/L)+BSA (5.0 µg/mL) 3. ARS (2.0×10^{-4} mol/L)+BSA (10 µg/mL).

and protein greatly improves the backward light scattering signal (Figure 12.2). The detection limit of BSA and HSA can be as low as 9.59×10^{-3} and 9.51×10^{-3} mg/L, respectively [32].

12.2.2 Nano probes

Nanoparticles are a new type of light scattering spectral probe with the development of nanotechnology. Nanoparticles have large specific area, which are able to adsorb electric components to its surface to form a charged layer, whereas dispersed particles exist steadily in a medium with colloidal form. When there is a protein carries charge opposite to the nanoparticles in a medium, they would attract together to form lager aggregates. Small-sized nanoparticles usually have weaker light scattering signals, whereas aggregates have higher scattering cross section, producing a obvious enhancement light scattering signal.

There have been many reports on detection of proteins by nanoparticles. Their principles are similar. First, colloidal solution of nanoparticles and proteins with opposite charges at an optimum pH environment are mixed. And then, they form large aggregates by electrostatic attraction and produce enhanced scattering signal. As a kind of light scattering probe, nanoparticles are easier to modify compared with ordinary small organic molecular dyes. Even for the same kind of nanoparticle, different modifications can be made through different modification methods for the determination and analysis of different types of proteins.

Currently, nanoparticle scattering probes for protein detection mainly include nanoparticles of metal and their compounds, such as gold, silver, iron, copper oxide, zinc sulfide, cadmium selenide, silver sulfide, and cadmium sulfide [33_39]. In addition, there are polymer nanoparticles, such as polystyrene, polyvinyl alcohol ketogenic particles [40–42], and some special materials, such as fullerenes and carbon nanotubes [43]. The principle of protein analysis with nanoparticle probe has a similar principle with small organic molecules. Nanoparticles that are modified with a charged molecule or group, form aggregates through electrostatic interaction with proteins [44].

With the development of nucleic acids, aptamers for different targets are gradually selected for protein detection. For example, Figures 12.3 and 12.4 exhibit the application of aptamer for lysozyme [4] and thrombin detection [6].

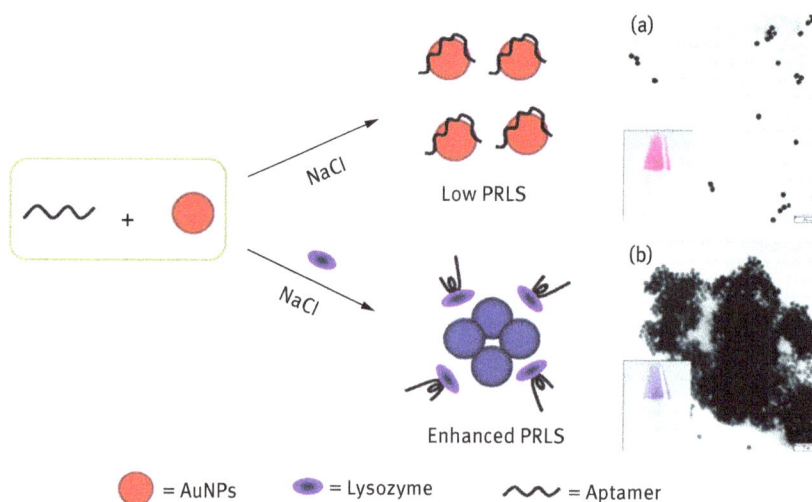

Figure 12.3: Diagram of aptamer application for lysozyme detection [4].

As an alkaline enzyme that can hydrolyze mucopolysaccharide in a pathogenic bacterium, lysozyme has antibacterial, anti-inflammatory, and antivirus activities. Lysozyme aptamer is a single-stranded DNA that can adsorb on the surface of gold nanoparticles to improve the stability of the gold nanoparticles. When NaCl exists, gold nanoparticles do not aggregate, thus they present a weaker scattering signal. When lysozyme exists, aptamer changes from a single-strand to a G-Quadruplex structure. At this time, aptamer no longer adsorbs on the surface of gold nanoparticle, and thus the gold nanoparticles aggregate in the presence of NaCl to produce stronger scattering light. By comparing the changes of light scattering intensity, a method for lysozyme detection can be establish [4].

Thrombin is an enzyme with good biological activity that plays important role in cascade reactions of anticoagulation and coagulation of blood. Figure 12.4 is a scheme of aptamer application for thrombin detection. In this method, a thrombin aptamer sequence was designed inside a hairpin DNA strand. Upon the addition of thrombin, the thrombin aptamer inside the hairpin DNA interacted specifically with thrombin, and the conformation of hairpin DNA would change. Then, a DNA strand, which was complementary to the hairpin DNA was introduced and the produced double-stranded DNA would decrease with the amount of thrombin added. When an RLS probe, methylene blue dye, which can only embed a double-stranded DNA rather than a single-stranded DNA, was added, the RLS intensity would present various response values based on different amounts of thrombin [6].

Compared with small organic molecular dye probes, light scattering detection of proteins using nanoparticles probes have different analytical parameter and mechanism. When small organic molecular dye scattering probe is used, as small organic molecular dyes are much smaller that proteins, aggregates forms by stacking of dyes on proteins. The amount of aggregates has a good linear relation with the protein and probe concentration. For nanoparticle probes, nanoparticles are much bigger than protein or have comparable size. And also, nanoparticles disperses as colloid that is different from small organic molecules, which directly dissolve in a solution. So, nanoparticles solutions are not much stable and are easy to aggregate.

So, when nanoparticle solutions are used for protein analysis, on the one hand, protein forms aggregate with nanoparticles through electrostatic interaction, and on the other hand, it destroys stability of nanoparticles while forming aggregates. In this process, charged proteins induced nanoparticle aggregates, thus enhancing the light scattering signal. In actual application, nanoparticle concentration is quite important. When the concentration of nanoparticles colloid is low, the nanoparticles are rather stable that the addition of protein is not easy to induce the aggregate of nanoparticles. However, when the particle concentration is high, the aggregate is easier. In addition, there is also differences in nanoparticle aggregate and ordinary small organic molecules aggregate: The protein induced nanoparticle aggregates are large and unstable in a solution. The light scattering intensity of the aggregates

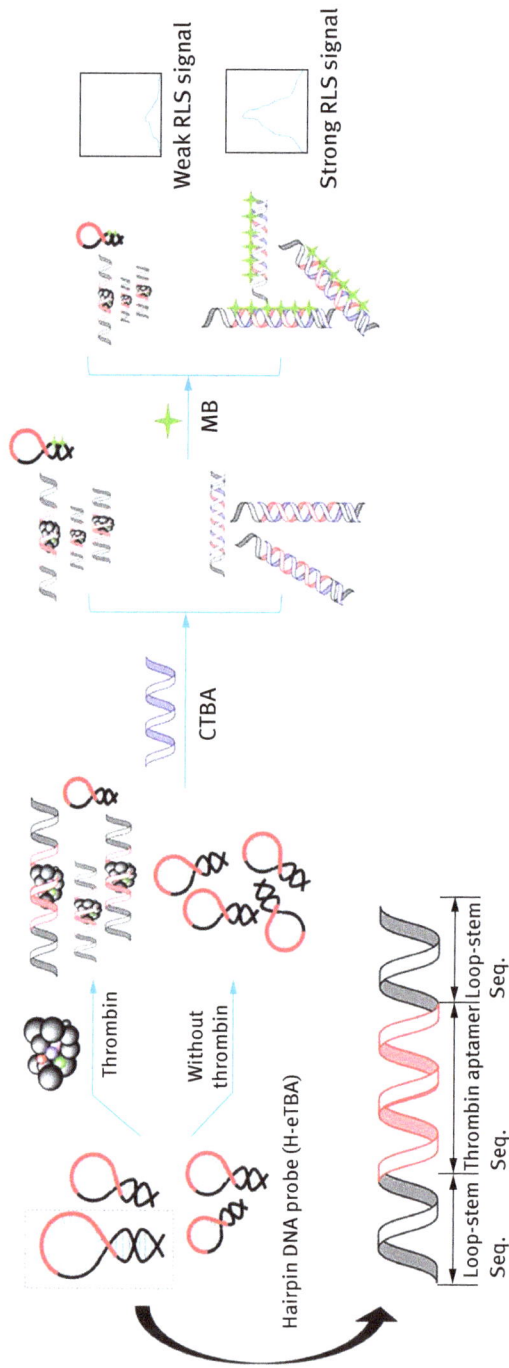

Figure 12.4: Diagram of aptamer application for thrombin detection [6].

solution would decrease with the sedimentation of aggregates after a while. So, in an actual application, light scattering spectral measurement needs to be performed in a short time after the formation of aggregates.

12.2.3 Other scattering probes

The charged micelle of a surfactant in a solution can aggregate into large-sized aggregates through an electrostatic attraction with proteins of opposite charges. It can be used for protein detection as there is a linear relationship between the light scattering signal enhancement and protein concentration [45–48]. For example, negative charged micelles formed by anionic surfactants, such as SDS and SDBS can interact with positive charged proteins in acid condition. The detection of protein is in the range of 0.01–10 µg/mL. Some polymers carrying charges, such as polymethacrylic acid and polydiallyldimethylammonium chloride [49–51]; ion associates, such as $[PdI_4]^{2-}$ [52]; inorganic molecules or ions, such as ammonium sulfate and aluminum ion [53–56], can also be used for the detection of protein by similar light scattering strategies.

12.3 Factors influencing protein light scattering analysis

12.3.1 Medium acidity

In the principle of protein light scattering analysis, formation of larger aggregates with a large scattering cross section is a key point. The formation is usually based on an electrostatic interaction, so, the acidity of the system is a key factor for protein detection as it directly decides the happening of an aggregation.

Usually, the environment acidity for scattering enhancement is at which the protein and the probes are carry opposite charges. This is because protein is an amphoteric electrolyte with different charges in different acidity. When pH is lower than the isoelectric point, then the protein has positive charges; when pH is higher than the isoelectric point, protein has negative charges; so does the probe, the pK_a value and other parameters of small organic molecular probes need to be considered. Take an example of anionic dye for BSA detection. As the isoelectric point of BSA is 4.6, the optimal acidity condition is at pH 3.98 when bromophenol blue is used as a probe [57]; whereas when Congo red is used as a probe, the best acidity condition is at pH 2.75 [58]. It can be observed that the acidity of the solution directly influences the electrostatic interaction between the probe and the protein.

If nanoparticle is used to detect protein, then protein aggregate through electrostatic interaction of nanoparticle surface-adsorbed molecules. So, the protein isoelectric point and the charge carries of the nanoparticles should be taken into consideration. For example, when trisodium citrate-capped gold nanoparticle is used to detect BSA, as BSA has an isoelectric point of 4.7; while pK_{a1} of trisodium citrate is 3.18, pK_{a2} is 4.74, the chosen pH is 4–5. In this range, BSA has small amounts of positive charges, while citric acid exists in the form of acidic group, bringing negative charges to the surface of the gold nanoparticle, thus there is a stronger electrostatic interaction between BSA and gold nanoparticles.

When gold and silver nanoparticles are used as probes, the covalent and hydrophobic interaction of the protein with silver and gold need to be considered. Protein surface has large number of thiol groups that is able to bind to the surface of gold and silver through Au–S or Ag–S bond. So, in some analytical systems for protein detection, the acidic environment of proteins and nanoparticles is around the protein's isoelectric point; where protein has no or only slight charges, then the protein is adsorbed to the nanoparticle surface with covalent interaction. The reagent adding sequence always has a larger influence on the degree of scattering enhancement because the aggregation depends on the electrostatic attraction that are related to charge carries of proteins and probes. For example, if proteins and probes interact under non-neutral conditions, the buffer needs to be mixed with probe before the addition of protein, that can create the most suitable acidic condition for the combination.

12.3.2 Probe type and concentration

Different types of light scattering probes have different charges in a different environment because of their various charged groups. So, when different probes are used to detect proteins, the sensitivity and detection ranges are quite different.

Probes have distinct light scattering properties. For example, small organic molecular probes have weak light scattering properties, producing an apparent enhancement after aggregate. While for some nanoparticle probes, light scattering signal changes before and after aggregation are rather complex. For small-sized nanoparticle probes, the scattering is weak; there is apparent enhancement after aggregation. While for nanoparticles that have strong light scattering intensity, the aggregation may lead sedimentation of bigger aggregates that decrease the whole scattering intensity of the solution on the contrary.

Probe concentration mainly affects the formation of aggregates. Using small organic molecules as an example, at a certain protein concentration when the concentration of small organic molecular dye increases, larger protein aggregates can be formed to obtain a more sensitive signal enhancement, and wider linear detection range. To produce aggregates, excessive small organic molecular dye probes are used, dye concentration is much more higher than that of proteins; the obtained sensitivity increases with the dye concentration. While the trend has a critical concentration, sensitivity reduces when concentration is higher than the value. Because too high concentration of the dye produces stronger molecular absorption, the absorption produces an effect similar to the inner-filter effect of fluorescence, leading to a self-quenching of light scatting intensity. Similar situations occur in the analysis system with scattering probe of nanoparticles. When nanoparticle concentration is higher, proteins can induce nanoparticle aggregates, whereas when nanoparticle probe concentration is lower, it is easy to stable in a medium, reducing the possibility of an aggregate.

12.3.3 Factors interfering the detection of protein light scattering

As described earlier, the analysis of protein light scattering is mainly based on enhancement of aggregate light scattering with a large scattering cross section based on electrostatic interaction. So, the factors interfering the chemical or particle electricity can directly influence the aggregate formation. For example, surfactants with opposite charges and small organic molecules can compete with probes to destroy aggregate formation. In addition, aggregations formed by proteins and common small organic molecules do not have characteristic light scattering peaks and the light scattering spectrum is wide in the range from UV to near IR region. In a practical application of protein detection, the light scattering spectrum of other particles, aggregates and sedimentations would possibly overlap the spectrum of protein aggregates that clearly influence the detection of protein.

12.4 Light scattering immunoassay

Although protein light scattering analysis is simple, rapid, and sensitive, this method has the poor selectivity. A specific reaction between protein–small molecule or an antigen–antibody immune response can effectively solve poor selectivity. So, when light scattering analysis is used for an immunoassay, selectivity can be well improved.

Immunoassay is an important aspect in bioanalytical chemistry, that is based on the immune reaction of antigens and antibodies, small-molecule haptin and antibodies, and small-molecule acceptor and protein aptamers. Immune reaction has a good selectivity that light scattering immunoassay has both high sensitivity and good selectivity. Usually, light scattering immunoassay involves label-free and labeled light scattering immunoassays.

A label-free light scattering immunoassay utilizes a light scattering signal that changes after an antigen (Ag) and an antibody (Ab) combine to form an antigen–antibody reaction for detection. A labeled light scattering immunoassay needs to modify the probe to an antigen (or antibody); through their immune reaction, the light scattering signal of probe changes, so the detection of an immune reaction can be realized.

12.4.1 Label-free light scattering immunoassay

12.4.1.1 Light scattering turbidity method

Immune scattering turbidity method is the simplest light scattering turbidity method that has been widely practiced in clinical detection. It is usually called light scattering turbidity method or simply turbidity method.

The basic principle is, when antigen and antibody form an immune complex, when antigen and antibody form an immune complex, a promotion reagent (such as polyethylene glycol) is added to the complex to induce the formation of bigger aggregates, and thus, the turbidity of the solution increased. If antibody concentration is fixed, the amount of the formed immune complex increases with the increasing of antigen in the sample. Thus, the turbidity of solution increases. At this time, the detected light scattering intensity is in direct proportion to the concentration of the complex if a certain wavelength of light was illuminated to the solution. The light scattering intensity is also closely related to various physical factors, such as adding time of the antigen and the antibody, intensity and wavelength of the light source, signal measuring angle, and so on.

Although the traditional light scattering turbidity method is sensitive, simple, and rapid and has been widely used in clinical immunology analysis, the method still has flaws. For inducing aggregation of the immune complex, the promotion reagent needs to use. However, the using promotion reagent can also bring obvious changes in the activity of the immune complex itself and cause irreversible damage to the immune sample. So, in a physiological environment, in-situ immune response of an antigen–antibody reaction has attracted interest of researchers.

To obtain a accurate and sensitive signal, light scattering measurement was conducted on the synchronous scanning of samples with a common fluorescence spectrophotometer. This method can represent light scattering properties of a sample in a wide wavelength range to obtain more light scattering information from the spectrum. For quantitative analysis, the maximum scattering wavelength can be chosen as the detection wavelength to obtain a sensitive detection signal.

12.4.1.2 Solid-phase radioimmunoassay

In addition to immunoassay and light scattering detection in a solution, immunoassay can also carries on solid phases by fixing antigen (or antibody) on solid phases of slides, and then incubating the slides in a solution with antigen (or antibody) to form immune complexes. Afterward, the light scattering signal from the slide is obviously increased compared to single fixation of antigen (or antibody) on a slide. By using slide for solid-surface light scattering immunoassay, there is higher sensitivity while effectively reducing the sample amount. The studies of light scattering detection on a solid surface create a foundation for solid-phase surface microarray immune scattering analysis.

A label-free immune scattering analysis can enhance the immune reaction scattering signal with a probe. For example, silver nanoparticles have strong light scattering signal and can be used as an admirable scattering probe, whereas silver nanoparticles and proteins have a good affinity; at certain pH environment, proteins can be steadily adsorbed to the surface of the silver nanoparticles.

Different proteins can be adsorbed to silver nanoparticles because of their structure difference. Based on this principle, first, the antigen on the slide, then the antibody binds on the slide through an immune reaction, and then the slide is dipped into the solution with silver nanoparticles for incubation. As antigen and antibody have different adsorption ability to silver nanoparticles, adjusting incubation time and pH, silver nanoparticles can be adsorbed on the slide with an antibody, rather than the slide containing antigen. In this way, when an antibody is added to the slide through an immune reaction, nanoparticles can be adsorbed to the slide, and immune reaction can be detected through the scattering signal of silver [59].

12.4.1.3 Pasmon resonance light scattering immunoassay

Another label-free immunoassay is based on gold nanoparticle's localized surface plasmon resonance (LSPR) scattering properties. This method is similar to surface plasmon resonance (SPR) immunoassay. The difference is that LSPR happens on the surface of metal nanoparticles and signal is the light scattering from nanoparticles [60]. So, the adsorbing of antigen (or antibody) on the surface of a metal nanoparticle after immune reactions, would change the dielectric coefficient of metal nanoparticles that change the LSPR properties. So, in LSPR immunoassay, the LSPR absorption and scattering spectral peak would shift with the amount of antigen (or antibody) on the surface of metal nanoparticles. As the shifting degree of peaks is related to the combined thickness of a protein on the surface of metal nanoparticle, the relation between the peak shift and the combined antigen (or antibody) concentration can be obtained.

In LSPR immunoassay, nanoparticle scattering probe was first fixed on the surface of a slide and antigen (or antibody) is adsorbed onto the nanoparticle for detection. Different methods can be used to detect LSPR signal. For example, by using of a common LSPR spectrophotometer, scattering signal from large amounts of nanoparticles in an immobilization area can be detected. If micro-spectroscopy system is used to detect LSPR scattering spectrum of single metal nanoparticle, the influence of immunoreaction on a single particle LSPR scattering can be observed and a highly sensitive detection of antigen can be achieved. Currently, both the methods are widely used and various single nanoparticle-based LSPR scattering methods have been established.

In addition, study of metal nanoparticle probes is going. Metal nanoparticles of different their or materials have various LSPR properties when used as LSPR probes, because their dielectric coefficient changes that leads to different shifting degree of the spectrum. There was a report that by using gold nanorod as an immunoassay probe, the obtained sensitivity was 100 times higher than that with spherical gold nanoparticles [61].

12.4.2 Labeled light scattering immunoassay

Labeled light scattering immunoassay is based on labeling a probe to an antigen (or antibody). A immune reaction process or immunoassay can be achieved by detection the light scattering property of the probe after immune reaction. Metal nanoparticles are the most widely used light scattering probes.

Gold and silver nanoparticles are the most common metal nanoscattering probes. Its detailed description is provided in Chapter 8. Gold and silver nanoparticles are the most popular probes for light scattering immunoassay, not only for their excellent plasmon resonance absorption and scattering properties, but also for their good biocompatibility and easy labeling. For example, gold nanoparticle labeled antigen can be prepared easily by through simple mixing of antigen and nanoparticles, because proteins have large amount of thiol and amino groups that can adsorbed on the surface of gold and silver nanoparticle through covalent interaction and hydrophobic interaction easily. This labeling method does not need chemical coupling or modification of antigen (or antibody), labeling sure of the bioactivity of antigen (or antibody).

In terms of a reaction method, immune reactions can be divided into two types: (1) sandwich immunoassay and (2) competitive immunoassay. In terms of scattering signal acquisition, immune reactions can be divided into three types: (1) Directly synthesizing large-sized nanoparticles with strong scattering for immune labeling and analysis (2) use of small-sized nanoparticles with weak scattering for labeling of antigen (or antibody) to produce enhanced scattering signal for analysis after the aggregation of nanoparticles; and (3) use of catalytic property of labeled nanoparticles to produce large nanoparticles with strong scattering for detection. In the following text, the book will introduce the different types of light scattering immunoassay in terms of light scattering signal acquisition.

12.4.2.1 Direct light scattering labels

Direct light scattering label method is similar to common labeling fluorescence spectroscopy. Usually, nanoprobes probe with strong light scattering signals is synthesized and labeled with an antigen or an antibody, the light scattering signal from the probe is used for immune detection. This method is simple, whereas normally, not many nanoparticle probes are available for direct labeling.

For example, when a gold nanoparticle is used as a probe, to equip it with strong scattering signal, the size of nanoparticles should be above 50 nm. Compared with gold nanoparticles, silver nanoparticles have better scattering properties, showing apparent scattering properties at 30–40 nm. While compared with gold nanoparticles, silver nanoparticles have the worse stability and biocompatibility, limiting their application as a direct label scattering probe.

In recent reports, gold and silver nanoparticles above 40 nm are usually used as an light scattering immune probe. The authors of this book established a light scattering immunoassay by using silver nanoparticles as probe. Figure 12.5 shows the sandwich analysis of two antibodies and detection of one antigen: one antibody is fixed on the solid surface to capture an antigen, another antibody is labeled with a scattering probe for antigen detection after recognition [62]. Although the sandwich analysis has a good selectivity, the experimental process is complex and an important step for fixing of antibody on a solid surface to capture the antigen should be considered. There are many alternative solid surfaces, such as microporous polyethylene, glass, magnetic nanoparticles, and so on. Compared with homogeneous immune response, this method is complex. However, it has important research value in terms of immune scattering chip and microarray analysis studies [63].

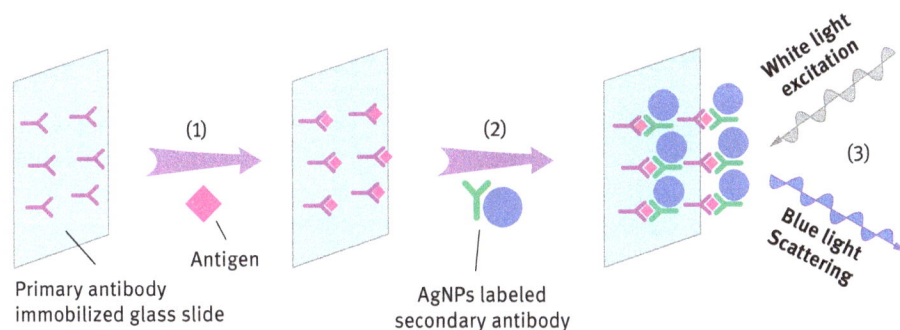

Figure 12.5: Labeled AgNPs on a slide for immunoassay [62].

Direct light scattering label method has laid foundation for exploring a new type of scattering probe. Thus, gold and silver nanoparticles can be used like fluorescence probes for labeling specific substances. The further applications in detection and recognition of specific proteins, such as biomarkers in living cells and tissues, would expand the using of light scattering probes in bio-imaging application.

12.4.2.2 Nanoparticle aggregate

As described in Chapter 7, nanoparticle aggregate is one of the most common and widely used light scattering analysis methods. Since the time it was developed, most scattering analysis methods are connected with an "aggregate," which means enhancement of the signal system. Immune reaction of antibody and antigen forms immune complexes with good specificity; it is only performed in specific antibody and antigen reactions. So, the use of light scattering signal produced by aggregating for immune reaction can effectively avoid poor selectivity of light scattering analysis method.

Simple immune response system based on nanoparticle aggregates can be used to obtain enhanced scattering signal. For example, usually, immune reaction between an antigen and a small-sized gold nanoparticles probe labeled antibody would induce the aggregation of nanoparticles probe and produce a strong light scattering signal [64–67].

Nanoparticle aggregation can be confirmed with a dynamic light scattering apparatus. By calculating the light scattering fluctuation of nanoparticles solution, dynamic light scattering technology is used to determine particle size distribution. After aggregation, Brownian movement of the aggregates slows down and the calculated particle "diameter" based on the detected dynamic light scattering signal enlarges. The changes in the diameter have a certain linear relation with the nanoparticle aggregation degree, which can be indirectly used for the detection of substance concentration that causes nanoparticle aggregates [68].

12.4.2.3 Nanocatalysis

Nanocatalysis utilizes an antigen- or an antibody-labeled nanomaterial as a catalyst to produce nanoparticles with strong light scattering for signal amplification. Usually, small-sized gold nanoparticle is used as a label to modify the antigen or antibody. Afterward, catalytic action of gold nanoparticle is used for producing large particles with enhanced scattering signals. The most common nanocatalytic scattering method is the silver-enhanced scattering analysis, which utilizes gold nanoparticles as a catalyst for growth of large sized silver nanoparticles in silver nitrate solution. The strong light scattering signal from the produced silver nanoparticles can be detected with equipment or can be observed with naked eyes [69–71]. In addition, gold nanoparticles can also be used as a catalyst for growth of other nanoparticles, for example, growth of copper oxide nanoparticles in copper sulfate solution [72].

The key factor to nanocatalyzed immune light scattering analysis is to establish a simple relationship between the amount of the labeled gold nanoparticle and the produced scattering intensity. The more the amount of gold nanoparticles, the more large-sized nanoparticles would be produced. Thus, the light scattering signal of the system will be linearly increased with the amount of gold nanoparticles. So, in this immunoassay method, a gold nanoparticles probe labeled sandwich strategy is usually used for analysis. The sandwich strategy has three steps. Firstly, a primary-antibody is immobilized on a solid surface (such as slide surface and magnetic nanoparticle), then the immobilized primary-antibody captured the antigen, finally a gold-labeled secondary-antibody recognize the antigen. After elution or magnetic separation, a gold-labeled antibody is obtained, which was later utilized for catalytic production of large nanoparticles. As the amount of gold labeled antibody is related to the concentration of antigen, the obtained light scattering signal from produced bigger particles also has a relationship with the antigen concentration.

Common immunoassay strategy can also be used for establishing a nanocatalysis scattering method in the similar way. In this strategy, a gold nanoparticles-labeled antibody is firstly mixed with the antigen for immunoreaction. Then, the unreacted gold nanoparticles are separated by centrifugation as the immune complex and gold nanoparticles has different dispersing property in solution. Finally, the separated gold nanoparticles are used for catalytic production of large nanoparticles. As the amount of the separated gold nanoparticles is related to the concentration of antigen, a light scattering method for antigen detection can be established based on the nanocatalysis [72,73].

12.4.3 Light scattering immune analysis based on aptamer

Aptamer is a DNA or RNA that combines targets with high affinity and high specificity. Its function is similar to antibodies but have more advantages: better stability, ability of voluminous production, and with no rejection reactions [74]. As described earlier, aptamer has been widely applied in small molecules and protein biomacromolecule detection, such as lysozyme and thrombin. With the same strategies, aptamer can also be used for immunoassay [75].

Prostate specific antigen (PSA) is a single-stranded glycoprotein secreted from epithelial cells of the prostate gland and ducts. It is a marker for differentiate benign and malignant prostate diseases in routine clinical practice and also an important indicator of postoperative diagnosis of prostate patients. Chen [76] used aptamer to establish PSA detection method (Figure 12.6). PSA aptamer is single-stranded DNA that adsorbs on the surface of a gold nanoparticle to improve its stability. When a PSA is present, the aptamer recognizes the PSA and no longer adsorbs onto the surface of gold nanoparticles. Thus, the gold nanoparticles are not stable at higher salt concentration. So, when KCl is added, gold nanoparticles aggregate, producing a strong signal, which can be used to detect PSA.

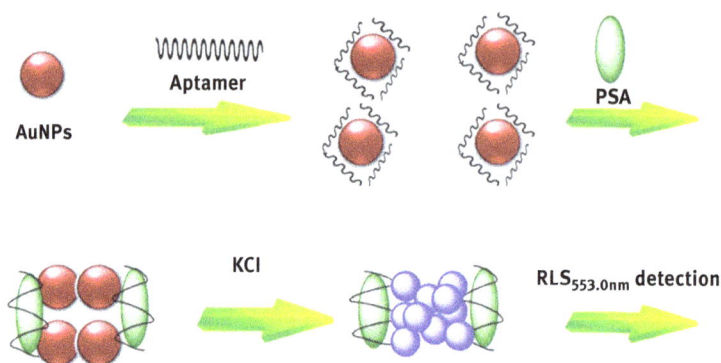

Figure 12.6: Aptamer application in immunoassay [76].

References

[1] Dong L J, Jia R P, Li Q F, et al. Quantitative determination of proteins by the Rayleigh light-scattering technique after optimization of the derivation reaction with Arsenazo-DBS. Anal. Chim. Acta, 2002, 459: 313–322.

[2] Feng P, Huang C Z, Li Y F. Direct quantification of human serum albumin in human blood serum without separation of g-globulin by the total internal reflected resonance light scattering of thorium-sodium dodecylbenzene sulfonate at water/tetrachloromethane interface. Anal. Biochem., 2002, 308: 83–89.

[3] Truong P L, Choi S P, Sim S J. Amplification of resonant Rayleigh light scattering response using immunogold colloids for detection of lysozyme. Small, 2013, 9: 3485–3492.

[4] Wang X, Xu Y, Xu X, et al. Direct determination of urinary lysozyme using surface plasmon resonance light-scattering of gold nanoparticles. Talanta, 2010, 82: 693–697.

[5] Wang X, Xu Y, Chen Y, et al. The gold-nanoparticle-based surface plasmon resonance light scattering and visual DNA aptasensor for lysozyme. Anal. Bioanal. Chem., 2011, 400: 2085–2091.

[6] Huang X P, Li Y Q, Huang X, et al. Combining a loop-stem aptamer sequence with methylene blue: a simple assay for thrombin detection by resonance light scattering technique. RSC Adv., 2015, 5: 30268–30274.

[7] Su M, Ma L, Li T, et al. A microarray-based resonance light scattering assay for detecting thrombin generation in human plasma by gold nanoparticle probes. Anal. Methods, 2013, 5: 5895–5898.

[8] Zhang H J, Lu Y H, Long Y J, et al. An aptamer-functionalized gold nanoparticle biosensor for the detection of prion protein. Anal. Methods, 2014, 6: 2982–2987.

[9] Huang X X, Long Y J, Zhang H J, et al. Gold nanoparticles as a probe for prion determination via resonance light scattering method. Anal. Sci., 2012, 28: 475–479.

[10] Verdian-Doghaei A, Housaindokht M R. Spectroscopic study of the interaction of insulin and its aptamer -sensitive optical detection of insulin. J. Lumines., 2015, 159: 1–8.

[11] Liu B S, Zhang H Y, Zhang H L, et al. The high-sensitivity determination of protein concentration by the enhancement of resonance light scattering of chlorophenol red. Spectrosc. Spectr. Anal., 2003, 23(2): 229–231.

[12] Li Y F, Huang C Z, Li M. A resonance light-scattering determination of proteins with Fast Green FCF. Anal. Sci., 2002, 18(2): 177–181.

[13]   Feng N C, H e S P, Zhang J, et al. Resonance light scattering study on interaction of solochrome cyanine R with protein and light scattering determination of trace protein. Spectrosc. Spectr. Anal., 2004, 24(2): 194–196.

[14]   Wei Y J, Li K A, Tong S Y. Protein-chromazurine S system elastic light scattering and its preliminary application. Chinese J. Chem., 1998, 56: 290–297.

[15]   Ma C Q, Li K A, Tong S Y. Microdetermination of proteins by resonance light scattering spectroscopy with tetraiodophenol sulfonphthalein. Fresen. J. Anal. Chem., 1997, 57(7): 915–920.

[16]   Liu S P, Liu Q. Resonance Rayleigh-scattering method for the determination of proteins with some monoazo dyes of chromotropic acid. Anal. Sci., 2001, 17(2): 239–242.

[17]   Liu S P, Yang R, Liu Q. Resonance Rayleigh method for the determination of proteins with orange G. Anal. Sci., 2001, 17(2): 243–247.

[18]   Wang Y T, Li K A, Tong S Y, et al. A novel protein assay with an Azo dye using Rayleigh light scattering technique. Anal. Lett., 2000, 133(2): 221–235.

[19]   Li S W, Li N, Zhao F L, et al. Luminescence properties of Pr^{3+} and energy transfer characteristics of $Pr^{3+} \rightarrow Gd^{3+}$ in $CaSiO_3$. Spectrosc. Spectr. Anal., 2002, 22(4): 619–622.

[20]   Li S W, Li N, Zhao F L, et al. Use eriochrome black T as RLS probe for protein detection. Chinese J. Anal. Chem., 2002, 30(6): 732–735.

[21]   Li Q F, Chen X G, Zhang H Y, et al. Microdetermination of proteins using m-carboxychlorophosphonazo as detection probe by enhanced resonance light scattering spectroscopy. Fresen. J. Anal. Chem., 2000, 368(7): 715–719.

[22]   Li Q F, Liu S H, Zhang H Y, et al. Microdetermination of proteins by enhanced resonance light scattering spectroscopy of m- acetylchlorophosphonazo. Anal. Lett., 2001, 34(7): 1133–1142.

[23]   Yang C X, Li Y F, Huang C Z. Determination of total protein content in human serum samples with Fast Red VR by resonance light scattering technique. Anal. Lett., 2002, 35(12): 1945–1957.

[24]   Yang C X, Li Y F, Huang C Z. Determination of proteins with Fast Red VR by a corrected resonance light-scattering technique. Anal. Sci., 2003, 19(2): 211–215.

[25]   Yang C X, Li Y F, Huang C Z. Determination of total protein content in human blood plasma samples with Ponceau G by resonance light scattering technique. Chinese J. Anal. Chem., 2003, 31(2): 148–152.

[26]   Huang C Z, Yang C X, Li Y F. Determination of proteins with Ponceau G by compensating for the molecular absorption decreased resonance light scattering signals. Anal. Lett., 2003, 36(8): 1557–1571.

[27]   Chen Y J, Yang J H, Wu X, et al. Resonance light scattering technique for the determination of proteins with resorcinol yellow and OP. Talanta, 2002, 58(5): 869–874.

[28]   Su X D, Li S W, Wang X H, et al. Beryllon RLS spectrum for protein detection. Instr. Anal. Monitor., 2003, 19(3): 35–36, 46.

[29]   Chen X L, Li D H, Zhu Q Z, et al. Determination of proteins at nanogram levels by a resonance light-scattering technique with tetra-substituted sulphonated aluminum phthalocyanine. Talanta, 2001, 53: 1205–1210.

[30]   Wang L, Li Y X, Zhu C Q, et al. Determination of proteins based on their resonance light scattering enhancement effect on manganese-tetrasulfonato phthalocyanine. Microchim. Acta, 2003, 143(4): 275–279.

[31]   Wang L, Wang Z L, Li S W. Simple fluorescence resonance scattering spectrometry for protein detection. Chemical Res. Appl., 2001, 13(5): 591–593.

[32]   Zhong H, Li N, Zhao F L, et al. Determination of proteins with Alizarin Red S by Rayleigh light scattering technique. Talanta, 2004, 62(1): 37–42.

[33]   Zhang S H, Fan Y S, Feng S, et al. Microdetermination of proteins by resonance light scattering technique based on aggregation of ferric nanoparticles. Spectrochim. Acta A, 2009, 72: 748–752.

[34] Huang K J, Wei C Y, Shi Y M, et al. The investigation of the interactions between CdSe quantum dots and human serum albumin by resonance Rayleigh scattering and second-order scattering spectra. Spectrochim. Acta A, 2010, 75: 1031–1035.

[35] Zhang S H, Fan Y S, Feng S. Resonance light scattering method for microdetermination of proteins based on aggregation of Au nanoparticles. Anal. Lett., 2011, 44: 656–666.

[36] Peng L, Li C Y, Xue H B, et al. Ultrasensitive determination of serum albumin using resonance light scattering based on ZnS-polyacrylic acid nanoparticles. Spectrochim. Acta A, 2010, 75: 1497–1500.

[37] Fang B, Gao Y, Li M, et al. Application of functionalized Ag nanoparticles for the determination of proteins at nanogram levels using the resonance light scattering method. Microchim. Acta, 2004, 147: 81–86.

[38] Wang L, Xu F, Zhou Y, et al. Preparation and application of MS-M^{2+} nanoparticles as a novel resonance light-scattering probe. Spectrochim. Acta A, 2004, 60: 2141–2145.

[39] Pan H C, Tao X C, Mao C J, et al. Aminopolycarboxyl-modified Ag2S nanoparticles: Synthesis, characterization and resonance light scattering sensing for bovine serum albumin. Talanta, 2007, 71: 276–281.

[40] Wang L, Chen H, Li L, et al. Quantitative determination of proteins at nanogram levels by the resonance light-scattering technique with macromolecules nanoparticles of PS–AA Spectrochim. Acta A, 2004, 60: 747–750.

[41] Zhou Y, She S, Zhang L, et al. Determination of proteins at nanogram levels using the resonance light scattering technique with a novel PVAK nanoparticle. Microchim. Acta, 2005, 149: 151–156.

[42] Chen H, Xu F, Hong S, et al. Quantitative determination of proteins at nanogram levels by the resonance light-scattering technique with composite nanoparticles of CdS/PAA. Spectrochim. Acta A, 2006, 65: 428–432.

[43] Zhao G C, Zhang P, Wei X W, et al. Determination of proteins with fullerol by a resonance light scattering technique. Anal. Biochem., 2004, 334: 297–302.

[44] Yguerabide J, Yguerabide E E. Resonance light scattering particles as ultrasensitive labels for detection of analytes in a wide range of applications. J. Cell. Biochem., 2001, 84: 71–81.

[45] Lu W, Feng P, Li Y F, et al. Determination of proteins based on their resonance light scattering enhancement effect on sodium dodecyl benzene sulfonate. Anal. Lett., 2002, 35: 227–238.

[46] Liu R, Yang J, Sun C, et al. Resonance light-scattering method for the determination of BSA and HAS with sodium dodecyl benzene sulfonate or sodium lauryl sulfate. Anal. Bioanal. Chem., 2003, 377: 375–379.

[47] Gao D J, He N, Tian Y, et al. Determination of bovine serum albumin using resonance light scattering technique with sodium dodecylbenzene sulphonate-cetyltrimethylammonium bromide probe. Spectrochim. Acta A, 2007, 68: 573–577.

[48] Chen Z, Liu J, Liang Y, et al. Use of sodium lauroyl sarcosinate in a high-sensitivity protein assay by resonance light scattering technique. J. Biomol. Screen., 2006, 11: 400–406.

[49] Chen Y, Gao D, Tian Y, et al. Resonance light scattering technique for the determination of proteins with polymethacrylic acid(PMAA). Spectrochim. Acta A, 2007, 67: 1126–1130.

[50] Chen Y H, Tian Y, Gao D J, et al. Determination of serum albumin in the presence of poly (diallyldimethylammonium chloride) by resonance light scattering technique. Spectrochim. Acta A, 2007, 66: 1011–1015.

[51] Chen Z G, Liu J B, Chen X M, et al. Resonance light scattering spectroscopy of beta- cyclodextrin-sodium dodecylsulfate- protein ternary system and its analytical applications. Anal. Sci., 2007, 23: 1305–1310.

[52] Luo J G, Liu Z F, Kong L, et al. Study on the resonance Rayleigh scattering and resonance non-linear scattering spectra of ion-association complexes of [PdI$_4$]$^{2-}$-protein systems and their analytical applications. Luminescence, 2009, 24: 429–437.

[53] Li A, Zhao H, Jin L. Protein analysis with terbium (III) and sodium dodecyl sulphonate by a second-order scattering technique. Luminescence, 2007, 22: 9–14.

[54] Long X, Zhang C, Cheng J, et al. A novel method for study of the aggregation of protein induced by metal ion aluminum (III) using resonance Rayleigh scattering technique. Spectrochim. Acta A, 2008, 69: 71–77.

[55] Jiang Z L, Peng Z L, Liu S P. Resonance scattering spectral determination of HSA with a new scattering enhanced reagent of $K_3[Fe(CN)_6]$. Chin. J. Chem., 2002, 20: 1566–1572.

[56] Wu L H, Mu D, Liang F H, et al. Determination of protein via resonance light scattering technique with$(NH_4)_2SO_4$. Chem. Res. Chin. Univ., 2010, 26: 189–193.

[57] Ma C Q, Li K A, Tong S Y. Microdetermination of proteins by resonance light scattering spectroscopy with Bromophenol Blue. Anal. Biochem., 1996, 239: 86–91.

[58] Wu X, Sun S, Guo C, et al. Resonance light scattering technique for the determination of proteins with Congo Red and Triton X-100. Luminescence, 2006, 21: 56–61.

[59] Ling J, Li Y F, Huang C Z. A label-free visual immunoassay on solid-support with silver nano-particles as plasmon resonance scattering indicator. Anal. Biochem., 2008, 383: 168–173.

[60] Raschke G, Kowarik S, Franzl T, et al. Biomolecular recognition based on single gold nano-particle light scattering. Nano Lett., 2003, 3: 935–938.

[61] Nusz G J, Marinakos S M, Curry A C, et al. Label-free plasmonic detection of biomolecular binding by a single gold nanorod. Anal. Chem., 2008, 80: 984–989.

[62] Ling J, Li Y F, Huang C Z. Visual sandwich immunoassay system on the basis of plasmon resonance scattering signals of silver nanoparticles. Anal. Chem., 2009, 81: 1707–1714.

[63] Saviranta P, Okon R, Brinker A, et al. Evaluating sandwich immunoassays in microarray format in terms of the ambient analyte regime. Clin. Chem., 2004, 50: 1907–1920.

[64] Sánchez-Martínez M L, Aguilar-Caballos M P, Gómez-Hens A. Homogeneous immunoassay for soy protein determination in food samples using gold nanoparticles as labels and light scattering detection. Anal. Chim. Acta, 2009, 636: 58–62.

[65] Jiang Z, Sun S, Liang A, et al. Gold-labeled nanoparticle-based immunoresonance scattering spectral assay for trace apolipoprotein AI and apolipoprotein B. Clin. Chem., 2006, 52: 1389–1394.

[66] Jiang Z L, Sun S J, Liang A H, et al. A new immune resonance scattering spectral assay for trace fibrinogen with gold nanoparticle label. Anal. Chim. Acta, 2006, 571: 200–205.

[67] Du B, Li Z, Cheng Y. Homogeneous immunoassay based on aggregation of antibody-functionalized gold nanoparticles coupled with light scattering detection. Talanta, 2008, 75: 959–964.

[68] Liu X, Dai Q, Austin L, et al. A one-step homogeneous immunoassay for cancer biomarker detection using gold nanoparticle probes coupled with dynamic light scattering. J. Am. Chem. Soc., 2008, 130: 2780–2782.

[69] Storhoff J J, Marla S S, Bao P, et al. Gold nanoparticle-based detection of genomic DNA targets on microarrays using a novel optical detection system. Biosens. Bioelectron., 2004, 19: 875–883.

[70] Wang Z, Lee J, Cossins A R, et al. Microarray-based detection of protein binding and function-ality by gold nanoparticle probes. Anal. Chem., 2005, 77: 5770–5774.

[71] Xu X, Georganopoulou D G, Hill H D, et al. Homogeneous detection of nucleic acids based upon the light scattering properties of silver-coated nanoparticle probes. Anal. Chem., 2007, 79: 6650–6654.

[72] Jiang Z, Liao X, Deng A, et al. Catalytic effect of nanogold on Cu(II)-N_2H_4 reaction and its application to resonance scattering immunoassay. Anal. Chem., 2008, 80: 8681–8687.

[73] Jiang Z L, Huang Y J, Liang A H, et al. Resonance scattering detection of trace microalbumin using immunonanogold probe as the catalyst of Fehling reagent-glucose reaction. Biosens. Bioelectron., 2009, 24: 1674–1678.

[74] Tan W, Donovan M J, Jiang J. Aptamers from cell-based selection for bioanalytical applications. Chem. Rev., 2013, 113: 2842–2862.

[75] Soldevilla M M, Villanueva H, Pastor F. Aptamers: A feasible technology in cancer immunotherapy. J. Immunol. Res., 2016;2016: 1083738.

[76] Chen Z G, Lei Y L, Chen X, et al. An aptamer based resonance light scattering assay of prostate specific antigen. Biosens. Bioelectron., 2012, 36: 35–40.

Sai Jin Xiao, Cheng Zhi Huang

13 Light scattering spectrometry of bioparticles

Biological particles are microorganisms or microbes with a size of 0.5–1,000 μm, including bionts such as spores, bacteria, cells, bacteriophages, organelles such as mitochondria, lysosomes, and nuclei. Biological particles widely exist in the surrounding environment, closely related to production and life.

The light scattering signals widely exist in nature, and is extensively applied to the determination of the molecular weight and particle size of polymer and micro-particles in biophysics and biochemistry [1–3]. Biological particles have complex compositions; their structure and size vary with the inhomogeneous structure's refractive index. Their scattering property is rather complex. Thus, the scattering spectrum, intensity, and distribution are closely related to the cell's size, shape, plasmalemma, and interior structure as well as the equipment and its working conditions.

Even so, from optical principles of light and cell interaction, we obtained optical information with regard to light scattering of biological particles [4–13], and analyzed the relationship of the signal and particle's composition, size, shape, rigidity, homogeneity, and refractive index [14, 15]. Thus, dynamic light scattering and the static one (including resonance light scattering, polarized scattering, etc.) have made significant progress in terms of optical characterization of particles. For example, through study of microorganisms light scattering feature and its changes, bioparticle group distinguish, identification and detection was realized, providing vital information for life sciences, biomedicine, environmental testing and clinical testing.

This chapter will focus on the application of different light scattering techniques, including resonance light scattering (RLS), polarized light scatter-ing spectrum, light scattering imaging technique and so on in bioparticle analysis. It needs to be shown that the "RLS technique" is still used, as bioparticle characteristic absorption in UV-to-visible light region is not clear owing to complex "turbidity" in particle, while the signal acquisition method is the same as that in discussed Chapter 3 (Section 3.2) in order to keep equivalent wavelength of the incident light and emission light, and obtain signal through synchronized sweep.

https://doi.org/10.1515/9783110573138-013

13.1 RLS properties of bacteria

13.1.1 Bacteria and their determination

13.1.1.1 Bacteria

Bacteria are the main groups of living bodies, and are also the maximum number in all living things, as much as 5×10^{30}. Bacteria widely distribute in soil and water or commensalism with other living bodies. We carry a number of bacteria in our body and the number of epidermis bacteria is about 10 times that of the total human cells. Bacteria are like a double-edged sword, but different bacteria have different properties, which are both harmful and beneficial to the environment, human, animals, and plants. Some of these are as follows:

(1) Some bacteria are causative agents for animals and plants, causing human tetanus, typhoid, pneumonia, tuberculosis, gonorrhea, anthrax, syphilis, plague, sand eye, or plants leaf spot disease, fire blight, wilting, and so on.

(2) Bacterial fermentation is beneficial to human, for example, saccharomycetes and other fungus fermented food are widely used in cheese, pickles, soy sauce, vinegar, wine, and yogurt.

(3) Bacteria secrete multiple antibiotics, mainly used in biopharmaceuticals. For example, streptomycin is from *Streptomyces*.

(4) Bacteria degrade into some organic chemicals to remove pollution, for example, methanotroph is used to resolve trichloroethylene and tetrachloroethylene pollution.

Thus, bacteria greatly influence human activities. We often use bacteria as bioreactor, diminishing polluted substances while preparing inorganic, organic, or biopolymers that are otherwise difficult to prepare by regular reactions; they have wide applications in the biopharmaceutical field. Bacteria have irreplaceable role in some fields such as technology, economy, environment, health, and so on; thus, study and detection of bacteria have wider and deeper meaning.

13.1.1.2 Bacterial analysis and detection

On the basis of cell growth kinetics and bacterial metabolic regulation in culture process, regular bacterial analysis is to detect biomass (or bacterial concentration) in a bacterial culture experiment. The detection method includes filter weighing method [16], blood count method [17], conductivity method [18], fluorescence staining method [19], flow cytometry [20], turbidity method, plate counting method [21–23], and so on.

These methods have a number of flaws; the accuracy can be influenced by factors such as human factors, major error, long time, and fatigue operators, on the one hand, and sensitivity and accuracy are not enough, on the other hand. To establish a simpler, rapid, and sensitive analysis method, we would naturally consider applications of light scattering signals for bacterial bioparticles. Here, we will introduce a few typical bacterial light scattering analytical methods and some spectral features.

13.1.2 RLS analysis of *saccharomyces cerevisiae*

Saccharomyces cerevisiae is one of the earlier used unicellular eukaryotic microorganism, an ideal organism for drug selection with common biochemical mechanisms such as mammal cells, with multiple disease-related genes. The detection of microorganism's quantity and growth in a fermentation process is quite helpful for process control and product quality control.

Figure 13.1 shows the scattering and absorption spectra of beer yeast [24]. It can be seen from the figure that its maximum scattering peak is at 308.0 nm (Figure 13.1(A)), while the intensity enhances with yeast cell concentration with a linear relation at a certain concentration range; the detection limit is 4.94×10^2 cells/mL, lower than 10^3 cells/mL for detection limit of flow cytometry [24]. As the medium signal has weak intensity in 220.0–700.0 nm range, the intensity is nearly zero at 308.0 nm. Thus, at this wavelength, we need not consider the influence of medium background for the yeast detection; in addition, no centrifugal separation is needed, and bacteria can be directly detected.

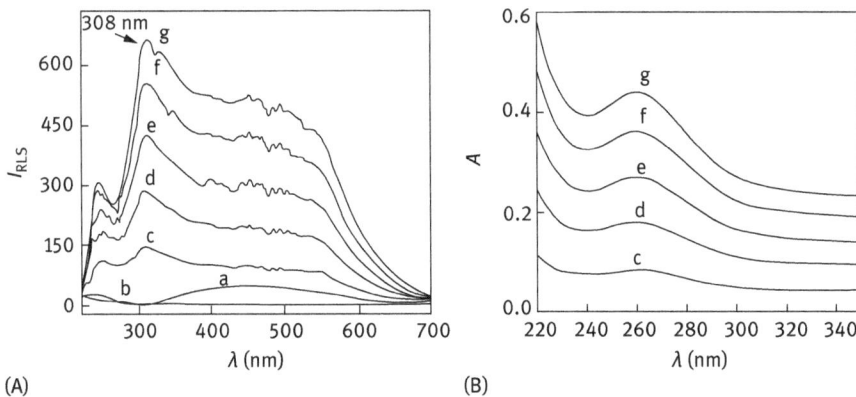

(A)   (B)

Figure 13.1: Yeast suspension scattering (A) and absorption spectra (B). a: Supernatant. b–g: Yeast concentration (cell number per mL); b: 0; c: 4.0×10^5; d: 8.0×10^5; e: 1.2×10^6; f: 1.6×10^6; g: 2.0×10^6.

It can be seen from yeast suspension (Figure 13.1(B)) that the yeast liquid is basically not influenced by the light absorption, with a characteristic absorption peak near 260 nm, which is mainly from cell protein and nucleic acid molecule absorption, then the intensity is not reduced because of inner-filter effect. This shows that the enhanced light scattering signal with an increase in yeast concentration is mainly reflected by the influence of particle size and number.

To better perform scattering spectral characterization and analysis on bacteria or particles of different sizes, shapes, and compositions, polycarboxylate microspheres are chosen for comparison. Beer yeast and polycarboxylate microspheres have no absorption in the visible light region, while they are quite different in their structure and composition: beer yeast has a fungal structure while the latter has a solid construction. Thus, the comparison of their absorbance and scattering spectra could explain properties and functions of the beer yeast. Figure 13.2 shows the absorbance and scattering spectra of beer yeast. When compared with Figure 13.1, we can find that they are quite similar in spite of their differences. The description is as follows.

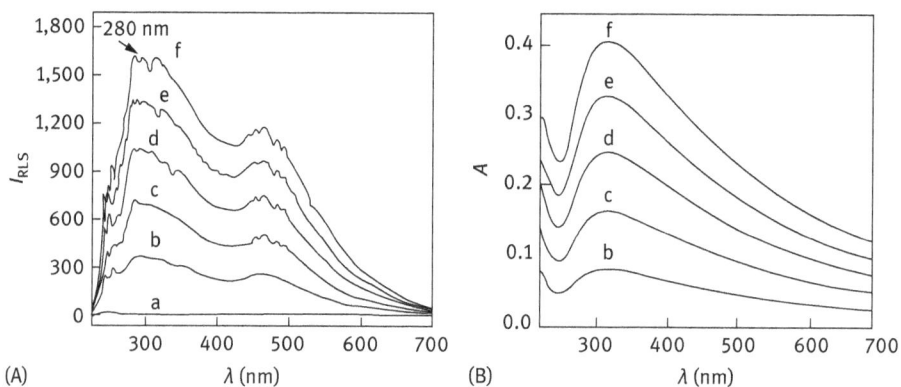

Figure 13.2: Polycarboxylate microsphere scattering (A) and absorption spectra (B). a–f: Polycarboxylate microsphere concentration (particle number per mL); a: 0; b: 1.08×10^7; c: 2.16×10^7; d: 3.24×10^7; e: 4.32×10^7; f: 5.40×10^7.

First, the maximum scattering peak has a different position than the absorption peak. The yeast cell's maximum scattering peak is at 308.0 nm, with a peak valley at 206 nm which is account for the absorption of protein and nucleic acid of cell at 260-280nm. For microsphere, its maximum scattering peak is at 280.0 nm, while the absorption peak is near 320 nm.

Secondly, as the yeast cell (about 10 μm) and microsphere (about 0.75 μm) have large size, obviously, their scattering signal features cannot be explained with the Rayleigh scattering theory, and so the Mie theory is needed. According to the Mie theory [14–15], the solution's scattering intensity is directly proportional to biquadrate of the medium's refractive index, while the light scattering section C_{sca} and the absorption section C_{abc} have no equivalent spectral profile; the amplitude increases as the sixth power and cubic power of the particle size, thereby increasing the particle size widens peak width and moving the wave toward the direction of longer wavelength.

Comparing Figures 13.1 and 13.2, it can be seen that they have different scattering and absorption spectral charts, while the beer yeast has larger size when compared with polycarboxylate microsphere; its maximum scattering peak is at 308 nm in longer wavelength when compared with that of polycarboxylate microsphere.

13.1.3 RLS analysis of *Escherichia coli* and citrus ulcers

E. coli is a normal flora in the intestinal tract of human and animals. In case of babies, *E. coli* enters the intestinal tract through swallowing and stays with him/her all their lives. When the immunity of the host decreases, or bacteria invade the parenteral tissues or an organ, it results in parenteral infection. Some *E.coli* has pathogenicity, and can directly cause their intestinal infection. *E. coli* is exhausted out of the body with feces, directly or indirectly pollute surrounding environment, water source, food and so on. Thus, analysis and detection of *E. coli* are quite meaningful.

E. coli has five Rayleigh scattering peaks, including 470, 510, and 730 nm (Figure 13.3), where the 470 nm Rayleigh scattering peak is mainly caused by the light source, and 510 and 730 nm peaks are resonance scattering peaks [25]. From Table 13.1, it can be seen that the *E.coli* concentration of $0.1 \times 10^8 - 38 \times 10^8$ number/ mL has a good linear relation with $I_{470\ nm}$, $I_{510\ nm}$, and $I_{730\ nm}$. On the basis of this, resonance light scattering method and growth dynamics research of *E. coli* can be built.

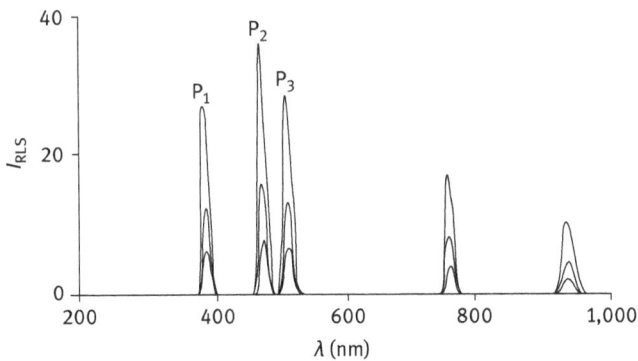

Figure 13.3: *E. coli* RLS spectra. P_1: λ_{ex}=390 nm; P_2: λ_{ex}=470 nm; P_3: λ_{ex}=510 nm. Concentration from up to down: 1.25×10^8/mL; 2.50×10^8/mL; and 5.00×10^8/mL.

Table 13.1: Linear relation of E. coli concentration (c) and RLS intensity.

Measuring wavelength (nm)	Linear range (×10^6/mL)	Regression equation	Linearly dependent coefficient	Detection limit (×10^6/mL)
470	7.400–3800	$I_{470nm}=0.71+2.77×10^{-8}c$	0.9998	7.400
510	7.400–3800	$I_{510nm}=0.97+2.17×10^{-8}c$	0.9997	7.400
730	7.400–3800	$I_{730nm}=1.08+1.35×10^{-8}c$	0.9987	7.400

Figure 13.3 shows the E.coli scattering spectra with Gaussian distribution. When different excitation wavelengths are used for E. coli, besides resonance peaks (the same as the excitation wavelength), there are some more scattering peaks, which are different from Raman scattering and Brilliouin scattering; there is a simple fraction, integer relation with the excitation light frequency (or wavelength).

The citrus ulcer RLS spectrum has obvious differences when compared with that of E. coli. The citrus ulcer spectrum has four resonance scattering peaks at 330, 425, 465, and 695 nm, as shown in Figure 13.4 [26]. Citrus ulcer has a good linear relation in $0.04 × 10^8$–$10.4 × 10^8$/mL range with an intensity of I_{330nm}, I_{425nm}, I_{465nm}, and I_{695nm}, which can be used for its quantitative analysis.

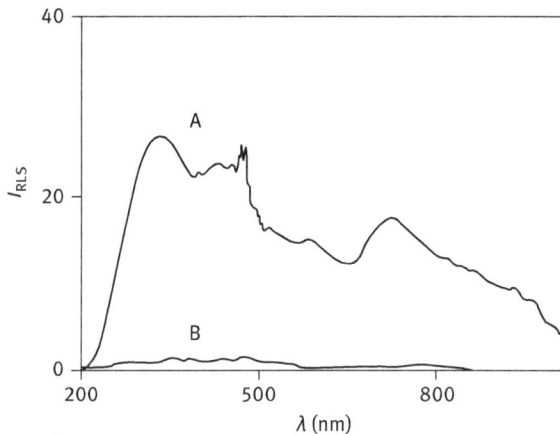

Figure 13.4: The RLS spectra of citrus ulcer. (A) Citrus ulcer of $5.22 × 10^8$; (B) distilled water.

13.2 Polarization resonance light scattering analysis of bioparticles

It is quite important to obtain microorganism-related information by characterization, identification, differentiation, and quantification of bioparticles. Current characterization and identification methods include optical spectrophotometry [26–29], flow cytometry [20, 25] different chromatography-mass spectrometry, capillary electrophoresis, surface resonance Raman technique, and so on. Although these methods have been applied, there are still a number of limitations. For example, the molecule selectivity of these methods may not have high microbial selectivity, and bacteria culture and separation may be needed which is time-consuming; and expensive apparatus and frequent operation may also be needed. Besides, cell growth conditions, time, and preparation method sometimes obviously change the cell component, and so repeatability is another problem. Thus, characterization and identification of microorganism still are limited in applications for its long analysis time, limited analysis amount, and so on.

The light polarization state is related to particle's size, structure, shape, refractive index, and so on. Since Tyndall found scattering light has polarization effect, the polarized scattering signal has been widely used in biology and medicine, especially, in the recent years. For example, a polarized scattering signal has been used for determination of the particle size, distribution in mammal cell [28], bacteria size [21, 29, 30], metal toxicity for bacteria [31], DNA superhelical structure density changing situation [32], protein and sodium dodecyl sulfate interaction [33]. On the basis of this, we put forward polarized synchronous light scattering, that is, by using fluorospectrophotometer with a polarizer, keeping excitation and emission wavelengths identical, and obtaining polarized light scattering signal through changing of polarizer' angel, eight microbiological particles are characterized [34, 35] (Table 13.2).

Table 13.2: Property description of six bacteria and two yeasts.

Bacteria name	Gram stain	Length (μm)	Width (μ)m	Shape
Bacillus subtilis	Positive	2.0–3.0	0.7–0.8	Rod shape
Bacillus thuringiensis	Positive	3–5	1.0–1.2	Rod shape
Bacillus megaterium	Positive	3.0–9.0	1.0–2.5	Rod shape
Staphylococcus aureus	Positive	0.8–1.0	0.8–1.0	Sphere
E. coli, BL-21	Negative	1.0–3.0	0.5	Rod shape
E. coli, DH5α	Negative	1.0–3.0	0.5	Rod shape
Saccharomyces cerevisiae	Fungus	3–10	1~5	Sphere or pillar
Schizosaccharomyces pombe	Fungus	10–20	6	Straight bar

As discussed in Section 13.1.2, we first obtained a resonance light scattering spectrum of these bacteria by an ordinary fluorospectrophotometer. When compared with an RLS spectrum, it can be found that these bioparticles have different features and a few similarities. For example, *S. cerevisiae* and *S. pombe* are similar in shape, while *S.*

pombe has better scattering ability than that of *S. cerevisiae* [35]. Apparently, their light scattering ability is related to their shape, size, and refractive index. *S. cerevisiae* is oral with a 10 µm size, while *S. pombe* is long rod with a long axis about 20 µm and a short axis about 6 µm.

B. subtilis, B. thuringiensi, B. megaterium, and *S. aureus* are similar [35]. They have similar spectra, while different from *S. cerevisiae* and *S. pombe,* they have different scattering abilities. Apparently, their light scattering ability is related to their shape, size, refractive index, and inner structure. *S. aureus* is spherical with a size of 0.8–1.0 µm, while the other three are rod bacteria with different sizes. It is interesting to note that *E. coli BL-21* and *E. coli DH5α* of different subspecies have quite different RLS spectra, [35] probably related to their gene expression or inner structure and plasma membrane.

It can be seen that a bacteria RLS spectrum is not enough to distinguish these bacteria, while its maximum scattering wavelength or other scattering wavelengths with a smaller interference can be used for quantitative detection of these cells (Table 13.3).

We use Mie theory to discuss the relationship light intensity of different polarized angle scattering, radiated by the polarized incident light, and different particles [34, 35]. Inspired by the principle of flow cytometry principle, we could distinguish cell groups based on forward light scattering and lateral light scattering related to cell size, shape, and inner structure. Figure 13.5 (left column) shows the scattering light signal of different polarized angles obtained by an ordinary fluorescence spectrophotometer with polarizer.

It can be seen from the figure that when compared with the RLS spectrum, the polarized light scattering spectrum could better distinguish the following bacteria and fungus. It can be seen from the polarized light scattering spectra that their characteristic scattering peaks are both in 400–550 nm for the above bacteria and fungus, with an increase in polarized angle; the intensity change trends for peaks at 550 nm are more obvious than that at 400 nm. Thus, the detection of polarized scattering signal helps obtain characteristic information of microorganism size, shape, and composition.

Further study of bioparticle-polarized angle scattering light, excited by the incident light of horizontal polarization, found that [35] when an incident light falls vertically on a bioparticle, the intensity is maximum at 0° polarized scattering angle, and then the intensity decreases with an increase in the angle. When the polarized angle is 90°, the intensity is nearly zero. By performing data analysis and disposal for bioparticle scattering intensity at different polarized angles (Figure 13.6), we find that fungus and bacteria can be well distinguished; while there are some repeatabilites for different bacterium, while *E. coli BL-21* and *B. megaterium, E. coli DH5α* and *S. aureus* are hard to distinguish. At this time, a synchronous resonance scattering spectrum and the shape of polarization scattering spectrum need to be combined for further differentiation.

Table 13.3: Calibration curves and analysis parameters of bacteria.

Bacteria type	Characteristic wavelength (nm)	Regression equation (c, cfu/mL)	Linear range (cfu/mL)	Detection limit (cfu/mL)	Relative coefficient (r)
S. cerevisiae	350	$I = 23.62+5.22\times10^{-4}c$	$2.0\times10^{4}-2.0\times10^{6}$	3.0×10^{3}	0.9999
S. pombe	350	$I = 1.70+1.22\times10^{-3}c$	$2.0\times10^{4}-2.0\times10^{6}$	3.0×10^{3}	0.9991
S. aureus	470	$I = -39.77+3.57\times10^{-5}c$	$1.7\times10^{5}-1.7\times10^{9}$	1.71×10^{4}	0.9995
E. coli BL-21	470	$I = 33.09+3.27\times10^{-5}c$	$4.0\times10^{5}-4.0\times10^{7}$	4.0×10^{4}	0.9994
E. coli DH5α	520	$I = 37.56+2.63\times10^{-6}c$	$7.0\times10^{5}-6.6\times10^{8}$	7.0×10^{4}	0.9995
B. subtilis	470	$I = 24.25+1.10\times10^{-5}c$	$1.5\times10^{5}-6.6\times10^{8}$	1.5×10^{4}	0.9997
B. thuringiensis	470	$I = -74.20+1.19\times10^{-4}c$	$3.3\times10^{5}-3.3\times10^{7}$	3.3×10^{4}	0.9987
B. megaterium	470	$I = 9.33+3.43\times10^{-5}c$	$5.0\times10^{5}-5.0\times10^{7}$	5.0×10^{4}	0.9998

Notes: RLS excitation and emission slit is 5.0 nm, and the photomultiplier voltage is 400 V.

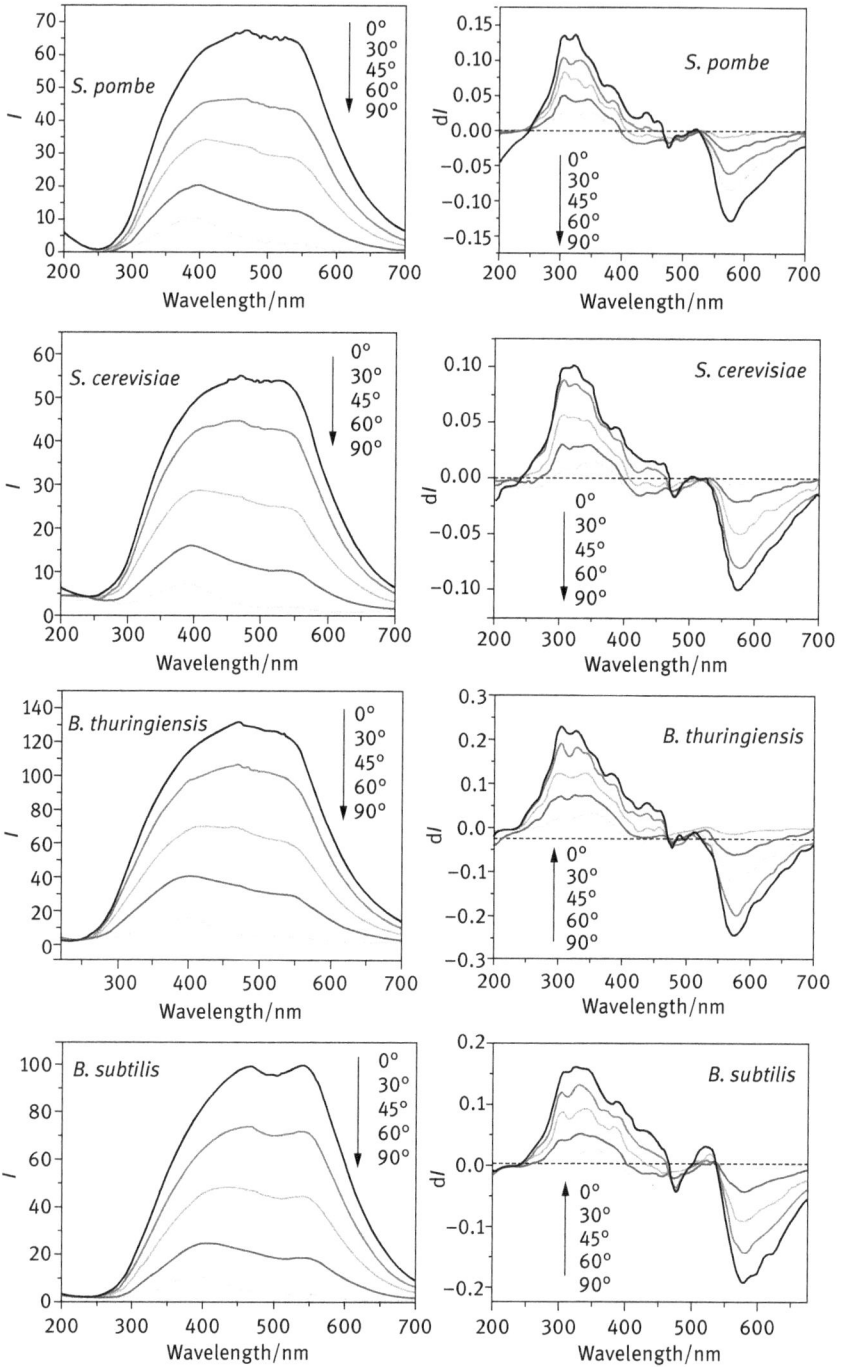

Figure 13.5: The polarized light scattering spectra (left column) and the first-order derivative RLS spectra (right column) of different microorganisms.

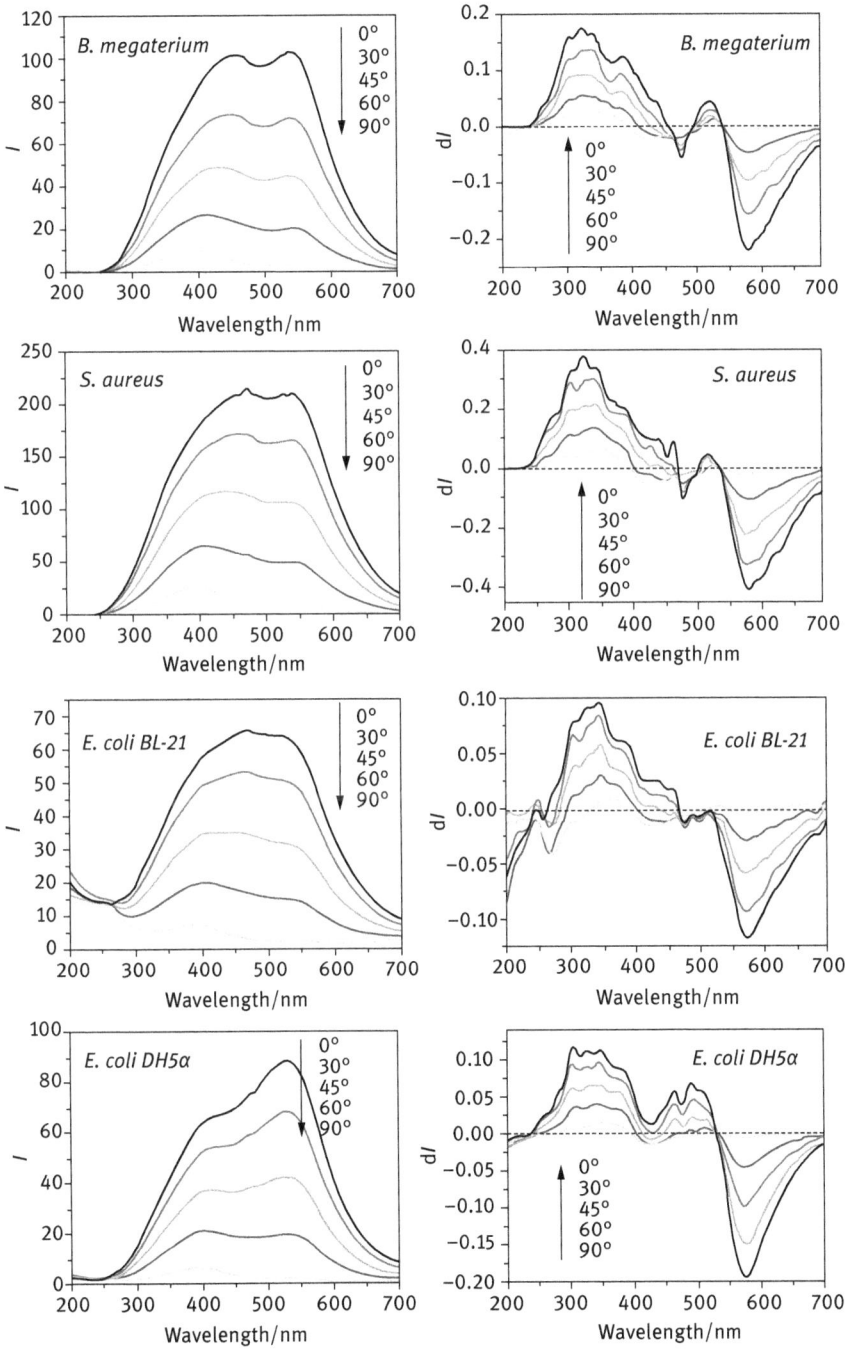

Figure 13.5: *continued*

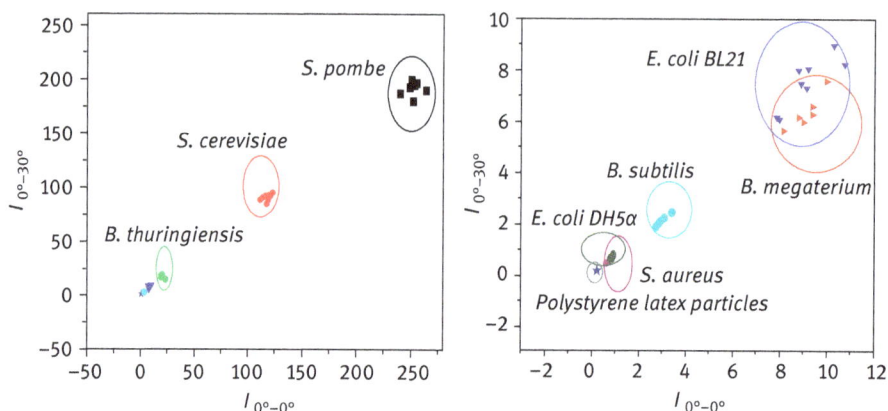

Figure 13.6: Differentiation of different microorganisms. Plotted with a scattering light intensity at 0° and 30° at 540 nm.

To better study the scattering spectrum of bioparticles, an ordinary fluorescence spectrophotometer is used to collect first derivative synchronous scattering spectrum of eight microorganisms listed in Table 13.2 [34, 35]. It can be seen from Figure 13.5 (right column) that the two spectra are quite different; they have positive peaks and negative ones, where the positive peak represents the enhanced synchronized light scattering signal, while the negative peak is the weakened synchronized light scattering signal.

Consider *S. cerevisiae* as an example, in 220–360 nm, it has many positive peaks, and 301.8 nm peak is the main characteristic peak; it has many negative peaks in 450~700 nm, and 580.2 nm peak is the main characteristic peak. Turning point of positive peaks and negative ones are the maximum synchronous light scattering signal at 350 nm. To better distinguish microorganism information, the authors introduced two parameters: the characteristic wavelength and distance of main positive peak and negative peak ($\lambda^+ - \lambda^-$), as listed in Table 13.4. Different microorganisms have different

Table 13.4: Bioparticle RLS and first derivative synchronous scattering spectra features.

Bacteria name	synchronous light scattering signal	Main positive peaks(λ^+, nm)	Main negative peaks(λ^-, nm)	$\lambda^+ - \lambda^-$(nm)[a]
S. pombe	350	314.6	575.8	261.2
S. cerevisiae	350	301.8	580.2	278.4
B. megaterium	470	297.6	587.0	289.4
B. subtilis	470	300.8	583.4	282.6
B. thuringiensis	470	313.4	569.2	256.8
S. aureu	470	302.6	593.4	290.8
E. coli. BL-21	470	300.8	583.0	282.2
E. coli DH5α	470	295.8	573.4	277.6

[a] The wavelength shift of main negative peaks and main positive peaks.

$\lambda^{+}-\lambda^{-}$, which may be related to microorganism's size, shape, refractive index, and inner structure.

13.3 Application of RLS in cell analysis

A cell is the basic unit of living as well as basic unit of physiological function. Mammalian cells are composed of cell membrane, cytoplasm, mitochondria, lysosomes, Golgi body, endoplasmic reticulum, nucleus and other subcellular structures. They play their special function to keep normal physiological function of a cell. Analyzing subcellular structure's light scattering properties has a reference value for real monitoring of cell skeleton and medium dynamic changes.

13.3.1 Interaction of light and cell

The light and cell interaction mechanism study has great theoretical guidance and promotion for the development of phototherapy; it is expected to provide a practical foundation for further establishment of a rapid, lossless histopathological assessment method. Scattering is caused by the refractive index difference. When we think about how a cell interacts with the light, a cell can be seen as an object covered by different index particles.

With cell's different physical composition and component size, different locations have different refractive indices. Organelles and suborganelles with different refractive indices are the main scatterers. A cell is also a main small-angle scatterer, widely used in the analysis as flow cytometry. While for in situ light scattering diagnostic technique, as a cell has the same refractive index with the surrounding cell or tissue, the cell has no great study value. At this time, specific organelle becomes an important potential scatterer; if the cell nucleus is a larger organelle in a cell, its size is closely related to the tumor development. The sizes of other scatterers in a cell, including mitochondria (0.5–1.5 µm), lysosome (0.5 µm), and peroxidase (0.5 µm), are related to light wavelength, showing that they could serve as a vital backward scatterer.

13.3.2 Light scattering analysis of cell

Till now, there has been no mathematical model for complex-cell scattering; there are still a number of challenges in establishing a relationship of the cell structure and scattering light. In cell scattering, the particle size is related to specific operations of optical imaging and electromagnetic methods. Although Mie theory has been widely used in scatterer analysis, a homogeneous sphere is needed as a cell model [36]. Another common method uses a covered sphere as the model [37], where the sphere and cover have different refractive indices, and can be used to model with homogeneous spherical nucleus.

Anomalous diffraction approximations, multipole solutions [39], and T-matrix computations [40] have also been used to predict cell scattering. These methods need finite geometry and refractive index hypothesis; this section will introduce light scattering of a whole cell and organelle.

Normally, in a cell and surrounding cells, tissues have the same refractive index. Thus, in the analysis of light scattering, although the study of overall cell light scattering does not have a great value, when a cell is separated from the surrounding cell or tissue, single cell scattering will play an important role in the analysis of its physiological properties.

13.3.2.1 Flow cytometer

A flow cytometer (FCM) is an equipment to represent cell's physical, chemical, and biological properties, and is able to acquire biophysical characteristics of diffused cells rapidly; it is normally used for automatic analysis and automatic sorting based on cell properties.

The basic principle of an FCM is to flow the cell through fluid of hydrodynamic focusing, and to obtain fluid light scattering signal and fluorescence signal when laser falls from a different angle. Signal gathering angle is to install a series of optical tester forward and side of incident light, each detector aims at intersection point of aimed flow and laser. A detector is for forward scattering signal, on the same line with laser, while there are detectors at vertical angle with the laser, and a detector is for side scattering signal, and the others are for fluorescence signal. Thus, when a suspension particle passes through the intersection point, each detector records forward and side scattering signals. If the fluorescence probe is used or cell autofluorescence is acquired, then we use the obtained scattering signal and fluorescence signal to represent physical and chemical properties of each particle.

Forward scattering is related to cell size, while side scattering is closely related to cell nucleus shape, particle type, or membrane roughness level (Figures 13.7 and 13.8). Hence, different size particles and cell samples of different containers produce different forward and side light scattering intensities through the detector, which can be used to analyze the sample's particle size, inner particle, and other information.

For example, when anticoagulation blood sample is added in an FCM, each cell in blood goes through the detector as an individual cell particle; their forward and side light scattering signals are collected by the detector. We plot forward scattering

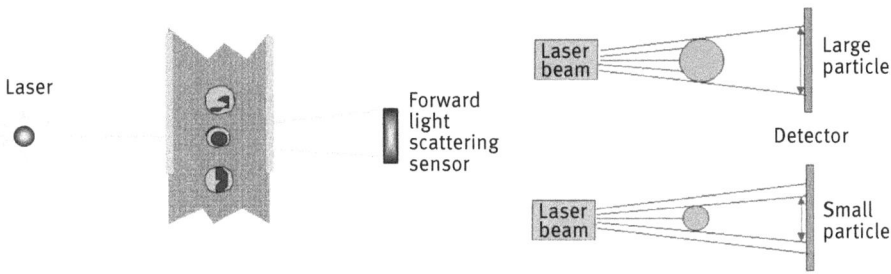

Figure 13.7: Forward scattering analysis and particle diameter relation with an FCM.

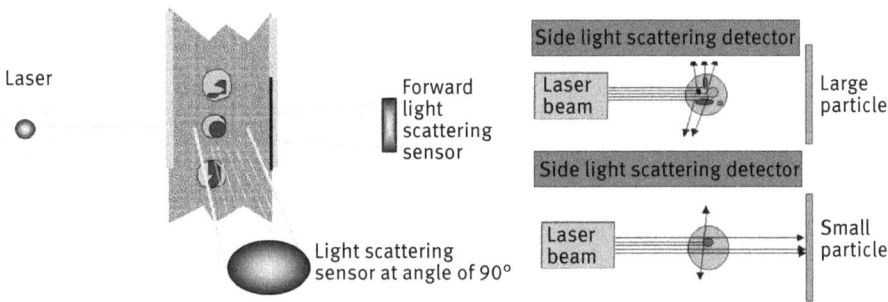

Figure 13.8: Side scattering analysis and particle size relation with an FCM.

signal intensity as the y-coordinate and side signal intensity as the x-coordinate to obtain Figure 13.9. It can be seen that there are three groups of particles distinguished by the scattering intensity. Based on blood cell component, it can be known that lymphocyte has minimum particle and container, and then there is monocyte, neutrophile granulocyte has largest particle size and rich container.

13.3.2.2 Highly sensitive FCM

Figures 13.7–13.9 show regular FCM, widely applied in life science research and clinical examination. With the advancement of equipment science, highly sensitive flow cytometer (HSFCM) came into existence. With features of rapidness and fastness, it has wide application prospects in single nanoparticle multiple parameter characterization, single virus bacteria, subcellular organelle, and other bionanoparticle biochemical tests. For example, Professor Yan's group, from Xiamen University, adopted fluid dynamic focusing technology to condense the sample's flow diameter to a few micrometers and increase its laser detection area and

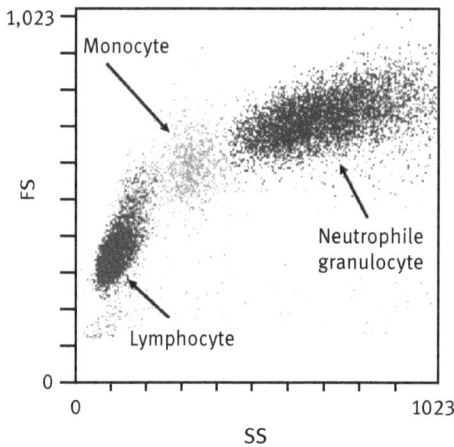

Figure 13.9: Flow cytometry analysis of the blood sample.

upgrade; they studied HSFCM, which greatly reduced the background signal, and successfully applied in the detection of DNA section length, characterization of single nanoparticle multiple parameter, determination of concentration, and rapid identification of pathogenic bacterium. The signal-to-noise ratio for the fluorescence detection of single algae red protein has reached 1.7; a high concentration sample detection ratio reaches over hundred times per second, and the analysis speed is greatly enhanced.

13.3.2.3 RLS analysis of classic cells

Besides FCM, an ordinary fluorescence spectrophotometer has also been applied in cell scattering analysis. The study of complete blood cell and rabbit erythrocyte scattering spectrum shows that when the excitation and emission wavelengths are consistent, their RLS spectra can be obtained [41, 42]. Rabbit erythrocyte has five RLS peaks at 308, 468, 560, 600, and 936 nm; the strongest peak wavelength is at 468 nm [42]. Complete blood cell and erythrocyte have RLS peaks at 310, 470, 560, and 600 nm [41]; leukocyte and serum have two weaker RLS peaks at 210 and 470 nm.

13.3.3 RLS analysis of mitochondria

Mitochondria quantity in cell is a mark of cellular metabolism. The more vigorous the cellular metabolism, the more mitochondria there are. As the main function of mitochondria is to transfer oxidization energy into ATP for other processes, it can be seen as "cell energy factory." In addition, mitochondria plays an important role

in metabolism reactions such as cell apoptosis, glutamate acceptor induced excitatory toxicity effect, cell proliferation, cell oxidation reduction state adjustment,
heme synthesis, cholesterol synthesis, heat production (to keep warm), and so on.
If some genes related to mitochondria had gene mutation, mitochondria diseases
can be caused. Thus, it is important to study properties and functions of
mitochondria.

13.3.3.1 RLS of different morphological mitochondria

The morphology of mitochondria changes, with expansion and condensation,
and has a close relationship with living body functions and pathology [43]. The
latest study shows that mitochondrial volume disorders may be caused by
opening of permeation hole of the oxidizing agent. When an ion channel
opens or expands to break mitochondria membrane, cytochrome is released,
causing cell necrosis or apoptosis [44]. Besides forward scattering reflects cell
volume, Mourant et al. mentioned that cell light scattering is mainly decided by
the inner organelle size and composition rather than the cell's integral size [45].
In 1998, the Mourant group tested optical properties of a complete cell, revealing the scatterer distribution with a particle size of 0.2–1 μm [45]. Their later
study confirmed that the main scatterer's particle size is smaller than the cell
nucleus; mitochondria around 0.5 μm is the main scatterer [46, 47].

For a long time, mitochondria needs to be separated from cell in advance of its
light scattering analysis, then property is detected separately [48]. The separated
mitochondria suspension can be used with a spectrophotometer or an FCM, and the
obtained information is for mitochondria in a normal and condensed state [49, 50].
Except for use of mitochondria size changes, light scattering technique is applied in
its permeability transition study [51–53], it can also be used for mitochondria shape
changes in apoptosis process [54–57].

Of the mentioned studies, in vitro mitochondria have been the main method
used for the analysis of optical shape. Mitochondria need to be separated from the
cell in advance; there has been a few optical measurement reports about mitochondria shape changes in a complete cell. Until 2002, Boustany et al. used an
optical scattering imaging microscope for the characterization of monolayer epidermal cells mitochondria, coincubated with calcium ion [58]. In their study, the
scattering spectrophotometer and microscope composed light scattering imaging,
where the scattering images of a monolayer cell could directly code scatterer's
large angle–narrow angle ratio in full vision [58]. Although there are other scatterers in cytoplasm, early studies have shown that mitochondria made more significant contributions to cell scattering. The method can be used for in-cell
mitochondria analysis. The authors used the facility to represent light scattering
images of polystyrene bead suspension; the result was consistent with

theoretically predicted results, where scattering image rates linearly decreased with a decrease in bead particles [58]. Then they used a biological system as the study object for calcium ion-induced mitochondria shape relation with the scattering image ratio; they further compared observed scattering changes with ellipsoid theoretical light scattering [58].

13.3.3.2 Mitochondria distribution in cell

Besides light scattering relation with mitochondria morphagy, one of main research hotspots about light scattering study is mitochondria distribution in cell. Studies have shown that, in normal cell, mitochondria concentrate near the nucleus, while mitochondria distribute randomly in cancer cell [59]. Cancer cell disposed by anticancer drugs, mitochondria concentration near nucleus can be observed [60, 61]. Thus, a label-free and noninvasive light scattering technique will play its role in the diagnosis of cancer.

Finite-difference time domain is the most common method for mitochondria light scattering study. In 2009, Sua and coworkers established a cell model of nucleus area and random distribution [62]. They detected and compared the two cell models, and used three-dimensional finite-difference time domain to simulate scattering of two types and detecting mitochondria distribution in models. Of course, real living body is quite complex rather than a simple sphere structure, but a combined and interacted network. Sua's study brought the first step in mitochondria distribution influence on light scattering.

13.3.4 RLS analysis of lysosome

Normally, in a study of mitochondria light scattering, lysosome distribution to scattered light is assumed to be negligible. Until in 2007, Foster et al. dyed cell with high extinction coefficient lysosome position dye and used an absorption sphere model to obtain lysosome size, distribution, and contribution to cell light scattering [63]. As the model only identifies lysosome refractive index, the group redesigned their experiment to verify the accuracy of lysosome distribution to scattering. *N*-aspartyl chlorin e6 (Npe6) is a common photosensitizer, positioned in lysosome [64]. Research has shown that if the cell is preincubated with a lysosome-positioned photosensitizer, after suitable optical excitation, it causes oxygen tension for lysosome, and then lysosome starts photodamage [64–66]. On the basis of this, Wilson and Foster used angularly resolved light scattering measurements to detect lysosome ablation, normal EMT6 cell lysosome particle size distribution, and contribution to cell light scattering [67].

13.4 Applications of light scattering in tumor analysis

In the past few decades, many technologies have been developed to determine physiological status with cell and tissue light scattering properties, including direct image technique [68, 69], elastic scattering spectroscopy [70], and photon migration method [71]. Although cell and tissue light scattering is related to their form and biochemical structure, the relation is not clear. Thus, although elastic scattering spectroscopy has been detected differences between normal tissue and pathological tissue, what specific physical and chemical changes these differences represent are still unclear.

As it still lacks quantitative model for microscopic observation prediction of elastic scattering changes, it is unable to evaluate effectiveness of spectral diagnosis, or evaluate system fluctuation effect on the analysis result. Although there are still problems to be solved in cell and tissue analysis, scientific workers have successfully taken the first step. This section will focus on the applications of light scattering technique in tumor analysis.

A tumor is cell abnormal lesions, normally including benign and malignant tumors. The former has slow growth speed with a smoother surface, and it does not invade adjacent normal tissues; the latter is called cancer, malignant tumors, and its mechanism is caused by out-of-control growth and proliferation mechanism. As cancer has malignant germs without being adjusted, it invades (soaks) adjacent normal tissues locally, and even transfers to other parts of the body through circulatory or lymphatic systems in the body, for distant metastases.

13.4.1 Light scattering properties of tissue

Tissue is a medium with strong scattering properties. At most wavelengths, its light scattering ability is far higher than the light absorption ability, as shown in Figure 13.10 [72]. For different tissue features, an average distance of light permeation through tissue is 0.02–0.05 cm [73], and an average distance of infrared absorbing is over 0.2 cm; the average distance reduces with the wavelength, and it can be as small as 0.01 cm. The blood absorption peak is in the visible light region [74].

The example shows a possible travel path in tissue with two forms of light absorption and light scattering, while scattering takes relatively major proportion. Scattering usually changes the direction of light path with an acute angle.

The tissue feature is expressed as the scattering parameter (μ_s), which is the reciprocal of the scattering average distance when light passes through the tissue; absorption parameter (μ_a) is the reciprocal of operation distance before absorption. Delivery of light in the tissue is related to the scattering angle; when light has

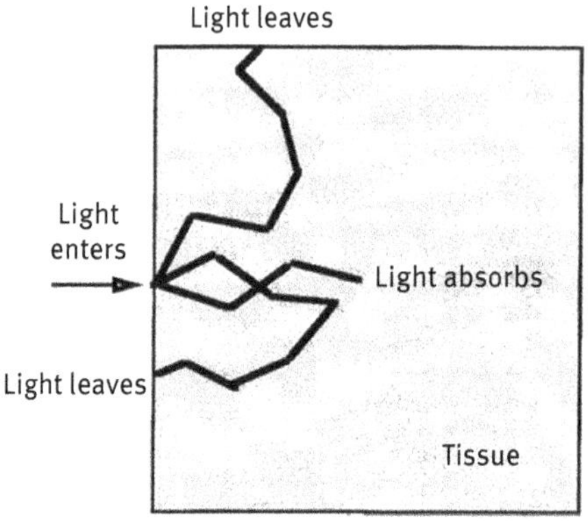

Figure 13.10: The light travel path in a tissue.

scattering at an angle θ, it can be expressed in terms of phase function $P(\theta)$ and scattering angle cosine:

$$g = \int P(\theta) \cos \theta d\Omega \qquad (13.1)$$

For a medium with high forward scattering ability, g is close to the value 1, while for isotropic scattering, $g=0$. The tissue's g value is 0.80–0.97. The scattering parameter is related to total forward scattering, and it can be expressed as follows:

$$\mu_s' = \mu_s(1-g) \qquad (13.2)$$

where μ_s' is usually used for tissue scattering. μ_s' is adjacent to direct recognition of the scattering medium; if an analyte (as milk solution) has strong scattering, it has a higher μ_s' value. Otherwise, for an analyte with weak scattering (as skim milk), it has a lower μ_s' value.

    Related tissue feature study now mainly focuses on two aspects: one is the light scattering properties depend on the wavelength, closely related to the shape of the light scatterer; and the second is tissue polarization properties (as relative position of electric field direction and light propagation direction). For example, when light propagates in the Z-axis direction while the electric field vibrates with X-axis or Y-axis, light has linear polarization. For different scatterers, light scattering varies light polarization to different degrees.

    Freyer analyzed the wavelength dependency and polarization-dependent light scattering properties of cell suspension and found that there are obvious differences between light scattering properties of tumor and nontumor cells, caused by average particle changes of scattering center with parameter of dozen nanometers [72]. The results show that light scattering technology is possible to be used for noninvasive cancer diagnosis. Later, many science researchers started to explore the feasibility of applying light scattering in tumor diagnosis.

13.4.2 Light scattering properties of cancer tissue

Changes of cell nucleus include enlarged nucleus, asymmetric nucleus form, increased DNA content, breakage of primary chromatin, and the most important indices of abnormal epidermis cell growth [75–77]. DNA multiple is also main factors for disease evaluation and treatment prediction [78, 79]. Optical diagnosis technology could detect epidermis cell structure without living tissue detection, obtain nucleus quantitative information without invasion, and apply to precancerous lesions at early stage of diagnosis. Although uniform spherical scatterer is used as model, Mie theory can be used to obtain quantitative morphological information with light scattering analysis and polarized scattering technology [77, 80, 81]. As cells are complex and inhomogeneous, Mie scatterers cannot explain all of them. Thus, scientific researchers spared no effort in establishing light scattering property model and method for normal and cancer tissues.

Studies have shown that there are obvious changes in light scattering properties of normal and cancer tissues [72, 77, 82–94]. One of the distinct changes is the increase of scattering section in cancer tissue and cells. In 2003, Richards-Kortum group [77] found in their study of angle-dependent light scattering properties with normal cells and Cervical epithelia cancer cells, and the angle-dependent light scattering properties of different depth of Cervical epithelia cells is different. The scattering section of Cervical epithelia cancer cells increase considerably. From epithelium to the middle layer, the scattering section gets smaller while the section increases for basal layer cells. Scattering changing trends of epidermis cells can also be observed in cancer tissue-detached confocal imaging. Backman et al. detected backward scattering spectrum, angle, polarized ratio, and other properties in living body tissue and cells to obtain their unique quantitative scattering information; they further preliminarily discussed applications of their method in early stage precancerous changes diagnosis. With a model of mouse with tumor colony, they found that the density of red blood cells, fractal dimension, and average roundness of subcellular structure all had significant changes in early stage of cancer [86].

After concertation of epithelial cell, nucleus increases and light scattering properties change with that (Figure 13.11). On the basis of this, the resolution is improved to construct a nanometer level recognition in the overall area. Figure 13.12 shows a special scattering spectrum to detect cancerous nucleus changes with the technique; based on its changes, a new light scattering imaging diagnostic unit is developed to judge whether there is concertation in the cell [87, 88]. To lead instrument probe into the body, constant probe with organ in wall to record scattering spectrum of epithelial cell, which will not damage tissue at all. Figure 13.12 shows in situ reflect of wider

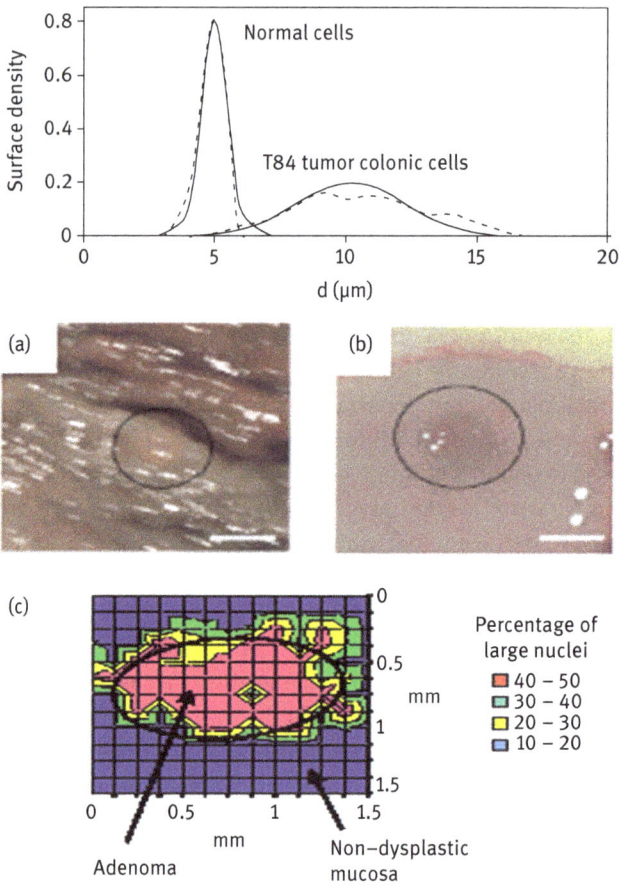

Figure 13.11: Light scattering comparison of tumor cell and normal cells, adenomatous polyp backward scattering photo and data analysis obtained based on this [88].

nucleic structure changes of tissue without tissue excision, where (c) and (d) are normal and abnormal development of four epithelial cells, respectively.

Compared with traditional cancer diagnosis methods, such as electrocardiogram, imaging, and biopsy, an optical technique has advantages of fast speed, no need of tranquilizer, and noninvasion in cancer diagnosis. In situ online cancer diagnosis study has indicated obvious differences of light scattering properties in cancer cells or tissues, and normal cells or tissues. There are not many studies for potential factors of these changes, which are the main challenges of light scattering technique for clinical cancer diagnosis applications. We believe that a rapid, noninvasive light scattering tumor diagnosis method will play an important role in cancer study and diagnosis.

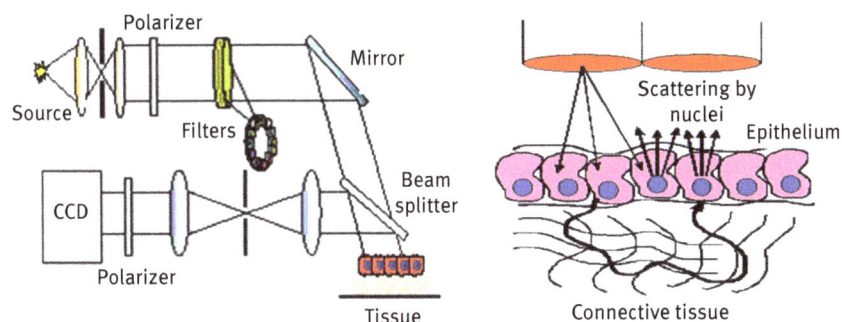

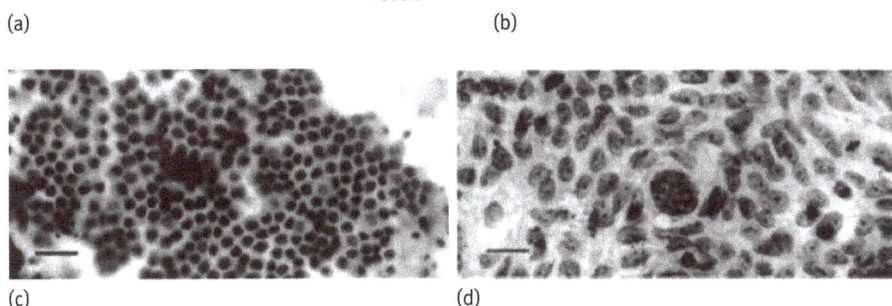

Figure 13.12: Backward light scattering imaging diagnostic unit and its application in cancer tissue section [87]. (a) The optical path of backward light scattering imaging diagnostic unit; (b) backward light scattering propagation of tissue; (c) backward light scattering imaging of normal cells; (d) backward light scattering imaging of cancer cells.

References

[1] Tanford C. Physical chemistry of macromolecules. New York and London: John Wiley and Sons, Inc., 1961, Vol. 275.

[2] Schure M R, Pallkar S A. Accuracy estimation of multiangle light scattering detectors utilized for polydisperse particle characterization with field-flow fractionation techniques: A simulation study. Anal. Chem., 2002, 74(3): 684–695.

[3] Andersson M, Wittgren B, Wahlund, K G. Ultrahigh molar mass component detected in ethyl-hydroxyethyl cellulose by asymmetrical flow field-flow fractionation coupled to multiangle light scattering. Anal. Chem., 2001, 73(20): 4852–4861.

[4] Kano K, Fukuda K, Wakami H, et al. Factors influencing self-aggregation tendencies of cationic porphyrins in aqueous solution. J. Am. Chem. Soc., 2000, 122(31): 7494–7502.

[5] Luca G D, Romeo A, Scolaro L M. Role of counteranions in acid-induced aggregation of isomeric tetrapyridylporphyrins in organic solvents. J. Phys. Chem. B, 2005, 109(15): 7149–7158.

[6] Kubát P, Lang K, Zelinger Z, et al. Aggregation and photophysical properties of water-soluble sapphyrins. Chem. Phys. Lett., 2004, 395: 82–86.

[7] Togash D M, Costa S M B, Sobral A J F N, et al. Self-aggregation of lipophilic porphyrins in reverse micelles of aerosol OT. J. Phys. Chem. B, 2004, 108(31): 11344–11356.

[8] Huang C Z, Li Y F, Deng S Y, et al. Assembly of methylene blue on nucleic acid template as studied by resonance light-scattering technique and determination of nucleic acids of nano-gram. Bull. Chem. Soc. Jpn., 1999, 72: 1501–1508.

[9]   Bao P, Frutos A G, Greef C, et al. High-sensitivity detection of DNA hybridization on microarrays using resonance light scattering. Anal. Chem., 2002, 74(8): 1792–1797.

[10] Huang C Z, Li K A, Tong S Y. Determination of nucleic acids by a resonance light-scattering technique with α,β,γ,δ-tetrakis [4- (trimethylammoniumyl) phenyl]porphine. Anal. Chem., 1996, 68(13): 2259–2263.

[11] Huang C Z, Li K A, Tong S Y. Determination of nanograms of nucleic acids by their enhancement effect on the resonance light scattering of the cobalt (II)/4-[(5-chloro-2-pyridyl)azo]-1,3-diaminobenzene complex. Anal. Chem., 1997, 69(3): 514–520.

[12] Chen Y J, Yang J H, Wu X, et al. Resonance light scattering technique for the determination of proteins with resorcinol yellow and OP. Talanta, 2002, 58(5): 869–874.

[13] Huang C Z, Li Y F, Feng P. Determination of proteins with α, β, γ, δ-tetrakis(4-sulfophenyl) porphine by measuring the enhanced resonance light scattering at the air/liquid interface. Anal. Chem. Acta, 2001, 443: 73–80.

[14] Yguerabide J, Yguerabide E E. Light-scattering submicroscopic particles as highly fluorescent analogs and their use as tracer labels in clinical and biological applications. Anal. Biochem., 1998, 262: 137–156.

[15] Yguerabide J, Yguerabide E E. Light-scattering submicroscopic particles as highly fluorescent analogs and their use as tracer labels in clinical and biological applications. Anal. Biochem., 1998, 262: 157–176.

[16] Zhou D Q. Text book of microbiology. Beijing: Higher Education Press, 2002.

[17] Du L X, et al. Experimental technology of industrial microbiology. *Tianjin:Tianjin Science and Technology Press*, 1992.

[18] Zhao J, Dai H, Tan W S. Effect of inner designs of spinner flasks on serum-free cultivation of insect cells in suspension. Chinese J. Biotechnol., 2000, 16(6): 755–758.

[19] Kepner R L Jr, Pratt J R. Use of fluorochromes for direct enumeration of total bacteria in environmental samples: Past and present. Microbiol. Mol. Biol Rev., 1994, 58(4): 603–615.

[20] Vives-Rego J, Lebaron P, Caron G N. Current and future applications of flow cytometry in aquatic microbiology. FEMS Microbiol. Rev., 2000, 24: 429–448.

[21] Bronk B V, Merwe W P V D, Stanley M. In vivo measure of average bacterial cell size from a polarized light scattering function. Cytometry, 1992, 13: 155–162.

[22] Bronk B V, Druger S D. Light scattering calculations exploring sensitivity of depolarization ratio to shape changes. II. Single rod-shaped vegetative bacteria in air. Appl. Opt., 2009, 48(30): 5655–5663.

[23] Katz, A, Alimova, A, Xu, M, et al. Bacteria size determination by elastic light scattering. IEEE J. Select.Top. Quant. Electron., 2003, 9(2): 277–287.

[24] Chen S F, Huang C Z, Tan K J. Light scattering method for rapid sensitive determination and characterization of beer yeast. Chemical J. Chinese U., 2006, 27(6): 1023–1027.

[25] Malacrinò P, Zapparoli G, Torriani S, et al. Rapid detection of viable yeasts and bacteria in wine by flow cytometry. J. Microbiol. Meth., 2001, 45(2): 127–134.

[26] Liao Z H, Luo Y H, Jiang Z L, et al. Resonance scattering spectroscopic study of Colibacillus. Anal. test. tech. instr., 2003, 9(2): 65–69.

[27] Xie J Y, Jiang Z L. RLS spectrum of orange ulcer germs. Chinese J. Phys. Chem., 2001, 17(5): 406–411.

[28] Bartlett M, Huang G, Larcom L. Measurement of particle size – in mammalian cells in vitro by use of polarized light spectroscopy. Appl. Opt., 2004, 43(6): 1296–1307.

[29] Bronk B V, Druger S D, Czege J. Measuring diameters of rod-shaped bacteria in vivo with polarized light scattering. Biophys. J., 1995, 69(3): 1170–1177.

[30] Van de Merwe W P, Li Z Z, Bronk B V. Polarized light scattering for rapid observation of bacterial size changes. Biophys. J., 1997, 73(1): 500–506.

[31] Bronk B V, Li Z Z, Czeg J. Polarized light scattering as a rapid and sensitive assay for metal toxicity to bacteria. J. Appl. Toxicol., 2001, 21(2): 107–113.

[32] Nicolini C, Diaspro A, Bertolotto M. Changes in DNA superhelical density monitored by polarized light scattering. Biophys. Res. Commun., 1991, 177(3): 1313–1318.

[33] Zhao X H, Huang C Z. Light scattering polarization properties of protein interacting with sodium dodecyl sulfate. Chinese Sci. Bull., 2006, 51(23): 2729–2732.

[34] Wang L J, Huang C Z. Polarized synchronous light scattering used for differentiating microorganisms. J. Southw. Univ. (Nat. Sci. Ed.), 2007, 29(7): 28–31.

[35] Huang C Z, Chen S F. Quantitation and differentiation of bioparticles based on the measurements of light-scattering signals with a common spectrofluorometer. J. Phys. Chem. B, 2008, 112(37): 11785–11793.

[36] Wilson J D, Foster T H. Mie theory interpretations of light scattering from intact cells. Opt. Lett., 2005, 30: 2442–2444.

[37] Brunsting A, Mullaney P. Light scattering from coated spheres: Model for biological cells. Appl. Opt., 1972, 11: 675–680.

[38] Streekstra G, Hoekstra A, Nijhof E, et al. Light scattering by red blood cells in ektacytometry: Fraunhofer versus anomalous diffraction. Appl. Opt., 1993, 32, 2266–2272.

[39] Videen G, Ngo D. Light scattering multiple solution for a cell. J. Biomed. Opt., 1998, 3: 213–220.

[40] Nilsson A, Alsholm P, Karlsson A, et al. T-matrix computations of light scattering by red blood cells. Appl. Opt., 1998, 37, 2735–2748.

[41] Xie J Y, Jiang Z L, Zhong F X, et al. Spectral properties of blood cell. Anal. Test Tech. Instr., 2000, 6(3): 178–181.

[42] Yang M M. Jiang Z L. Resonance scattering spectral study of rabbit red cell. Spectrosc. Spectr. Anal., 2002, 22(2): 204–208.

[43] Hackenbrock C R. Ultrastructural bases for metabolically linked mechanical activity in mitochondria. J. Cell Biol., 1966, 30: 269–297.

[44] Green D R, Reed J C. Mitochondria and apoptosis. Science, 1998, 281: 1309–1313.

[45] Mourant J R, Freyer J P, Hielscher A H, et al. Mechanisms of light scattering from biological cells relevant to noninvasive optical-tissue diagnosis. Appl. Opt., 1998, 37: 3586–3593.

[46] Mourant J R, Johnson T M, Carpenter S, et al. Polarized angular dependent spectroscopy of epithelial cells and epithelial cell nuclei to determine the size scale of scattering structures. J. Biomed. Opt., 2002, 7: 378–387.

[47] Mourant J R, Johnson T M, Freyer J P. Characterizing mammalian cells and cell phantoms by polarized backscattering fiberoptic measurements. Appl. Opt., 2001, 28: 5114–5123.

[48] Packer L. Metabolic and structural states of mitochondria. J. Biol. Chem., 1960, 235, 242–249.

[49] Hackenbrock C R. Ultrastructural bases for metabolically linked mechanical activity in mitochondria. J. Cell Biol., 1966, 30: 269–297.

[50] Hunter D R, Haworth R A, Southward J H. Relationship between configuration, function, and permeability in calcium-treated mitochondria. J. Biol. Chem., 1976, 251: 5069–5077.

[51] Hoek J B, Farber J L, Thomas A P, et al. Calcium ion-dependent signaling and mitochondrial dysfunction: mitochondrial calcium uptake during hormonal stimulation in intact liver cells and its implication for the mitochondrial permeability transition. Biochim. Biophys. Acta, 1995, 1271: 93–102.

[52] Kristal B S, Dubinsky J M. Mitochondrial permeability transition in the central nervous system: Induction by calcium cycling dependent and-independent pathways. J. Neurochem., 1997, 69: 524–538.

[53] Scorrano L, Petronilli V, Bernardi P. On the voltage dependence of the mitochondrial permeability transition pore. J. Biol. Chem., 1997, 272: 12295–12299.

[54] Vander-Heiden M G, Chandel N S, Williamson E K, et al. Bcl-xL regulates the membrane potential and volume homeostasis of mitochondria. Cell, 1997, 91: 627–637.

[55] Zamzami N, Susin S A, Marchetti P, et al. Mitochondrial control of apoptosis. J. Exp. Med., 1996, 183: 1533–1544.

[56] Finucane D M, Bossy-Wetzel E, Waterhouse N J, et al. Bax-induced caspase activation and apoptosis via cytochrome c release from mitochondria is inhibitable by Bcl-xL. J. Biol. Chem., 1999, 274: 2225–2233.

[57] Narita M, Shimizu S, Ito T, et al. Bax interacts with the permeability transition pore to induce permeability transition and cytochrome c release in isolated mitochondria. Proc. Natl. Acad. Sci. USA, 1998, 95: 14681–14686.

[58] Boustany N N, Drezek R, Thakor N V. Calcium-induced alterations in mitochondrial morphology quantified in situ with optical scattering imaging. Biophys. J., 2002, 83: 1691–1700.

[59] Gourley P L, Hendricks J K, McDonald A E, et al. Ultrafast nanolaser flow device for detecting cancer in single cells. Biomed. Microdevices, 2005, 7(4): 331–339.

[60] Thomas W D, Zhang X D, Franco A V, et al. TNF-related apoptosis-inducing ligand induced apoptosis of melanoma is associated with changes in mitochondrial membrane potential and perinuclear clustering of mitochondria. J. Immunol., 2000, 165(10): 5613–5620.

[61] Liu L Y, Vo A, Liu G Q, et al. Distinct structural domains within C19ORF5 support association with stabilized microtubules and mitochondrial aggregation and genome destruction. Cancer Res., 2005, 65(10): 4191–4201.

[62] Sua X T, Singha K, Rozmusa W, et al. Light scattering characterization of mitochondrial aggregation in single cells. Opt. Express, 2009, 17(16): 13381–13388.

[63] Wilson J D, Cottrell W J, Foster T H. Index-of-refractiondependent sub-cellular light scattering observed with organelle specific dyes. J. Biomed. Opt., 2007, 12: 014010.

[64] Reiners J J Jr, Caruso J A, Mathieu P, et al. Release of cytochrome c and activation of procaspase-9 following lysosomal photo damage involves Bid cleavage. Cell Death Differ., 2002, 9: 934–944.

[65] Lin C W, Shulok J R, Kirley S D, et al. Photodynamic destruction of lysosomes mediated by Nile blue photosensitizers. Photochem. Photobiol., 1993, 58: 81–91.

[66] Georgakoudi I, Foster T H. Effects of the subcellular redistribution of two Nile blue derivatives on photodynamic oxygen consumption. Photochem. Photobiol., 1998, 68: 115–122.

[67] Wilson J D, Foster T H. Characterization of lysosomal contribution to whole-cell light scattering by organelle ablation. J. Biomed. Opt., 2007, 12(3): 030503.

[68] Rajadhyaksha M, Grossman M, Esterwitz D, et al. In-vivo confocal scanning laser microscopy of human skin: Melanin provides strong contrast. J. Invest. Dermatol., 1995, 104: 946–952.

[69] Izaat J, Hee M, Huang D, et al. Micron-resolution biomedical engineering with optical coherence tomography. Opt. Photon. News, 1993, 4: 14–19.

[70] Mourant J, Bigio I, Boyer J, et al. Spectroscopic diagnosis of bladder cancer with elastic light scattering. Lasers Surg. Med., 1995, 17: 350–357.

[71] Fishkin J, Coquoz O, Anderson E, et al. Frequency-domain photon migration measurements of normal and malignant tissue optical properties in a human subject. Appl. Opt., 1997, 36: 10–20.

[72] Mourant J R, Hielscher A H, Eick A A, et al. Evidence of intrinsic differences in the light scattering properties of tumorigenic and nontumorigenic cells. Cancer(Cancer Cytopathol), 1998, 84: 366–374.

[73] Cheong W-F, Prahl S A, Welch A J. A review of the optical properties of biological tissues. IEEE J. Quant. Elect., 1990, 26: 2166–2185.

[74] Liu H, Chance B, Hielscher A H, et al. Influence of blood vessels on the measurement of hemoglobin oxygenation as determined by time-resolved reflectance spectroscopy. Med. Phys., 1995, 22: 1209–1217.

[75] Thiran J, Macq B. Morphological feature extraction for the classification of digital images of cancerous tissues. IEEE Trans. Biomed. Eng., 1996, 43(10): 1011–1020.

[76] Koss L G. Diagnostic cytology and its histopathologic bases. Philadelphia: Lippincott, 1992.

[77] Drezek R, Guillaud M, Collier T, et al. Light scattering from cervical cells throughout neoplastic progression: Influence of nuclear morphology, DNA content, and chromatin texture. J. Biomed. Opt., 2003, 8(1): 7–16.

[78] Erhardt K, Auer G, Bjorkholm E, et al. Prognostic significance of nuclear DNA content in serious ovarian tumors. Cancer Res., 1984, 44: 2198–2202.

[79] Mourant J R, Canpolat M, Brocker C, et al. Light scattering from cells: The contribution of the nucleus and the effects of proliferative status. J. Biomed. Opt., 2000, 5(2): 131–137.

[80] Gurjar R S, Backman V, Perelman L T, et al. Imaging human epithelial properties with polarized light-scattering spectroscopy. Nat. Med., 2001, 7(11): 1245–1248.

[81] Sokolov K, Drezek R A, Gossage K, et al. Reflectance spectroscopy with polarized light: Is it sensitive to cellular and nuclear morphology? Opt. Express., 1999, 5(13): 302–317.

[82] Ramachandran J, Powers T M, Carpenter S, et al. Light scattering and microarchitectural differences between tumorigenic and non-tumorigenic cell models of tissue. Opt. Express., 2007, 15(7): 4039–4053.

[83] Arifler D, Guillaud M, Malpica A, et al. Light scattering from normal and dysplastic cervical cells at different epithelial depths: Finite-difference time-domain modeling with a perfectly matched layer boundary condition. J. Biomed. Opt., 2003, 8(3): 484–494.

[84] Mourant J R, Bocklage T J, Powers T M, et al. Detection of cervical intraepithelial neoplasias and cancers in cervical tissue by in vivo light scattering. J. Low Genit. Tract. Dis., 2009, 13(4): 216–223.

[85] Choi W, Yu C C, Fang-Yen C, et al. Field-based angle-resolved light-scattering study of single live cells. Opt. Lett., 2008, 33(14): 1596–1598.

[86] Kim Y L, Liu Y, Wali R K, et al. Simultaneous measurement of angular and spectral properties of light scattering for characterization of tissue microarchitecture and its alteration in early precancer. IEEE J. Sel. Top. Quant. Electron., 2003, 9(2): 243–256.

[87] Gurjar R S, Backman V, Perelman L T, et al. Imaging human epithelial properties with polarized light-scattering spectroscopy. Nat. Med., 2001, 7: 1245–1248.

[88] Backman V, Wallace M B, Perelman L T, et al. Detection of preinvasive cancer cells. Nature, 2000, 406: 35–36.

[89] Jacques S L, Ramella-Roman J C, Lee K. Imaging skin pathology with polarized light. J. Biomed. Opt., 2002, 7(3): 329–340.

[90] Jacques S L, Roman J R, Lee K. Imaging superficial tissues with polarized light. Lasers Surg. Med., 2000, 26: 119–129.

[91] Hielscher A H, Mourant J R, Bigio I J. Influence of particle size and concentration on the diffuse backscattering of polarized light from tissue phantoms and biological cell suspensions. Appl. Opt., 1997, 36(1): 125–135.

[92] Liu Y, Kim Y L, Li X, et al. Investigation of depth selectivity of polarization gating for tissue characterization. Opt. Express., 2005, 13(2): 601–611.

[93] Drezek R, Dunn A, Richards-Kortum R. Light scattering from cells: Finite-difference time-domain simulations and goniometric measurements. Appl. Opt., 1999, 38(16): 3651–3661.

[94] Ghosh N, Patel H S, Gupta P K. Depolarization of light in tissue phantoms-effect of a distribution in the size of scatterers. Opt. Express., 2003, 11(18): 2198–2005.

Jian Wang, Ke Jun Tan

14 Aquatic environmental light scattering spectrometry

Water is the source of life. The production and living activities of human beings can't do without water. There are less available freshwater resources, especially usable freshwater on the Earth's surface. So the monitoring of water environment and testing of water quality are important. There are many factors influencing the water quality, including heavy metals and organic pollutants. In this chapter, we will introduce light scattering analysis of pollutants in water environment, including heavy metals and typical organic pollutants.

14.1 Light scattering measurement of typical metal ions in water environment

When heavy metals, such as Hg, Cd, Pb, Cr, and metalloid As are too high in our body, their presence will seriously damage our health, so their detection and analysis are very essential.

14.1.1 Mercury ions

Mercury is a nonessential element of the human body. When the human body is often exposed to a small amount of mercury, it can cause chronic mercury poisoning, which is characterized by headache, dizziness, numbness and pain of the limbs, muscle tremor, dyskinesia and other neurogenic symptoms. Too much inhalation of mercury vapors can cause acute mercury poisoning, bringing serious illness symptoms, such as hepatitis, nephritis, proteinuria, and uremia. Minamata disease first discovered in 1993 in Kumamoto prefecture, Kyushu, Japan, was due to high concentration of mercury. Therefore, it is necessary to determine mercury in the environment, especially the Chinese herbal medicines around the thermometer factory.

Table 14.1 presents the currently reported methods for light scattering analysis of Hg^{2+}. It can be divided into the following three types.

The first method is based on large-particle associates of mercury ions and other ions [1], such as ion associates of mercury ion and cation, or large-particle associates of mercury anion complex and cation with electrostatic or hydrophobic interactions. These ion associates can enhance the scattering signal, which can be used for

https://doi.org/10.1515/9783110573138-014

Table 14.1: Light scattering spectrometry for mercury ions.

System	Linear application	Detection limit	Application	References
AuNPs–T_n–Hg^{2+}	4.0×10^{-8}–6.0×10^{-7} mol/L	1.0 nmol/L	Tap water and lake water	[2]
AuNPs–T_n–Hg^{2+}	0.044–0.50 µmol/L	0.013 µmol/L	Environmental water samples	[7]
Herring sperm DNA– AuNPs –Hg^{2+}	3.3–3333.3 nmol/L	2.5 nmol/L	Water samples	[8]
Hg^{2+}–AuNPs–stannous chloride	0.5–50 µmol/L	0.23 µmol/L	Water samples	[9]
AuNPs–S^{2-}–Hg^{2+}	0.025–0.25 µmol/L	0.013 µmol/L	Water samples	[5]
Au–ssDNA–T_n–Hg^{2+}	1.3–1667 nmol/L	0.7 nmol/L	Water samples	[10]
Hg^{2+}–Au–ssDNA–NH_2OH–Cu^{2+}	0.1–400 nmol/L	0.03 nmol/L	Water samples	[10]
AuNPs–VC–Hg^{2+}	2.0–80 µmol/L	2.0 µmol/L	–	[11]
AuNPs–Sn^{2+}–Hg^{2+}	0.1–30 µmol/L	0.051 µmol/L	Wastewater	[12]
CuSNPs–Hg	0.01–100 µmol/L	0.003 µmol/L	Water	[13]
Magnetic nanoparticle–T_n–Hg^{2+}	1.0–80.0 pmol/L	500 fmol/L	Tap water, river water, and lake water	[14]
Mercury (II)–dithiazone	0.028–6.140 µg/mL	0.012 µg/mL	Water samples	[15]
Iodomercurate potassium–hexadecylpyri-dine bromide	15–200 µg/L	5 µg/L	Environmental water samples and synthetic water samples	[16]
Mercury–thioacetamide	0.2–10.0 µmol/L	0.02 µmol/L	Water samples	[17]
Methyl triphenyl bromide–potassium iodide–mercury	0.04–1.8 µg/mL	1.3 ng/mL	synthetic water samples, surface water, and industrial sewage	[18]

detecting mercury ions. The advantage of this method is that mercury ions interact with different anions and cations, and the disadvantage is that this method based on electrostatic reaction has poor selectivity and can be easily interfered by other ions.

The second method is based on dispersion–aggregation system to enhance light scattering signal with nanoparticles. Nano-materials development has greatly promoted the measurement of mercury ions [2, 3]. The basic principle is that the nanoparticles are dispersed without mercury and the scattering signal is low, but mercury can induce nanoparticles aggregation and enhance the scattering signal, which can be used for the detection of mercury. The most common example is to stabilize the nanopparticles with poly T single-stranded DNA. When no mercury ions exist, the nanoparticles are dispersed, the solution is red and the scattering signal is low. When the mercury ions exist, mercury ion and T-base form a steady $T–Hg^{2+}–T$ double-stranded structure; at this time, DNA no long stabilizes nanoparticles. When salt is added, nanoparticles aggregate and the solution changes to blue color with enhanced scattering signal, mercury ion can be quantitatively detected based on the relationship between the scattering signal and the concentration of mercury ion. Because T-base can specifically bind mercury ions, the selectivity of the method is better, and other ions do not interfere with the determination of Hg. Based on the stable double-stranded structure of $T-Hg^{2+}-$ T, many methods have been developed for the determination of mercury ions [3, 4]. In addition, scientists modified the thiol group on the surface of a nano-particles. With the interaction of mercury ions and the thiol group, a nanoparticle aggregate is induced [5, 6] to enchance the scattering signal, which can also be used to detect mercury ions.

Prof. Jiang Z L and his group developed one new method for the detection of mercury ions (Figure 14.1) [10]. They first modified ssDNA on 10 nm gold nanoparticle

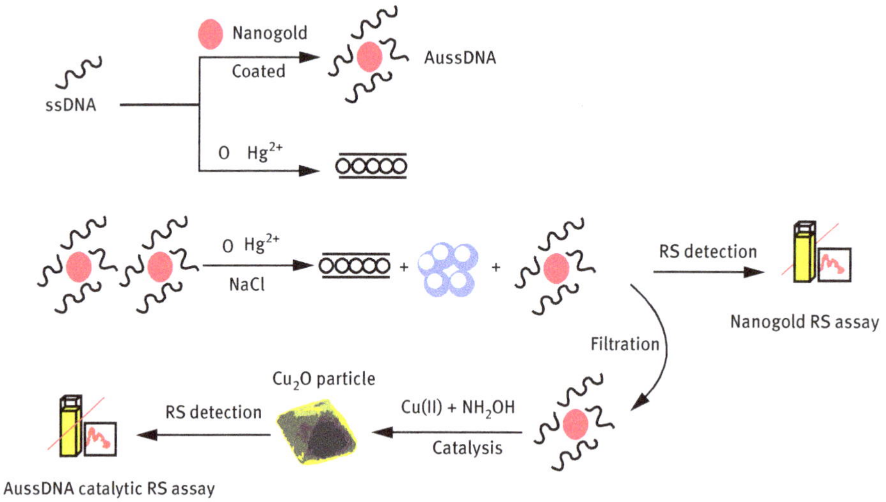

AussDNA catalytic RS assay

Figure 14.1: RLS method for the detection of mercury ions with gold nanoparticles as a probe [7].

to obtain Au–ssDNA probe. When NaCl exists, Hg^{2+} ion and Au–ssDNA interact to form a stable $T–Hg^{2+}–T$ mismatch double-stranded structure, inducing the aggregation of gold nanoparticles with resonance light scattering (RLS) enhancement at 540 nm. Hg^{2+} has a linear relation with the enhanced RLS intensity, and the linear range is 1.3–1,667 nmol/L, and the detection limit of Hg^{2+} is 0.7 nmol/L.

They then removed the large gold nanoparticle aggregates through a filter membrane, and the excess Au-ssDNA filtered liquor can catalyze $NH_2OH/Cu^{2+}–$ EDTA complex to produce a new Cu_2O particle at 60°C. Excessive amount of Au-ssDNA gradually reduces with the increase in mercury ions, and then gradually reducing Cu_2O particle scattering intensity at 602 nm. The reduced scattering intensity $\Delta I_{602\ nm}$ has a linear response in the range of 0.1–400 nmol/L; Hg^{2+}ion detection limit is 0.03 nmol/L. This aptamer-modified gold nanoparticle catalyzed scattering method for mercury detection in water samples has advantages of high sensitivity, selectivity, and simplicity.

The third method is based on the interaction between mercury and noble metal nanoparticle resulting in the nanoparticle form changes and light scattering signal changes. For example, vitamin C can reduce Hg^{2+} into Hg^0, which adsorbs on the surface of the nanoparticle, binding with gold to change the nanoparticle form, and then the particle size gradually increases, resulting in the scattering signal gradually enhances, which can be used for the detection of mercury ion (Figure 14.2) [11]. Based on the reduction of Sn^{2+}, Hg^{2+} is converted into Hg^0 inducing light scattering signal enhancement, which it can be used for the detection of mercury ion [12].

In general, the developments of nano-material and nanotechnology improve the sensitivity and selectivity of detection method and benefit for visual detection.

14.1.2 Cadmium ions

Cadmium ion is one heavy metal ion with strong toxicity, which is able to accumulate in the body with half-life period as long as 20–40 years, causing cumulative poisoning and seriously influencing human health [19], leading to symptoms, such as renal failure, osteoporosis, diabetes, hypertension, and so on. Currently, there are many methods for the detection of cadmium ion, such as fluorescence method [20], colorimetry [21, 22], scattering method [23], and so on.

Scattering method for cadmium ion detection is based on the formation of large-particle associates that cause scattering signal enhancement (Table 14.2). Cadmium ions and anions (as I^-, S^{2-}) form an anion complex, which interacts with a large-molecule cation or surfactant to form an ion associate, resulting in the scattering signal enhancement. The current application of scattering method is for the detection of cadmium ion in water sample.

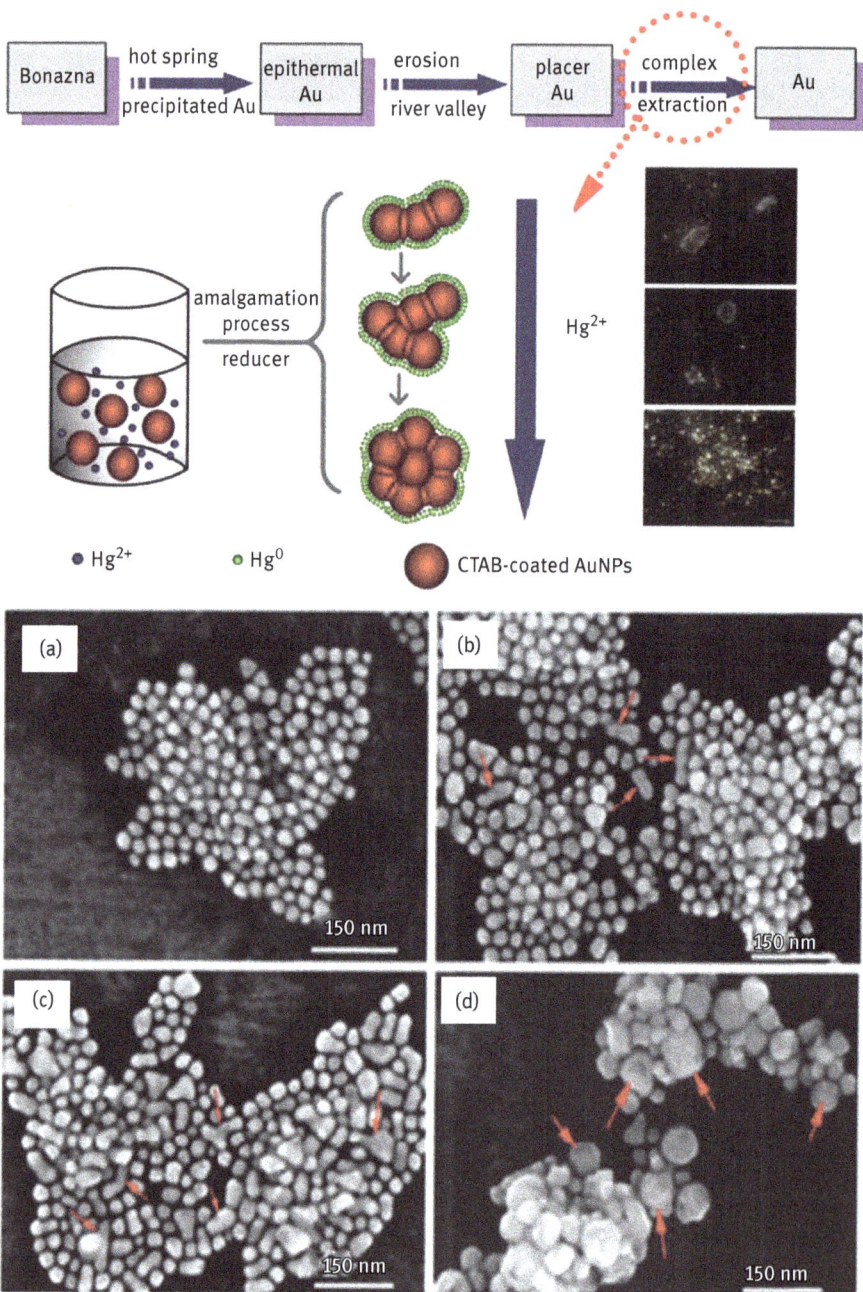

Figure 14.2: Detection of mercury ion based on changes in the shape of gold nanoparticles [11].

Table 14.2: Light scattering spectrometry of cadmium ions.

System	Linear range	Detection limit	Application	Reference
Cd–iodine–acridine red	0–80 µg/L	–	High purity zinc, zinc sulfate, grain, and tap water	[23]
Cd (II)– phenanthroline bromophenol blue	0–3.2 mg/L	4.50 µg/L	Synthetic sample and water	[24]
Cd^{2+}–PAN–PVA	0.017–1.0 µg/mL	5.30 ng/mL	Water	[25]
Cd^{2+}–1, 10-phenanthro-line–perfluorooctane sul-fonic acid	0.08–4.0 µg/mL	0.6 ng/mL	Tap water and lab wastewater	[26]
Cd^{2+}–phenanthroline–bright yellow	0–10 µg/L	0.086 µg/L	Feed	[27]
Cd^{2+}–KBr–phenanthroline	0~0.15 µg/mL	0.012 µg/mL	Ginger	[28]
Cd^{2+}–S^{2-}–OP	0.0–20.0 µg/mL	$6.33×10^{-3}$ µg/mL	Water samples	[29]
Cd^{2+}–S^{2-}–Triton X-100	0.067–20.0 µg/mL	20.11 ng/mL	Water sample	[30]
Cd^{2+}–I^-–CV	0.697–43.0 ng/mL	0.210 ng/mL	Water sample	[31]
Cd^{2+}–I^-–Rhodamine	0–1.0 µg/25mL	0.60 ng/mL	Pure zinc	[32]

14.1.3 Lead ions

Lead is mainly used for lead–acid battery. With the development of vehicle industry, the amount of lead for battery gradually increased. Lead vapor and dust easily enters into the human body through the respiratory tract and the esophagus. Lead is a heavy metal element, which is harmful for our health, accumulates in living bodies, affecting nerve, hematopoiesis, digestion, reproduction and development, bones, and so on. It is most serious harm to the hematopoictic system of bone marrow and nervous system. Lead and lead oxide dissolves in the blood causing lead poisoning and damaging the central nervous system, resulting in abnormal intelligence and behavior [33]. At present the detection methods of lead include colorimetric method [34], fluorescence method [35], and scattering method [36].

The main idea of scattering method is based on the interaction between lead ion with anion and cation to form large-particle associates. The analytes are mainly in the water sample.

Fu developed a RLS method for lead ion detection with thrombin–aptamer probe (Figure 14.3) [44]. Thrombin–aptamer probe has a flexible, irregular, and a curved structure with weaker scattering signal. When thrombin–aptamer probe combines with lead ion, the structure becomes a G-quadruplex structure with enhanced scattering signal. Based on above, the detection method of lead ion has been developed.

Prof. Jiang Z L and his group developed a visual detection method of lead ions with the use of specific binding of aptamer and lead ions (Figure 14.4) [46]. First, they synthesized AuPd nanoparticles, and then twined flexible aptamer on the

Table 14.3: RLS method for lead ions.

System	Linear application	Detection limit	Application	References
Pb^{2+}–NaTBP–PEG	0.03–1.0 µg/mL	2.6 ng/mL	River water	[36]
Pb^{2+}–I^-–TPB	0.008–6 µg/10mL	0.78 ng/mL	Environmental water sample	[37]
Pb^{2+}–I^-–CTAB	0–10.0 ng/mL	0.07 ng/mL	Food	[38]
	1.0–40.0 ng/mL	0.74 ng/mL	Environmental water sample	[39]
Pb^{2+}–KI–RhB 6G	0.04–0.20 µg/mL	0.032 µg/mL	Environmental water sample	[40]
	0.04–0.2 ng/mL	3.5×10^{-2} ng/mL	Tap water	[41]
Pb^{2+}–8–hydroxyquino-line 5–sulfonic acid	0.0801–12 µmol/L	0.240 nmol/L	Environmental water sample	[42]
Pb^{2+}–BSA–Congo red	0.0–7.0 µg/mL	0.358 µg/mL	Water samples	[43]
Pb^{2+}–Na_2S	0-12.0ug/mL	5.73 ng/mL	Environmental water sample	[28]
Pb^{2+}–TBA	1.0–120.0 nmol/L	0.90 nmol/L	–	[44]
DTC–AgNP–Pb^{2+}	0.01–60 µmol/L	4 nmol/L	Tap water, river water, and lake water	[45]
Pb^{2+}–Au–ssDNA–$H_2PO_2^-$–Ni^{2+}	0.20–42 nmol/L	0.1 nmol/L	Water samples	[46]

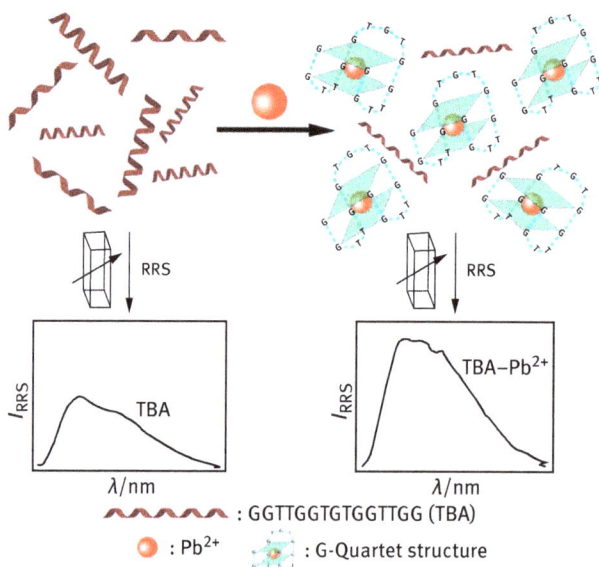

Figure 14.3: RLS method for the detection of lead ion with thrombin–aptamer as probe [44].

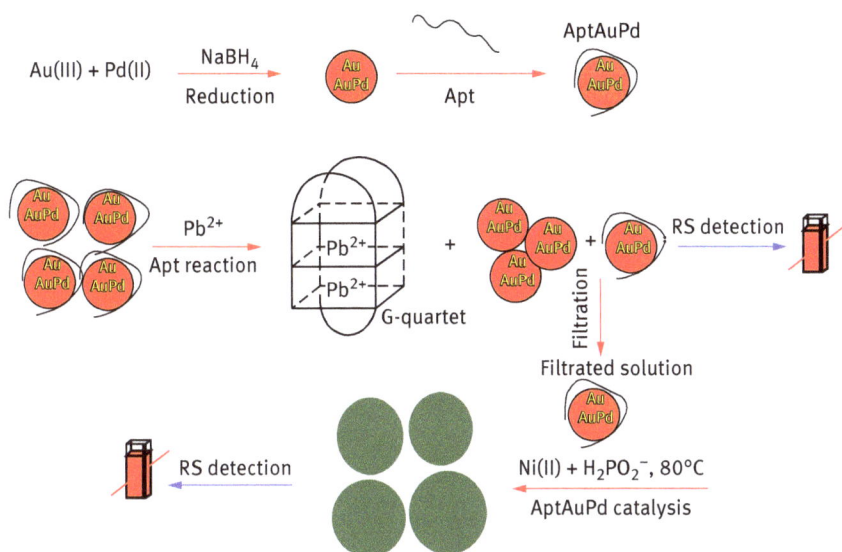

Figure 14.4: RLS combination method for thrombin–aptamer and lead ion for the detection of lead ion detection [46].

nanoparticle surface. After aptamer binding with lead ions, the flexible aptamer becomes a rigid G-quartet structure, ripping-off from the nanoparticle surface. And then filtered to obtain unbonded nanoparticles, which is used as a catalyst for Ni nanoparticle. By detecting the scattering signal changes of Ni nanoparticle, we can detect the concentration of lead ion. This method is similar to that method developed by Jiang's group for the detection of mercury ion [8].

14.1.4 Chromium ions

Chromium is one essential trace element. The lack of chromium can cause atherosclerosis, whereas excessive chromium is toxic. Researches show that hexavalent chromium is the most toxic, trivalent chromium comes second. Some researchers think that hexavalent chromium causes lung cancer, so there is a lot of literature for the determination of hexavalent chromium [47].

The scattering method for the determination of chromium ion is based on the interaction between anion complex of chromium ion and I^- with cation to form an associate enhancing scattering signal. There are a few literatures available for the detection of chromium ion with nanoparticles as light scattering probe (Table 14.4). Figure 14.5 is a scheme of RLS detection of chromiu ion. Chromium ion can corrdinate with citrate bound on the surface of silver nanoparticle, which reduced from silver ion by citric acid [48, 49] to induce nanoparticle aggregate, resulting in light scattering signal enhancement. It can be used for the detection of chromium ion.

Table 14.4: Scattering method for chromium ion detection.

System	Linear range	Detection limit	Application	References
AuNPs–Cr^{3+}	1.0–10.0 pmol/L	1.0 pmol/L	Water sample	[48]
AuNPs–Cr^{3+}	0.5–10.0 µmol/L	0.13 µmol/L	Cell imaging	[49]
Cr(VI)–KI–acridine orange	0.008–0.48 mg/L	3.8 µg/L	Environmental water	[50]
Cr (VI)– KI–Rh B	30–80 µg/L	2.44 µg/L	Electroplating wastewater	[51, 52]
Cr (VI)–KI–Rh 6G	1.0×10^{-7}–20×10^{-7} mol/L 3.1×10^{-3}~ 0.35 µg/mL	-3.1 µg/L	Water	[53, 54]
Cr (VI)–KI–Janus green	0.01–0.2 mg/L	2.0 µg/L	Synthetic water samples	[55]
Cr (VI)–I$_3^-$–RhB	10.0–500 µg/L	2.2 µg/L	Environmental water	[56]
Cr(VI)–iodide–starch	34–400 µg/L	6.7 µg/L	Herbicide	[57]
Cr (VI)– KI–safranin T	0.01–0.32 µg/mL	4.0 ng/mL	Water sample	[58]

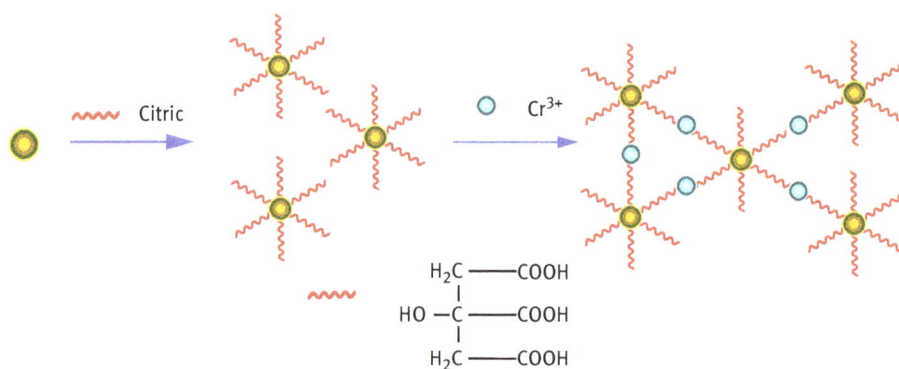

Figure 14.5: Scheme of RLS detection of chromium ion based on the complexation between citrate with chromium ion [48].

14.1.5 Arsenic ions

Arsenic is an essential trace element, but excessive amount of arsenic is poisonous. The toxicity of trivalent arsenic is stronger than that of pentavalent arsenic, which is mostly present as arsenic trioxide (also known as "Pishuang" in Chinese). Arsenic poisoning can be detected by measuring arsenic content in blood, urine, and hair. The recent determination methods of arsenic include electrochemical process [59], dynamic light scattering [60] and RLS method [61, 62], but the scattering system is still limited in ion associates.

Li G R used iodine green and heteropolyacid anion $[As_2(Mo_2O_7)_{12}]_{14}^-$ to form ion associate in sulfuric acid medium for the detection of arsenic content in hair with high sensitivity and low detection limit [63].

14.2 Light scattering detection of small organic molecules in water environment

14.2.1 Surfactant

Surfactant is also called "industrial monosodium glutamate", which is used in the fields of textiles, printing and dyeing, washing, cosmetics, pesticides, mining, oil, and so on. When the surfactant reaches a certain concentration in water, it causes damage to not only various living bodies in water but also public health through the food chain. So monitoring the surfactant in water is quite necessary [64].

Table 14.5 lists surfactant detection with scattering method mainly based on the formation of associated large particles. The interaction between surfactant with proteins, nucleic acids, dye and chromogenic agent and other scattering probes by electrostatic, hydrophobic, and hydrogen bond forces can form associates and

Table 14.5: Scattering methods for surfactant detection.

Probe	Surfactant	Linear range	Detection limit	Application	References
Basic fuchsin	SDBS	0.1–56µmol/L	1.86pmol/L	Environmental water samples	[65]
Chlorpromazine hydrochloride	SDBS	0.2–12 mg/L	47µg/L	Environmental water samples	[66]
Chlorpromazine hydrochloride	SDS	0.4–15 mg/L	106µg/L	Environmental water samples	[66]
Chlorpromazine hydrochloride	SLS	0.4–20 mg/L	117µg/L	Environmental water samples	[66]
Chlorpromazine hydrochloride	SDBS	0.09–10 mg/L	18µg/L	Environmental water samples	[67]
Chlorpromazine hydrochloride	SDS	0.15–15 mg/L	46µg/L	Environmental water samples	[67]
Chlorpromazine hydrochloride	SLS	0.67–12.5 mg/L	200µg/L	Environmental water samples	[67]
Ethyl violet	SDBS	0.0011–2.5 mg/L	1.1µg/L	Environmental water samples	[68]
Ethyl violet	SDS	0.0025–5.0 mg/L	2.5µg/L	Environmental water samples	[68]

(continued)

Table 14.5 (Continued)

Probe	Surfactant	Linear range	Detection limit	Application	References
Ethyl violet	SLS	0.27–3.0 mg/L	270 µg/L	Environmental water samples	[68]
Janus green	SDBS	0–50 mg/L	17.4 µg/L	Environmental water samples	[69]
Thionine	SDBS	0.05–3.0 mg/L	17 µg/L	Environmental water samples	[70]
Thionine	SDS	0.5–3.0 mg/L	166 µg/L	Environmental water samples	[70]
Thionine	SLS	1.0–5.0 mg/L	571 µg/L	Environmental water samples	[70]
RhB 6G	SDBS	0.02–5.6 ug/L	0.006 µg/L	Environmental water samples	[71]
RhB 6G	SDS	0.00002–0.014 mg/L	0.005 µg/L	Environmental water samples	[71]
Victoria Blue B	SDBS	0.08–3 mg/L	13 µg/L	Environmental water samples	[72]
AO	SDBS	0.028–8.71 mg/L	8.36 µg/L	Environmental water samples	[73]
RhB B	SDS	0.0–80.0 mg/L	23 µg/L	Environmental water samples	[74]
RhB B	SDBS	0.0–36.0 mg/L	38 µg/L	Environmental water samples	[74]
Janus Green	SDBS	0–3.48 µg/mL	178 ng/mL	Environmental water samples	[75]
Azoviolet	CTAMB	0.4–4.8 µmol/L	38 nmol/L	Water sample	[76]
Azoviolet	Zeph	0.2–6.0 µmol/L	21 nmol/L	Water sample	[76]
HAuCl$_4$	CTMAB	0.8–15 µmol/L	40.1 nmol/L	Environmental water samples	[77]
HAuCl$_4$	CTMAC	0.1–15 µmol/L	40.6 nmol/L	Environmental water samples	[77]
HAuCl$_4$	CPB	0.25–10 µmol/L	47.2 nmol/L	Environmental water samples	[77]
Eosin Y	CTAB	0–40 mg/L	0.012 mg/L	Water sample	[78]
Bromophenol blue	CTMAB	0–1.85 mg/L	0.43 µg/L	Synthetic water	[79]
Silver nanoparticle	CTMAB	5.0–500 nmol/L	3.8 nmol/L	Environmental water samples	[80]
K$_2$HgI$_4$	CPC	0.03–7.5 mg/L	6.7 µg/L	Detergent	[81]
Orange II	CPB	0.5–20 µmol/L	1.91 nmol/L	Life wastewater	[82]
Methyl orange	CPB	1–50 µmol/L	7.61 nmol/L	Life wastewater	[82]

Table 14.5 (Continued)

Probe	Surfactant	Linear range	Detection limit	Application	References
Golden orange I	CPB	1–50 µmol/L	0.691 nmol/L	Life wastewater	[82]
Ca^{2+}–Qu	CPC	0.1–19 µmol/L	0.05 µmol/L	Industrial wastewater	[83]
Ca^{2+}–Qu	CTAB	0.1–20 µmol/L	0.05 µmol/L	Industrial wastewater	[83]
AgI_2	CTMAB	0.2–5.0 µmol/L	80 nmol/L	Synthetic water sample	[84]
NaTPB	CDBAC	0.01–0.8 mg/L	3.2 µg/L	Environmental water samples	[85]
NaTPB	CPB	0.02–0.5 mg/L	5.6 µg/L	Environmental water samples	[85]
NaTPB	CTMAB	0.02–0.8 mg/L	4.7 µg/L	Environmental water samples	[85]
NaTPB	TPB	0.01–0.6 mg/L	3.0 µg/L	Environmental water samples	[85]
NaTPB	Zeph	0.01–0.8 mg/L	2.2 µg/L	Environmental water samples	[85]
Bromothymol blue	CPB	0.03–1.0 mg/L	3 µg/L	Synthetic sample, actual water sample	[86]
Bromothymol blue	CPB	0–4.0 mg/L	3.1–9.8 µg/L	Water sample	[87]
Bromocresol green	Zeph	0–3.5 mg/L	3.1–9.8 µg/L	Water sample	[87]
Bromocresol green	CTAB	0–3.5 mg/L	3.1–9.8 µg/L	Water sample	[87]
Eosin Y	Zeph	0–0.2 mg/L	4.3 µg/L	—	[88]
Eosin Y	CTMAB	0–0.2 mg/L	5.02 µg/L	—	[88]
Eosin Y	CPC	0–0.2 mg/L	4.54 µg/L	—	[88]
Eosin Y	CPB	0–0.2 mg/L	5.46 µg/L	—	[88]
Eosin Y	TPB	0–0.2 mg/L	5.73 µg/L	Synthetic water	[88]
H_2PtCl_6	CTMAC	1.5–25 µmol/L	1.38 µmol/L	Environmental water samples	[89]
H_2PtCl_6	CTMAB	2.5–17.5 µmol/L	0.637 µmol/L	Environmental water samples	[89]
H_2PtCl_6	CPB	1–17.5 µmol/L	0.739 µmol/L	Environmental water samples	[89]
Naphthol black 12B	Zeph	0.32–14.4 µmol/L	88 nmol/L	Water sample	[90]

remarkably enhance the light scattering signal, which can be used for surfactant detection. This method has poor specificity. Usually, one probe can be used for the detection of several surfactants.

14.2.2 Persistent organic pollutants

Persistent organic pollutants (POPs) refer to natural or synthesized organic pollutants with long half-life period that exist in the environment for a long time. They can accumulate through the food chain causing serious harm to health and environment. POPs have high toxicity, persistence, bioaccumulation, and hydrophilic hydrophobicity. Currently, determined POPs include polychlorinated biphenyls (PCBs), chlordane, hexachlorobenzene (HCB), tetrachlorodibenzo-p-dioxin (PCDDs), and perfluorochemical [91–93].

Polychlorinated biphenyls (PCBs) are one type of the representative POPs, with 209 isomers and features of persistence, long-term residual bioaccumulation, semivolatility, and high toxicity. It has long-distance transfer in the atmosphere and gets deposited in the soil. So, detection of PCBSs in the environment is necessary. Table 14.6

Table 14.6: Light scattering spectral detection of persistent organic pollutant.

Detection object	on system	Linear range	Detection limit	Application	Reference
PCB_{77}	$AgNPs–PCB_{77}$	40–1000 ng/mL	6.3 ng/mL	Synthetic sample	[94]
PCB_{52}	$AgNPs–PCB_{52}$	90–1000 ng/mL	33 ng/mL	Synthetic sample	[94]
PCB_{28}	$AgNPs–PCB_{28}$	80–1000 ng/mL	26 ng/mL	Synthetic sample	[94]
PFOS	QhCl–PFOS	0.10–50.0 µmol/L	9.88 nmol/L	Water sample and serum	[95]
PFOS	RhB 6G–PFOS	0.10–30.0 µmol/L	9.95 nmol/L	Environmental water sample	[96]
PFOS	BSA–PFOS	0.2–25 µmol/L	20.0 nmol/L	Environmental water sample	[97]
PCB_{77}	Berberine hydrochloride $-PCB_{77}$	6.8×10^{-2}–1.5 µmol/L	15 nmol/L	Environmental water sample	[98]
PCB_{77}	CdTe quantum dot-PCB_{77}	0.068–3.4 µmol/L	21 nmol/L	—	[99]
PCB_{52}	CdTe quantum dot-PCB_{52}	0.2–10 µmol/L	150 nmol/L	—	[99]
PFOS	BSA-PFOS	0.2–25.0 µmol/L	20.0 nmol/L	Environmental water sample	[100]
PFOS	Brilliant cresyl blue -PFOS	0.64–30.0 µmol/L	63.8 nmol/L	Water sample	[101]
PFOS	RhB –PFOS	0.17–10 µmol/L	17 nmol/L	Environmental water sample	[102]

lists light scattering analysis method of POPs in recent years. One of them use silver nanoparticles as a light scattering probe for the detection of PCBs based on the enhanced scattering signal of AgNPs–PCBs aggregate resulted from the interaction between PCBs and AgNPs [94].

Perfluorooctane sulphonate (PFOS) and perfluorooctanoic acid (PFOA) are another kind of POPs after organochlorine pesticides and dioxins, and they are all anionic surfactants with fluorine that are widely used in industrial production and in living consumption. PFOS and PFOA are final converted products of perfluorochemicals [101], and their molecular structures are shown in Figure 14.6. As fluorine has maximum electronegativity, it has a strong polarity for C–F bond, which is one of the highest-energy covalent bond that equips perfluorochemicals with good thermal stability and chemical inertness and makes they are able to tolerate sunlight and high temperature. It is hard to degrade even with microbial and higher animal metabolism [102]. For example, PFOS does not decompose even it is boiled for 1 hour in a concentrated nitric acid solution, and only splits with high-temperature incineration [103, 104]. PFOS has long-distance migration ability and bioaccumulation, which brings great threat to humans and animals [105, 106].

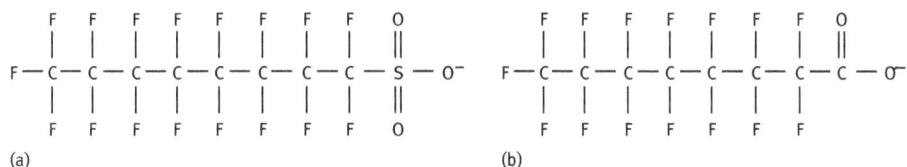

Figure 14.6: Molecular structural formula of (a) PFOS and (b) PFOA.

Currently, scattering assays of PFOS/PFOA are mainly based on their associations with scattering probes, which can cause enhanced scattering signal, so that the detections of PFOS/PFOA are realized. Prof. Tan K J and his group developed highly sensitive methods for the detection of PFOS based on the interaction between PFOS with quinine hydrochloride, rhodamine 6G, ethyl violet, and so on, to form an ion associate, which caused enhanced scattering signal [96–98].

14.2.3 Phenols and other small organic molecules

There are multiple phenolic pollutants in environment, mainly are phenol, cresol, and hydroquinone. Phenols are widely used in fertilizer, paint, rubber, synthetic resin, phenolic plastics, leather making, papermaking, pesticides, spices, dyes, and other industries. Phenol is the main phenolic pollutant of wastewater discharged from coking, producing gas and refining industries, which pollutes the environment. Phenolic substance brings denaturation and deactivation of proteins in cell protoplasm with carcinogenesis, mutation, and reproductive toxicity [107–109]. In water environment, phenol influences the growth and development of living bodies in water [110–112]. For this, Environmental Protection Agency (EPA) mentioned in phenol standard regulation with a phenolic concentration of 2.56 mg/L, chronic toxicity occurs for freshwater aquatic organisms. 3.5 mg/L is the concentration limit for humans, and 0.3 mg/L is the concentration limit to keep river water from any foul smell. Chinese *Surface water environmental quality standard* (GB 3838–2002) regulates that the maximum allowable concentration of volatile phenol is V type water. Chinese *Standards for drinking water quality* (GB 5749–2006) regulates that volatile phenolic substances (on the basis of phenol) should not exceed 0.002 mg/L, *Integrated wastewater discharge standard* (GB 8978–1996) regulates the standard for first level: 0.3 mg/L, second level: 0.4 mg/L, and third level: 1.0 mg/L. Therefore, the analysis of phenol is of great significance to the development of life science and environmental science.

The light scattering assay of phenols is based on the molecular aggregate formation of a system, which enhances the scattering signal for detection. Sun et al. used hydrogen peroxide and horseradish peroxidase to oxidize phenol into benzoquinone; benzoquinone forms molecular aggregates with graphene quantum dots through electrostatic interaction to produce an enhanced RLS signal. The intensity has a linear relation with phenol concentration, so the phenol scattering analysis method was established (Figure 14.7) [113].

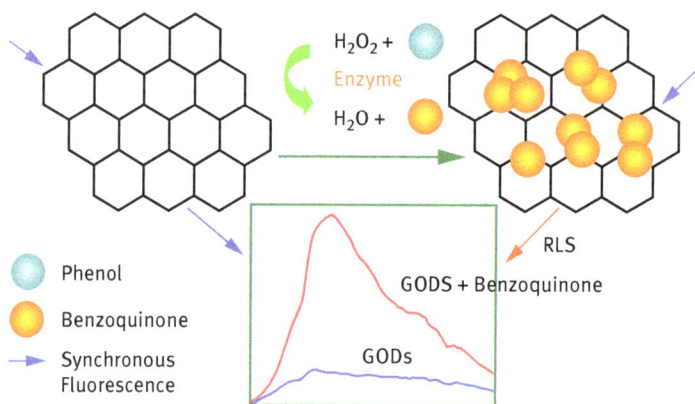

Figure 14.7: Diagram of graphene quantum dots for phenol detection [113].

As presented in Table 14.7, RLS method is also widely used in other small organic molecules for pollutant detection.

Table 14.7: Light scattering analyses of organic pollutants in environment.

System	Linear range	Detection limit	Application	References
GQDs–phenol	0.06–2.16 μmol/L	22 nmol/L	Environmental water sample	[113]
Apt–AuNPs–biphenol A	3.33–333.33 ng/mL	0.012 ng/mL	Commodity	[114]
C_6H_5OH–H_2SO_4–$KBrO_3$– XO–CTMAB	0.0161–1.36 μg/mL	4.87 ng/mL	Environmental water sample	[115]
KBrO–acridine orange– naphthol	0.141–28.0 μmol/L	42.2 nmol/L	Environmental water sample	[116]
Phosphorus molybdenum– malachite green	0.18–5.0 μg/mL	55 ng/mL	Environmental water sample	[117]
HSA–Sudan red I	0.10–2.80 μg/mL	100 ng/mL	Environmental water sample	[118]
Methyl violet–WO_4^{2-}	0.25–7.40 μg/mL	76.3 ng/mL	—	[119]
Ethyl violet–WO_4^{2-}	0.14~3.0 μg/mL	40.5 ng/mL	—	[119]
Crystal violet–WO_4^{2-}	0.10–4.80 μg/mL	31.4 ng/mL	—	[119]
Methyl green–WO_4^{2-}	0.17–3.80 μg/mL	50.2 ng/mL	—	[119]
Iodine green–WO_4^{2-}	0.22–5.80 μg/mL	67.1 ng/mL	—	[119]
Malachite green–WO_4^{2-}	0.06–2.8 μg/mL	18.1 ng/mL	Fish meat	[119]
Brilliant green–WO_4^{2-}	0.08–4.80 μg/mL	22.6 ng/mL	—	[119]
Methylene blue–SDBS	0.004–3.0 μg/mL	1.2 ng/mL	Human serum	[120]
Methylene blue -SDBS	0.008~3.0 μg/mL	1.4 ng/mL	—	[120]
Methylene blue –SDBS	0.006–3.0 μg/mL	1.7 ng/mL	—	[120]
Potassium iodide mercury– paraqua	0.01–13.0 μg/mL	8.0 ng/mL	Rice	[121]
Aniline–nitrite –N,N- dimethylaniline	0.02–0.2 μg/mL	19.2 ng/mL	Environmental water sample	[122]
Ethyl violet (EV) and 1-OHP	0.004–0.982 μg/mL	1.2 ng/mL	Urine	[123]

14.3 Turbidity analysis

Water in the environment becomes turbid because of the presence of soil, sand, tiny organic and inorganic matter, soluble and colored organic chemicals, plankton and other microorganisms, and other mechanical impurities. To evaluate water quality, the concept of turbidity was proposed [124–127].

14.3.1 Turbidity

Turbidity is expressed in light scattering signal of turbid system. As light scattering signal is dependent on the incident wavelength, scattering particle size, and medium, we will discuss turbidity on the basis of size of scattering particles.

(1) When particles in a suspension are smaller than the wavelength of light, scattering light obeys the Rayleigh scattering rule:

$$I_r = \frac{K}{\lambda^4} NV^2 \left(\frac{n_1^2 - n_2^2}{n_1^2 + n_2^2}\right)^2 I_0 \qquad (14.1)$$

where I_r is the scattering light intensity, K is the parameter, N is the particle number in an unit volume, V is the particle volume, λ is the incident light wavelength, and I_0 is the incident light intensity, whereas n_1, n_2 are the refractive indexes of water and particle, respectively.

In equation (14.1), if a particle is homogeneous, V is the constant; it can be further concluded the light scattering intensity I_r is directly proportional to the suspension concentration:

$$I_r = K_R N I_0 \qquad (14.2)$$

where K_R is the Rayleigh scattering coefficient:

$$K_R = \frac{K}{\lambda^4} NV^2 \left(\frac{n_1^2 - n_2^2}{n_1^2 + n_2^2}\right)^2 \qquad (14.3)$$

(2) When a suspension particle is larger than the wavelength of light, the scattering light obeys the Mie scattering rule:

$$I_M = K_M A N I_0 \qquad (14.4)$$

where I_M is the scattered light intensity, A is the particle surface area, K_M is the scattering coefficient of Mie scattering. If the surface area A is constant, then light intensity I_M is in direct proportion to the suspension concentration.

Combining Rayleigh scattering with Mie scattering, when the incoherent light within a certain wavelength passes through an homogeneous solution with uniform particles, equation (14.2) and (14.4) can be expressed in a general equation, as shown in equation (14.5), light intensity is directly proportional to the suspension concentration.

$$I_S = K_S N I_x \qquad (14.5)$$

where K_S is the scattering coefficient, N is the concentration of the solution, and I_x is the incident light intensity at scattering position.

14.3.2 Measurement of turbidity

Based on equation (14.5), three methods (Figure 14.8) has been developed for detecting turbidity, such as vertical scattering, forward and backward scattering according to different angles between scattering light and incident light.

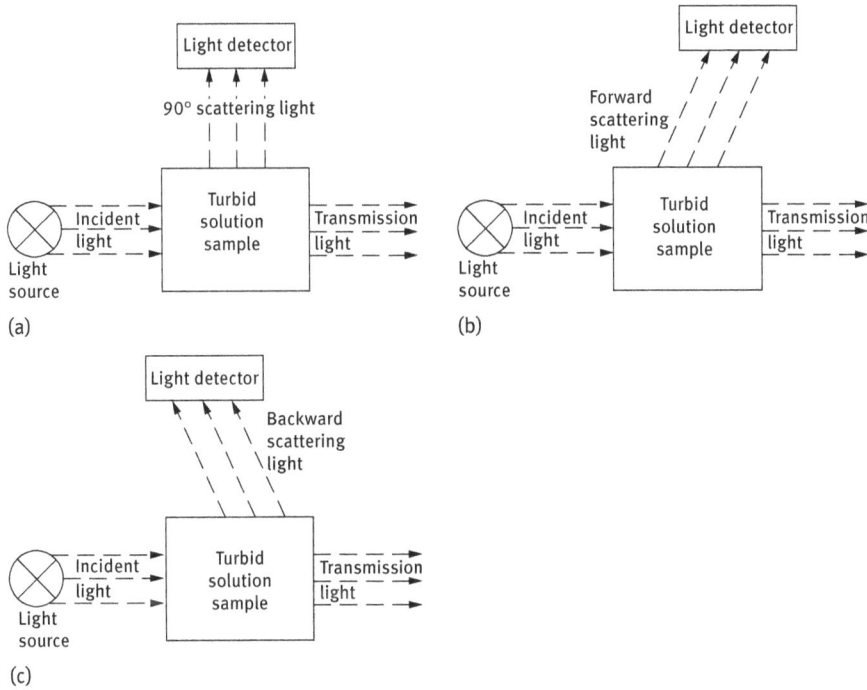

Figure 14.8: The diagram of turbidity detection. (a) Vertical scattering (b) Forward scattering (c) Backward scattering.

In the above-mentioned three methods, backward light scattering technique can be used for highly sensitive detection of multiple substances with simple equipment and are widely applied for the turbidity analysis in tap water and river water. Especially, when the angle between incident light and the scattering light is 180°, a fiber can be used to deliver incident light and scattering light; scattering probe is connected to the main engine through an optical fiber, which is can be detected on online real-time long-distance turbidity measurement in different environment [128].

In addition, surface scattering turbidity measurement (Figure 14.9 [a]), underwater scattering light measurement (Figure 14.9 [b]) and other measurement methods were developed. And then portable, desktop and online turbidity meters were designed successively.

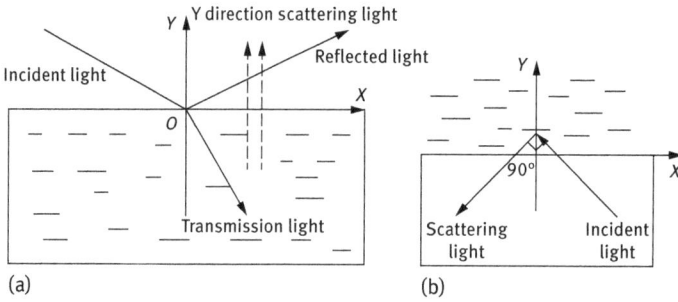

Figure 14.9: Diagram of principle of turbidity detection of surface scattering type (a) and underwater scattering light detection (b).

Although the 90° detector reduces the influence of stray light, comparing various advantages and disadvantages of above methods, Tan K J used backward light scattering units as shown in Figure 14.10 to develop a backward light scattering turbidity meter and successfully applied it for the detection of chlorine in human urine [129].

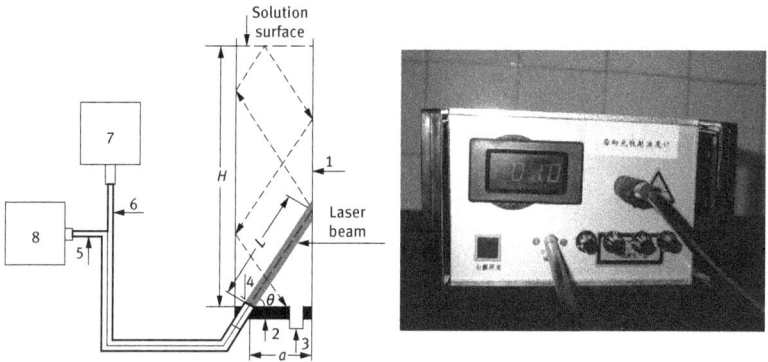

Figure 14.10: Scheme of 180° backward light scattering method and turbidity meter [129]. 1. Optical probe; 2. foundation; 3. leakage hole; 4. optical fiber inlet; 5. backward scattering signal delivery light fiber; 6. laser light transmitting optical fiber; 7. xenon light; 8. photomultiplier. θ is the incident light angle(°); L is the light scattering optical path(mm); and H is the solution height (mm).

References

[1]   Liu S P, Liu Z F, Li M. Double scattering spectra of mercury (II)-thiocyanate – rhodamine dye systems and their analytical applications. Chinese J. Anal. Chem., 1996, 24(5): 501–505.

[2]   Liu Z D, Li Y F, Ling J, et al. A localized surface plasmon resonance light-scattering assay of mercury(II) on the basis of Hg^{2+}-DNA complex induced aggregation of gold nanoparticles. Environ. Sci. Technol., 2009, 43: 5022–5027.

[3]   Gao Z F, Song W W, Luo H Q, et al. Detection of mercury ions(II) based on non-cross-linking aggregation of double-stranded DNA modified gold nanoparticles by resonance Rayleigh scattering method. Biosens. Bioelectron., 2015, 65: 360–365.

[4]   Li L, Wen Y, Xu L, et al. Development of mercury(II) ion biosensors based on mercury-specific oligonucleotide probes. Biosens. Bioelectron., 2016, 75: 433–445.

[5]   Fan Y, Long Y F, Li Y F. A sensitive resonance light scattering spectrometry of trace Hg^{2+} with sulfur ion modified gold nanoparticles. Anal. Chim. Acta, 2009, 653: 207–211.

[6]   Su D, Yang X, Xia Q, et al. Colorimetric detection of Hg^{2+} using thioctic acid functionalized gold nanoparticles. RSC Adv., 2013, 3: 24618–24624.

[7]   Zhou B, Shi L F, Liu L, etc. Resonance light scattering detection of mercury(II) ion using unlabelled AuNPs. App. Chem. Ind., 2012, 41(02): 344–346.

[8]   Ling S M, Li J F, Liang A H, etc. Resonance scattering detection of trace Hg^{2+} using herring sperm NDA modified nanogold. Spectrosc. Spectr. Anal., 2010, 30: 486–488.

[9]   Huang D F. Resonance scattering spectroscopic study on ion association system of potassium mercuric bromide-benzethoniumchloride and its analytical application. Chinese J. Anal. Lab., 2005, 24: 48–51.

[10] Jiang Z, Fan Y, Chen M, et al. Resonance scattering spectral detection of trace Hg^{2+} using aptamer-modified nanogold as probe and nanocatalyst. Anal. Chem., 2009, 81(13): 5439–5445.

[11] Qi W J, Wang Y, Wang J, et al. Light scattering investigations on mercury ion induced amalgamation of gold nanoparticles in aqueous medium. Sci. China-Chem., 2012, 55(7): 1445–1450.

[12] Zhang P, Chen S, Kang Y, et al. Trace mercury ion determination based on the highly selective redox reaction between stannous ion and mercury ion enhanced by gold nanoparticles. Spectrochim. Acta A, 2012, 99(24): 347–352.

[13] Yue Q, Shen T, Wang J, et al. A reusable biosensor for detecting mercury(II) at the subpicomolar level based on "turn-on" resonance light scattering. Chem. Commun., 2013, 49: 1750–1752.

[14] He Q, Long Y F. Determination of trace mercury based on gold nanoparticles-stannous chloride system by resonance light scattering technique. Chinese J. Anal. Chem., 2009, 37: 290–290.

[15] Liu T L, Jian H X. Resonance scattering determination of mercury in environmental waters. Guangdong Trace Elem. Sci., 2010, 17: 61–64.

[16] Huang D F. Resonance scattering spectrometry of the reaction product of potassium mercuric iodide and CPB and its analytical application. Phys. Chem. Inspect. (Chem. Classif.), 2006, 42: 237–239.

[17] C, Wang Y Y, Li H, Tan K J. Determination of trace mercury with thioacetamide in environmental water samples by resonance light scatting. Chinese J. Anal. Chem., 2011, 29(2): 233–237.

[18] Cui F L, Qin L X, Li F, etc. Resonance light scattering spectrum of methyltriphenylphosphonium bromide-potassium iodide-mercury system and its application. Metall. Anal., 2009, 9: 41–44.

[19] Pang S Q, Wang Z K, Xu J W, etc. Recent advance in analytical methods of cadmium. J. Food Eng., 2008, 2: 62–65.

[20] Liu Z, Zhang C, He W, et al. A highly sensitive ratiometric fluorescent probe for Cd^{2+} detection in aqueous solution and living cells. Chem. Commun., 2010, 46: 6138–6140.

[21] Xue Y, Zhao H, Wu Z, et al. Colorimetric detection of Cd^{2+} using gold nanoparticles cofunctionalized with 6-mercaptonicotinic acid and L-cysteine. Analyst, 2011, 136: 3725–3730.

[22] Wang A J, Guo H, Zhang M, et al. Sensitive and selective colorimetric detection of cadmium(II) using gold nanoparticles modified with 4-amino-3-hydrazino-5-mercapto-1,2,4-triazole. Microchim. Acta, 2013, 180(11–12): 1051–1057.

[23] Liu X D, Liu S P, Huang C Z. Recent advance in analytical methods of cadmium. Chinese J. Anal. Chem., 1998, 26: 93–96.

[24] Zeng W, Li S Q. A study on the resonance light scattering Spectra of cadmium phenanthorline bromophenol blue system and its analytical application. J. Anal. Sci., 2007, 23: 205–208.

[25] Han Z H, Lv C Y, Yang S Y. Resonance Rayleigh scattering method for the determination of trace cadmiun in water with PAN in the presence of polyvinyl alcohol. J. Anal. Sci., 2006, 22: 77–79.

[26] Yuan R, Tan K J. Determination of trace cadmium by resonance light scattering technique with 1, 10-phenanthroline-perfluorooctane sulfonate system. Environ. Chem., 2012, 31: 885–889.

[27] Li Y M, Shi P F, Li R Y. Determination of cadmium in feed by resonance light scattering technique. China Feed, 2012(20): 40–42.

[28] Gao X Y, Gao G Y, Zhang N, etc. Determination of trace cadmium in ginger by resonance light scattering method with microwave digestion procedure. J. China Condiment. 2014, (02): 92–96.

[29] Zeng M. Determination of trace lead and cadmium in environmental water by resonance light scattering method after separation with sulfhydryl cotton. Metall. Anal., 2007, 27: 44–47.

[30] Han Z H, Lv C Y, Fu S F, etc. Determination of trace cadmium in environmental water by resonance light scattering method. Chinese J. Health Lab. Technol., 2005, 15: 20–22.

[31] Yang H X, Li G R, He A T, Tan G H. Determination of cadmium with the ternary complex of Cd^{2+}-I^-CV by resonance ray leigh scattering. Chinese J. Health Lab. Technol., 2007, 17: 1366–1402.

[32] Liu S P, Liu Z F, Luo H Q. Resonance Rayleigh scattering method for the determination of trace amounts of cadmium with iodide–rhodamine dye systems. Anal. Chim. Acta, 2000, 407: 255–260.

[33] Zheng C Y, Hu M S. The harm of lead and the role of Zinc in children's growth and development. Guangdong Trace Elem. Sci., 2001, 8: 10–13.

[34] Chai F, Wang C, Wang T, et al. Colorimetric detection of Pb^{2+} using glutathione functionalized gold nanoparticles. ACS Appl. Mater. Interf., 2010, 2: 1466–1470.

[35] Beqa L, Singh A K, Khan S A, et al. Gold nanoparticle-based simple colorimetric and ultra-sensitive dynamic light scattering assay for the selective detection of Pb(II) from paints, plastics, and water samples. ACS Appl. Mater. Interf., 2011, 3: 668–673.

[36] Tan K J, Huang C Z, Huang Y M. Determination of lead in environmental water by a backward light scattering technique. Talanta, 2006, 70: 116–121.

[37] Nan H J, Liu Z F, Liu S P, etc. A Study on the resonance Rayleigh scattering of lead(II)-iodide-tetradecyl pyridinium bromide system and its analytical application. Chinese J. Spectrosc. Lab., 2008, 25: 249–253.

[38] Gao X Y, Gao G Y, Kong X X, etc. Rapid determination of trace lead in food by resonance light scattering method. J. Oil Food, 2014, 22(2): 76–79.

[39] Zhong M H. Determination of trace lead (II) in environmental water by resonance light scattering method. Guangdong Trace Elem. Sci., 2005, 12: 65–67.

[40] Luo D C, Liu J F. Determination of micro lead content in environmental water with system of Pb-KI-rhodamine 6G by resonance Rayleigh scattering method. J. Chem. Reag., 2007, 29: 547–548.

[41] Wang D, Luo H Q, Li N B. Resonance Rayleigh scattering method for the determination of trace lead with iodide-rhodamine 6G system. Environ. Chem., 2005, 24: 97–100.

[42] Qian Q M, Tang X X, Zhang Y L, etc. Resonance light scattering detection of Pb^{2+} using 8-hydroxyquinoline-5-sulphonic acid. Appl. Chem. Ind., 2013, 42(1): 168–170.

[43] Wang X H. On application of resonance light scattering in water environmental monitoring. Chengdu: Master's Thesis of Sichuan Normal University, 2005.

[44] Fu X B, Qu F, Li N. B, et al. A label-free thrombin binding aptamer as a probe for highly sensitive and selective detection of lead (II) ions by a resonance Rayleigh scattering method. Analyst, 2012, 137: 1097–1099.

[45] Cao H, Wei M, Chen Z, et al. Dithiocarbamate-capped silver nanoparticles as a resonance light scattering probe for simultaneous detection of lead (II) ions and cysteine. Analyst, 2013, 138: 2420–2426.

[46] Jiang Z L, Fan Y Y, Liang A H, et al. Resonance scattering spectral detection of trace Pb^{2+} using aptamer-modified AuPd nanoalloy as probe. Plasmonics, 2010, 5: 375–381.

[47] Chen W, Cao F, Zheng W, et al. Detection of the nanomolar level of total Cr[(iii)and(vi)] by functionalized gold nanoparticles and a smartphone with the assistance of theoretical calculation models. Nanoscale, 2015, 7: 2042–2049.

[48] Chen M, Cai H H, Yang F, et al. Highly sensitive detection of chromium (III) ions by resonance Rayleigh scattering enhanced by gold nanoparticles. Spectrochim. Acta A, 2014, 118: 776–781.

[49] Wu T, Liu C, Tan K J, Hu P P, Huang C Z. Highly selective light scattering imaging of chromium (III) in living cells with silver nanoparticles. Anal. Bioanal. Chem., 2010, 397: 1273–1279.

[50] He D X, Li G R. Determination of trace chromium (VI) in environmental water by resonance light-scattering. J. Instr. Anal., 2004, 23: 89–91.

[51] Zhong M H. Determination of chromium (VI) in electroplating wastewater by resonance light scattering. Electroplat. Pollut. Contr., 2006, 26: 36–37.

[52] Pan Q S, Li J Q. Determination of the content of chromium in the kind plant of ramie by ion resonance scattering method. J. Wuhan University Sci. Eng., 2007, 20: 14–18.

[53] Liang A H, Jiang Z L, Hang S Y, etc. Resonance scattering determination of trace Cr(VI) in water with Rhodaine 6G. Environ. Pollut. Contr., 2004, 26: 472–475.

[54] Wang Z L, Zhang X S, Luo Y J, etc. Synthesis of 2-[benzothiazolylazo]-7-(4-carboxylphenylazo)-chromotropic acid and its color reaction with iron (III)., Metallurg. Anal 2003, 23: 4–5, 3.

[55] He D X, Li G R, Chen Y S, etc. Determination of trace Chromium(VI) in water samples by resonance light-scattering. J. Anal. Sci., 2006, 22: 336–338.

[56] Shen J S, Li X R. Determination of Cr(VI) and Cr(III) in environmental water by resonance scattering with the ion association complex of Rhodamine B and I_3^-. Chinese J. Anal. Chem., 2001, 29: 944–946.

[57] Li N, Liu Y W, Zhou H Y, etc. Determination of trace chromium in traditional Chinese medicines by resonance light scattering. Chinese J. Anal. Lab., 2007, 26: 85–87.

[58] Chen L H, Liu F G. Determination of chromium in water by resonance light scattering. Chinese J. Health Lab. Technol., 2006, 16: 205–206.

[59] Dai X, Nekrassova O, Hyde M E, et al. Anodic stripping voltammetry of arsenic (III) using gold nanoparticle-modified electrodes. Anal. Chem., 2004, 76: 5924–5929.

[60] Kalluri J R, Arbneshi T, Khan S A, et al. Use of gold nanoparticles in a simple colorimetric and ultrasensitive dynamic light scattering assay: Selective detection of arsenic in groundwater. Angew. Chem. Int. Ed., 2009, 48: 9668–9671.

[61] Liu Y F, Li G R, Tan G H. Resonance Rayleigh scattering spectra of Arsenic(V)-Molybdate-Iodine green in systems and their analytical application. Guangdong Trace Elem. Sci., 2010, 17: 60–65.

[62] Huang Y L, Xu H, Liu H. Determination of urinary arsenic by resonance Rayleigh scattering method. Appl. Chem. Ind., 2010, 1: 124–126.

[63] Li G, Liu Y, Wang Y. Determination of arsenic in hair by resonance Rayleigh scattering method with iodine green-molybdate system. Anal. Lett., 2009, 42(1): 94–107.

[64] Wang B H, Zhang X J, Ji W, etc. Hazard analysis of surfactants in ecosystem. Chem. Ind. Eng. Progr., 2007, 26(9): 1263–1268.

[65] Zhai H Y, Hu D L, Huang C H. Resonance scattering determination of anionic surfactant with fuchsin basic. Chem. Bull., 2013, 76(7): 663–665.

[66] Yang Q L, Liu Z F, Lu Q M, etc. Second-order Scattering and frequency doubling scattering method for the determination of some anionic surfactants with chlorpromazine hydrochloride as the probe. Chinese J. Anal. Chem., 2007, 35(6): 797–802.

[67] Yang Q L, Liu Z F, Lu Q M, etc. Resonance Rayleigh scattering method for determination of some Anionic surfactants with Chlorpromazine hydrochloride as probe. Chem. J. Chinese Univ., 2006, 27(12): 2281–2284.

[68] Chen S, Liu S P, Luo H Q. Resonance Rayleigh scattering spectra of ethyl violet-anionic surfactant systems and their analytical application. Chinese J. Anal. Chem., 2004, 32(1): 19–24.

[69] Li M L, Wang Y S, Xiao X L. Second-order scattering and frequency-doubling scattering of anionic surfactants-Janus green system and their analytical applications. Chinese J. Anal. Lab., 2004, 23(12): 39–41.

[70] Wang Z Y, Xu H. Resonance Rayleigh scattering spectrometric determination of anionic surfactants based on the reaction with thionin. Phys. Chem. Inspect. (Chem. Classif.), 2012, 48(11): 1331–1334, 1340.

[71] Chen X H, Lyu C Y, Liu H T, etc. Resonance Rayleigh scattering spectra of Rhodamine 6G-Anionic surfactant systems and their analytical application. J. Anal. Sci., 2009, 25(3): 272–276.

[72] Chen Z G, Peng Y R, Xie F, et al. Determination of anionic surfactant in surface water by resonance light-scattering technology. Int. J. Environ. Anal. Chem., 2010, 90(7): 573–585.

[73] Xiao X, Wang Y, Chen Z, et al. Resonance light scattering method for the determination of anionic surfactant with acridine orange. Spectrochim. Acta A, 2008, 71(2): 398–402.

[74] Yuan J, Wang X H, Xie J L, etc. RLS determination of anionic surfactants with Rhodamine B. Phys. Chem. Inspect (Chem. Classif), 2006, 42: 24–26.

[75] Xiao X L, Wang Y S, Li M L, etc. Detection of anionic surfactants in water by resonance light scattering method using janus green B. Chinese J. Health Lab. Technol., 2004, 14(3): *290–291*.

[76] Yang C X, Li Y F, Huang C Z. Determination of cationic surfactants in water samples by their enhanced resonance light scattering with Azoviolet. Anal. Bioanal. Chem., 2002, 374(5): 868–872.

[77] Wang X, Wu L H, Chen Y H, etc. Determination of cationic surfactants with chloroauric acid for resonance light scattering. Chinese J. Anal. Chem., 2011, 39(12): 1882–1886.

[78] Yao H C, Chen J, Xu E J, etc. Determination of cetyltrimethylammonium bromide in water with eosin Y resonance light scattering method. China Surfactant Deterg. Cosmet., 2011, 41(5): 381–383.

[79] He Z X, Liu Q T, Li M L. Resonance Rayleigh light scattering spectra of interaction of cationic surfactant with bromphenolblue and its analytical application. Chinese J. Health Lab. Technol., 2005, 15(3): 295–296.

[80] Ouyang Z Z, Bai S, Long Y F. Determination of trace cetyltrimethylammonium bromide using silver nanoparticles by RLS spectra. Chem. Res. Appl., 2014(11): 1771–1774.

[81] Gao J Y, Zhang B, Li S W, etc. Research on determination of cetylpyridinium chloride by resonance light scattering spectra. Chem. Res. Appl., 2008, 20(1): 90–92.

[82] Zhai H Y, Deng Y, Ruan S Q. Resonance scattering determination of cationic surfactants with Acidic monoazo dyes. J. Anal. Sci., 2010, 26(4): 455–458.

[83] Gan N, Li T H, Xu W M, etc. Resonance Rayleigh scattering determination of cationic surfactant in natural waters based on formation of Ca^{2+}– quercitrin- surfactant complex. China Surfactant Deterg. Cosmet., 2007, 37(1): 54–57.

[84] Cheng Y Y, Jiang Z L, Liang A H. A new spectral method for the determination of cationic surfactant based on the resonance scattering effect of AgI_2. association particles. Spectrosc. Spectr. Anal., 2006, 26(10): 1888–1890.

[85] Wang M X, Liu Z F, Hu X L, etc. Resonance Rayleigh Scattering Spectra of Interaction of Sodium Tetraphenylboron with cationic surfactants and their analytical application. Chinese J. Anal. Chem., 2005, 33(10): 1427–1430.

[86] Jiang T, Zhang G W, Wang A P, et al. Determination of cetylpyridinium bromide by dual wavelength resonance light scattering ratiometry. J. Instr. Anal., 2008, 27(9): 984–987.

[87] Yang Z, Liu S P, Liu Z F, et al. Resonance Rayleigh scattering spectra of interaction of trace amounts of cationic surfactants with some acidic triphenyl methane dyes and their analytical applications. Chem. J. Chinese Univ., 2004, 25(6): 1040–1042.

[88] Liu S P, Liu Z F. Double scattering spectra of Cationic Surfaetant-Eosin Y system and their analytical applications. Chinese J. Anal. Chem., 1996, 24(6): 665–668.

[89] Wang X, Study on the determination of surfactants and drugs by light scattering spectrometry. Changchun: Jilin University Master's thesis, 2011.

[90] Feng S, Wang J, Fan J. Determination of a cationic surfactant with naphthalene black 12B by the resonance light scattering technique. Ann. Di Chim., 2006, 96(5–6): 293–300.

[91] Yu G, Huang J, Zhang P Y. Persistent organic pollutants: one of the important global environmental problems. J. Environ. Protect., 2001(4): 37–39.

[92] Li J G, Zhao Y F, Wu Y N. A review for the body buden of persistent organic pollutants in China. Environ. Chem., 2011, 30(1): 5–19.

[93] Wang Y H, Cai Y, Jiang G B. Research processes of persistent organic pollutants (POPs) newly listed and candidate POPs in stockholm convention. Chinese Sci.: Chem., 2010, 40(2): 99–123.

[94] Zhang F, Wu X, Zhan J. Study on the resonance light scattering spectra of the interaction of quinine dihydrochloride with perfluorooctane sulfonate and its analytical applications. Luminescence, 2011, 26: 656–661.

[95] Wu F, Tan K J, Liu Z D. RLS spectrum study and application of quinine hydrochloride and perfluorooctane sulfonate system. Chinese Sci. Chem., 2011, 41(7): 1163–1169.

[96] Tan K J, Wu F. Rapid detection method of PFOS in environment water sample based on resonance light scattering technique. China, 201110065276.4.2011- 03-18

[97] Wu F, Zhu J, Tan K J. Resonance light scattering spectra of perfluorooctane sulfonate-protein system and its analytical application. Chinese J. Anal. Chem., 2012, 29(8): 969–973.

[98] Zhang F J. Application of spectral analysis technique for the determination of polychlorinated biphenyls (PCBs). Jinan: Shandong University Master's thesis, 2011.

[99] Zhang F J, Wu X, Li L. A ratiometry of light scattering and fluorescence emission for the determination of polychlorinated biphenyls. The 5th Shanghai International Analytical Chemistry Forum Collected Works, 2010-09-15, 74–75.

[100] Wu F, Zhu J, Tan K J. Resonance light scattering spectra of perfluorooctane sulfonate-protein system and its analytical application. Chinese J. Anal. Chem., 2012, 29(8): 1163–1169.

[101] Olsen G W, Burris J M, Mandel J H, et al. Serum perfluorooctane sulfonate and hepatic and lipid clinical chemistry tests in fluorochemical production employees. J. Occup. Environ. Med., 1999, 41: 799–806.

[102] Renner R. Scotchgard ban highlights unknowns. Environ. Sci. Tech., 2000, 34: 371A-373A.

[103] Guo R, Cai Y Q, Jiang G B, etc. Current research of perfluorooctane sulfonate. Progr. Chem., 2006, 18: 808–813.

[104] Jia C X, Pan G, Chen H: Sorption and desorption behavior of perfluorooctane sulfonate on the natural sediments. Acta Scientiae Circumstantiae, 2006, 26: 1611–1617.

[105] Loos R, Locoro G, Huber T, et al. Analysis of perfluorooctanoate (PFOA) and other perfluorinated compounds (PFCs) in the River Po watershed in N-Italy. Chemosphere, 2008, 71: 306–313.

[106] Giesy J P, Kannan K. Perfluorochemical surfactants in the environment. Environ. Sci. Tech., 2002, 36: 146A–152A.

[107] do Céu Silva M, Gaspar J, Silva I D, et al. Induction of chromosomal aberrations by phenolic compounds: possible role of reactive oxygen species. Mutat. Res., 2003, 540(1): 29–42.

[108] Bukowska B, Kowalska S. Phenol and catechol induce prehemolytic and hemolytic changes in human erythrocytes. Toxicol. Lett., 2004, 152(1): 73–84.

[109] Sun L W, Qu H H, Wu D, etc. Study on DNA damage of several different biomaterials induced by hudroxybenzene compounds. Environ. Chem., 2003, 22: 390–393.

[110] Kong Z M, Zhang G D, Sun L W. Study on DNA damage of spermary cell in mice induced by environmental endocrine disrupting chemicals. Environ. Pollut. Prev., 2002, 24: 76–78.
[111] Li J H, Yuan J, Ni X, etc. Effects of phenol on the grownth of spirulina maxima. Acta Hydrobiologica Sinica, 2001, 25: 294–296.
[112] Gao H, Zhang S H, Xiong D Q, etc. Study on acute oxieities of pheonl and anilien to two marine organism. Mar. Environ. Sci., 2006, 25: 33–36.
[113] Su R L, Wang Y, Ni Y N, et al. Graphene quantum dots and the resonance light scattering technique for trace analysis of phenol in different water samples. Talanta, 2014, 125: 341–346.
[114] Yao D M, Liang A H, Yin W Q, et al. Resonance light scattering determination of trace bisphenol A with signal amplification by aptamer–nanogold catalysis. Luminescence, 2014, 29: 516–521.
[115] Tan G H. The new methods for determination of phenol and hydropuinone in environmental water samples. Hengyang: Nanhua University Master's thesis, 2007.
[116] Chen Y S, Yang H M, He A T, etc. Determination of trace naphthols in water by resonance light scattering. Chinese,J. Spectrosc. Lab. 2009, 26: 324–327.
[117] Fan X, Lyu C Y, Liu Y M, etc. Determination of trace malachite green byresonance Rayleigh – scattering method. Chinese J. Health Lab. Technol., 2006, 16: 1158–1159.
[118] Huang M Y, Peng X J, Li Y Z, etc. Detection of sudan I in hot chilli and related products by resonance light scattering method. Mod. Prev. Med., 2007, 34: 528–529.
[119] Yu L H, Liu Z F, Hu X L, etc. Resonance Rayleigh scattering method for the determination of some basic triphenylmethane dyes using tungstate as a probe with wolframate radical. Chinese Sci. Bull., 2009, 54: 3283–3290.
[120] Liu S, Liu S P, Xie L S, etc. Determination of methylene blue by a resonance light scattering technique with dodecy ibenzene sulfonic acid sodiumsalt. Chinese J. Anal. Chem., 2009, 26: 966–970.
[121] Zeng M, Ouyang X, Chen M K, etc. Determination of paraquat by resonance Rayleigh scattering method. Chinese J. Anal. Chem., 2008, 36: 112–115.
[122] Wu X H, Liang L N, Dong C Z, etc. Resonance light scattering determination of aniline in water based on diazotization-coupling reaction. Chinese J. Pub. Health, 2008, 24: 1406–1407.
[123] Ouyang Y F, Wang Y S, Mi X W, etc. Determination of trace 1-Hydroxypyrene in human urine with ethyl violet by resonance light scattering method using sodium dodecy 1-benzene sulfonate as sensitizer. J. Instr. Anal., 2007, 26: 214–220.
[124] Zhang H T. Standard examination method annotation for drinking water. Chongqing: Chongqing University Press, 1993.
[125] Ravi J. Turbidity Measurement. ISA Transactions, 1993, 32: 397–405.
[126] Suk N S, Guo Q, Psuty N P. Feasibility of using a turbidimeter to quantify suspended solids concentration in a tidal saltmarsh creek. Estuar. Coast Shelf Sci., 1998, 46: 383–391.
[127] Steven A S. A simple method of correction for forward Rayleigh scattering in turbidity measurement. Appl. Optics, 1993, 32: 4646–4651.
[128] Tan K J, Huang C Z, Li Y F. Highly sensitive multifunctional analyzer based on backward light scattering technology. China: 2005100205337.
[129] Tan K J, Li Y F, Huang C Z, et al. Determination of chlorine in human urine by detecting backscattering signals with a new optical assembly. Chinese Chem. Lett., 2006, 17(5): 679–682.

Xiao Bing Pang, Yan Yin

15 Atmospheric environmental light scattering spectrometry

Atmospheric light scattering is an important natural phenomenon occurring in the earth's atmosphere, which weakens the direct radiation of the sun's light, changes the balance of the earth's solar radiation, and leads to global climate change. The atmospheric scattering of the sun's light forms the blue sky. Without atmospheric light scattering, the sky would be dark. When the light wave collides and interacts with scattering particles, such as atmospheric molecules and aerosols, scattering particles re-emit light in all directions. The intensity of the light and polarization properties are related to concentration, size and morphology of particles. The scientists have developed a series of aerosol detector based on these relations for the study of atmospheric aerosol. This chapter will briefly introduce the natural phenomena of atmospheric light scattering, the earth's atmosphere and air pollution, and then discuss aerosol detection instrument based on the principle of light scattering.

15.1 Atmospheric light scattering phenomenon

As discussed in Chapter 1, the clouds at noon appear white or grey for the existence of lots of water drops or ice crystal and other larger particles and their particle sizes are comparable to the wavelengths of the solar spectrum to occur Mie scattering. As the intensity of Mie scattering has no obvious relation with the incident wavelength, the scattered light still appears white of the sunlight, and the clouds in the sky usually appear white. If the clouds are too thick, the scattered light cannot pass through the cloud, which appear grey in the sky.

The blue sky on a sunny day is owing to Rayleigh scattering when the sunlight passes through the atmosphere. As the atmospheric particle size is far smaller than the wavelength of the sunlight, the intensity of Rayleigh scattering is in reverse proportion to the wavelength biquadrate, and so the red light at longer wavelength has weaker scattering intensity compared with that of the purple light at shorter wavelength. The smaller proportion of red light at longer wavelength is scattered, and so most of the light is directly radiated to the ground, while the shorter wavelength purple light and blue light are mostly scattered by the atmosphere, leading to the blue sky.

Scattering of atmospheric molecule reduces significantly with the density of atmosphere and height, and thus the color of the sky changes as well. For example, at a height of 8 km, 11 km, 13 km and 21 km, the sky appears cyan, dark cyan, dark purple and black purple, respectively. In the higher sky, the air

https://doi.org/10.1515/9783110573138-015

has become very thin, so that the scattering is weak, and the sky appears quite dark. Because there is no light scattering from the atmosphere, the sky on the moon is dark at night or day.

15.1.1 Earth atmosphere

Atmosphere of the earth can be divided into troposphere, stratosphere, mesosphere, thermosphere, and exosphere.

The troposphere is the bottom of the atmosphere, at a height of 10–15 kilometers, and its remarkable feature is that the temperature decreases with height, and the convection in the vertical direction is rapid and obvious. The troposphere contains 90% of the mass in the atmosphere, so most of the atmospheric chemistry occurs in the troposphere. Besides, all the weather phenomena on the earth occur in the troposphere, so it is closely related to human activities, and it is the main area for the atmospheric observation and aerial remote sensing.

The stratosphere is located above tropopause to the height of 45–55 km in the atmosphere, and contains an ozone layer of about 20 km thickness. The ozone layer protects life on the surface of earth by absorbing ultraviolet light from the sun. The stratospheric temperature enhances with the increase of altitude, so the convection is very slow in the vertical direction.

Mesosphere is directly above stratosphere and extends from 80 to 90 km. In this atmosphere, air is quite thin. The temperature decreases as the height increases, and there is a strong vertical convection movement. It is the coldest area in the atmosphere, and the temperature can be as low as -80 °C.

Thermosphere is directly above mesosphere. N_2 and O_2 absorb a lot of radiation at short wavelength, so the temperature is higher. Due to the effect of the sun and cosmic rays, most of the air components at the top of the mesosphere and the bottom of the thermosphere are ionized into charged particles with higher density, so it is also called the ionosphere. The ionosphere can reflect electromagnetic waves emitted from the ground, and plays an important role in radio communication on the ground.

Exosphere extends in the sky over 500 km, so that gas molecules have enough energy to overcome gravity and escape from the atmosphere.

15.1.2 Atmospheric environment

The atmosphere is a gas protective layer formed by the gravitational force of the earth, which protects life on the earth and provides life elements such as oxygen, water and all kinds of elements to the earth's creatures. The atmosphere is related to survival of people, directly influencing human health and future development. The atmosphere is composed of clean air, vapor, aerosol, and so on. When they change, the environment of the atmosphere is deteriorated. The change in atmospheric compositions directly influences quality of air, local weather, and global climate.

15.1.2.1 Dry air

The clean atmosphere beyond water vapor is usually called the dry air. with main components of nitrogen (volume rate 78.1%), oxygen (20.9%), argon (0.9%), and trace gas (0.1%). The trace gas includes nitrogen dioxide, methane, nitrous oxide, ozone, sulfur dioxide, nitrogen oxide, volatile organic compounds, neon, helium, and so on. Although the trace gas only takes up a small amount of air, it greatly influences the atmosphere. For example, nitrogen dioxide, methane and nitrous oxide are essential gases causing global greenhouse effect, and sulfur dioxide, nitrogen oxides, volatile organic gases and ozone are main pollutant gases in the cities.

15.1.2.2 Water vapor

The volume of water vapor in the atmosphere increases from 0 to 4% with the change in area and time, which can be lower than 1% in cold and dry north pole, or can be as high as 4% in humid tropics. Thus, vapor concentration is an important parameter for the weather report. Water is earth's most vital factor, which exists in gas, liquid, or solid three forms depending on the atmosphere. Thus, water vapor is essential for biogeochemical cycles and atmospheric cloud physics.

15.1.2.3 Aerosol

Aerosol is a small solid or liquid particle in the atmosphere, which is formed mainly from direct release of volcanic eruptions, dust storms, forest fires, plant emissions (such as spore and pollen), marine droplet, fossil fuel combustion, direct release during land use process, or secondary formation of a gas (such as SO_2, NH_3, isoprene, or other compounds) during atmospheric photo-oxidation process; cloud droplets, ice crystals, rain, and snow and other particles, composed of water and ice, are also a part of aerosol.

Aerosol directly influences earth's radiation balance through scattering and absorption of light radiation or indirectly affect the balance by changing cloud formation through cloud condensation nuclei, and indirectly influences the balance, thus affecting global climate, local weather, and atmospheric visibility [1–2], which are quite harmful for human health. When aerosol concentration is too high, it results in a lot of diseases, such as pneumonia, bronchitis, heart failure, and stroke. Besides, aerosol also provides an interface for atmospheric heterogeneous chemical reactions, promoting atmospheric photo-oxidation process. It is of great importance to investigate atmospheric aerosols.

The light scattering intensity of aerosol is closely related to its concentration, particle size, and shape, and so light scattering is important for the study of aerosol. Optical particle counting technology based on light scattering principle has been widely applied to various aerosol detectors. The following sections will introduce the atmospheric light scattering instrument (Sections 15.3–15.9) on the basis of an optical particle counter (Section 15.2), which will serve as a reference as how to construct a new instrument for the analysis of light scattering spectrum.

15.2 Optical particle counter

An optical particle counter is used to detect the number of aerosol particles and the distribution of particle size in an atmospheric unit volume. Its basic principle is to directly radiate a test chamber with a high-energy light source (a laser or light emitting diode). When the particle passes through the chamber to produce light scattering, a small light impulse is formed because of the signal transfer, amplification and identification, thus the counting of a large number of electric pulses can be performed.

Particle counters are widely used in atmospheric chemistry simulation to study aerosol formation mechanism and observe aerosol formation, evolution and aging process [3]. Therefore, optical particle counters have been used for the study of aerosol and air pollution as well as for monitoring the quality of indoor air. Figure 15.1(a) shows

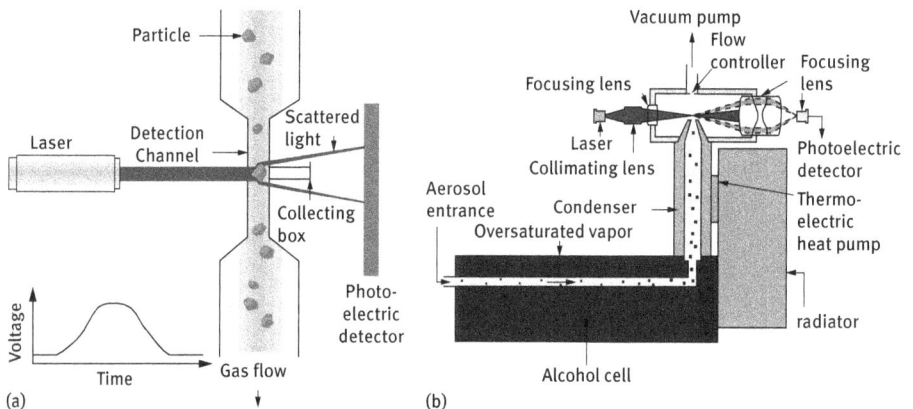

Figure 15.1: The working principle of a particle counter based on the light scattering technique: (a) basic principle (from Wikipedia; the content is translated and revised by the authors). (b) TSI company condensation particle counter optical principle (from TSI website; the content is translated and revised by the authors).

the optical schematic of the light scattering particle counter. The sensitivity of the particle counter based on light scattering method is 0.05 μm or larger. A condensation nuclei counter reduces the detection sensitivity to a nanometer scale. The particles are changed into nucleation centers when the supersaturated gases (such as ethanol gas) are expanded into droplets. Then, condensation nuclei counter measures and calculates these aerosol particles (Figure 15.1(b)). A famous international aerosol producer company (TSI) uses the above technique with a detection range of $1-10^7$ cm^{-3} and a particle size range of 10–1,000 nm.

An optical particle counter can perform continuous monitoring for site measurement, and it has a compact and firm structure that requires minimum maintenance. However, an optical particle counter has low sensitivity and resolution in the detection of particle size, aerosol spectrometry is often needed in some high-precision measurements.

According to equation (2.38), when the scattering particle diameter d is far larger than the incident wavelength (λ), the scattering intensity (I) can be given by a simplified scattering equation:

$$I = I_0 K(n, \theta)d^2 \tag{15.1}$$

where, I is the scattering intensity, and K is a function of refractive index (n) and scattering angle (θ). To retain most aerosol particles in the Mie scattering range compared with light source wavelength, most optical particle counters use visible or infrared light source (500–1,100 nm). When an instrument is used to detect a cloud or water droplet with larger particles, the simplified equation (15.1) can be applicable, while for smaller aerosol, the Rayleigh scattering equation is more suitable [4].

To more sensitively detect a single aerosol particle signal, most counters use high-intensity laser as the light source. As shown in Figure 15.2, to detect smaller particles, the intensity of laser and the sensitivity of the detector need to be improved to detect small particles following Rayleigh's Law (below 100 nm) with an in-chamber laser technique. The light scattering intensity of the particles produced by the chamber laser detector is over 100 times stronger than the standard laser detector using the same power, which significantly improves the sensitivity of the instrument.

An ultrahigh sensitive aerosol spectrometer (UHSAS) and single-particle soot photometer (SP2) usually adopt this kind of an in-chamber solid laser technique, while most of the ordinary optical particle counters use simple laser, such as diode laser. As light intensity from one 20 μm particle is 10 order magnitude than that from one 50 nm particle, far exceeding linear detection range of the single detector. So some instruments can choose to measure particles in different ranges, and some instruments have multiple detectors to expand the particle size detection range by multiple signal amplification systems.

(a)

(b)

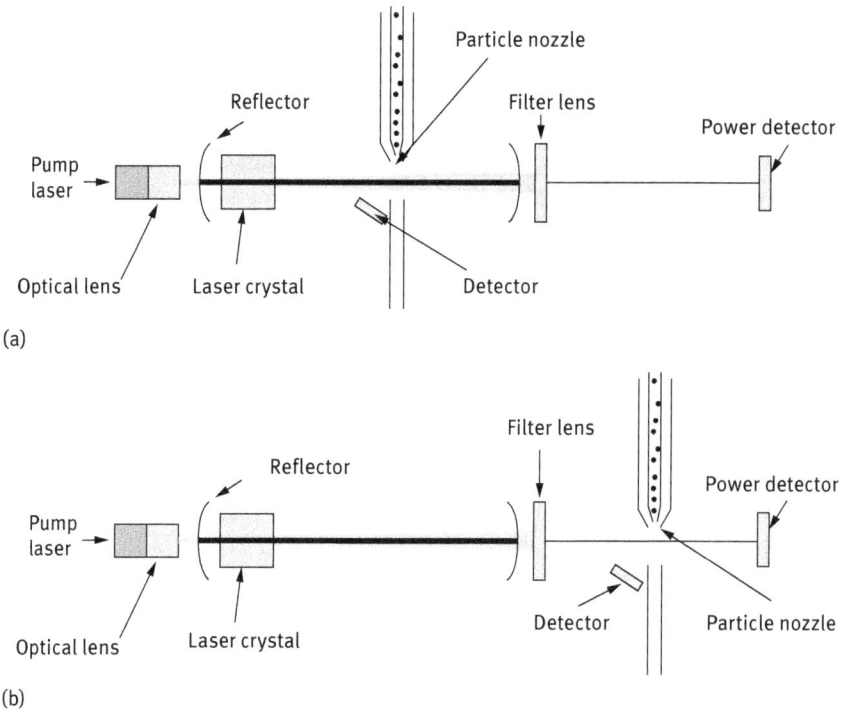

Figure 15.2: The working principle schematic diagram of an in-chamber optical particle counter (a) and an ordinary optical particle counter (b).

15.3 Aerosol spectrometer

The aerosol spectrometer is different from the low precision optical particle counter, but using the laser to irradiate the aerosol, and detect the particle size and concentration by measuring the scattering light intensity produced by the aerosol. It has the function of high resolution for particle size detection and counting, so that it is not too complicated to measure the particle size according to Mie scattering.

Laser is the key component of an aerosol spectrometer and different companies use different lasers. For example, the Ultra-High Sensitivity Aerosol Spectrometer (UHSAS) produced by Droplet Measurement Technologies (DMT) company uses laser of a semiconductor diode pump (Nd^{3+}:$YLiF_4$ solid laser). The basic space mode is a internal chamber laser at 1054 nm with an power of 1 kilowatts. The laser reflectivity can achieve as high as 0.99999. The direction of the laser emission is perpendicular to the direction of the particle flow. The scattering signals perpendicular both to the direction of the particle flow and the laser direction are used for detection (Figure 15.3).

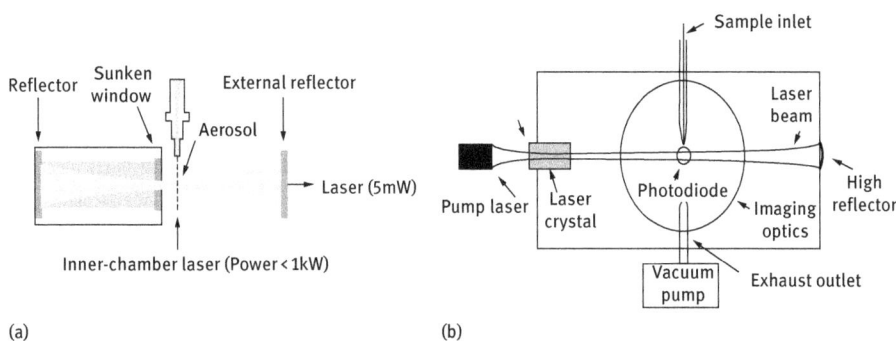

Figure 15.3: The working principle schematic diagram of an aerosol spectrometer. (a) An innner-chamber laser (the photo is from TSI website; the content is translated and revised by the authors) and (b) the working principle schematic diagram of DMT aerosol spectrometer laser (the photo is from DMT website; the content is translated and revised by the authors). An aerosol spectrometer is composed of five subsystems as follows: 1) the main optical system: its function is to generate laser, detect the scattered light generated by aerosol, and provide protective shell for optical system and aerosol sampling parts. 2) an air flow system: its function is to bring aerosol samples into the optical detectable zone to control the flow rate. 3) the simulated electronic signal system: its function is to amplify and deal with the light scattering signals of particles; 4) Digital electronic system: its function is to analyze particle signals, classify binary signals according to user specified binary drawings, and communicate with the monitoring/controlling instrument of computer. and 5) the user computer: its function is to control the instrument and collect and report data through the computer program.

The particle size of the aerosol measured by UHSAS is 0.06–1.0 μm, and the maximum particle counting rate is 3000/second. Studies have found that a highly sensitive aerosol spectrometer detected particle concentration is consistent with that of condensation particle counter, and the particle size is consistent with that of the scanning mobility particle sizer [5].

15.4 Aerosol mass spectrometer

The principle of light scattering for the detection of particle sizes is widely adopted by aerosol mass spectrometer (AMS). The AMS is one of the most advanced instrument for aerosol research, which can be used for the detection of particle chemical component (organic compounds, sulfates, nitrates, ammonium salts, and chlorine salts) and on-line detection of the particle size distribution to obtain chemical composition and size distribution data of atmospheric aerosol with high temporal resolution.

In the aerosol mass spectrometry with dual laser for thermal analysis, the atmospheric particles are pumped into the instrument and the scattered light is generated when irradiated by two consecutive luminescent lasers. The scattered light is converted into a electrical signal by photomultiplier to trigger a Nd doped yttrium aluminum garnet (Nd:YAG) laser for thermal desorption of particulates. As the distance between the two lasers is constant, time and speed of the particles moving through the distance can be calculated, and so the particle aerodynamic diameter is obtained. As the velocity of the particle is decided by its aerodynamic diameter, particles of different sizes can be selected by setting the laser light time of the thermal desorption. Only the particles with suitable particle size can be thermally desorbed by the Nd:YAG laser with accurate light out time and their chemical components are heated and eventually be detected by mass spectrometry. Most other particles with different sizes will miss the opportunity to be thermally desorbed by laser because of their different flight speeds [6].

The AMS produced by TSI adopts this technology (Figure 15.4), which could real-time monitor the particle size and chemical components of nondissolvable aerosols. The AMS could detect chemical components of every single particle <30 nm, thereby significantly improving our understanding of the nucleation mechanism. A movable aerosol is usually used for ground, boats, planes and global study, thus further improving our understanding of aerosol spatial distribution; similar to other series of coupling AMS technologies, this also allows in situ detection of chemical particle effective density, refractive index, volatility, and cloud activation properties [7]. However, the currently adopted dual-laser thermal desorption AMS has stopped production.

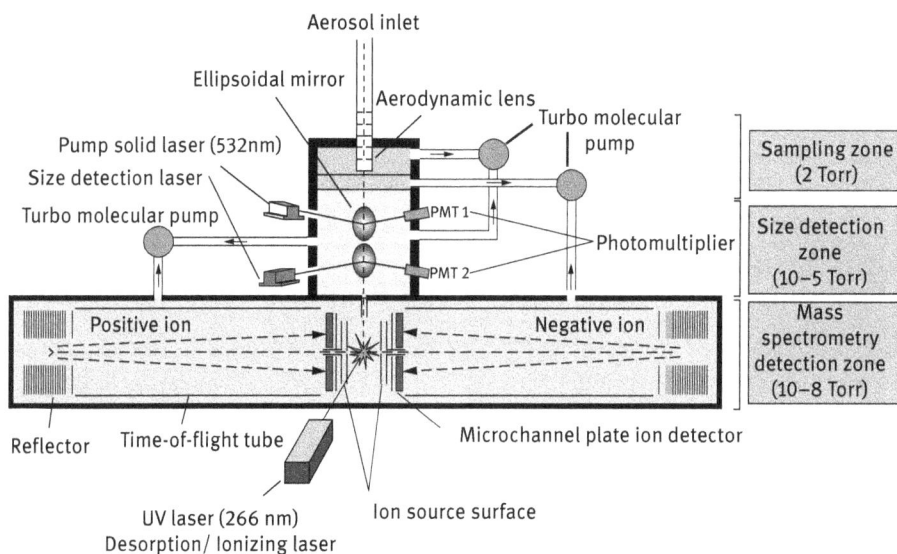

Figure 15.4: The working principle schematic diagram of aerosol mass spectrometer. The photo is from TSI website (the content is translated and revised by the authors).

15.5 Aerodynamic particle sizer

An aerodynamic particle sizer (APS) uses the above mentioned dual-laser method to detect particle's aerodynamic size. APS produced by TSI Company adopts the above mentioned technology: when the test particle enters the detection zone, it passes through two close laser beams to produce scattering of light (Figure 15.5); each particle produces a double-peak signal, and the time difference between peaks is time-of-flight. The instrument can achieve a resolution of 4 nanoseconds, which can calculate the particle's exact velocity and obtain aerodynamic particle size. The amplitude of the peak signal is light scattering intensity, but the light scattering intensity is not always a reliable indicator of particle size. Thus, APS provides two kinds of measurement methods: aerodynamic diameter and light scattering intensity.

The APS uses flight of time to the real-time measured particle aerodynamic diameter, which is not influenced by the refractive index or Mie scattering index and is able to detect an aerodynamic particle size of 0.5–20 μm and relative light scattering intensity of particle in the range of 0.37–20 μm.

In recent years, the APS with a UV laser source (355 nm) is used to study the particle size distribution and concentration of atmospheric bioaerosol particles (including bacteria, spores, and pollen), and is helpful to understand the air microbial pollution and the contribution of bioaerosols to atmospheric processes,

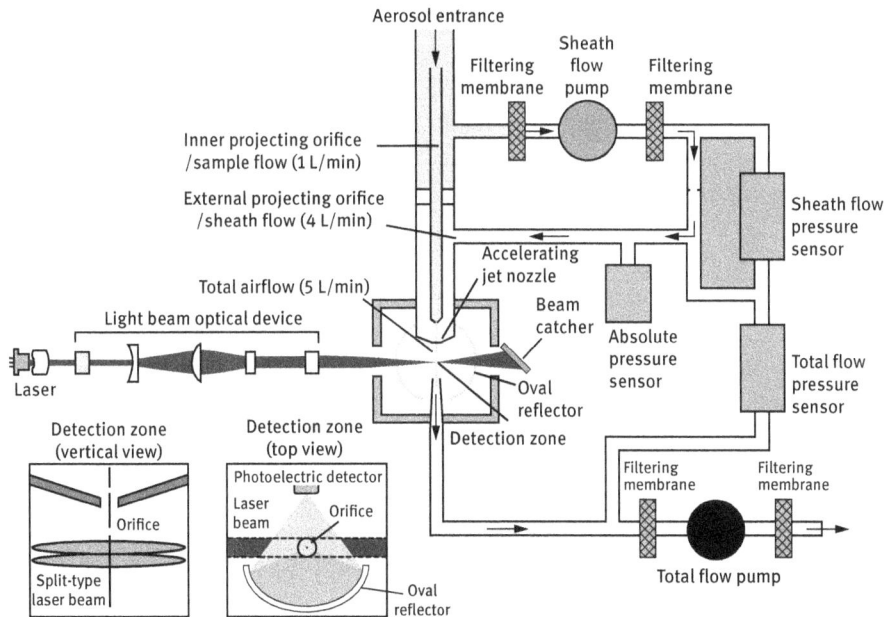

Figure 15.5: The working principle schematic diagram for the aerodynamic particle sizer produced by TSI company. Photo is from TSI website (the content is translated and revised by the authors).

including clouds and precipitation [8]. The APS is not only used for the detection of aerosol, but is also an important tool to simulate atmospheric chemical reactions, and observe formation of aerosol, evolution, and aging process [9].

15.6 Single-particle soot photometer

An single-particle soot photometer (SP2) is a tool to directly measure the aerosol's single-particle charcoal (soot), which is suitable to detect soot, snow, or water black charcoal. Black charcoal particles are essential in the atmosphere for they strongly absorb irradiation from the sun and ground. Soot particles have smaller proportion in aerosol with dramatic effect of aerosol optical thickness.

As a highly absorbing particle, black carbon has the potential to alter the atmospheric temperature distribution and the radiation balance of the earth / gas system. If black carbon appears on cloud particulates, they will affect the energy absorbed by clouds, thereby changing the temperature and life of clouds. If black carbon particles are deposited on ice surfaces such as glaciers or polar ice caps, this will affect the energy absorbed by these ice surfaces, thereby promoting the melting rate of ice. Black charcoal is also considered to be a cloud of coagulant tuberculosis. The traditional black carbon instrument can infer the content of soot by measuring light absorption. These instruments are not specially used for the measurement of black carbon, and the indirect measurement is influenced by many factors such as refractive index, humidity and so on. The measurement can not provide the mixed state information of the single particle black carbon.

SP2 is a highly specific instrument for determining black charcoal. It uses a laser induced incandescent instrument to measure the carbon black mass of a single particle, which provides the quality of black charcoal in each particle, but does not distinguish the mixed or fractal dimensions of the black charcoal. SP2 can also provide particle size distribution, that is to say, it can provide optical dimensions of individual particles containing carbon black and determine the ratio of particles to black carbon. This makes SP2 an important tool to study the black carbon concentration in atmosphere, mixing process and atmospheric transport. When combined with other measurements, the properties of the particles can be further understood.

SP2, from USA DMT Company, uses a high-power in-chamber Nd:YAG laser (1064 nm) as the light The scattering of single aerosol is produced when particles in the air enter into the laser beam, which will be detected to indicate the mixed state and particle size of the black carbon in the single particle, and to detect the number and mass concentration of non black carbon, and to record the scattering light and time evolution. When the black carbon enters the laser, it is heated to gasification (3500 K) to release incandescent light. The light intensity is only related to the mass of black carbon rather than its shape or mixed state in the particle [10].

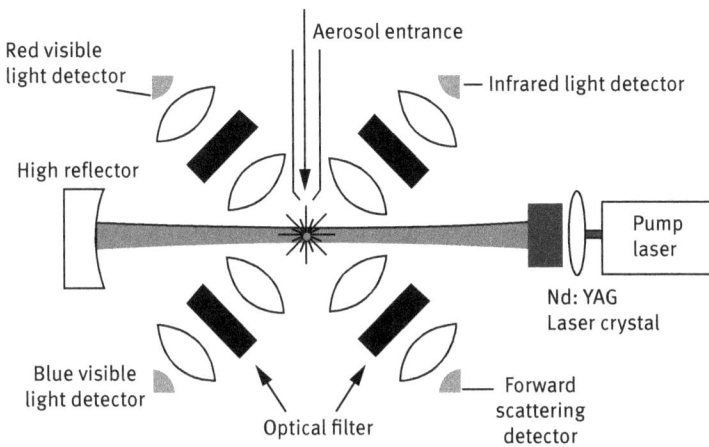

Figure 15.6: Schematic diagram of optical photometer detection principle for single particle soot. The four detectors are for scattering signal of different wavelength range. The photos are from DMT website (the content is translated and revised by the authors).

Figure 15.6 shows schematic diagram and main optical components of laser. The particle size range detected by the instrument is 200–1500 nm, and the quality range of black carbon is $(1–300)×10^{-15}$ g/particle. The particle concentration range is 0 – 10000 particulate/cm^3. When the boiling temperature of particles is high enough, they will be heated to a certain temperature and will release incandescent light. Two sensitive detectors with different wavelength band widths (350–800 and 630–800 nm) are used to detect the incandescent light, whose intensity is proportional to the burning quality of materials. Black carbon (coal particle) is the material with richest incandescent property in the atmosphere. Other incandescentable materials include metals, but their boiling points of these materials are different from that of black charcoal, and so they are easily differentiated from black carbon in SP2 measurement. Black charcoal and other materials can still be incandescent with their intensity directly proportion to that of burning material mass. But if other materials are evaporated before the black carbon reaches the incandescence, the material will slightly delay the occurrence of black carbon incandescence.

SP2 uses four different detectors at the 5 MHz sampling frequency. The scattering and incandescent pulse signal will be released when each particle is detected, and sysytem can obtain a fixed data points (about 200) from the above-mentioned four detectors. Besides, a number of instrument environmental parameters, such as temperature, pressure, flow rate, and detector voltage, are recorded per second. The instrument data process provides the intensity of scattered light, the intensity of incandescence of two channels, as well as the time delay between various signals. These parameters can be used to determine particle size, incandescent temperature, and black carbon quality. In addition, SP2 can determine the black carbon mixing state of a single particle by the ratio of light scattering to incandescent signal or from the time delay between scattering and white hot signals. The research in this field is novel, and the research on how to accurately determine the mixing state from these measurements is still in progress [12].

The instrument uses four optical detectors. One detector uses optical filters to only transmit 1064 nm of radiation, which will detect the scattered light signals of all particles, and this detector is identified as the data in channel 0. Two photomultiplier (PMT) detectors are employed to measure the visible regions of incandescent signals, and use optical filters to pass through wideband light (400–650nm) and narrowband light (610-650nm), which are determined as data in channel 1 and 2. The two detectors can accurately detect the element carbon through the color temperature signal ratio of burning particles. The fourth detector is for forward scattering signal to analyze burning particle mixture state and coating volume [13].

15.7 Cloud particle spectrometer with polarization detection

In situ differentiation and measurement of a small water droplets and crystal with a diameter lesser than 50 μm is still a challenge. Lack of such measurement hinders the understanding of the process of ice forming clouds. The polarized cloud particle spectrometer (Cloud Particle Spectrometer with Polarization Detection, CPSPD) produced by DMT company in the United States makes use of the light scattering of the cloud particles through the laser beam. By measuring the focused forward and backward scattering light, the size of the cloud particles and the thermodynamics phase (the liquid or ice of the single cloud particle in the optical diameter range from 2 to 50 μm) are obtained. Cloud particle's optical equivalent diameter is calculated from forward scattering light.

The CPSPD uses a 680-nm linearly polarized laser as the light source with four detectors for 8°–34° forward scattering and 146°–172° backward scattering (Figure 15.7). The particle's water equivalent optical diameter (rather than the geometric diameter) is determined by the forward scattering signal. The CPSPD detects the particle size of 0.6–50 μm by light intensity. The backward scattering light passes through the beam to direct light into two independent detectors. A detector is to calculate the intensity of total backward scattering; this intensity can determine the shape of the particle. The second detector is to calculate the intensity of the vertically polarized light; the rate of that light with the parallel polarized light is the polarization ratio.

Similar to the laser radar polarization, for a spherical droplet, the polarization ratio is nearly zero. A detector in the forward direction is used to direct the position of the particle. An optical mask of a qualifying detector prohibits scattering of light over

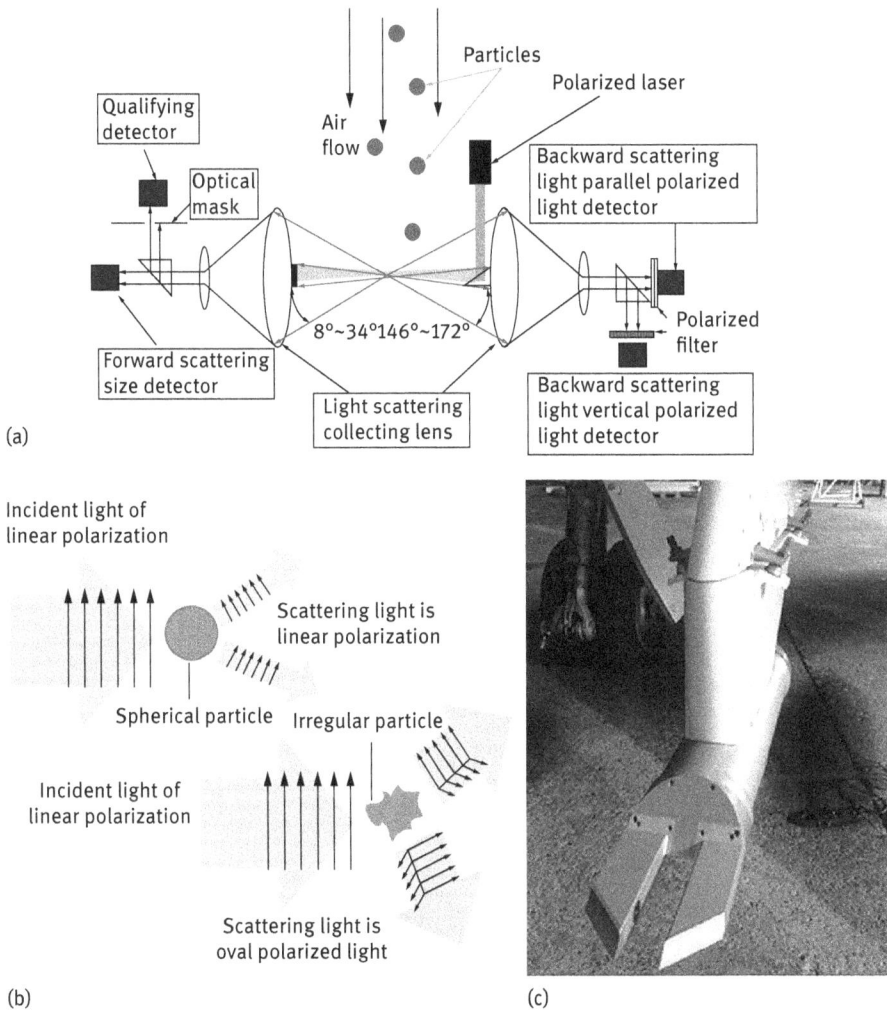

(a)

(b)

(c)

Figure 15.7: The working principle of cloud particle spectrometer with polarization detection. (a) The main optical parts and scattering light path: (b) Schematic diagram of the interaction between linearly polarized light with spherical and non-spherical particles. (c) The object installed in the aircraft. From DMT website (the content is translated and revised by the authors).

0.75 nm particle away from the measurement center. Particles in focus ±0.55 mm range will be detected by qualifying detector. The beam splitter separates the forward scattering light into two parts, of which 70% passes to the qualifying detector, the other 30% passes to the forward scattering detector, and the peak amplitude of the forward scattered light signal is recorded and used as the particle count signal.

Some researchers used the polarized light detection of CPSPD to distinguish nonspherical dust single particle of different types to understand whether the dust of specific source position has unique properties based on its optical properties, which can be used for the on-line dust type detection. In addition, these optical scattering differences provide evidence for the potential difference of climate change by the aerosol [14]. The spectrometer detects the shape of non spherical particles by measuring the scattered light of a single cloud particle and analyzing the variation of orthogonal polarized light component. Normally, spherical particle's scattered light and the incident light are on the same polarized surface. The scattered light of the non-spherical particles is partially polarized on the same plane of the incident light, and the other will be in the vertical direction of the incident plane. The instrument can detect ice particles in the mixed phase cloud of 2-50 μm. This new function can provide an important for the understanding of ice formation in the cloud [15].

15.8 Backscatter cloud probe

Backscatter cloud probe (BCP), as shown in Figure 15.8(a), delivers to various directions through laser beam scattering light, in which the scattered light at 143°, 169° (156°±13°) is transmitted to the cloud probe 143°, 169°(156°±13°) after optical lens concentration, these photons are introduced to photoelectric detector, and light pulse transfers into electric pulse, then sent to signal disposal for magnification and digitization.

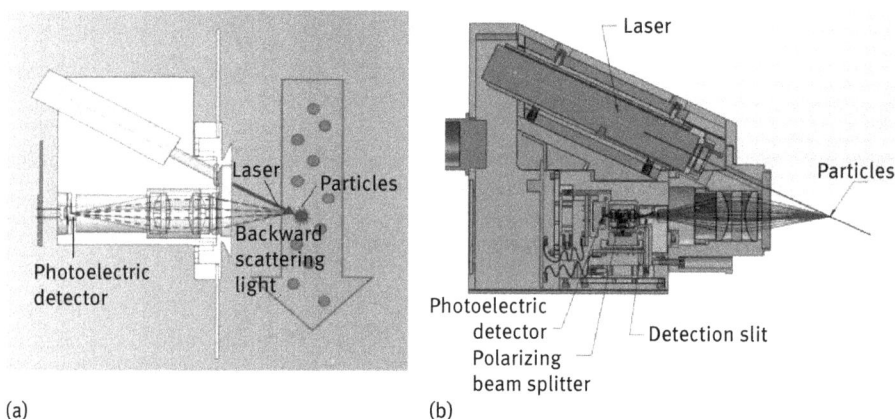

(a)   (b)

Figure 15.8: (a) The working principle of backscatter cloud probe. (b) The working optical principle of backscatter cloud probe with polarization. The figures are from DMT website (the content is translated and revised by the authors).

The Mie theory helps to calculate the scattering signal peak value and determines size of the cloud. The detection time interval can be determined by the program in the equipment. The BCP sends cloud particle size distribution with set frequency, which could be used to determine cloud particle number and concentration [16], while BCP cannot distinguish cloud particle type, such as water droplet, ice crystal, dust or volcanic ash. Backscatter cloud probe with polarization (BCPD) can realize the above-mentioned functions [17]. The light scattering generated by the cloud particles is collected at a three-dimensional angle of 136.5° and 173.5°, and the backscattered light is collected and guided to the detection slit. The light passes through the slit and continues through the polarization beam splitter, in which the polarized light component of the P direction is guided to a light detector and the other S direction is detected by other optical detectors (Figure 15.8(b)).

A photodetector converts the scattered light of each particle into electrical pulse, which are then sent to the signal processor for amplification and digitalization. As a spherical particle mainly produces P polarized light, that is to say, the direction of polarized light is in the same plane as that of the incident light, while a nonspherical or crystal particle has scattered light, and P scattered light is more than S polarized light and S polarized light is in vertical direction to the incident light surface. Herein, P polarization signal is used to determine the particle size distribution, that is, the number of particles per optical diameter.

15.9 Cloud droplet probe

Cloud droplet probe (CDP) is installed under the wing of a plane for counting the cloud droplets and aeronautically detecting the particles (Figure 15.9). By measuring

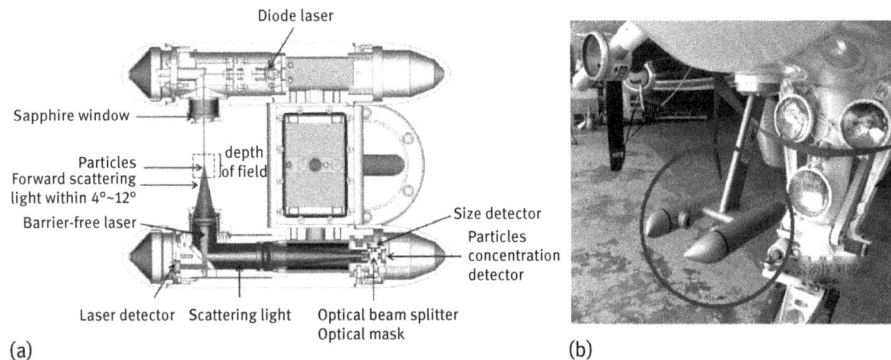

Figure 15.9: (a) The optical principle of cloud droplet detector. (b) Cloud droplet detector installed under the wing of aircraft for cloud droplet detection. The photos are from DMT website (the content is translated and revised by the authors).

the light scattering pulse generated by cloud droplets passing through the laser beam, the particle size of cloud droplets is measured in the range of 2–50 μm. The CDP uses a laser beam (658 nm) to radiate the particle to produce scattering light. The CDP detects forward scattering of light; and the intensity of the scattered light depends on the size of the cloud droplet, component and shape, which can be used to detect the size and concentration of cloud droplet. The sampling frequency is 1–10 Hz. It needs to be demonstrated that the instrument is only suitable to detect the water droplets and is unsuitable for ice crystal measurement. For precise measurement, CDP only measures cloud droplets passing through homogeneous regions of laser intensity, known as depth of field. When the aircraft speed is 200 m/s, the measuring space is 48 cm^3. As the particle passes through the laser beam, the light is scattered in all directions (Figure 15.9), while the CDP collects only the forward scattered light at an angle of 4°–12° (The photon in the range of 0°–4° is considered to be part of the laser, which is easy to burn out the photodetector). The collected light is directed to a 50/50 beam splitter and and finally to a pair of photodetector [18]. The two detectors are called "particle size instrument" and "qualifying instrument", respectively. Only scattered signal of a cloud droplet in the laser field depth range can reach the qualifying instrument.

CDP has been widely applied in the study of raindrop particle size in the cloud, However, when the high cloud concentration is over 500 /cm^3 or the speed of airborne survey is higher than 200 m/s, the probe result has obvious error. This instrument is similar to forward scattering spectrometer probe, while CDP works in a higher-speed flight.

15.10 Applications of light scattering in atmospheric remote sensing

On the basis of the atmospheric light scattering principle, the laser radar has been widely used in the analysis and monitoring of atmospheric remote sensing. To exhibit application prospects of a light scattering instrument, we will simply introduce the structure and working principle of a laser radar.

15.10.1 Working principle of light detection and ranging

Light detection and ranging (LIDAR) abbreviated as RDA. It is a remote sensing technology for optical detection and distance measurement, which use laser beam scattering, absorption, frequency shift and broadening of the spectrum for remote sensing detection in the upper atmosphere. Its principle is that laser radar sends laser pulses to the atmosphere and receives the backward scattering light signals produced by the atmosphere particles (molecules, atoms, ions, aerosol particles, etc.), and determines the position of the atmosphere by measuring the time difference between the emitted laser and the received scattering light. It detects the density or concentration of the atmosphere by measuring the intensity of scattered light. By using lasers at various wavelengths and diverse types of scattering, it also measures different atmospheric components, atmospheric optical properties, meteorological parameters and so on.

A laser radar includes a laser emission system, a signal receiving system, and an information processing system, which works in two modes of operation using pulse or continuous wave. Based on the principle of light scattering, laser radar can be divided into Doppler, Mie scattering, Rayleigh scattering, Raman scattering, Brillouin scattering and other lidar.

There are two types of lidar based on measurement: ground-based LIDAR and space-based LIDAR (Figure 15.1). Ground-based LIDAR is that laser emits from the ground to study the atmosphere by measuring backscattered light from the atmosphere. While the space based LIDAR is transmits laser from an aircraft or satellite, and to study atmospheric composition by measuring the backscattered light of the atmosphere. For example, the CALIPSO satellite laser LIDAR from USA can observe the aerosol generated by natural and human activities and even thin clouds that can not be observed by the human eye. Scientists can obtain the height, total and type of clouds and aerosols through the global three-dimensional perspective of the CALIPSO, which helps to better understand the atmosphere and climate change.

Since lidar is of great importance, the following major lidars will be introduced.

15.10.2 Doppler LIDAR

Doppler LIDAR is mainly used to detect wind's speed and direction. It uses near infrared laser to detect the moving aerosols in the air. Aerosols are ubiquitous in the low troposphere and are ideal tracers for atmospheric wind speed and direction. A moving aerosols scattering light frequency has a Doppler frequency movement when compared with the original laser frequency, and movement is in direct proportion to their speed, and so that the change in direction of the wind can be observed. Doppler LIDAR is most effective in the absence of rain, which can observe the weather phenomena such as wind direction and wind speed that can not be observed by the naked eyes. Doppler laser LIDAR is quite accurate in the detection of speed of wind (as precise as 10 cm/s).

The working principle of Doppler laser LIDAR is the Doppler effect; the relationship of an observer and an emission source frequency is given as

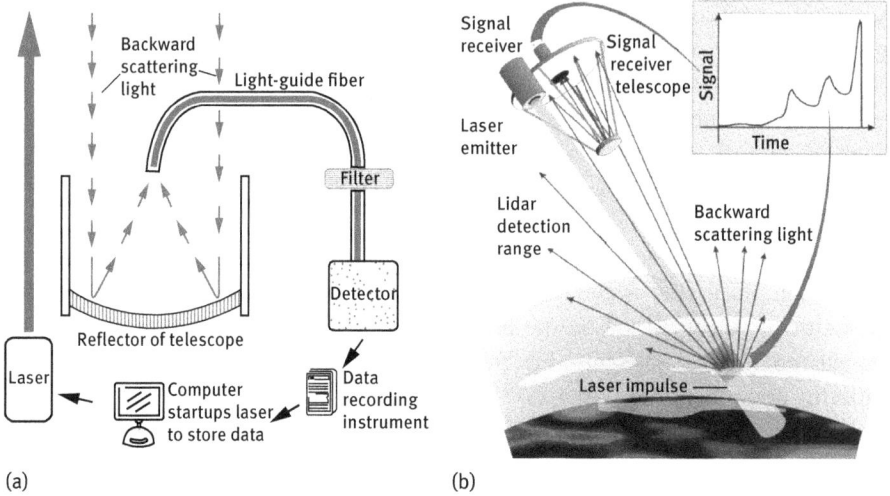

(a)

(b)

Figure 15.10: Working principle of light detection and ranging. (a) The working principle schematic diagram of ground-based LIDAR. The figure from National Oceanic and Atmospheric Administration website (the content is translated by the authors). (b) The working principle schematic diagram of satellite-based LIDAR. The figure is from European Space Agency website (the content is translated by the authors).

$$v' = \frac{v \pm v_0}{v \pm v_s} v \qquad (15.2)$$

Equation (15.2) is consistent with equation (1.2), where v' the observed frequency, and here is the frequency observed by LIDAR; v is the original emission frequency of the emission source in the medium, and here is the frequency of the emission frequency from the laser LIDAR. v is the traveling speed of the wave in the medium, and here is the velocity of laser propagation. And v_0 is the velocity of the observer relative to the medium. If the observer moves closer to the source, the operative symbol is "+"; otherwise, the symbol is "–". Herein, the observer is LIDAR. v_s is the moving speed of the source relative to the medium. If the observer moves closer to the source, the operative symbol is "–"; otherwise, the symbol is "+". Here v_s is the moving speed of the aerosol relative to the laser LIDAR.

According to equation (15.2), if the aerosol moves away from the direction of LIDAR. If the aerosol moves away from the direction of the laser radar, the light scattering wavelength of the aerosol increases; if the aerosol moves toward the laser LIDAR, light scattering wavelength of the aerosol decreases.

In general, the laser LIDAR is capable to scan the upper hemisphere, so it can draw a three-dimensional map of the turbulence in the atmospheric boundary layer. When the laser scanner points to the vertical direction, it can provide the variation of the vertical velocity with the height and time. LIDAR sends the laser beam to the air, and receives aerosol backward scattering light, and analyzes atmospheric characters, and calculates the velocity of aerosol based on the principle that the backward scattering light of aerosol is proportional to velocity. Since the Doppler laser radar can scan from different directions and measure the wind speed in multiple directions, the information of wind direction and velocity can be obtained by combining the above information (Figure 15.11).

15.10.3 Differential absorption LIDAR

Differential absorption LIDAR (DIAL) laser radar simultaneously emits two or more closely spaced wavelengths of laser when measuring the concentration of atmospheric components one laser wavelength (λ_{on}) is on the absorption spectrum line of the gas molecule, and another laser wavelength (λ_{off}) is close to the absorption line but is not located in the absorption spectrum range of the test gas molecule. As the two wavelengths are close, light scattering of atmospheric molecule and particle is basically equal. However, the difference in the backward scattered light intensity at the end of the two wavelengths is attributable to the absorption of the λ_{on} laser in the gas molecule, so the ratio of the backward scattered light intensity of the two or more wavelengths can be used to determine the concentration of the atmospheric composition.

As shown in Figure 15.12, vapor has many dispersed absorption spectral lines. When the atmospheric water vapor concentration is measured by DIAL, the laser emitted two very close wavelengths of laser pulses. One wavelength located on the water vapor absorption spectrum line is called λ_{on}, the other is far away from the absorption spectrum band called λ_{off}. If the two waves are close to each other, the scattering of light from the atmosphere and particulates is basically the same at the two wavelengths. The difference of the backward scattered light intensity between these two wavelengths is entirely due to the absorption of water molecules to λ_{on}. Therefore, the measurement of the backscattering ratio at two wavelengths based on the function of distance can be used to calculate the concentration distribution of water vapor.

When measuring atmospheric ozone, DIAL does not require precise laser frequency, because ozone has a wide absorption band (about 200 nm), which is different from the narrow spectral line of water vapor. This requires a sufficient wavelength interval between the selected λ_{on} and λ_{off} to ensure a significant difference of ozone at two wavelengths. However, due to the different scattering of aerosol from these two wavelengths, the uncertainty caused will lead to error in measurement. The difference in aerosol scattering between λ_{on} and λ_{off} can be deduced from third measurements of a longer wavelength λ_A. Thus, DIAL detects ozone concentration with three wavelengths to obtain a good precision. The component concentration is directly proportional to the light scattering intensity ratio of DIAL at λ_{on} and λ_{off} (Figure 15.12(B)); specific calculus formula can be obtained from the reference [21].

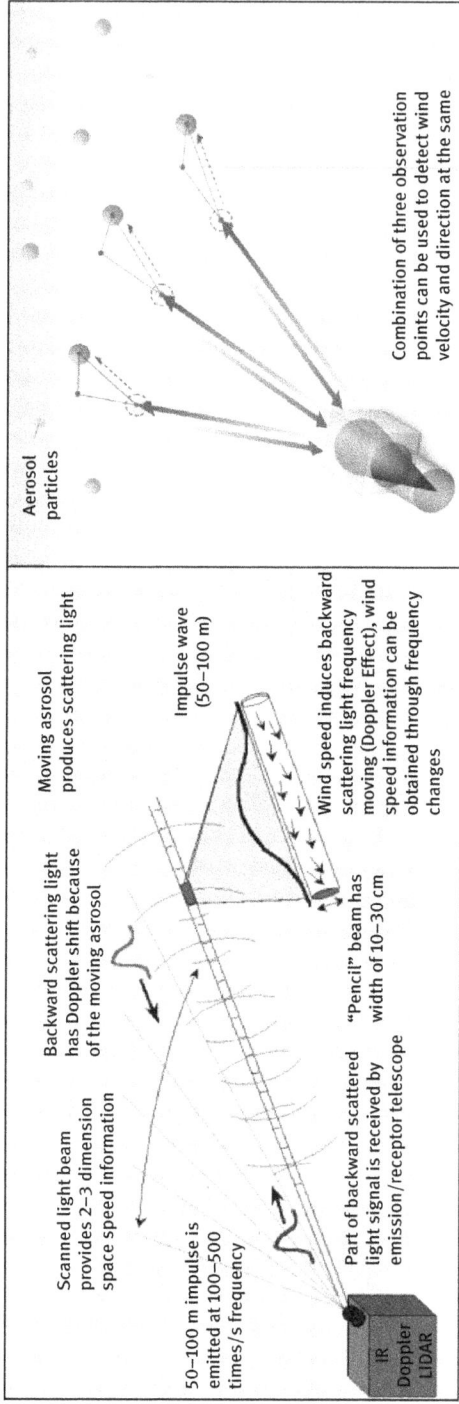

Figure 15.11: The working principle schematic diagram of Doppler RIDAR. Doppler RIDAR emits laser and receives backward scattered light of light frequency changes to detect wind velocity. The photos are from HK observatory (the content is translated and revised by the authors).

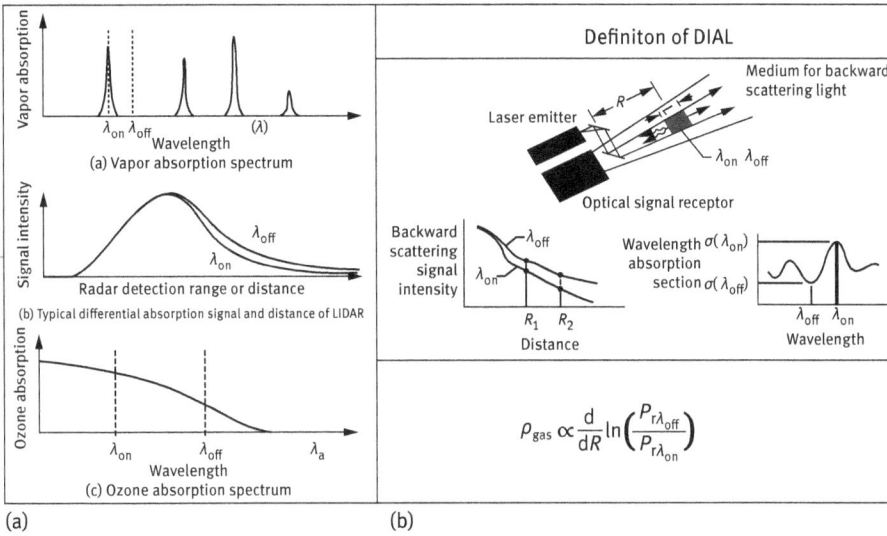

Figure 15.12: Differential absorption LIDAR: Choose two wavelengths and differentiate components of test substance based on the distinct molecule absorption at the above two wavelengths. (B) The working principle schematic diagram of differential absorption LIDAR. The atmospheric composition concentration of test gas (ρ_{gas}) is proportional to the ratio of light scattering intensity of λ_{off} and λ_{on}. (Photos are from NASA website; the content is translated and revised by the authors.)

DIAL has the ability of distance resolution, which helps to detect atmospheric pollution at varying distances, and is widely used for discharge monitoring of pollutants such as greenhouse gas, ozone, volatile organic compounds, aerosol, and so on.

15.10.4 Cloud physics LIDAR

Cloud physics LIDAR (CPL) is a backscatter LIDAR for simultaneous operation at three wavelengths (1,064, 532, and 355 nm). The main function of CPL is to provide a multi-wavelength, high spatial and temporal resolution measurement chart for cirrus, subvisual cirrus and aerosol [22]. The vertical resolution of CPL is 30 m, and the horizontal resolution is usually 200 m. The CPL has the vertical measurement resolution of 30 m and a horizontal measurement resolution of 200 m. The CPL instrument is mainly used to detect the 180° backscatter parameter. The basic measurement derives multiple data results, including time–height section images, layer boundaries of cloud and aerosol, cloud optical thickness, aerosol layer and planetary boundary layer, and extinction distribution.

CPL provides a series of cloud's physical information, which includes the following data:

(1) Provides cloud distribution figure with a vertical measurement resolution of 30 m and horizontal measurement resolution of 200 m, observed at wavelength of 1,064, 532, and 355 nm.
(2) aerosol boundary layer and plume distribution observed at three wavelengths.
(3) Determines the morphology of the cloud at 1,064 nm depolarization ratio (such as ice or water).
(4) Determines the size distribution of cloud particles through the observational result at 532 nm.
(5) Directly detects the optical thickness of cirrus cloud through the observation result at 355 nm.

15.10.5 Airborne Raman ozone and temperature LIDAR

Airborne Raman ozone and temperature LIDAR (AROTEL) instrument includes a light source, detector, and data processor. The light source uses two kinds of lasers: XeCl excimer laser emitting at 308 nm without polarization (200 mJ, 200 Hz) and Nd:YAG laser emits 50 Hz of 355 nm (350 mJ), 532 nm (600 mJ) and 1064 nm (1.2 J) three wavelengths. The data will be collected through photon counting and simulation detection techniques. The data will be stored at a resolution of two minutes, and the ozone, temperature and aerosol distribution obtained after the data processing will be displayed on the plane closed circuit television.

The ozone determination is based on differential absorption LIDAR technology, using lights at two different wavelengths: one is ozone's strong absorption light wave at 308 nm and the other is ozone's minimum absorption at 355 nm. Due to the simultaneous generation of Rayleigh and Raman scattering signals from atmospheric molecules, the four transmitting wavelengths will be collected when the two emission wavelengths are transmitted. The Raman signal is used to detect ozone in the presence of aerosols or clouds. By eliminating the influence of the instrument parameters, such as detection efficiency and the size of the telescope, the expression of the ozone concentration can be obtained through the two signal slope differece between absorbing the ratio of the absorption wavelength and the non absorption wavelength. Data are collected every 2 minutes. The vertical resolution of 0.5–1.5 km can be obtained, depending on the altitude and integration time, and the ozone concentration distribution in the 35–40 km range near the plane can be obtained. With a detection precision over 10%, while the accuracy depends on elevation and collecting time.

The temperature measurement is mainly based on the atmospheric temperature distribution curve over 40 km above the aircraft, and the intensity of the backscattering signal is measured by Rayleigh and Raman scattering. The intensity of the signal is proportional to the atmospheric density. The atmospheric density and temperature can be obtained from the ideal gas law from the hydrostatic equilibrium system. Rayleigh scattering is suitable for vertical profile of atmosphere temperature higher than 25~80 km, while Raman scattering is applicable to the atmosphere region above the aircraft up to 25 km [9]. Because the returned signal is very strong, Rayleigh scattering is the preferred signal. But there are aerosols in the atmosphere below 25 km, and the aerosol seriously interferes with the Rayleigh scattering signal, and Raman scattering is necessary to avoid effected of aerosol scattering. The temperature distribution time resolution is 2 minutes (the horizontal scanning depth is 24 km), vertical resolution is 0.5–2 km, and the temperature measurement precision is 1–2°C [23].

15.10.6 Aerosol LIDAR

Aerosol LIDAR mainly detects the backward scattering signal of aerosol and/or cloud at 532 and 1,064 nm and depolarization of aerosol/cloud at 532 nm; the backward scattering signal at 532 nm and 1064 nm provides information about aerosol/cloud particle spatial concentration distribution. Comparison of aerosol / cloud scattering at two wavelengths can provide information of particle size, and the depolarization effect of particles can provide indicative information of particle morphology.

Aerosol LIDAR is an additional instrument of Airborne Raman Ozone and Temperature LIDAR (AROTEL). The aerosol measuring light source is a Continuum 9050 Nd:YAG laser, with the emission energy is about 600 mJ (1064 nm), 250 mJ (532 nm) and 350 mJ (355 nm), respectively. The AROTEL emits 308 nm laser and measures the atmospheric ozone concentration and atmospheric temperature distribution using molecular and Raman scattering from 355 nm and 308 nm beams. The back-scattered light at all wavelengths is obtained from adjustable field of view Newton telescope with a diameter of 16 inches.

Under the subsequent optical component, the ultraviolet (UV) signal is separated from the 532 nm and 1064 nm signals by the two color beam splitter, and the UV signal is guided to the AROTEL receiver component, but signals at 532 nm and 1064 nm are led to the aerosol LIDAR receiver component. In aerosol LIDAR receiving component, a rotating shutter stops the strong signal at 532 and 1,064 nm near the field to reduce the weaker signal distortion at a higher amplitude. The 532 and 1,064 nm signals are separated by a dichroic beam splitter, and the 532 nm signal is further separated by a polarized beam splitter by an orthogonal polarization component. A computer controlled half-wave plate is rotated in front of the polarization beam splitter, so that the 532 nm signal is separated into parallel polarization and vertical polarization relative to the polarization of the laser pulse. Signals at 532 and 1064 nm as well as the two polarization states at 532 nm are transmitted to the aerosol LIDAR data acquisition machine through optical fiber.

In order to more accurately measure the signal in its entire dynamic range, it is guided to two separate detectors, of which 10% signal to one detector and the 90%

signal is transmitted to another detector. The reflection signal at 532 nm is measured by a photomultiplier, and reflection signal at 1064 nm is measured by a photodiode. Because the intensity of the high light signal is strong enough, all data are obtained in simulation form, and the 12 bit simulation-to-digital converter is adopted. The instrument can operate under the light conditions during the daytime and nighttime, but the quality of data is slightly reduced during daytime.

References

[1]   Ryerson T B, Andrews A E, Angevine W M, et al. The 2010 California research at the Nexus of air quality and climate change (CalNex) field study. J. Geophys. Res. Atmos., 2013, 118(11): 5830–5866.

[2]   Change I P O C. Climate change 2013: The physical science basis. Fifth Assessment Report (AR5), 2014.

[3]   Wang X, Liu T, Bernard F, et al. Design and characterization of a smog chamber for studying gas-phase chemical mechanisms and aerosol formation. Atmos. Meas. Tech., 2014, 7(1): 301–313.

[4]   Glantschnig W J, Chen S H. Light scattering from water droplets in the geometrical optics approximation. Appl. Opt., 1981, 20(14): 2499–2509.

[5]   Cai Y, Montague D C, Mooiweer-Bryan W, et al. Performance characteristics of the ultra high sensitivity aerosol spectrometer for particles between 55 and 800 nm: Laboratory and field studies. J. Aerosol Sci., 2008, 39(9): 759–769.

[6]   Allen J O, Fergenson D P, Gard E E, et al. Single-particle detection efficiencies of aerosol time-of-flight mass spectrometry during the North Atlantic marine boundary layer experiment. Environ. Sci. Tech., 2000, 34(1): 211–217.

[7]   Banta R M, Pichugina Y L, Kelley N D, et al. Wind energy meteorology: insight into wind properties in the Turbine-Rotor layer of the atmosphere from high-resolution Doppler LIDAR. B. Am. Meteorol. Soc., 2013, 94(6): 883–902.

[8]   Leskinen A, Yli-Pirilä P, Kuuspalo K, et al. Characterization and testing of a new environmental chamber designed for emission aging studies. Atmos. Meas. Tech. Discuss., 2014, 7(6): 5921–5951.

[9]   Stephens M, Turner N, Sandberg J. Particle identification by laser-induced incandescence in a solid-state laser cavity. Appl. Opt., 2003, 42(19): 3726–3736.

[10]   Moteki N, Kondo Y. Effects of mixing state on black carbon measurements by laser-induced incandescence. Aerosol. Sci. Tech., 2007, 41(4): 398–417.

[11]   Schwarz J P, Gao R S, Fahey D W, et al. Single-particle measurements of midlatitude black carbon and light-scattering aerosols from the boundary layer to the lower stratosphere. J. Geophys. Res. Atmos., 2006, 111(D16): D16207.

[12]   Gao R S, Schwarz J P, Kelly K K, et al. A novel method for estimating light-scattering properties of soot aerosols using a modified single-particle soot photometer. Aerosol. Sci. Tech., 2007, 41(2): 125–135.

[13]   Browell E V. Differential absorption lidar sensing of ozone. Proceedings of the IEEE, 1989, 77(3): 419–432.

[14]   Baumgardner D, Newton R, Krämer M, et al. The cloud particle spectrometer with polarization detection (CPSPD): A next generation open-path cloud probe for distinguishing liquid cloud droplets from ice crystals. Atmos. Res., 2014, 142(0): 2–14.

[15] Alvarez R J, Senff C J, Langford A O, et al. Development and application of a compact, tunable, solid-state airborne ozone lidar system for boundary layer profiling. J. Atmos. and Ocean. Tech., 2011, 28(10): 1258–1272.

[16] Langford A O, Senff C J, Alvarez R J, et al. Long-range transport of ozone from the Los Angeles Basin: A case study. Geophys. Res. Lett., 2010, 37(6): L06807.

[17] Lance S, Brock C A, Rogers D, et al. Water droplet calibration of the cloud droplet probe (CDP) and in-flight performance in liquid, ice and mixed-phase clouds during ARCPAC. Atmos. Meas. Tech., 2010, 3(6): 1683–1706.

[18] Gallagher M W, Connolly P J, Crawford I, et al. Observations and modelling of microphysical variability, aggregation and sedimentation in tropical anvil cirrus outflow regions. Atmos. Chem. Phys., 2012, 12(14): 6609–6628.

[19] Crosier J, Bower K N, Choularton T W, et al. Observations of ice multiplication in a weakly convective cell embedded in supercooled mid-level stratus. Atmos. Chem. Phys., 2011, 11(1): 257–273.

[20] Bruneau D, Gibert F, Flamant P H, et al. Complementary study of differential absorption lidar optimization in direct and heterodyne detections. Appl. Opt., 2006, 45(20): 4898–4908.

[21] Yost C R, Minnis P, Ayers J K, et al. Comparison of GOES-retrieved and in situ measurements of deep convective anvil cloud microphysical properties during the tropical composition, cloud and climate coupling experiment (TC4). J. Geophys. Res. Atmos., 2010, 115(D10), D00J06.

[22] Heaps W S, Burris J. Airborne Raman lidar. Appl. Opt., 1996, 35(36): 7128–7135.

[23] Burris J, Heaps W, Gary B, et al. Lidar temperature measurements during the Tropical Ozone Transport Experiment (TOTE)/Vortex Ozone Transport Experiment (VOTE) mission. J. Geophys. Res. Atmos., 1998, 103(D3): 3505–3510.

Index

https://doi.org/10.1515/9783110573138-016

www.ingramcontent.com/pod-product-compliance
Lightning Source LLC
Chambersburg PA
CBHW080656220326
41598CB00033B/5221